# Benthic Biodiversity from Microbes to Multicellular Organisms and Its Functional Interplays

# Benthic Biodiversity from Microbes to Multicellular Organisms and Its Functional Interplays

Guest Editors

**Mayya Gogina**
**Judith Piontek**

MDPI

Basel • Beijing • Wuhan • Barcelona • Belgrade • Novi Sad • Cluj • Manchester

*Guest Editors*
Mayya Gogina
Biological Oceanography
Leibniz Institute for Baltic Sea
Research Warnemünde
Rostock
Germany

Judith Piontek
Biological Oceanography
Leibniz Institute for Baltic Sea
Research Warnemünde
Rostock
Germany

*Editorial Office*
MDPI AG
Grosspeteranlage 5
4052 Basel, Switzerland

This is a reprint of the Special Issue, published open access by the journal *Biology* (ISSN 2079-7737), freely accessible at: https://www.mdpi.com/journal/biology/special_issues/benthic_biodiversity.

For citation purposes, cite each article independently as indicated on the article page online and as indicated below:

| Lastname, A.A.; Lastname, B.B. Article Title. *Journal Name* **Year**, *Volume Number*, Page Range. |
|---|

**ISBN 978-3-7258-4353-4 (Hbk)**
**ISBN 978-3-7258-4354-1 (PDF)**
**https://doi.org/10.3390/books978-3-7258-4354-1**

Cover image courtesy of Mayya Gogina

# Contents


**About the Editors** . . . . . . . . . . . . . . . . . . . . . . . . . . . . . . . . . . . . . . . . . . . . . . **vii**

**Preface** . . . . . . . . . . . . . . . . . . . . . . . . . . . . . . . . . . . . . . . . . . . . . . . . . . . . . **ix**

**Mayya Gogina and Judith Piontek**
Beyond the Surface: The Peculiar World of Benthic Biodiversity, from Microbes to Multicellular Life and Their Ecosystem Roles
Reprinted from: *Biology* **2025**, *14*, 368, https://doi.org/10.3390/biology14040368 . . . . . . . . . **1**

**Mayya Gogina, Judith Rahel Renz, Stefan Forster and Michael L. Zettler**
Benthic Macrofauna Community Bioirrigation Potential (BIPc): Regional Map and Utility Validation for the South-Western Baltic Sea
Reprinted from: *Biology* **2022**, *11*, 1085, https://doi.org/10.3390/biology11071085 . . . . . . . . . **5**

**Irini Tsikopoulou, Christopher J. Smith, Konstantia Nadia Papadopoulou and Melanie C. Austen**
Linking Species Functional Traits to Specific Biogeochemical Processes under Trawling Pressure
Reprinted from: *Biology* **2022**, *11*, 1378, https://doi.org/10.3390/biology11101378 . . . . . . . . . **22**

**Olivia Dixon, Johanna Gammal, Dana Clark, Joanne I. Ellis and Conrad A. Pilditch**
Estimating Effects of Sea Level Rise on Benthic Biodiversity and Ecosystem Functioning in a Large Meso-Tidal Coastal Lagoon
Reprinted from: *Biology* **2023**, *12*, 105, https://doi.org/10.3390/biology12010105 . . . . . . . . . **35**

**Natalia Anna Miernik, Urszula Janas and Halina Kendzierska**
Role of Macrofaunal Communities in the Vistula River Plume, the Baltic Sea—Bioturbation and Bioirrigation Potential
Reprinted from: *Biology* **2023**, *12*, 147, https://doi.org/10.3390/biology12020147 . . . . . . . . . **54**

**Sichen Zheng, Tianshi Zhang, Kang Tu, Li Li, Zhihong Liu, Biao Wu, et al.**
Population Genetics of Manila Clam (*Ruditapes philippinarum*) in China Inferred from Microsatellite Markers
Reprinted from: *Biology* **2023**, *12*, 557, https://doi.org/10.3390/biology12040557 . . . . . . . . . **76**

**Huiting Wu, Yang Zhang, Anfeng Chen and Thomas L. Stubbs**
A Highly Diverse Olenekian Brachiopod Fauna from the Nanpanjiang Basin, South China, and Its Implications for the Early Triassic Biotic Recovery
Reprinted from: *Biology* **2023**, *12*, 622, https://doi.org/10.3390/biology12040622 . . . . . . . . . **90**

**Katharina Romoth, Alexander Darr, Svenja Papenmeier, Michael L. Zettler and Mayya Gogina**
Substrate Heterogeneity as a Trigger for Species Diversity in Marine Benthic Assemblages
Reprinted from: *Biology* **2023**, *12*, 825, https://doi.org/10.3390/biology12060825 . . . . . . . . . **107**

**Aleksandra Anatolyevna Istomina, Avianna Fayazovna Zhukovskaya, Andrey Nikolaevich Mazeika, Ekaterina Andreevna Barsova, Victor Pavlovich Chelomin, Marina Alexandrovna Mazur, et al.**
The Relationship between Lifespan of Marine Bivalves and Their Fatty Acids of Mitochondria Lipids
Reprinted from: *Biology* **2023**, *12*, 837, https://doi.org/10.3390/biology12060837 . . . . . . . . . **123**

**Maria Sachs, Manon Dünn and Hartmut Arndt**
Benthic Heterotrophic Protist Communities of the Southern Baltic Analyzed with the Help of Curated Metabarcoding Studies
Reprinted from: *Biology* **2023**, *12*, 1010, https://doi.org/10.3390/biology12071010 . . . . . . . . . **136**

**Florian Baletaud, Gaël Lecellier, Antoine Gilbert, Laëtitia Mathon, Jean-Marie Côme, Tony Dejean, et al.**
Comparing Seamounts and Coral Reefs with eDNA and BRUVS Reveals Oases and Refuges on Shallow Seamounts
Reprinted from: *Biology* **2023**, *12*, 1446, https://doi.org/10.3390/biology12111446 . . . . . . . . . **155**

**Denise Marx, Agata Feldens, Svenja Papenmeier, Peter Feldens, Alexander Darr, Michael L. Zettler and Kathrin Heinicke**
Habitats and Biotopes in the German Baltic Sea
Reprinted from: *Biology* **2024**, *13*, 6, https://doi.org/10.3390/biology13010006 . . . . . . . . . . . **176**

**Mayya Gogina, Sarah Joy Hahn, Ramona Ohde, Angelika Brandt, Stefan Forster, Ingrid Kröncke, et al.**
Baseline Inventory of Benthic Macrofauna in German Marine Protected Areas (2020–2022) before Closure for Bottom-Contact Fishing
Reprinted from: *Biology* **2024**, *13*, 389, https://doi.org/10.3390/biology13060389 . . . . . . . . . **208**

**Natalia Anna Gintowt, Halina Kendzierska and Urszula Janas**
Seasonal Dynamics of Benthic Infauna Communities in *Zostera marina* Meadows: Effects of Plant Density Gradients
Reprinted from: *Biology* **2025**, *14*, 153, https://doi.org/10.3390/biology14020153 . . . . . . . . . **237**

# About the Editors

**Mayya Gogina**

Researcher in the Department of Biological Oceanography at the Leibniz Institute for Baltic Sea Research Warnemünde (IOW) in Rostock, Germany. Her work focuses on marine benthic ecology, with particular interests in species distribution modelling, habitat mapping, and the structures and functions of macrofauna communities in the Baltic Sea. She investigates biogeographic patterns, biological traits, and key ecosystem processes, such as bioturbation and bioirrigation. Her research also explores how biodiversity responds to environmental change, climate impacts, and human activities like bottom-trawl fishing.

**Judith Piontek**

Scientist in the Department of Biological Oceanography at the Leibniz Institute for Baltic Sea Research Warnemünde (IOW) in Rostock, Germany. Her research centers on marine microbiology, with a focus on the structures and functions of bacterial communities in both pelagic and benthic environments. She investigates how microbes mediate biogeochemical processes and the role of bacteria in carbon remineralization within marine ecosystems.

# Preface

Benthic biodiversity plays an important regulatory role in marine ecosystem functioning, including the transport of energy, solutes, and materials within sediments and across the sediment–water interface. Biological interactions within and between components often expose co-structures and may superimpose on the relationships between biodiversity and its abiotic drivers, potentially masking or even reversing them. Due to limited accessibility, the seafloor environment still hides many of its inhabitants and their functions from us, even in the most studied regions.

Progress made in global and regional analyses has revealed the potential controlling mechanisms of large-scale benthic biodiversity patterns. Changes in benthic communities along natural gradients have been the focus of ecological research for decades. However, with a strong bias toward macrozoobenthos and biogeochemical elemental cycling, the structures and functions of other organism groups living in and on sediment, including prokaryotes, protists, microphytobenthos, and meiofauna, are largely understudied. Particularly, the monitoring of benthic habitats is necessary to expand our insights into the underlying mechanisms of spatial and temporal variability. Elucidating the interactions between the different components of the benthic biota and parameterizing the implications of biodiversity changes for ecosystem functions along natural gradients and in response to anthropogenic disturbance remain continuing challenges.

This Special Issue aimed to explore how changes in the diversity of benthic communities—from prokaryotes to macrofauna—and the interactions between these groups influence community metabolism, biogeochemical fluxes, and transport processes. We encouraged observational, experimental, and modelling studies that reflect the latest advances in the field, and this collection represents the outcome of these efforts.

We sincerely thank MDPI *Biology* and especially our Assistant Editor, Ms. Annie Ji, for her highly professional, thoughtful, and kind support throughout this process. Annie, it was a pleasure working with you—your attentive communication, care, and clarity are truly appreciated.

**Mayya Gogina and Judith Piontek**
*Guest Editors*

biology

MDPI

*Editorial*

# Beyond the Surface: The Peculiar World of Benthic Biodiversity, from Microbes to Multicellular Life and Their Ecosystem Roles

**Mayya Gogina * and Judith Piontek**

Leibniz Institute for Baltic Sea Research Warnemünde, Seestrasse 15, D-18119 Rostock, Germany; judith.piontek@io-warnemuende.de
* Correspondence: mayya.gogina@io-warnemuende.de

Below the water column depths, marine sediments harbor a vibrant tapestry of life that underpins a variety of ecological balances. From the tiniest microbes to complex multicellular organisms, benthic communities play pivotal roles in nutrient cycling, energy flow, and habitat formation. Understanding these intricate ecosystems is paramount, especially as they face intensive evolutionary fluctuation and natural succession. Changing climate conditions and the increasing influence of other threats from human activities are putting pressure on communities in benthic marine ecosystems and require adaptations within short periods of time.

In this Special Issue, we wanted to delve into the multifaceted world of benthic biodiversity, exploring the dynamic interactions among its diverse communities and their impacts on marine ecosystem functioning. We were lucky to collect 13 articles, totaling 252 pages, that have been co-authored by 65 researchers from 8 countries.

The contributions span a broad spectrum of topics, beginning with microbial communities that form the foundation of benthic ecosystems. Advanced metabarcoding techniques have unveiled the astonishing diversity and versatility of benthic protists, shedding light on their roles in biogeochemical processes and food webs and addressing their resilience to environmental perturbations. Transitioning to meiofauna and macrofauna, several studies investigate species distribution patterns in response to natural gradients and anthropogenic pressures like bottom trawling, sea level rise, or organic and pollutant influences from the river plume, suggesting species that suffer most from these perturbations and spotting those that are tolerant. These findings underscore the sensitivity of benthic organisms to habitat alterations, emphasizing the need for comprehensive monitoring and conservation strategies.

Interdisciplinary approaches featured in this issue highlight the functional interplays between different benthic compartments, but also the need for more integrated assessments. For instance, research on bioturbation and bioirrigation demonstrates how macrofaunal activities enhance microbial processes, facilitating nutrient exchange across the sediment–water interface. Such interactions are crucial for maintaining ecosystem health and productivity. However, establishing robust quantitative relationships to improve predictive capacity remains a challenge.

The contributions in this issue primarily focus on observational and modeling approaches to understand benthic biodiversity distribution and ecosystem functioning (Table 1). Purely experimental studies were unfortunately lacking. We therefore advocate for more of them in the future, as they can reveal not only species-specific responses but also potential cascading effects on community structure and ecosystem services. Such insights remain missing for many organism groups but are vital for predicting future changes and formulating adaptive management plans.

**Table 1.** Summary of 13 papers included in this Special Issue, distinguishing between organism groups studied and methodologies and highlighting key findings.

| Reference | Focal Organism Group | Methodological Approach | Key Ecosystem Function Finding |
|---|---|---|---|
| [1] | Macrofauna | Observational, Modeling, Experimental | The Bioirrigation Potential index (BIPc) derived from observational macrofauna data correlates with irrigation rates, confirming that macrofaunal activity significantly enhances the efficiency of solute transfer in benthic environments. |
| [2] | Macrofauna | Observational, Experimental | Bottom trawling shifts benthic communities toward opportunistic, deposit-feeding species that do not bioturbate. This disturbance does not affect oxygen consumption (site- and season-dependent) but reduces carbon mineralization due to the removal of reactive surface sediment. The decline in community complexity and bioturbation leads to decreased sediment oxygenation, a reduction in carbon mineralization, and higher organic carbon concentrations in the sediment. |
| [3] | Macrofauna | Observational | Sea level rise will modify the seafloor macrofauna communities in estuaries and subsequently alter the related ecosystem functions. |
| [4] | Macrofauna | Observational | With increasing depth and decreasing influence from the river plume, species density and related sediment mixing and bioirrigation decreased. Highest macrofauna diversity was observed in the upper first cm of sediment, but the highest biomass was in deeper (6 cm) depth layers. |
| [5] | Macrofauna | Observational | Using microsatellite markers, the genetic diversity of commercially important clams was shown to be influenced by breeding mode and revealed no genetic isolation of populations by distance. |
| [6] | Macrofauna, brachiopods | Observational, paleo | Based on precise age constraints from conodont biostratigraphy and quantitative brachiopod data, a previous underestimate of the diversity of Olenekian brachiopod fauna diversity is suggested, and brachiopod recovery in the studied section is attributed to the latest Spathian period. |
| [7] | Macrofauna | Observational | Field assessments document a direct link between macrofaunal diversity and sediment structure. Small patches of different soft sediment types are associated with elevated species richness and a higher rate of occurrence of rare species. |
| [8] | Macrofauna | Observational | Research on marine bivalves suggested that lipid matrix membranes of the mitochondria of long-lived species are less sensitive to in vitro-initiated peroxidation compared to species with shorter life spans. |
| [9] | Microbes, protists | Observational | Metabarcoding-based assessment of benthic protists, which act as controllers of bacterial and microphytobenthos production and contribute significantly to the carbon flux, suggested community response to salinity, sediment properties, and oxygen. |
| [10] | Fish (demersal and benthopelagic), megafauna | Observational | Field surveys using environmental DNA metabarcoding (eDNA) and baited video (BRUVS) revealed no diversity hotspots for fish at seamounts. Shallower seamounts, as biomass oases and refuges for threatened megafauna, were spotted as protection priorities. |

**Table 1.** *Cont.*

| Reference | Focal Organism Group | Methodological Approach | Key Ecosystem Function Finding |
|---|---|---|---|
| [11] | Macrofauna | Observational, Modeling | Field geological and biological surveys and modeling were used to map the distribution of seabed habitats and biotopes and their inhabitants. |
| [12] | Macrofauna | Observational | Field surveys compare recent benthic macrofauna biodiversity in German Marine Protected Areas of the Baltic and the North Sea along environmental gradients and decreasing bottom trawling intensity. |
| [13] | Macrofauna | Observational | Seagrass (regardless of density) positively affects macrozoobenthic communities and their functioning, indicating meadows as key biotopes that can support biogeochemical processes in coastal zones more effectively than bare sands. |

To serve as a quick reference for readers to navigate the diverse topics covered in this Special Issue and to synthesize the wealth of information presented, Table 1 categorizes the 13 published studies based on their focal organism groups (microbes, meiofauna, macrofauna, fish) and methodological approaches (observational, experimental, modeling) and outlines key findings related to ecosystem functions. In conclusion, this compilation of research offers a snapshot of some current advancements in benthic biodiversity studies. It underscores the importance of integrative approaches to unravel the complexities of marine ecosystems and informs strategies for their understanding and preservation amidst a rapidly changing world.

Additionally, Figure 1 illustrates schematically the expected interconnections between different benthic biodiversity components, depicting how microbial, meiofaunal, and macrofaunal interactions might drive or alter essential processes like nutrient cycling, carbon remineralization, deoxygenation, and overall energy transfer.

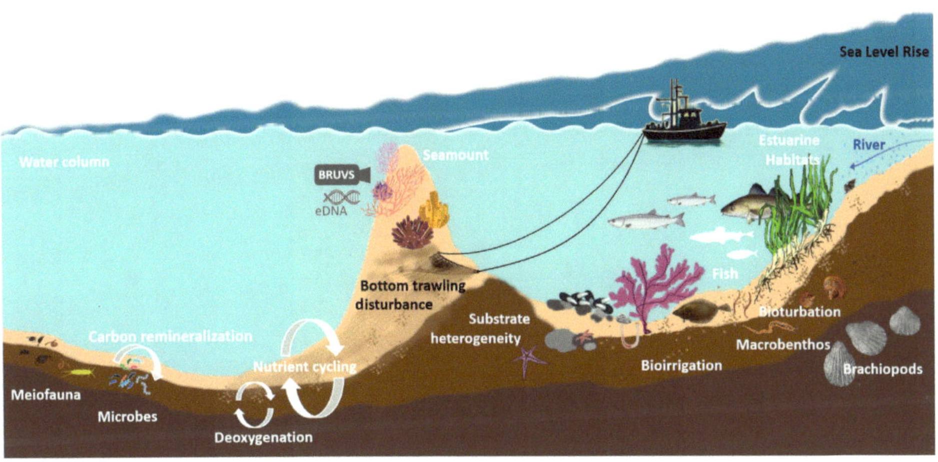


**Figure 1.** Schematic illustration of the expected interconnections between different benthic biodiversity components and processes (labeled in white), exemplary pressures (labeled in black), as well as some currently actively developed monitoring methods (shown in gray).

**Conflicts of Interest:** The authors declare no conflict of interest.

## References

1. Gogina, M.; Renz, J.R.; Forster, S.; Zettler, M.L. Benthic Macrofauna Community Bioirrigation Potential (BIPc): Regional Map and Utility Validation for the South-Western Baltic Sea. *Biology* **2022**, *11*, 1085. [CrossRef] [PubMed]
2. Tsikopoulou, I.; Smith, C.J.; Papadopoulou, K.N.; Austen, M.C. Linking Species Functional Traits to Specific Biogeochemical Processes under Trawling Pressure. *Biology* **2022**, *11*, 1378. [CrossRef] [PubMed]
3. Dixon, O.; Gammal, J.; Clark, D.; Ellis, J.I.; Pilditch, C.A. Estimating Effects of Sea Level Rise on Benthic Biodiversity and Ecosystem Functioning in a Large Meso-Tidal Coastal Lagoon. *Biology* **2023**, *12*, 105. [CrossRef] [PubMed]
4. Miernik, N.A.; Janas, U.; Kendzierska, H. Role of Macrofaunal Communities in the Vistula River Plume, the Baltic Sea—Bioturbation and Bioirrigation Potential. *Biology* **2023**, *12*, 147. [CrossRef] [PubMed]
5. Zheng, S.; Zhang, T.; Tu, K.; Li, L.; Liu, Z.; Wu, B.; Zhou, L.; Sun, X. Population Genetics of Manila Clam (*Ruditapes philippinarum*) in China Inferred from Microsatellite Markers. *Biology* **2023**, *12*, 557. [CrossRef] [PubMed]
6. Wu, H.; Zhang, Y.; Chen, A.; Stubbs, T.L. A Highly Diverse Olenekian Brachiopod Fauna from the Nanpanjiang Basin, South China, and Its Implications for the Early Triassic Biotic Recovery. *Biology* **2023**, *12*, 622. [CrossRef] [PubMed]
7. Romoth, K.; Darr, A.; Papenmeier, S.; Zettler, M.L.; Gogina, M. Substrate Heterogeneity as a Trigger for Species Diversity in Marine Benthic Assemblages. *Biology* **2023**, *12*, 825. [CrossRef] [PubMed]
8. Istomina, A.A.; Zhukovskaya, A.F.; Mazeika, A.N.; Barsova, E.A.; Chelomin, V.P.; Mazur, M.A.; Elovskaya, O.A.; Mazur, A.A.; Dovzhenko, N.V.; Fedorets, Y.V.; et al. The Relationship between Lifespan of Marine Bivalves and Their Fatty Acids of Mitochondria Lipids. *Biology* **2023**, *12*, 837. [CrossRef] [PubMed]
9. Sachs, M.; Dünn, M.; Arndt, H. Benthic Heterotrophic Protist Communities of the Southern Baltic Analyzed with the Help of Curated Metabarcoding Studies. *Biology* **2023**, *12*, 1010. [CrossRef] [PubMed]
10. Baletaud, F.; Lecellier, G.; Gilbert, A.; Mathon, L.; Côme, J.-M.; Dejean, T.; Dumas, M.; Fiat, S.; Vigliola, L. Comparing Seamounts and Coral Reefs with eDNA and BRUVS Reveals Oases and Refuges on Shallow Seamounts. *Biology* **2023**, *12*, 1446. [CrossRef] [PubMed]
11. Marx, D.; Feldens, A.; Papenmeier, S.; Feldens, P.; Darr, A.; Zettler, M.L.; Heinicke, K. Habitats and Biotopes in the German Baltic Sea. *Biology* **2024**, *13*, 6. [CrossRef] [PubMed]
12. Gogina, M.; Hahn, S.J.; Ohde, R.; Brandt, A.; Forster, S.; Kröncke, I.; Powilleit, M.; Romoth, K.; Sonnewald, M.; Zettler, M.L. Baseline Inventory of Benthic Macrofauna in German Marine Protected Areas (2020–2022) before Closure for Bottom-Contact Fishing. *Biology* **2024**, *13*, 389. [CrossRef] [PubMed]
13. Gintowt, N.A.; Kendzierska, H.; Janas, U. Seasonal Dynamics of Benthic Infauna Communities in Zostera marina Meadows: Effects of Plant Density Gradients. *Biology* **2025**, *14*, 153. [CrossRef] [PubMed]

*biology*

MDPI

*Article*

# Benthic Macrofauna Community Bioirrigation Potential (BIPc): Regional Map and Utility Validation for the South-Western Baltic Sea

**Mayya Gogina [1,*], Judith Rahel Renz [2], Stefan Forster [2] and Michael L. Zettler [1]**

1 Leibniz Institute for Baltic Sea Research, Seestraße 15, 18119 Rostock, Germany; michael.zettler@io-warnemuende.de

2 Marine Biology, Institute of Biological Sciences, University of Rostock, Albert-Einstein-Straße 3, 18059 Rostock, Germany; judith-rahel@gmx.de (J.R.R.); stefan.forster@uni-rostock.de (S.F.)

* Correspondence: mayya.gogina@io-warnemuende.de; Tel.: +49-381-519-7393

**Simple Summary:** The sediments on the seafloor are inhabited by multiple macroscopic organisms such as shells and worms, which, among other things, influence the biogeochemical cycling by flushing the near-bottom water through their gangways. This is called bioirrigation, one of key processes in the functioning of marine sediments. The density of animals, in addition to the features (or traits) of each species, define their specific contributions to this process. Measuring the intensity of this rather dynamic process in nature is difficult and costly; therefore, the available direct observations are too scarce for large-scale assessments. However, such assessments are essential for broadening our understanding of ecosystem functioning, and of the role that biodiversity plays in it. To address this shortage of observational data, a traits-based index "BIPc" that expresses the bioirrigation potential, based on available data on sediment-dwelling animals, comes into play. In this paper, we focus on the performance of the BIPc index in the south-western Baltic Sea, and on how it changes in space and time. The results support the usefulness of this index, but also highlight its existing limitations. Modelled distribution map layers of the bioirrigation potential and scores for 120 key species required for index calculation are made available for reuse.

**Abstract:** Benthic community bioirrigation potential (BIPc), an index developed to quantify the anticipated capacity of macrofauna to influence the solute exchange at the sediment–water interface, was calculated for the south-western Baltic Sea. This index can be regarded as an effect trait that is useful for predicting ecosystem processes impacted by animal burrow ventilation. The special feature, and presumably an advantage, of BIPc, compared to alternative recently developed benthic macrofauna-based bioirrigation indices, lies in its ability to distinguish the taxa-specific score values between diffusion- and advection-dominated sediment systems. The usefulness of the BIPc index was compared against the estimates of the well-established community bioturbation potential index (BPc). The BIPc index displayed a moderately but significantly stronger correlation with estimates of irrigation rates derived from tracer experiments. Using a random forest machine learning approach and a number of available relevant environmental predictor layers, we have modelled and mapped the spatial differences in this ecosystem functioning expression. The key species contributing to bioirrigation potential in the study area were identified. The interannual variation in BIPc was assessed on a small exemplary dataset. The scores required to calculate the index, that were assigned to 120 taxa dominating abundance and biomass in the region, are provided for reuse. The utility, temporal variability and uncertainty of the distribution estimate are discussed.

**Keywords:** benthic organisms; ecosystem functioning; irrigation; trait-based index; solute transport; sediment–water interface; mapping; species distribution model

## 1. Introduction

Both fluid and particle transport affect the physical properties of sediment; however, Aller [1] suggested that, in terms of weight, water pumping is about 100 times greater compared to bioturbation-driven particle transport. Biogeochemistry and microbial community structures of sediments in the coastal seas are influenced by the irrigation activity of large populations of burrowing macroinfauna [2]. Burrowing benthic macrofauna is a classic example of ecosystem engineers having a major impact on ecosystem functioning by influencing the sediment matrix and pore solutes in the aquatic sediments. This impact is disproportionally large compared to the abundance and biomass of macrofauna [3]. Furthermore, according to global estimates, macrofauna dominates the biomass on the continental shelf [4].

Importantly, the effects of bioturbation and, in particular, burrow ventilation, on oxygen uptake differ in diffusion- and advection-dominated systems, i.e., in fine grained, muddy sediments with low permeability compared to coarse grained, sandy sediments with high permeability [5].

Using a resazurin tracer (suitable for decoupling animal respiration and inorganic oxygen consumption from microbial respiration), microcosm experiments in lake sediments revealed that the transport of fluids into the sediment due to the activities of chironomid larvae enhanced sediment respiration by a factor of 2.5 in the diffusion-dominated sediment [6]. A study using the same tracer in marine sediments of an advection-dominated system did not detect any bioturbation-caused change in the total oxygen uptake, but evidenced a significant increase in biologically mediated oxygen uptake [3]. Other freshwater studies suggested that bioirrigation by macrozoobenthos can be responsible for an enhancement of sediment respiration ranging from 17 to 360% [7,8].

The constant transport of solutes caused by the activity of sediment-dwelling macrofauna modify the habitat and influence the availability of resources. These allogenic engineers exert a major influence on the biogeochemistry of aquatic sediments by altering the microstructure, oxidating solute species, water pumping through the sediment and enhancing bacterial activity [3,9]. Though the effects of benthos are mainly surficial, limited to a few centimetres above the sediment surface and a few decimetres below it, this sediment–water interface is both biologically active and chemically reactive [10]. The reasons that justify the approach of focusing on the effects of macrofauna in soft sediments are: macrofauna, as covered by largely available data, seem to comprise the most potent modifiers, and soft-bottom habitats form the bulk of the seafloor in the Baltic Sea region [9]. Despite a recently emerging global database covering bioturbation intensity, ventilation rate, and the mixing depth measurements of marine sediments [11], there is still only sporadic and rare data on bioirrigation rates directly measured in experimental set-ups and biodiffusion coefficients derived from bio-mixing models [12,13]. Toussaint et al. [14] concluded that both biotic and abiotic factors are required to explain the variability in oxygen consumption, total mineralisation, and nitrification and denitrification estimates, as macrofaunal activities have different effects across habitats. Using the BIPc index calculated on the base of biotic data, it could be considered to proportionally vary the bioirrigation in generic models for marine sediment biogeochemistry (such as ERSEM, e.g., [15]), and eventually to more accurately assess the effects on fluxes on different scales.

Bioirrigation index can be regarded as an effect trait that is useful for predicting ecosystem processes rates, particularly in cohesive sediments. Several such indices have been developed recently (e.g., [16]). Here, we focus on the bioirrigation potential of the benthic community (BIPc), developed and described by Renz et al. [17], as a proxy for burrow ventilation by fauna, specifically using data from the Baltic Sea. Environmental steering distinguishing diffusion- and advection-dominated systems could result in a more effective interpretation of what this statistic actually means. Measurements of irrigation do not typically discriminate between physical and faunal burrow ventilation, and it is therefore difficult to predict any organism's impact on actual ventilation rates based on trait scores alone, especially in permeable sediments. In contrast to the IPc index proposed by Wrede et al. [16], the BIPc index proposed by Renz et al. [17] was developed with a view to distinguishing between diffusive and advective environmental settings by recording different scores according to burrow type and feeding type of those two types of benthic systems.

However, compared with related well-developed bioturbation potential, which has already been assessed and mapped for various regions [18–20], the classification of sediments according to their bioirrigation potential is a very recent endeavour. Despite already emerging applications, for example in the assessment of the degradation of ecosystem functions in response to sediment contamination [21], its applicability should be further explored [22].

Interestingly, despite agreement on the importance of burrow ventilation in fuelling oxic mineralisation and nitrification processes [23,24], in the recent studies, the IPc index developed by Wrede et al. [16] was not selected as an explanatory variable for corresponding oxygen or nitrate fluxes, and indicated no correlation to the measured irrigation rate [14,22]. This mismatch could suggest that an index is not an accurate estimation of burrow ventilation rate. Instead, it was found to correlate more strongly to the burrow ventilation depth [22].

As a critical comment for IPc, that also remains valid for BIPc, both indices do not account for any temporal dynamics of faunal activity. This is crucial for the discontinuous and fitful ventilation and short temporal scale at which the stimulation of oxic mineralisation (the one biogeochemical process that is expected to be most strongly linked to burrow ventilation) takes place, particularly as electron acceptors that are transported downwards are rapidly consumed. Toussaint et al. [14] therefore suggested that an index that would account for temporal dynamic could be a more useful proxy for biogeochemical processes, but to our knowledge no such index currently exists.

Here, we estimate the index and provide the map of BIPc for the western Baltic Sea that can serve as basis for association with other metrics of ecosystem functioning. They can also be useful for predicting and scaling up anthropogenic impacts on ecosystem functions.

## 2. Materials and Methods

### 2.1. Study Area

The semi-enclosed brackish Baltic Sea is connected to the North Sea by two narrow and shallow Danish straits (the Belt Sea and the Sound). Its environmental conditions, highly stratified by strong vertical salinity and temperature gradients, are driven by restricted water exchanges through the straits, discharge of fresh waters from the rivers, and specific topography. Halocline, controlled by freshwater runoff, wind-induced mixing and advection [25], occurs at 10 m to 30 m depth in shallower parts. The study area in the south-western Baltic Sea (Figure 1) comprises 14,800 $km^2$ and has an average depth of 19 m. Shallow seafloor habitats along the shore and on top of the offshore glacial elevations are characterised by patches of rocks, till, gravel and coarser sands. With increasing water depth, substrates become finer, and organic-rich muddy sediments prevail in the basins and in the deeper parts of the trenches [26]. The main natural abiotic drivers of species richness and composition of benthic macrofauna communities in the area are near-bottom salinity and oxygen conditions [27]. There is a strong salinity gradient with values declining from 20 to 25 in the western part of Kiel Bay towards 7 in the eastern-most part of the study area in the Pomeranian Bay, with the highest temporal variability in salinity occurring in the western part. Oxygen depletion that negatively affects the diversity and density of soft-bottom fauna [28] is irregularly observed in the deeper regions of the Kiel Bay, the Bay of Mecklenburg and in the Arkona Basin [27].

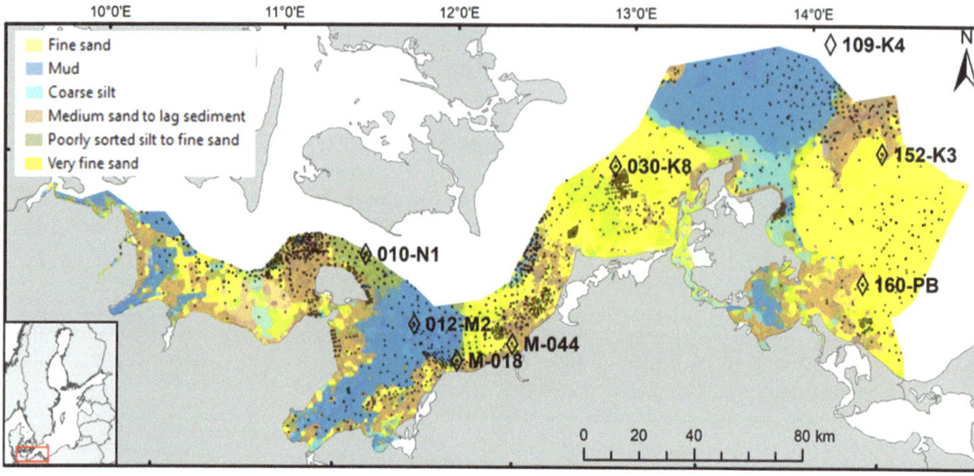


**Figure 1.** Map of the south-western Baltic Sea depicting the positions of the stations used as the reference dataset in this study (sampled in 2000–2020, shown by small grey dots), as well as long-term monitoring stations used to assess BIPc variability (shown by rhombus). The red rectangle indicates the location of the study area on a map of the Baltic Sea (lower left corner).

## *2.2. Biological Dataset and Environmental Predictors*

The biological data used is this case study covers 2170 sampling events. The positions of the stations are plotted in Figure 1. It comprises the data from the German part of the Baltic Sea described in Gogina et al. [20] updated for the period 1999–2020.

At each station, three replicate benthic samples were collected with 0.1 $m^2$ van Veen grab and washed through a 1 mm sieve. Any animals remaining were preserved on board in a 4% buffered formaldehyde–sea water solution. The retained material was sorted in a laboratory with a stereomicroscope and identified to the lowest possible taxonomic level, and the taxonomy was harmonised following the World Register of Marine Species (WoRMS).

The scores required to calculate the community bioirrigation potential, which were assigned to 120 taxa dominating abundance and biomass in the region, are included in the Supplementary Table S1. Taxa with defined species scores (for feeding, burrow type and depth to calculate BIPc) covered 93.5% of AFDW biomass and 88.9% of abundance in the area. Within the subset of data from eight monitoring stations analysed to characterise temporal variability, taxa that were covered by the scores list were responsible for at least 97% of summed abundance and at least 99% of summed AFDW biomass at each of the stations.

An interplay of physical, chemical, and biological components has a direct influence on the habitats and community structure, thereby shaping its bioirrigation potential. Available full-coverage layers for environmental variable for the German part of the Baltic Sea listed in Gogina et al. [20] were used as predictors for this study. For the final spatial distribution model that demonstrated the best performance, the selected predictors were: mean and standard deviation (SD) of salinity; mean inorganic suspended particle matter (SPM); bottom shear stress in Pa; age of water mass since the last contact with the surface; mean near-bottom oxygen concentration and SD of summer temperature of near-bottom water (modelled with the resolution of 600 × 600 $m^2$ [29,30]); bathymetry and sediment median grain size [31,32]; detritus concentration near the bottom (μmol/l) modelled with a 1 nm resolution [33,34]; and % of total organic content in surface sediments [35]. Pairs of predictors were tested for collinearity, and other highly correlated independent variables (those that indicated within any pair Pearson correlation r > 0.90, p level 0.05, and had the lower predictive power) were omitted from the analysis to ovoid overfitting (for variables and correlation matrices, see Supplementary Table S2). A larger set of other predictors was also tested, but did not enter the final model, including ice thickness as well as near-bottom salinity, temperature and oxygen modelled as described in Neumann et al. [36].

### 2.3. Calculating Community Bioirrigation Potential (BIPc)

To quantify the potential for solute exchange at the sediment–water interface, community bioirrigation potential (BIPc; [17]) was calculated for the south-western Baltic Sea. In order to account for different underlying physical processes in mud and sand, BIPc applies different scores for advective systems (here attributed to medium sand and coarser sediment types as classified by Tauber [32], and for diffusive benthic system (very fine and fine sand sediments, all other muddy and less permeable sediment types). The scores were assigned to pre-selected dominating 120 taxa, based on existing literature and expert judgement.

To calculate BIPc, the mean individual biomass (expressed by the relation $B_i/A_i$, where $B_i$ is an ash free dry weight in g $m^{-2}$, and Ai is abundance in ind. $m^{-2}$) of each species within a sample is multiplied by the relevant scores for the trait categories feeding type ($FT_i$), burrow type ($BT_i$) and depth ($L_i$), and they are weighted in turn by species abundance as given in the following equation. Afterwards, the results are summed up across all species present in the sample at a particular station (Equation (1)):

$$BIP_c = \sum_{i=1}^{n} \sqrt{\frac{B_i}{A_i}} \times A_i \times FT_i \times BT_i \times L_i \quad (1)$$

In those cases where trait categories were irrelevant or negligible with regard to solute exchange across the sediment–water interface (e.g., epifauna), a score of "zero" was assigned.

### 2.4. Assessing Temporal Variability in BIPc

Data from 8 long-term monitoring stations was used to assess the temporal variability in BIPc (Figure 1, Table 1). With regard to the decisions taken for distinguishing between diffusive and adjective systems, it is worth noting specific approach used for the long-term monitoring stations of the University of Rostock located in fine sand near the M-018 in the Bay of Mecklenburg. For all practical purposes of bioirrigation estimates, this location is known as a diffusion-driven site [37]. Since the total organic content and fine fraction at 010-N1 are even higher than at M-018, this station was also considered as a diffusion-driven system, despite the somewhat higher median grain size. Referring to this argumentation, all very fine and fine sand sediments were approximated as dominated by diffusion-driven processes.

### 2.5. Modelling Spatial Distribution of BIPc and Validating Index Performance

In order to obtain a full coverage BIPc map, we applied a Random Forest machine learning algorithm [38] in the R package "RandomForest" [39,40] to the data described in Section 2.2. Random Forest (RF) is a method based on an ensemble of randomly constructed decision trees, and unlike simple spatial interpolation methods, it helps to account for the variation in distribution driven by fine-scale habitat changes even where sampling density is not sufficiently high [20]. Calculated community bioirrigation potential BIPc was treated as the response variable, whereas environmental variables served as predictors. The log10 (x + 1) transformations was applied to BIPc values prior fitting the RF models. The number of trees was set to 1000; 1 to 5 variables per node were tested, and subsequently the best-performing model based on the highest % of variance explained was selected.

The predictive accuracy of the final model was assessed by computing the non-parametric Kendall's $\tau$ rank correlation between modelled and observed BIPc values. The importance of predictors in explaining the spatial distribution of BIPc in the final model was estimated using %IncMSE (the increase in mean squared error of the final prediction as a result of random shuffling of a particular variable, estimated with the out-of-cross-validation, considered as a robust and informative measure).

**Table 1.** Average values of sediment and environmental variables measured at monitoring stations, observed in the period 2000–2020. In brackets, values of standard deviation (SD) are reported. For station M-044 marked with *, SD could not always be defined, as multiple measurements of sediment and environmental variables were missing; this station was only sampled in the years 2000–2008.

| **Monitoring Station** | **010-N1** | **012-M2** | **M-018** | **M-044 *** | **030-K8** | **109-K4** | **152-K3** | **160-PB** |
|---|---|---|---|---|---|---|---|---|
| Diffusive (D) vs. advective (A) | D | D | D | A | A | D | A | A |
| Number of sampling events | 22 | 24 | 35 | 9 | 21 | 21 | 20 | 17 |
| Median grain size (μm) | 146<br>(91) | 21<br>(7) | 108<br>(41) | 197<br>(67) | 224<br>(6) | 19<br>(7) | 218<br>(17) | 191<br>(11) |
| Fraction finer 63 μm (%) | 34.2<br>(27.7) | 84.5<br>(22.7) | 26.5<br>(16.2) | 0.0<br>(-) | 1.8<br>(1.9) | 84.3<br>(24.7) | 2.5<br>(3.3) | 1.8<br>(2.6) |
| Fraction coarser 2000 μm (%) | 0.28<br>(0.92) | 0.84<br>(3.36) | 0.04<br>(0.15) | 0.00<br>(-) | 0.01<br>(0.02) | 0.61<br>(2.29) | 0.84<br>(1.23) | 0.03<br>(0.1) |
| Sorting (phi) | 1.64<br>(0.74) | 1.76<br>(0.47) | 1.31<br>(0.42) | 0.56<br>(-) | 0.49<br>(0.12) | 1.67<br>(0.43) | 0.60<br>(0.14) | 0.50<br>(0.14) |
| Skewness (phi) | −0.40<br>(0.61) | −0.37<br>(0.44) | −0.49<br>(0.48) | 0.04<br>(-) | −0.16<br>(0.11) | −0.46<br>(0.32) | −0.07<br>(0.23) | −0.04<br>(0.12) |
| Total organic content (%) | 2.97<br>(0.85) | 9.27<br>(1.13) | 1.58<br>(0.36) | 0.01<br>(-) | 0.29<br>(0.1) | 12.05<br>(1.55) | 0.44<br>(0.18) | 0.25<br>(0.1) |
| Oxygen (near bottom) (ml/l) | 5.07<br>(1.47) | 4.84<br>(1.73) | 5.26<br>(1.75) | 7.79<br>(0.84) | 6.34<br>(0.53) | 3.84<br>(1.03) | 5.45<br>(1.28) | 7.15<br>(1.69) |
| Salinity (near bottom) | 21.1<br>(2.5) | 20.0<br>(2.6) | 18.9<br>(2.1) | 14.91<br>(1.7) | 12.6<br>(3.2) | 18.0<br>(2.6) | 10.7<br>(2.2) | 8.1<br>(1.4) |
| Water depth (m) | 28.1 | 24.3 | 20.3 | 10.8 | 22.6 | 47.4 | 30.6 | 14.3 |
| Latitude (WGS 84), N | 54°33.08′ | 54°18.86′ | 54°10.99′ | 54°12.94′ | 54°43.41′ | 55°0.01′ | 54°37.96′ | 54°14.41′ |
| Longitude (WGS 84), E | 11°19.17′ | 11°33′ | 11°46.01′ | 12°5.14′ | 12°47.02′ | 14°4.96′ | 14°16.96′ | 14°4.11′ |

To explore the degree of redundancy or usefulness of the BIPc index, we have compared the resulting BIPc layer against the latest estimated distribution of well-established community bioturbation potential index BPc [18] in the study area. The distribution of BPc, used as a reference here, was modelled with RF [20]. BPc values were log10 (x + 1) transformed prior entering the Random Forest model as response variable, and 24 environmental layers were used as predictors. The number of trees was set to 500, best performing model considered 3 features at each split point and explained 51.4% of variance (for more details see [20]). A spatial overlay of predicted hot and cold spots of each of two indices, BIPc and BPc, was analysed.

Values of BIPc and BPc estimated based on macrofauna data collected with van Veen grab were also tested for correlation with mean total solute fluxes at the sediment–water interface assessed using incubated cores from Lipka [41]. In this case, total fluxes from the sediments were determined from the oxic phase of incubation experiments and they denoted the fluxes of dissolved chemical species by both advection of water (e.g., by organisms, via hydro-irrigation or due to relocation of pore-water and particles in surface sediments by human activities) and simple molecular diffusion across the sediment–water interface.

A distance-based Redundancy analysis (dbRDA; [42]) based on Euclidean distance was also used to extract and summarise the variation in (log-tarnsformed) BIPc explained by environmental predictors.

Taking advantage of the available data on the vertical distribution of benthic macrofauna in sediment cores collected by multicorer in the Fehmarn Belt in June 2020, we compared the BIPc values obtained when taking into account the observed distribution depth of macrofaunal individuals in the sediment, with that assumed based on literature.

## 3. Results

### 3.1. Key Species

A few dominant bivalve species such as *Mya areanaria* and *Arctica islandica*, as well as polychaetes *Marenzelleria viridis*, *Scoloplos armiger* and *Hediste diversicolor*, indicated the highest contribution to the overall BIPc in the study area. Differences in the relative contribution of a few key taxa to the summed BIPc were observed between sediment types (Table 2).

**Table 2.** Key species contributing to BIPc overall and top five key taxa listed per sediment type.

| Key Taxa and Contribution to Overall Total BIPc | | Key Taxa and Contribution to Total per Sediment Type | | | |
|---|---|---|---|---|---|
| **overall** | | **Mud** | | **Medium sand** | |
| *Mya arenaria* | 22.40% | *Arctica islandica* | 22.90% | *Marenzelleria viridis* | 19.80% |
| *Marenzelleria viridis* | 18.80% | *Scoloplos armiger* | 11.70% | *Mya arenaria* | 18.10% |
| *Arctica islandica* | 9.90% | *Terebellides stroemii* | 10.50% | *Arctica islandica* | 13.40% |
| *Scoloplos armiger* | 7.70% | *Macoma balthica* | 10.10% | *Scoloplos armiger* | 10.00% |
| *Hediste diversicolor* | 7.40% | *Lagis koreni* | 5.20% | *Pygospio elegans* | 5.00% |
| *Pygospio elegans* | 4.70% | **Fine sand** | | **Coarse sand** | |
| *Macoma balthica* | 3.90% | *Mya arenaria* | 31.80% | *Arctica islandica* | 15.70% |
| *Arenicola marina* | 2.30% | *Marenzelleria viridis* | 19.80% | *Marenzelleria viridis* | 15.60% |
| *Astarte borealis* | 2.30% | *Hediste diversicolor* | 10.00% | *Scoloplos armiger* | 10.80% |
| *Heteromastus filiformis* | 1.60% | *Arctica islandica* | 6.70% | *Mya arenaria* | 7.40% |
| **Sum** | 81% | *Scoloplos armiger* | 5.70% | *Pygospio elegans* | 7.40% |

### 3.2. Temporal Variability in BIPc

The temporal variability in BIPc observed in the period 2000–2020 at eight monitoring stations (with four stations each located in diffusion and advection dominated systems) is plotted in Figure 2. Despite significant temporal fluctuations (see also Supplementary Figure S1 for the plot showing the time-series of BIPc values), the results of pairwise tests indicated that similar magnitudes of the BIPc occur at stations with similar habitat types regardless of salinity (see also Table 1). BIPc in advection-driven systems is quantified significantly higher than in diffusion-driven systems, which is partly explained by the definition of scores for each system. Advection-dominated stations showed values of (untransformed) BIPc that were, on average, five times higher, and they were also prone to higher temporal fluctuations (Tukey multiple comparisons of means p adjusted 0.041).

### 3.3. Spatial Distribution of BIPc and Comparison with Patterns in BPc

The best fitted model explained 59.29% of the observed variance in BIPc. Kendall's $\tau$ between the modelled and observed values was 0.65 ($p < 0.05$). The importance of regional drivers of BIPc distribution, estimated as relative importance of predictors in the final RF model, is displayed in Figure 3. Median grain size was the most important predictor for spatial distribution of BIPc in the study area, followed by total organic content in sediments, inorganic suspended particle matter, temperature variability and water depth. The large effect size of sediment grain size on the potential bioirrigation activity of the organisms is not surprising since this parameter determined the delineation between diffusion- and advection-dominated sites.

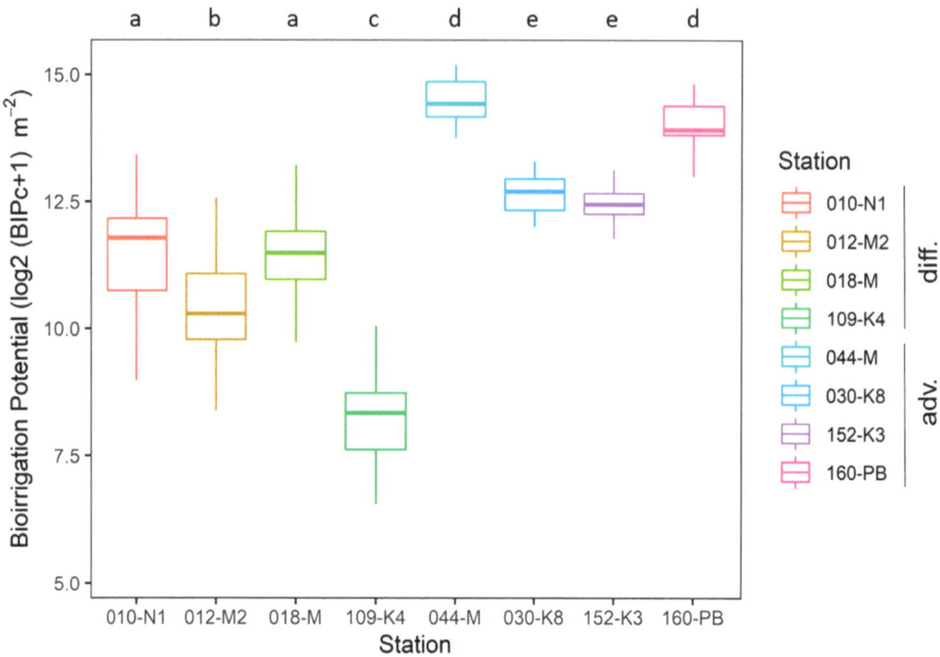


**Figure 2.** Boxplots showing the variability in BIPc over the years (two decades) at eight monitoring stations. Stations are first sorted according to the type of system (diffusion- or advection-driven) and then according to decreasing salinity. Subsets sharing the same letter above the plots are not significantly different (based on ANOVA and Tukey honestly significant difference test). The locations of the stations are indicated in Figure 1.

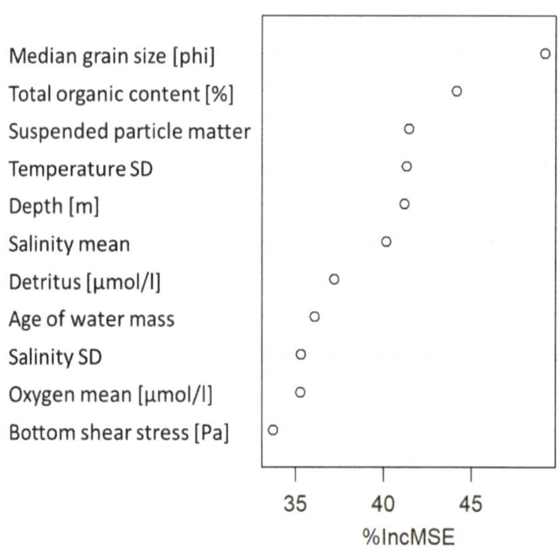


**Figure 3.** Importance of predictors driving regional distribution of BIPc in the final RF model quantified as the Mean Decrease Accuracy (%IncMSE).

Strangely, dbRDA results suggested that monthly oxygen means explain most of the variability in BIPc (56.6%) out of all the tested modelled predictor variables. However, in the Random Forest multiannual oxygen mean was chosen, and it was quantified as one of the less strong predictors, compared to, for example, sediment parameters, temperature or salinity variability. On the other hand, in support of the high importance of total organic content predictor variable obtained from the large-scale estimate, dbRDA performed on measured predictors, solely available as point data, showed that total organic content had the highest explanatory power, clarifying along 49.3% of BIPc variability.

Interestingly, models fitted separately for BIPc values based solely on BIPdiff or BIPadv scores showed lower performance, indirectly indicating the relevance of distinguishing between differences in two sediment systems.

A comparison between the BIPc map generated here (Figure 4a) with that of BPc obtained on the basis of a similar dataset (Figure 4, see [20] for more details on modelling BPc distribution), suggested that BIPc pattern is mostly similar to that of BPc, but not identical. An overlay of the two layers revealed that BIPc produced higher relative scores than BPc in the deeper parts of the Bay of Lübeck, on soft sediments around Adler Ground, to the east of Kadet Trench, in the central part of the Greifswald Lagoon and in the south/western part of Pomeranian Bay, whereas larger patches of lower relative scores were observed in Kiel Bay, north off Zingst and at intermediate depths of Arkona Basin (Figure 4b).

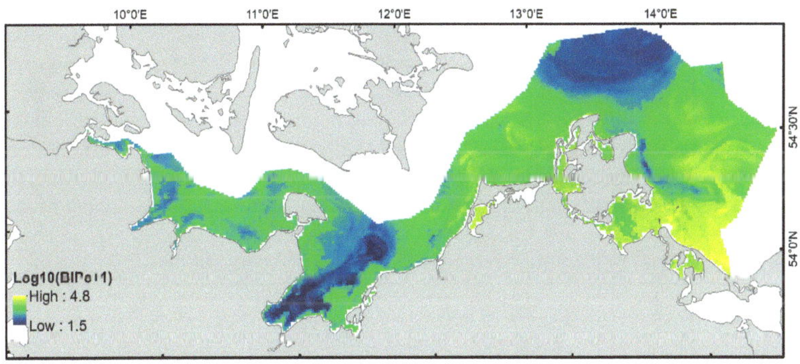


(**a**)

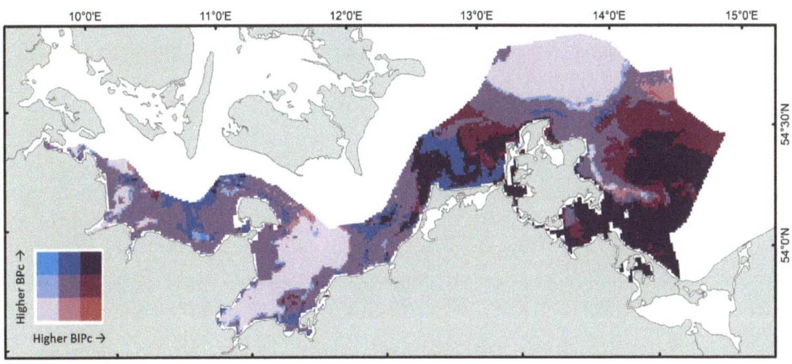


(**b**)

**Figure 4.** Predicted spatial distribution of BIPc (log-transformed) in the south-western Baltic Sea (**a**). Bivariate map of bioturbation and bioirrigation potentials hotspots in the south-western Baltic Sea (**b**). The distribution of BPc is a modelling result obtained using the same method and reported in Gogina et al. [20]. Red areas indicate relatively higher scores of BIPc compared to BPc, whereas blue areas are solely hotspots of BPc, but not of BIPc.

## 3.4. Validating BIPc Index Sensitivity to Changes in Solute Fluxes

There are some comparisons indicating that the patterns shown by BIPc distribution may not be unrealistic. Powilleit and Forster [43] reported bioturbation and bioirrigation rates from Pommeranian Bight. Reported bioirrigation indicated the highest rates ever measured using the tracer NaBr in this marine area, supporting the findings in Figure 4a. Furthermore, the spatial pattern of the tracer in pore waters also showed a positive association with BIPc values. A correlation of index with bioirrigation constants calculated for various depth intervals reported in Powilleit and Forster [43] was not significant, but this is not surprising due to the very low sample size. The strength of linear association was higher for deeper sediment depth intervals. In particular, the linear relationship of BIPc with bioirrigation constants was at its weakest for surface sediment layer (0–5 cm) and steepest for the bioirrigation values derived for 10–15 cm sediment interval (see Supplementary Figure S2a for more details). The NaBr tracer indicates the spatial proximity of introduced overlying water with irregularities in nutrient distribution, a fact described in many core sectioning studies on an overall averaged scale (e.g., [1]). However, quantifying the exact effect of faunal solute (or particle) transport on specific locally observed interface fluxes remains difficult.

We have tested BIPc index values against irrigation constants estimated in an experiment conducted in April 2018, where bromide tracer was used in 14 incubated cores (inner ⌀ = 10 cm) of fine and very fine sand sediment collected in the Oder Bank and in Greifswald Lagoon (sediment median grain sizes were 0.197 mm and 0.074 mm, respectively). After incubation, those cores were sliced and sieved to analyse the inhabiting macrofauna (unpublished own data). No significant linear correlation was found between derived bioirrigation parameters and any of the major macrofaunal parameters (abundance, biomass) or calculated functional indices (BIPc and BPc) using Pearson's correlation coefficient (Supplementary Figure S2b). Spearman's rank correlation suggested that, among the biotic parameters considered, BIPc calculated based on fauna present in cores and with use of diffusive scores indicated the strongest significant association with irrigated bromide amount (Table 3). The significance of this correlation was eliminated following the use of advective scores for those fine sediment cores. BPc derived from the same faunal data also significantly correlated with dissolved tracer irrigation amount; however, the association was weaker.

An analysis of BIPc values obtained when taking into account the real observed depth of vertical distribution of macrofaunal individuals (i.e., average depth of the sediment slice where the organisms were found in sediment cores) compared to BIPc values obtained with theoretical depth (i.e., when position or borrows depth reported in the literature was taken into account) revealed an obvious overestimation of BIPc values (Figure 5) in the latter case. This is not a surprise, as "theoretical" burrow depth taken from the literature often anticipates the maximum penetration depth known for any particular species. It also seems acceptable for this index that reflects the potential, i.e., the latent capacity of organisms to burrow and ventilate at a certain sediment depth. However, of course individual organisms can have the position at any depth between the declared lowest horizon and the sediment surface, subject to environmental settings, interspecific interactions and food availability. Therefore, the index calculated for different seafloor areas inhabited by similar communities may not mirror the differences in the realised effects on rates of processes.

**Table 3.** Spearman correlation coefficients (above the diagonal) with significance levels (*p*-value below the diagonal) calculated between bioirrigation intensity measured in 14 cores using bromide tracer (marked in green) and macrofauna parameters and functional indices BIPc and BPc (marked in blue). Significant correlation coefficients are shown in bold font. To calculate BIPc diff, the corresponding diffusion system scores are used for all cores, whereas for BIPc adv, advective scores are applied for fine sand sediment cores.

| Spearman Correlation Coefficients ($n$ = 14) / Corresponding $p$-Values | **Inventory Br mmol/m²** | **Irrigation L/(m² d) Entire Core Depth** | **Irrigation L/(m² d) in 2 to 10 cm Sediment Depth Layer** | **BIPc Diff** | **BIPc Diff in 2 to 10 cm Sediment Depth Layer** | **BIPc Adv (Adv Scores in Fine Sands)** | **BPc** | **BPc in 2 to 10 cm Sediment Depth Layer** | **Abundance, ind/m²** | **Wet Weight Biomass, g/m²** | **Ash Free Dry Weight Biomass, g/m²** | **Wet Weight Biomass in 2 to 10 cm Sediment Depth Layer, g/m²** |
|---|---|---|---|---|---|---|---|---|---|---|---|---|
| Inventory Br mmol/m² | | **0.896** | **0.866** | **0.723** | **0.569** | 0.473 | **0.582** | 0.437 | 0.207 | 0.446 | 0.477 | 0.389 |
| Irrigation L/(m² d) entire core depth | 0.000 | | **0.936** | **0.553** | 0.405 | 0.278 | 0.447 | 0.319 | 0.054 | 0.227 | 0.253 | 0.295 |
| Irrigation L/(m² d) in 2 to 10 cm sediment depth layer | 0.000 | 0.000 | | 0.509 | 0.328 | 0.264 | 0.434 | 0.282 | 0.068 | 0.225 | 0.264 | 0.311 |
| BIPc diff | 0.003 | 0.040 | 0.063 | | **0.789** | 0.516 | **0.895** | **0.789** | 0.169 | **0.763** | **0.780** | **0.578** |
| BIPc diff in 2 to 10 cm sediment depth layer | 0.034 | 0.151 | 0.252 | 0.001 | | 0.248 | **0.538** | **0.903** | −0.315 | **0.934** | **0.903** | **0.859** |
| BIPc (adv scores in fine sands) | 0.088 | 0.337 | 0.361 | 0.059 | 0.392 | | 0.415 | 0.257 | 0.453 | 0.231 | 0.253 | 0.125 |
| BPc | 0.029 | 0.109 | 0.121 | 0.000 | 0.047 | 0.140 | | **0.644** | 0.343 | **0.552** | **0.600** | 0.284 |
| BPc in 2 to 10 cm sediment depth layer | 0.118 | 0.266 | 0.329 | 0.001 | 0.000 | 0.375 | 0.013 | | −0.312 | **0.846** | **0.829** | **0.771** |
| Abundance, ind/m² | 0.478 | 0.854 | 0.816 | 0.563 | 0.273 | 0.104 | 0.230 | 0.277 | | −0.271 | −0.229 | −0.495 |
| Wet weight biomass, g/m² | 0.110 | 0.435 | 0.440 | 0.002 | 0.000 | 0.427 | 0.041 | 0.000 | 0.349 | | **0.991** | **0.890** |
| Ash free dry weight biomass, g/m² | 0.085 | 0.382 | 0.361 | 0.001 | 0.000 | 0.383 | 0.023 | 0.000 | 0.431 | 0.000 | | **0.873** |
| Wet weight biomass in 2 to 10 cm sediment depth layer, g/m² | 0.169 | 0.306 | 0.280 | 0.030 | 0.000 | 0.670 | 0.326 | 0.001 | 0.072 | 0.000 | 0.000 | |

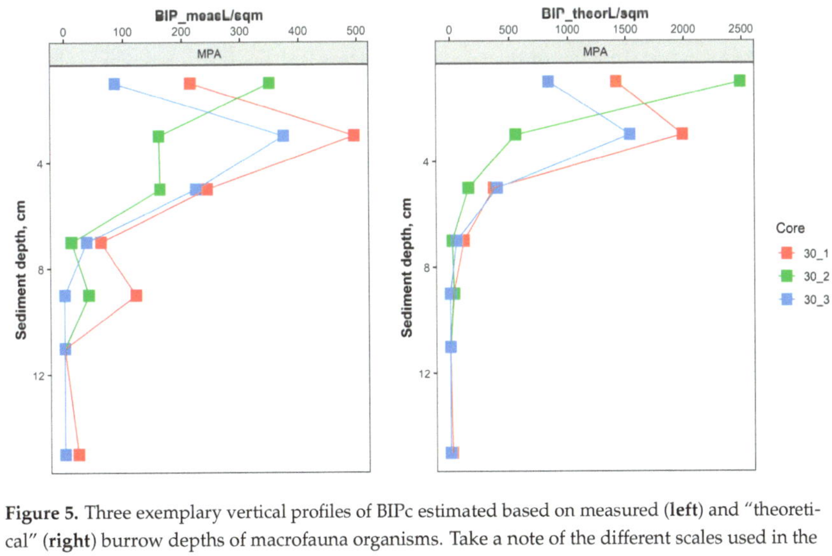


**Figure 5.** Three exemplary vertical profiles of BIPc estimated based on measured (**left**) and "theoretical" (**right**) burrow depths of macrofauna organisms. Take a note of the different scales used in the graphs.

Comparing BPc and BIPc with the measured ventilation rates (ml $h^{-1}$ ind.$^{-1}$, based on a mean rate over time inclusive of rest periods) and maximum reported irrigation depth in sediments (in cm), based on scars data (only $n = 4$) found in [11], suggested a somewhat stronger association and higher sensitivity of the latter index (where non-significant $R^2$ equalled 0.33 vs. 0.54 for ventilation rates, and 0.70 vs. 0.94 for irrigation depth, respectively).

The correlation between values of BIPc and BPc derived from grabs-based data and mean total solute fluxes from Lipka [41] was (positively) significant only for phosphate. This association was slightly stronger for BIPc than for BPc. Other Spearman rank correlation values were not significant for both indices. Moderate non-significant negative association with silica efflux was somewhat higher for BPc (Table 4). When the fluxes were averaged for multiple visits per station, the maximum silica efflux displayed a significant negative association with BIPc ($r = -0.89$, $p < 0.05$, $n = 5$).

**Table 4.** Spearman rank correlation coefficients between the estimated total effluxes from sediments obtained from Lipka [41] and trait-based bioturbation and bioirrigation indices, BPc and BIPc ($n = 12$). Significant values ($p < 0.05$) are highlighted in bold.

| | Oxygen | Silica | Ammonium | Phosphate | Manganese |
|---|---|---|---|---|---|
| BPc | −0.13 | −0.39 | 0.17 | 0.71 | 0.15 |
| BIPc | −0.20 | −0.33 | 0.18 | 0.73 | 0.18 |

## 4. Discussion

Organic matter remineralisation and nutrient regeneration, cycling of carbon, nitrogen, and metals are among the most important active processes in marine sedimentary habitats. They are often characterised by uni- and multivariate fluxes, which are known to be altered by bioirrigation of infauna [44,45]. Since the physical and hydrological properties of sediments are decisive in determining the magnitude of solute exchange in benthic ecosystems, BIPc accounts for differences in underlying physical processes in sand and mud by use of different traits scores for the same species [17]. Both bioirrigation and bioturbation are considered to have a desirable influence on those soft sediment types, as they are integral to a healthy soft-sediment ecosystem.

Suitable data on dissolved nutrients fluxes at the sediment–water interface and measured irrigation rates, which could be effectively used to validate the utility of bioirrigation indices, are scarcely available. Published global estimates, such as those based on relationships between bottom water oxygen and nitrogen loss [46], fluxes of phosphorus [47] and biogenic silica [48] appeared to be too coarse in spatial resolution when compared to our investigation. Apart from negligible sampling size, the weak relationship between estimated BIPc values in this study and irrigation volumes from Powilleit and Forster [43] (based on a handful of stations) could be associated with the poor performance of the index, or also with experimental artefacts arising during the measurements. The differences in association for different sediment layers indicated that correspondence between the Br transport and fauna is blurred close the sediment–water interface where, in addition to bioirrigation, molecular diffusion drives tracer fluxes most. For fine sand cores, the strongest association of BIPc index with tracer irrigation was found when using diffusive system scores for index estimation. The use of advective system scores to derive the index did not fit the measurements data. This suggests that burrow-linings in burrows constructed in more permeable sands may force more transport compared to diffusion, as assumed originally during the development of the BIPc index. For example, fertilisation and early larval development of lugworm *Arenicola marina* in sand occurs in U or J-shaped burrows. For ventilation and respiration, worms can pump water into those burrows down to 30 cm of sediment depth with rates as high as 430 mL per hour. Later larval development takes place enclosed in mucous tubes in the upper sediment layers [49,50]. The association of dissolved tracer irrigation with BPc was weaker. The increased total phosphate effluxes from sediments with higher bioirrigation suggested by the data from Lipka [41] and our BIPc results are in line with findings of Chaffin and Kane [51], who anticipated that bioirrigating fauna might be a source of internal phosphorus that explains the "trophic paradox" in lakes. These results are essential to justify our decision to assign fine sands to diffusion-driven systems in the performed analysis and, more importantly, to advocate for the usefulness of the BIPc index.

Using extensive benthic macrofauna data, including 2170 sampling events for over two decades, in combination with species distribution modelling methods, we have identified the potential key bioirrigating species and, for the first time, have mapped the distribution of expected bioirrigation hotspots in the south-western Baltic Sea. The resulting digital map layers are provided as Supplementary File S1. The distribution pattern for bioirrigation index BIPc largely resemble the spatial pattern of benthic community bioturbation potential index BPc [18], reported in earlier publications (e.g., [19,20]). This was also somewhat expected, since both traits-based indices are mostly determined by the same species dominating abundance and biomass, and by their attribution to functional groups that largely overlap in definition. Although the mathematical formulation is different, it is mostly the relative distribution of high and low values, and not the absolute units, that make up most of the utility for both indices to inform and support nature-based solutions. In the area west of the Fehmarn Belt, no hotspot for BIPc was observed, but one was observed for BPc. This could be a result of potential differences in processes driven by differences in community and the lower expected solute exchange in the diffusion-dominated system of that region.

It is important to acknowledge that, due to coupling of different processes, high process rates can also take place where measured effluxes are low. Miatta and Snelgrove [45] found that macrofauna explained up to 41% of the variation in benthic fluxes, whereas environmental variables only explained up to 19%, highlighting the importance of biodiversity for ecosystem functioning. However, their later results [52] suggested a relationship between resource availability and macrofaunal density, diversity, and taxonomic and trait composition. Nevertheless, those results also showed that organic matter remineralisation exhibited a more complex response, presumably reflecting variations in hydrodynamics and/or physical disturbances in heterogeneous continental margin habitats. In addition, remineralisation and bioirrigation do not necessarily show a strong linear relationship, as other drivers such as the supply of fresh and easily degradable organic matter, sedimentary settings, temperature and salinity shape these patterns [53]. This is particularly true when physical advections take over, in which case importance of bioirrigation as a contributor to functional processes, such as cycling of nutrients and metals, is expected to decrease.

However, although the relationship between bioirrigation potential and relevant processes on the scale of our study area remains to be examined, due to very limited availability of data covering the rate measurements of processes [11,41], the map of BIPc index can be considered a useful approximation of structuring activities in terms of the potential effects of organisms on solute transport, facilitation of transport of oxygen and excretory products [54].

Spatial distribution of another trait-based bioirrigation potential, IPc, was investigated in the North Sea sediments by Wrede et al. [16]. Based on a range of multifactorial experiments, the authors concluded that IPc distinctly improved the quantitative spatial assessment of bioirrigation activity. For the present study, no such extensive dataset on measured bioirrigation rates was available, and the presented validation is rather limited, providing nearly no statistically significant inferences. The results of the present study were also controversial in terms of the comparison of the strength of relations of BIPc and BPc to solute transports. Wrede et al. [55] demonstrated (based on generalised linear modelling results) that IPc provided a better estimation of phosphate, silicate, ammonium, nitrate and nitrite fluxes than community density, biomass or BPc. The authors suggested that predictive models of nutrient flux across the sediment–water interface will benefit greatly from incorporating macrofaunal irrigation behaviour by means of such indices. The BIPc index was developed particularly for the Baltic Sea realm [17]. In our results, an improved association of BIPc was suggested only for fluxes of phosphate and maximum recorded fluxes of silica.

Our results are prone to be more in agreement with the findings of Toussaint et al. [14], who found that BPc, and not IPc, as well as permeability, and not grain size or porosity, were mainly significantly involved in explaining biochemical process rates (and though the measured irrigation rate was also a frequent significant predictor, it did not correlate well with IPc). Based on the data analysed here, we have to conclude that adjustments distinguishing between diffusion and advection performed in estimates of BIPc do not seem to fully solve this crux. Data on permeability was unfortunately not available for our case study (but see [37]). Importantly, Toussaint et al. [14] concluded, and we confirm based on our results, that measuring processes is still essential, and their link with biological traits has not been not sufficiently studied to enable reliable and purely traits-based ecological assessments and future predictions [56]. In line with the findings of De Borger et al. [22] for IPc, here BIPc also showed a stronger association with burrow ventilation depths than with ventilation intensity or magnitude of fluxes. In order to achieve a firm predictive framework for the forecasting of relationships between bioturbators and sediment respiration in different environments, the present data need to be incorporated into existing biogeochemical models [6]. This will allow a more precise modelling of benthic-pelagic oxygen fluxes and even carbon retentions. A combination of tracer measurements with genomics was suggested as one possible way to better understand the microbial basis of differences between respiration in advection- and diffusion-dominated systems.

Both interannual variability and seasonality influence the intensity of bioirrigation [54]. Roskosch et al. [57] also hypothesised that seasonal variation caused by annual cycles of *Chironomus plumosus* would occur even if bioirrigation was measured under stable laboratory conditions. Thus, the long-term spatial distribution predicted here represents a climatic pattern which is not expected to necessarily demonstrate a strong association with momentary single measured fluxes or ventilation rates or depth estimates (for example with those few considered values included in the global database by Solan et al., [11]). Species scores for BIPc calculation parameterise by a single constant the changing sediment ventilation depth *or the rates* of the *dynamic process*, such as transport or mixing, that can also vary directionally. Naturally, macrobenthic communities that define BIPc respond to changes in the seabed and especially changes in sediment characteristics. Moreover, since we a priori introduce difference in BIPc as a function of sediment type, and then use sediment type as a predictor, the inevitable numerical gain of model accuracy seems to be predetermined. The propinquity of profiles to those expected from the distribution of macrofauna within the cores supports the influence of bioturbation and bioirrigation on elemental flows. Upper sediment layers with the highest BIP values usually contain the highest biochemical reaction rates within marine deposits [58]. However, these effects are not fully understood. Further measurements are required to explain the high natural variability, to understand anthropogenic effects on these processes, and finally to estimate the consequences of potential future changes for the ecosystem.

## 5. Conclusions

To conclude, our results suggested that using the BIPc index can help improve our understanding of the effects of bioirrigation on ecosystem functioning, but more evidence is required to provide a reliable synthesis of its pros and cons and to understand which ecological processes (that influence the fluxes of organic matter, nutrients and energy) it best approximates. Ultimately, only rate measurements in natural conditions can serve as cornerstones for index validation. Although statistical modelling may remain the best available tool for projecting future changes, we recognise its limitations and encourage more efforts to be made to gain a better mechanistic understanding of the causal relationships between bioirrigation and process rates on larger spatial scales. Further sensitivity studies and the development of trait-based indices, including the one discussed here, can bring us one step further to improving our ability to predict and manage biogeochemical functioning of intact and anthropogenically altered benthic communities, which is crucial for sustainable aquatic conservation and marine economy.

**Supplementary Materials:** The following are available online at https://www.mdpi.com/article/10.3390/biology11071085/s1, Table S1: Scores required to calculate BIPc for dominating taxa; Table S2: List of abiotic predictors and collinearity check results; Figure S1: Figure showing time-series of BIPc values at 8 monitoring stations; Figure S2: Correlations of BIPc values with measured bioirrigation rates. (a) Correlations of predicted BIPc values with bioirrigation constants resulting from bromide tracer experiments estimated for various depth intervals, as reported in Powilleit and Forster (2018). (b) Pearson correlation coefficients (above the diagonal) with significance levels ($p$-value below the diagonal) calculated between bioirrigation intensity measured with bromide tracer and macrofauna-based parameters; File S1: GIS layers of modelled BIPc distribution.

**Author Contributions:** Conceptualization, M.G. and M.L.Z.; methodology, M.G., S.F., M.L.Z. and J.R.R.; software, M.G.; validation, M.G., S.F. and M.L.Z.; formal analysis, M.G.; investigation, M.G.; resources, M.G., S.F. and M.L.Z.; data curation, M.G., S.F. and M.L.Z.; writing—original draft preparation, M.G.; writing—review and editing, M.G., S.F. and M.L.Z.; visualization, M.G.; supervision, M.G. and M.L.Z.; project administration, M.G., S.F. and M.L.Z.; funding acquisition, M.G., S.F. and M.L.Z. All authors have read and agreed to the published version of the manuscript.

**Funding:** This research was funded by the German Federal Ministry for Education and Research (Bundesministerium für Bildung und Forschung, BMBF) within its KÜNO projects SECOS, grant number 03F0666, and SECOS-Synthese, grant number 03F0738, and its DAM pilot mission project MGF Baltic Sea (MGF-Ostsee), grant number 03F0848A.

**Institutional Review Board Statement:** Not applicable.

**Informed Consent Statement:** Not applicable.

**Data Availability Statement:** Data is contained within the article or Supplementary Material.

**Acknowledgments:** We greatly acknowledge the valuable support of all IOW colleagues deployed in field sampling and laboratory analyses, and the captains and crews of RVs Elisabeth Mann Borgese, Poseidon, Alkor, and Maria S. Merian. We thank the two reviewers for inspiring comments that helped to improve the quality of the manuscript, and we are grateful to Irina Steinberg for correcting the English.

**Conflicts of Interest:** The authors declare no conflict of interest.

## References

1. Aller, R.C. Experimental studies of changes produced by deposit feeders on pore water, sediment, and overlying water chemistry. *Am. J. Sci.* **1978**, *278*, 1185–1234. [CrossRef]
2. Koretsky, C.M.; Meile, C.; Cappellen, P.V. Quantifying bioirrigation using ecological parameters: A stochastic approach. *Geochem. Trans.* **2002**, *3*, 17. [CrossRef] [PubMed]
3. Baranov, V. Influence of Bioturbation on Sediment Respiration in Advection and Diffusion Dominated Systems. Ph.D. Dissertation, Humboldt-Universität zu Berlin, Berlin, Germany, 8 February 2018. [CrossRef]
4. Wei, C.L.; Rowe, G.T.; Escobar-Briones, E.; Boetius, A.; Soltwedel, T. Global patterns and predictions of seafloor biomass using Random Forests. *PLoS ONE* **2010**, *5*, e15323. [CrossRef] [PubMed]
5. Meysman, F.J.R.; Galaktionov, E.S.; Gribsholt, B.; Middelburg, J.J. Bioirrigation in permeable sediments: Advective pore-water transport induced by burrow ventilation. *Limnol. Oceanogr.* **2006**, *51*, 142–156. [CrossRef]
6. Baranov, V.; Lewandowski, J.; Romeijn, P.; Singer, G.; Krause, S. Effects of bioirrigation of non-biting midges (Diptera: Chironomidae) on lake sediment respiration. *Sci. Rep.* **2016**, *6*, 27329. [CrossRef]
7. Granelli, W. Influence of Chironomus plumosus larvae on the oxygen-uptake of sediment. *Arch. Hydrobiol.* **1979**, *87*, 385–403.
8. Soster, F.M.; Matisoff, G.; Schloesser, D.W.; Edwards, W.J. Potential impact of Chironomus plumosus larvae on hypolimnetic oxygen in the central basin of Lake Erie. *J. Great Lakes Res.* **2015**, *41*, 348–357. [CrossRef]
9. McCall, P.L.; Tevesz, M.J. (Eds.) Sedimentary Geology 100. In *Animal-Sediment Relations: The Biogenic Alteration of Sediments*; Springer Science and Business Media: Berlin, Germany, 1985; pp. vii–xi.
10. Rhoads, D.C.; Boyer, L.F. The Effects of Marine Benthos on Physical Properties of Sediments. In *Animal-Sediment Relations*; McCall, P.L., Tevesz, M.J., Eds.; Topics in Geobiology; Springer: Boston, MA, USA, 1982; Volume 100. [CrossRef]

11. Solan, M.; Ward, E.R.; White, E.L.; Hibberd, E.E.; Cassidy, C.; Schuster, J.M.; Hale, R.; Godbold, J.A. Worldwide measurements of bioturbation intensity, ventilation rate, and the mixing depth of marine sediments. *Sci. Data* **2019**, *6*, 58. [CrossRef]
12. Soetaert, K.; Herman, P.M.; Middelburg, J.J.; Heip, C.; deStigter, H.S.; van Weering, T.C.E.; Epping, E.; Helder, W. Modelling 210Pb–derived mixing activity in ocean margin sediments: Diffusive versus nonlocal mixing. *J. Mar. Res.* **1996**, *54*, 1207–1227. [CrossRef]
13. Morys, C.; Forster, F.; Graf, G. Variabilty of bioturbation in various sediment types and on different spatial scales in the southwestern Baltic Sea. *Mar. Ecol. Prog. Ser.* **2016**, *557*, 31–49. [CrossRef]
14. Toussaint, E.; De Borger, E.; Braeckman, U.; De Backer, A.; Soetaert, K.; Vanaverbeke, J. Faunal and environmental drivers of carbon and nitrogen cycling along a permeability gradient in shallow North Sea sediments. *Sci. Total Environ.* **2021**, *767*, 144994. [CrossRef]
15. Butenschön, M.; Clark, J.; Aldridge, J.N.; Allen, J.I.; Artioli, Y.; Blackford, J.; Bruggeman, J.; Cazenave, P.; Ciavatta, S.; Kay, S.; et al. ERSEM 15.06: A generic model for marine biogeochemistry and the ecosystem dynamics of the lower trophic levels. *Geosci. Model Dev.* **2016**, *9*, 1293–1339. [CrossRef]
16. Wrede, A.; Beermann, J.; Dannheim, J.; Gutow, L.; Brey, T. Organism functional traits and ecosystem supporting services—A novel approach to predict bioirrigation. *Ecol. Indic.* **2018**, *91*, 737–743. [CrossRef]
17. Renz, J.R.; Powilleit, M.; Gogina, M.; Zettler, M.L.; Morys, C.; Forster, S. Community bioirrigation potential (BIPc), an index to quantify the potential for solute exchange at the sediment-water interface. *Mar. Environ. Res.* **2018**, *141*, 214–224. [CrossRef]
18. Solan, M.; Cardinale, B.J.; Downing, A.L.; Engelhardt, K.A.M.; Ruesink, J.L.; Srivastava, D.S. Extinction and ecosystem function in the marine benthos. *Science* **2004**, *306*, 1177–1180. [CrossRef]
19. Gogina, M.; Morys, C.; Forster, S.; Gräwe, U.; Friedland, R.; Zettler, M.L. Towards benthic ecosystemfunctioning maps: Quantifying bioturbation potential in the German part of the Baltic Sea. *Ecol. Ind.* **2017**, *73*, 574–588. [CrossRef]
20. Gogina, M.; Zettler, M.L.; Vanaverbeke, J.; Dannheim, J.; Van Hoey, G.; Desroy, N.; Wrede, A.; Reiss, H.; Degraer, S.; Van Lancker, V.; et al. Interregional comparison of benthic ecosystem functioning: Community bioturbation potential in four regions along the NE Atlantic shelf. *Ecol. Ind.* **2020**, *110*, 105945. [CrossRef]
21. Raymond, C.; Samuelsson, G.S.; Agrenius, S.; Schaanning, M.T.; Gunnarsson, J.S. Impaired benthic macrofauna function 4 years after sediment capping with activated carbon in the Grenland fjords, Norway. *Environ. Sci. Pollut. Res. Int.* **2021**, *28*, 16181–16197. [CrossRef]
22. De Borger, E.; Tiano, J.; Braeckman, U.; Ysebaert, T.; Soetaert, K. Biological and biogeochemical methods for estimating bioirrigation: A case study in the Oosterschelde estuary. *Biogeosciences* **2020**, *17*, 1701–1715. [CrossRef]
23. Mermillod-Blondin, F.; Rosenberg, R.; François-Carcaillet, F.; Norling, K.; Mauclaire, L. Influence of bioturbation by three benthic infaunal species on microbial communities and biogeochemical processes in marine sediment. *Aquat. Microb. Ecol.* **2004**, *36*, 271–284. [CrossRef]
24. Braeckman, U.; Provoost, P.; Gribsholt, B.; Van Gansbeke, D.; Middelburg, J.J.; Soetaert, K.; Vincx, M.; Vanaverbeke, J. Role of macrofauna functional traits and density in biogeochemical fluxes and bioturbation. *Mar. Ecol. Prog. Ser.* **2010**, *399*, 173–186. [CrossRef]
25. Dargahi, B.; Kolluru, V.; Cvetkovic, V. Multi-layered stratification in the Baltic Sea: Insight from a modelling study with reference to environmental conditions. *J. Mar. Sci. Eng.* **2017**, *5*, 2. [CrossRef]
26. Darr, A.; Gogina, M.; Zettler, M.L. Detecting hot-spots of bivalve biomass in the south-western Baltic Sea. *J. Mar. Syst.* **2014**, *134*, 69–80. [CrossRef]
27. Zettler, M.L.; Friedland, R.; Gogina, M.; Darr, A. Variation in benthic long–term data of transitional waters: Is interpretation more than speculation? *PLoS ONE* **2017**, *12*, e0175746. [CrossRef]
28. Vaquer-Sunyer, R.; Duarte, C.M. Thresholds of hypoxia for marine biodiversity. *Proc. Natl. Acad. Sci. USA* **2008**, *105*, 15452–15457. [CrossRef]
29. Klingbeil, K.; Mohammadi-Aragh, M.; Gräwe, U.; Burchard, H. Quantification of spurious dissipation and mixing—Discrete variance decay in a finite–volume framework. *Ocean Model.* **2014**, *81*, 49–64. [CrossRef]
30. Gräwe, U.; Naumann, M.; Mohrholz, V.; Burchard, H. Anatomizing one of the largest saltwater inflows into the Baltic Sea in December 2014. *J. Geophys. Res. Oceans* **2015**, *120*, 7676–7697. [CrossRef]
31. Tauber, F. Meeresbodenrelief in der Deutschen Ostsee/Seabed Relief in the German Baltic Sea. Bundesamt für Seeschifffahrt und Hydrographie. Hamburg, Germany. 2012. Available online: https://www.geoseaportal.de/ (accessed on 10 September 2016).
32. Tauber, F. Meeresbodensedimente in der Deutschen Ostsee/Seabed Sediments in the German Baltic Sea. Bundesamt für Seeschifffahrt und Hydrographie. Hamburg, Germany. 2012. Available online: https://www.geoseaportal.de/ (accessed on 10 September 2016).
33. Friedland, R.; Neumann, T.; Schernewski, G. Climate change and the Baltic Sea Action Plan: Model simulations on the future of the western Baltic Sea. *J. Mar. Syst.* **2012**, *105–108*, 175–186. [CrossRef]
34. Schernewski, G.; Friedland, R.; Carstens, M.; Hirt, U.; Leujak, W.; Nausch, G.; Neumann, T.; Petenati, T.; Sagert, S.; Wasmund, N.; et al. Implementation of European marine policy: New water quality targets for German Baltic water. *Mar. Policy* **2015**, *51*, 305–321. [CrossRef]
35. Leipe, T.; Tauber, F.; Vallius, H.; Virtasalo, J.; Uscinowicz, S.; Kowalski, N., Hille, S.; Lindgren, S.; Myllyvirta, T. Particulate organic carbon (POC) in surface sediments of the Baltic Sea. *Geo-Mar. Lett.* **2011**, *31*, 175–188. [CrossRef]

36. Neumann, T.; Koponen, S.; Attila, J.; Brockmann, C.; Kallio, K.; Kervinen, M.; Mazeran, C.; Müller, D.; Philipson, P.; Thulin, S.; et al. Optical model for the Baltic Sea with an explicit CDOM state variable: A case study with Model ERGOM (version 1.2). *Geosci. Model Dev.* **2021**, *14*, 5049–5062. [CrossRef]
37. Forster, S.; Khalili, A.; Kitlar, J. Variation of nonlocal irrigation in a subtidal benthic community. *J. Mar. Res.* **2003**, *61*, 335–357. [CrossRef]
38. Breiman, L. Random forests. *Mach. Learn.* **2001**, *45*, 5–32. [CrossRef]
39. Liaw, A.; Wiener, M. Classification and regression by randomForest. *R News* **2002**, *2*, 18–22.
40. R Core Team. *R: A Language and Environment for Statistical Computing*; R Foundation for Statistical Computing: Vienna, Austria, 2021; Available online: https://www.r-project.org/ (accessed on 1 November 2021).
41. Lipka, M. Current Biogeochemical Processes and Element Fluxes in Surface Sediments of Temperate Marginal Seas (Baltic Sea and Black Sea). Ph.D. Dissertation, Ernst-Moritz-Arndt-Universität, Greifswald, Germany, 2018. Available online: https://nbn-resolving.org/urn:nbn:de:gbv:9-opus-23761 (accessed on 21 May 2020).
42. Legendre, P.; Anderson, M.J. Distance-based redundancy analysis: Testing multispecies responses in multifactorial ecological experiments. *Ecol. Monogr.* **1999**, *69*, 1–24. [CrossRef]
43. Powilleit, M.; Forster, S. Continuous and High Transport of Particles and Solutes by Benthos in Coastal Eutrophic Sediments of the Pomeranian Bay. *Front. Mar. Sci.* **2018**, *5*, 472. [CrossRef]
44. Volkenborn, N.; Woodin, S.A.; Wethey, D.S.; Polerecky, L. Bioirrigation. In *Encyclopedia of Ocean Sciences*, 3rd ed.; Academic Press: Cambridge, MA, USA, 2019; Volume 2, pp. 663–670. [CrossRef]
45. Miatta, M.; Snelgrove, P.V.R. Benthic nutrient fluxes in deep-sea sediments within the Laurentian Channel MPA (eastern Canada): The relative roles of macrofauna, environment and sea pen octocorals. *Deep Sea Res. Part I* **2021**, *178*, 103655. [CrossRef]
46. Bohlen, L.; Dale, A.W.; Wallmann, K. Simple transfer functions for calculating benthic fixed nitrogen losses and C:N:P regeneration ratios in global biogeochemical models. *Glob. Biogeochem. Cycles* **2012**, *26*, GB3029. [CrossRef]
47. Wallmann, K. Phosphorus imbalance in the global ocean? *Glob. Biogeochem. Cycles* **2010**, *24*, GB4030. [CrossRef]
48. Dale, A.W.; Paul, K.M.; Clemens, D.; Scholz, F.; Schroller-Lomnitz, U.; Wallmann, K.; SGeilert, S.; Hensen, C.; Plass, A.; Liebetrau, V.; et al. Recycling and burial of biogenic silica in an open margin oxygen minimum zone. *Glob. Biogeochem. Cycles* **2021**, *35*, e2020GB006583. [CrossRef]
49. Baumfalk, Y.A. Heterogeneous grain size distribution in tidal flat sediment caused by bioturbation activity of *Arenicola marina* (Polychaeta). *Neth. J. Sea Res.* **1979**, *13*, 428–440. [CrossRef]
50. Fish, J.D.; Fish, S. (Eds.) *A Students Guide to the Seashore*; Cambridge University Press: Cambridge, UK, 1996.
51. Chaffin, J.D.; Kane, D.D. Burrowing mayfly (Ephemeroptera: Ephemeridae: *Hexagenia* spp.) bioturbation and bioirrigation: A source of internal phosphorus loading in Lake Erie. *J. Great Lakes Res.* **2010**, *36*, 57–63. [CrossRef]
52. Miatta, M.; Snelgrove, P.V.R. Sedimentary Organic Matter Shapes Macrofaunal Communities but Not Benthic Nutrient Fluxes in Contrasting Habitats Along the Northwest Atlantic Continental Margin. *Front. Mar. Sci.* **2021**, *8*, 756054. [CrossRef]
53. Piontek, J.; Endres, S.; Le Moigne, F.A.C.; Schartau, M.; Engel, A. Relevance of Nutrient-Limited Phytoplankton Production and Its Bacterial Remineralization for Carbon and Oxygen Fluxes in the Baltic Sea. *Front. Mar. Sci.* **2019**, *6*, 581. [CrossRef]
54. Norkko, J.; Shumway, S.E. (Eds.) Chapter 10. Bivalves as bioturbators and bioirrigators. In *Shellfish Aquaculture and the Environment*, 1st ed.; John Wiley and Sons, Inc.: Hoboken, NJ, USA, 2011; pp. 297–317.
55. Wrede, A.; Andresen, H.; Asmus, R.; Wiltshire, K.H.; Brey, T. Macrofaunal irrigation traits enhance predictability of nutrient fluxes across the sediment-water interface. *Mar. Ecol. Prog. Ser.* **2019**, *632*, 27–42. [CrossRef]
56. Garcia, C.; Solan, M.; Bolam, S.G.; Sivyer, D.; Parker, R.; Godbold, J.A. Exploration of multiple post-extinction compensatory scenarios improves the likelihood of determining the most realistic ecosystem future. *Environ. Res. Commun.* **2021**, *3*, 045001. [CrossRef]
57. Roskosch, A.; Hette, N.; Hupfer, M.; Lewandowski, J. Alteration of Chironomus plumosus ventilation activity and bioirrigation-mediated benthic fluxes by changes in temperature, oxygen concentration, and seasonal variations. *Freshw. Sci.* **2012**, *31*, 269–281. [CrossRef]
58. Zabel, M.; Hensen, C. Quantification and Regionalization of Benthic Reflux. In *Marine Geochemistry*; Springer: Berlin/Heidelberg, Germany, 2006. [CrossRef]

biology

MDPI

Article

# Linking Species Functional Traits to Specific Biogeochemical Processes under Trawling Pressure

**Irini Tsikopoulou [1,*], Christopher J. Smith [1], Konstantia Nadia Papadopoulou [1] and Melanie C. Austen [2]**

1 Hellenic Centre for Marine Research, P.O. Box 2214, 71003 Heraklion, Greece
2 School of Biological and Marine Sciences, University of Plymouth, Plymouth PL4 8AA, UK
* Correspondence: itsikopoulou@hcmr.gr; Tel.: +30-281-033-7727

**Simple Summary:** Bottom trawls when fishing move over large areas with different parts of the gears physically impacting the sea bottom, including the trawling wires, doors, ground rope and net. In this way, the trawl nets remove animals from bottom waters, the sediment surface and shallow sub-surface. The animals that live in the sea bottom with their activities and lifestyle play an important role in major ecosystem processes such as nutrient cycling. In this study, we investigated the relationship between species functional characteristics and ecosystem functions under trawling pressure. Our results indicated that under trawling, more opportunistic lifestyles and deposit feeders were associated with the ecosystem processes while in the undisturbed areas these processes were connected with bioturbating and burrowing species. Finding these links helps scientists and policy makers to better predict the impact of fishing disturbance on marine environment and set appropriate thresholds for marine ecosystem impacts.

**Abstract:** The impact of otter trawling on the relationship between functional traits of benthic invertebrates and specific biogeochemical processes were investigated in the oligotrophic Cretan Sea. The fishery is managed through a seasonal closure during the summer. During two seasons (winter and summer) replicate samples were taken from the field from a commercial trawl ground and an adjacent control area. Environmental parameters related to sediment biogeochemistry were measured including particulate organic carbon, sedimentary organic carbon, bottom water and sedimentary chlorophyll a and phaeopigment concentrations as well as benthic oxygen consumption. A significant impact of trawling was recorded only for bottom water chlorophyll and sedimentary organic carbon. Furthermore, the links between species traits and specific ecosystem processes were affected by trawling, highlighting the importance of unique functional modalities on ecosystem functioning. The traits that mostly influenced benthic biogeochemistry in the control sites were related to bioturbation and burrowing activities. In contrast, in the trawled sites, the associated traits were related to more opportunistic lifestyles and deposit feeding species that do not act as bioturbators. Thus, under trawling disturbance, this shift can decouple the species-sediment relations and affect nutrient cycling.

**Keywords:** oxygen flux; functional traits; trawling impact; seasonal fishery

## 1. Introduction

Grouping benthic invertebrate species according to their functional identity is a widespread approach that has led to an improved understanding of the role of these species within a community [1]. It also provides a useful tool for linking the ecosystem processes with specific functions as biological traits analysis can provide a direct link to certain functional properties related to life history, behavioral or morphological characteristics of species that drive ecosystem functioning [2]. A better understanding of the link between species functional traits and the processes related to specific ecosystem functions (i.e., nutrient cycling, oxygen consumption, denitrification) could help increase our ability to predict the impact of fishing disturbance on the benthic ecosystem functioning.

Demersal trawls have a large footprint on the seabed [3,4]. They move over large areas, with different parts of the gears physically impacting the sediment, including the trawling wires, doors, ground rope and net [5]. The actual impact on the seabed is therefore complex, partially turning over the sediment, partially reducing spatial heterogeneity and partially increasing it on different scales [6–8]. At the biological level, the trawl net removes animals from bottom waters, the sediment surface and shallow sub-surface. Impact studies have shown that the abundance of macrofauna and megafauna (infaunal and epifaunal) is generally reduced with corresponding changes in the community and trophic structure [6,9–12]. Smaller body-sized fauna, however, with fast life cycles may be more resistant to trawling [8,13,14].

Sediment resuspension and sediment-water nutrient exchanges may also be strongly influenced by the mechanical effects of trawling [15–20]. The effects of trawling on biogeochemistry depend on sediment characteristics, with stronger impacts occurring on muddy sediments compared to sand [17,21]. In addition, bioturbating organisms are particularly important as agents for irrigation and the movement of oxygen into the sediment [22,23]. With trawling impacting larger species in the sediment, continually impacted areas have a lower level of bioturbation and consequently different levels of sedimentary fluxes [8,12,13,18]. Trawling was also found to cause changes in oxygen regime, influencing the nitrogen cycle, as oxygen regulates both nitrification and denitrification [15,24].

To date, there are only few studies that have investigated the impact of trawling on the relationship between specific functional traits and biogeochemical processes [17,18,21,25–27], primarily in Northern waters (North Sea and Baltic Sea). In the upper few millimeters of the marine sediment, the oxidation of organic carbon controls the fluxes of oxygen and nutrients across the sediment-water interface, ultimately impacting primary productivity in the water column [28]. In the highly oligotrophic Eastern Mediterranean, where regeneration of nutrients from the seabed is likely to be extremely important for pelagic productivity and resultant increases in productivity may be at some distance from regeneration sites [29,30], it is essential to reveal the role of specific traits in biogeochemical processes that take place at the sediment-water interface.

Previous work in the study area has shown that macrofauna are impacted by trawling [6,31]. As a result, it is likely that the changes in the benthic community due to trawling will also have an impact on the community functional composition that regulate the biogeochemical processes in the sediment-water interface and this impact could be chronic, particularly in areas of the seabed that are regularly trawled. The fishery in Heraklion Bay off Crete is regulated through seasonal closures during the summer months when biological activity is likely to be highest. However, it is uncertain whether this closed season is sufficient to restore biological activity at the sediment-water interface. Taking these into account, we carried out a study to test the null hypothesis that trawl fishing affects the sediment biogeochemistry linkage to specific species functional traits. This hypothesis was tested by determining oxygen flux rates, chlorophyl and organic matter concentrations in bottom water and sediment, as well as species composition and functional traits in sediment cores collected inside and outside of trawled areas during both summer (closed season) and winter (open season).

## 2. Materials and Methods

### 2.1. Study Area

The study was carried out in and adjacent to one of the main commercial trawling lanes in Heraklion Bay (Island of Crete, southern Aegean, Figure 1). The area had been identified previously and work had been carried out there for some time with respect to trawling impacts on sediment characteristics and macrofaunal community structure [6,32]. The trawling lane follows the 200 m contour and narrows with the contours behind Dia Island. This natural constriction allowed for easy identification of the trawling lane and adjacent non-trawled control areas. The surface sediments of all the sampling areas were generally similar in terms of percentage composition. Sediments were predominantly silt (85–90%) with smaller fractions of clay and sand (mostly made up of shell fragments of *Turritellinella tricarinata*, *Aporrhais serresiana* and *Nucula* sp.). At the start of each of the sampling periods, 2 days were spent surveying with side scan sonar to verify the limits of the fishing lane (areas covered with trawl door marks). Four areas were identified for bottom sampling, two control areas to the south of the trawling lane (SOE, South Out East, 190 m depth; SOW, South Out West, 215 m depth) and two areas in the trawling lane (FLE, Fishing Lane East, 185 m depth; FLW, Fishing Lane West, 230 m depth) (Figure 1). Trawling takes place between the beginning of September and the end of May followed by an annual 4 month closed period (June–September). Based on observed trawling activity and knowledge of the gear, it was estimated that the intensity of trawling in the study area was 200% (total coverage 2 times per year). It should be noted that during all the sampling there were no trawlers in the vicinity.

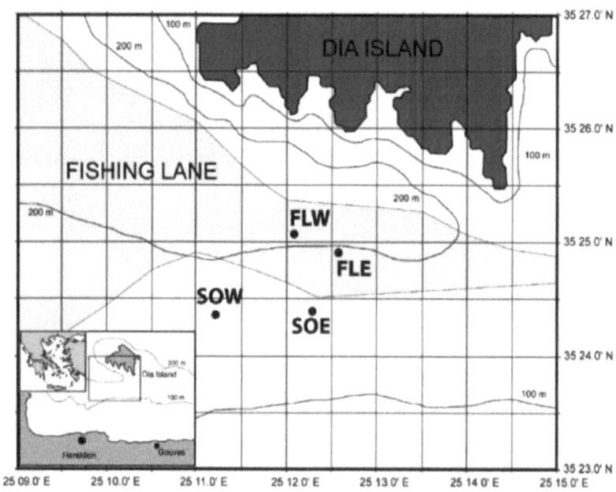


**Figure 1.** Study area showing trawling lane and control sites in Heraklion Bay, Crete (FLE, fishing lane east; FLW, fishing lane west, SOE, control area east; SOW, control area west).

### 2.2. Field Data Collection

Field work was carried out at the sampling sites in Heraklion Bay during July 2002 (summer) and March 2003 (winter). From each of the four sampling areas at Dia Island, 6 individual multicore drops were made (Bowers & Connely, Oban, UK, core diameter 10 cm, core length 40 cm). From each drop, two cores were selected on the basis of good penetration, undisturbed sediment fabric and clear water over the surface. One of these cores was stored in a cool box for use in the oxygen flux determination and the other was processed for sedimentary chemical parameters. This core was sub-sampled with smaller perspex core tubes and sectioned for granulometry (4 cm diam., 0–5 cm depth), organic carbon and chlorophyll concentrations (2 cm diam., 0–2 cm). Summarizing the sampling, there was a total of 6 replicates for an east and a west trawled and untrawled area in summer and winter.

### 2.3. Physico-Chemical Analyses

Water chlorophyll and phaeopigments were determined according to the fluorometric method of Yentsch and Menzel [33] using a TURNER 112 fluorometer. The analysis of water particulate organic carbon was performed according to Parsons et al. [34] by chromic acid wet oxidation. Sedimentary organic carbon was analysed by the wet oxidation method of Walkley and Black [35]. Chlorophyllous pigments in the sediment were determined according to Strickland and Parsons [36].

*2.4. Oxygen Flux Determination*

The oxygen flux set-up consisted of a cold sea water recycling system (15 degrees C: stable bottom water temperature at 200 m in the area) with water flowing from a chiller unit through a cooling coil in the bottom water reservoir tank and then into the core incubation tank and back into the chiller (Figure 2). The incubated cores stood upright with the cooling water covering the major part of the core tube. Each core tube was fitted with a special head containing an electrical stirring motor, with an opening allowing overflow, sampling and insertion of an oxygen and temperature probe. The vane of each stirrer was height adjusted for each core tube and was situated near to the top of the water column such that no vortices were formed. Turning speed was approximately 1 revolution per 1.5–2 s. A 12-channel peristaltic pump (Watson-Marlow, UK) fed a constant flow (1 mL per minute) of water from the bottom water reservoir tank to the cores. When the cores were full, they overflowed into the cooling tank. Oxygen concentrations were recorded from each of the cores and the reservoir tank after 24 h of incubation. Oxygen was measured using a dissolved oxygen meter and probe (WTW Oxi-330). Oxygen flux was estimated from:

$$Fx = (Ci - Co) \cdot Q/A,$$

where:

Fx flux of nutrient x micromol $m^{-2}$ $h^{-1}$)
CI concentration in the reservoir tank (μM)
Co concentration in the Core overlying water (μM)
Q flow of water through the core (l $h^{-1}$)
A area of the core ($m^2$)

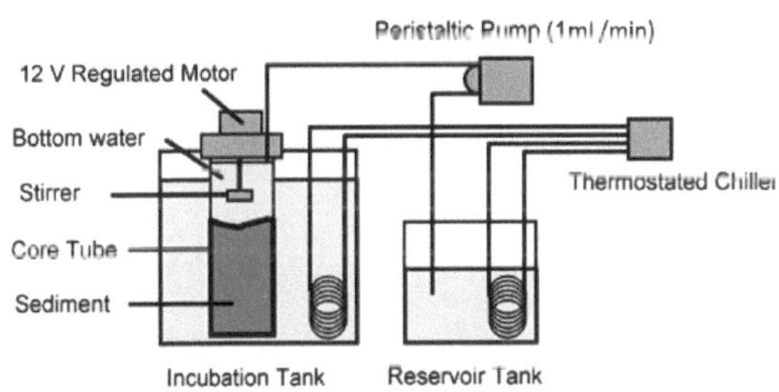


**Figure 2.** Experimental set-up for oxygen flux incubation.

Mean rates were estimated from the replicates from each sampling area.

*2.5. Macrofaunal Community and Biological Trait Analyses*

After the end of the experiment, cores were sieved through a 0.5 mm sieve for the determination of the benthic macroinvertebrate community. The fauna was sorted for species identification, enumeration and biomass measurement. A biological trait analysis (BTA) was conducted on the macrofaunal communities to determine their bioturbation attributes [37]. The biological response and effects traits considered in this study (7 traits, 27 modalities) describe the life history, morphological and behavioral characteristics of the benthic community [38] that may influence oxygen sediment-water exchanges and organic matter degradation (Table 1). Individual taxa were coded for the modalities of each trait using a fuzzy-coding procedure, which allows assessment of the affinity of a taxon to multiple categories. The trait scores were standardized for each species by re-coding the scores as percentage frequencies (Supplementary material Table S1). The species were classified into different functional categories based on information from a variety of literature sources [22,39–41] and databases (www.marlin.ac.uk/biotic; www.polytraits.lifewatchgreece.eu; accessed on 28 May 2021).

**Table 1.** Description of traits and trait modalities used in the biological trait analysis.

| | Trait | Modalities | Code | Trait Definition |
|---|---|---|---|---|
| Effect traits | Maximum body size (length) | <10 | S.10 | <10 mm |
| | | 11–20 | S.20 | 11–20 mm |
| | | 21–100 | S.100 | 21–100 mm |
| | Longevity (max) | <1 | L.1 | <1 year |
| | | 1–3 | L.3 | 1–3 years |
| | | 3–10 | L.10 | 3–10 years |
| | Feeding mode | Suspension & filter | FM.suspension | obtains food from water |
| | | Surface deposit | FM.sdeposit | including grazers |
| | | Sub-surface deposit | FM.subsdeposit | sub-surface deposit |
| | | Scavenger/opportunist | FM.scavenger | feeds upon dead animals |
| | | Surface predator | FM.predator | actively predates upon animals |
| | Bioturbation mode | Diffusive mixers | BM.diffusive | vertical and/or horizontal movement of sediment or particles, organisms with activities that result in a constant and random local sediment biomixing over short distances |
| | | Surface depositors | BM.deposition | deposition of particles at the sediment surface, species whose activities are restricted to <1–2 cm of the sediment |
| | | Upward conveyors | BM.upConveyors | upwards movement of particles resulting from biological activity, head down feeders that actively transport sediment to the sediment surface |
| | | Downward conveyors | BM.downConveyors | downwards movement of particles resulting from biological activity, head up feeders that actively transport sediment from the sediment surface |
| Response traits | Living habit | Tube-dwelling | LH.tube | builds a tube |
| | | Burrow-dwelling | LH.burrow | builds a burrow, includes mucus-lined burrows |
| | | Free-living | LH.free | freely moves around sediment/water |
| | | Inhabits crevices | LH.crevices | Inhabits crevices/holes/under stones |
| | Sediment position | Surface | Pos.S | surface dwellers |
| | | Infauna: 0–5 cm | Pos.5 | shallow-dwellers |
| | | Infauna: 6–10 cm | Pos.10 | buried deeper |
| | | Infauna: >10 cm | Pos.deep | deep-dwelling |
| | Mobility | Sessile | M.sessile | immobile, fixed in a place, stalked or not |
| | | Swim | M.swim | includes those which may stop swimming temporarily |
| | | Crawl/creep/climb | M.crawl | those which move above bed slowly |
| | | Burrower | M.burrower | Infers relatively low mobility |

Traits are categorized in effect and response traits. Response traits refer to functions related to the ability of the organism to survive and effect traits refer to how organisms influence the environment.

### 2.6. Data Analyses

To detect significant differences in environmental variables and oxygen fluxes between the samples, univariate statistical techniques were used. Specifically, three-way mixed ANOVA with trawling and season as fixed factors and site as nested factor within trawling was conducted to determine the effects of these factors on each of the variables measured (i.e., oxygen flux, organic carbon, chlorophyll a and phaeopigments). Residual analysis was performed to test for the assumptions of the three-way ANOVA. Normality was assessed using Shapiro Wilk's normality test and homogeneity of variances was assessed by Levene's test. Statistical significance was accepted at the $p < 0.025$ level for simple two-way interactions and effect of trawling at each group of the other factors.

To assess the relationship between species traits and benthic biogeochemical processes as well as the effect of trawling on this relationship, a fourth-corner analysis was performed separately for the disturbed (trawled) and the undisturbed sites and for the two seasons [24,42]. The fourth-corner analysis requires three different data tables: the R table, with measurements of the environmental variables (i.e., oxygen, silt and clay content, chlorophyll a, and organic carbon); the L table, constituted by the biomass of each species in each sample; and the Q table, composed of fuzzy-coded trait data for each species. Prior to fourth-corner analysis, a standardization was applied to environmental data and variables were checked for collinearity. The Hellinger transformation was also applied to species community data. The fourth-corner method evaluates the significance of bivariate associations (i.e., one single trait and one single environmental variable at a time) [43]. For the later analysis, 49,999 permutations were used in all randomization procedures and the false discovery rate method (FDR) was selected to adjust *p*-values for multiple testing. The fourth-corner method was carried out with the ade4 package in the R program (version 4.0.2) [43,44].

## 3. Results

### 3.1. Physico-Chemical Analyses

Figure 3 shows the bottom water and sedimentary parameters (no bottom water data for the easterly sampling sites during summer sampling for technical reasons). For chlorophyll a (Chl a), there was a statistically significant two-way interaction between season and trawling (Table 2). A significant effect of trawling was recorded for the summer sampling ($F = 57.1$, $p = 0.000$, lower values in the trawled area) but not for winter ($F = 2.29$, $p = 0.138$). No significant changes were recorded for the bottom water phaeopigments (Table 2). Particulate organic carbon in the bottom water varied significantly only between seasons ($F = 16.51$, $p = 0.003$, higher in winter).

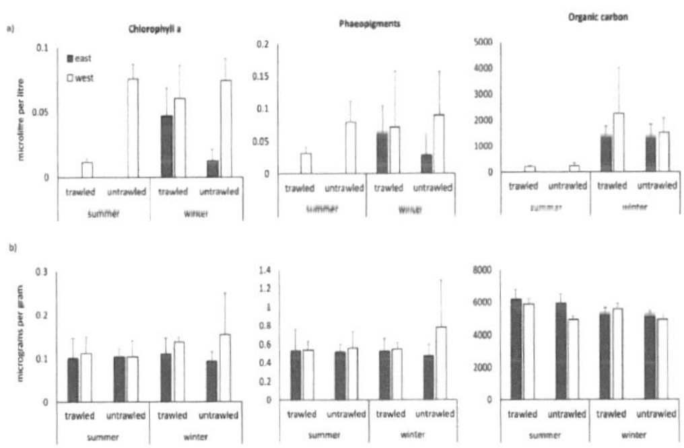


**Figure 3.** (**a**) Bottom water and (**b**) sedimentary chlorophyll a, phaeopigments and organic carbon from the Heraklion Bay trawled and untrawled sampling sites with mean and standard deviation in summer and winter. Grey bars indicate the eastern sites and white bars the western sites.

**Table 2.** Three-way ANOVA results for the comparisons of the bottom water and sedimentary environmental variables as well as oxygen flux for the factor trawling, season and site. Significant differences at $p < 0.05$ indicated with "*", at $p < 0.01$ indicated with "**" and non-significant indicated with "ns".

| | | Bottom Water | | | | | | Sediment | | | | | | Flux | |
|---|---|---|---|---|---|---|---|---|---|---|---|---|---|---|---|
| | | Chlorophyll a | | Phaeopigments | | Organic Carbon | | Chlorophyll a | | Phaeopigments | | Organic Carbon | | Oxygen | |
| Source of Variation | DF | F | *p* | F | *p* | F | *p* | F | *p* | F | *p* | F | *p* | F | *p* |
| trawling | 1 | 22.55 | ** | 1.40 | ns | 0.75 | ns | 0.02 | ns | 0.31 | ns | 116.96 | ** | 0.03 | ns |
| season | 1 | 17.93 | ** | 1.38 | ns | 16.51 | ** | 1.30 | ns | 0.23 | ns | 16.30 | ** | 33.84 | ** |
| site | 1 | - | - | - | - | | | 2.71 | ns | 2.04 | ns | 9.55 | * | 16.53 | ** |
| season × trawling | 1 | 19.80 | ** | 0.49 | ns | 0.92 | ns | 0.02 | ns | 0.47 | ns | 0.51 | ns | 3.55 | ns |
| trawling × site | 1 | - | - | - | - | - | - | 0.29 | ns | 1.68 | ns | 8.46 | * | 2.23 | ns |
| season × site | 1 | . | - | - | - | - | - | 3.62 | ns | 1.57 | ns | 8.85 | * | 2.23 | ns |
| trawling × season × site | 1 | . | - | - | - | - | - | 0.33 | ns | 0.46 | ns | 0.18 | ns | 0.53 | ns |

Sedimentary Chl values were around 0.1 microg g-1 sediment. There were no significant differences between any of the factors tested (Table 2). Sedimentary phaeopigment values ranged 0.75–1.25 microg g-1 sediment, with the exception of a peak mean value in the westerly fishing lane in winter. Trends were similar to that of Chl a and there were no significant differences between any of the factors tested. A two-way interaction between site and trawling was recorded for sedimentary organic carbon (Table 2). The simple main effect of trawling was significant for the westerly sites (F = 25.6, $p$ = 0.000) but not for the easterly sites (F = 1.72, $p$ = 0.197). Nevertheless, the same trend (higher organic carbon in the trawled area)—even if not statistically significant—was also found in the easterly sites (Figure 3).

Oxygen influxes were recorded in every case (Figure 4). Trawling did not significantly affect oxygen flux rates (Table 3). Significant differences were recorded for the factors site and season (Table 2).

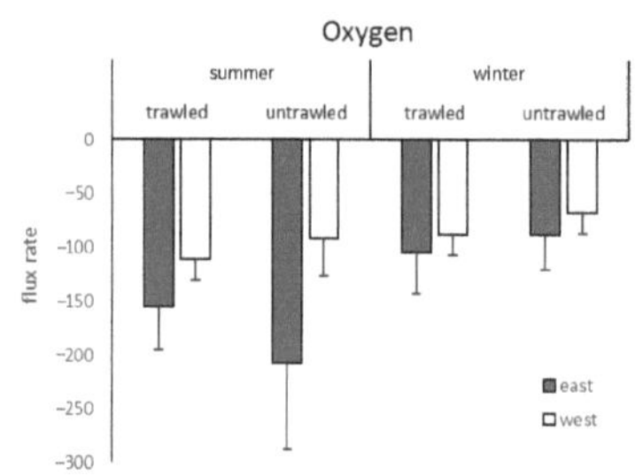


**Figure 4.** Oxygen flux rates (micromole per square metre per hour) from the trawled and untrawled sampling sites with mean and standard deviation in summer and winter. Grey bars indicate the eastern sites and white bars the western sites.

**Table 3.** Summary of the major functional groups related (positively: + and negatively: −) to specific biogeochemical processes in different trawling intensities and seasons. OC: organic carbon, POC: particulate organic carbon.

| Season | Trawling | | Description | Sediment Chl *a* | OC | Silt and Clay | Oxygen | Water Chl *a* | POC |
|---|---|---|---|---|---|---|---|---|---|
| summer | untrawled | Group 1 | long-lived infauna, burrowers and suspension feeding species | − | − | | − | | |
| winter | | Group 2 | sessile species that live in tubes and suspension feeders | − | − | | − | | |
| | | Group 3 | free living infauna that moves in the sediment, scavengers and predators | + | + | | + | | + |
| summer | trawled | Group 4 | deposit feeding species | | | − | | | |
| winter | | Group 5 | free living infauna that moves in the sediment, deposit feeders | | − | | − | + | + |
| | | Group 6 | Suspension feeding species | | | | | − | |

### 3.2. *Link between Species Traits and Biogeochemical Processes*

A total of 40 species were found in the samples after the end of the experiment. Their functional traits and abundances are shown in Supplementary material (Tables S1 and S2). The most abundant group was Polychaeta, followed by Bivalvia (mainly *Abra alba*), Sipuncula and Crustacea. Crustacea were not found in the cores from the trawled site nor during the trawling season (winter). The relationships between species functional traits and biogeochemical variables in different sites are summarized in Figure 5.

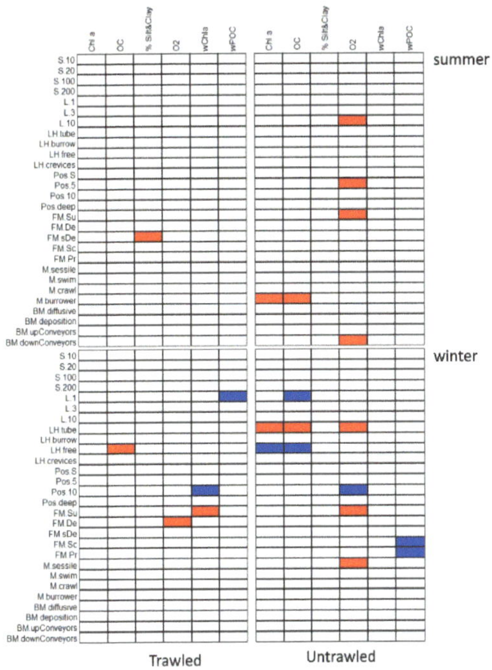


**Figure 5.** Representation of significant ($p < 0.05$) associations identified by the fourth-corner method. The blue shades indicate significant positive associations and red shades significant negative associations between traits and biogeochemical variables. Variables with no significant associations are shown in white. $p$ values were adjusted for multiple comparisons using the FDR procedure. Codes for traits are explained in Table 1. Chla: chlorophyll a, OC: organic carbon, %Silt&Clay: silt and clay percentage, $O_2$: oxygen, wChla: water column chlorophyll, wPOC: particulate organic carbon.

The significant associations (negative and positive) of species traits and biogeochemical variables describing either the bottom water or the sediment, were different in trawled and untrawled sites. More associations were recorded in the untrawled sites. The traits that are linked to biogeochemical processes in trawled sites versus control sites in both seasons are summarized in Table 3.

In the untrawled sites, biogeochemical processes are influenced by the more complex functional traits community consisting of free-living scavengers and predators, by sessile suspension feeders and also by burrowing species with high longevity that are oriented with their heads towards the sediment surface. These species transport sediment from the surface to deeper layers as they feed. In contrast, in the trawled sites biogeochemical processes were only related to short-lived species with high mobility either deposit feeders or suspension feeders.

## 4. Discussion

This study explored how trawling affects the linkage between functional traits of benthic macrofaunal species and specific biogeochemical processes such as oxygen consumption and organic matter degradation. In general, significant differences were detected only for bottom water Chl a and sediment OC concentrations between the trawled and control areas indicating that there was an impact of bottom fishing on benthic biogeochemistry. Nevertheless, oxygen consumption was not affected. In addition, the traits that are linked to specific ecosystem processes were different between trawled and undisturbed sites, highlighting the importance of unique functional traits in preserving ecosystem functioning.

The sediment and the adjacent bottom water environmental parameters could be directly affected by trawling in two ways, either removal by resuspension or constant exposure at the sediment surface by uncovering deeper sedimentary-locked carbon with continuous trawling, but also indirectly through species loss which can cause changes in ecosystem functioning [17,18,25,45,46]. As a result, differences in bottom water Chl a concentration and sedimentary organic carbon at the different sites can be explained by the aforementioned processes. Pusceddu et al. [47] in the northwestern Aegean, found that sedimentary organic carbon concentrations displayed a significant increase immediately after the initiation of the trawling season, probably from trawling-induced uplift from deeper sediment layers, but no significant short-term changes in sedimentary pigment levels. In accordance, Sciberras et al. [17] have linked bottom trawling to increased sediment Chl a and organic carbon and attributed this enhancement to a considerable reduction in bacterial biomass due to sediment resuspension that leads to a slow-down in the remineralization of the labile portion of organic matter within the sediment rather than to a loss in macrofaunal community bioturbation potential. Despite the findings of Sciberras et al. [17], macrofaunal species play also an important role in controlling the levels of OC within the sediment via bioturbation [48]. Thus, the difference in OC between the trawled and untrawled areas may be related to differences in macrofauna functional traits. Specifically, a decline in community complexity and bioturbation capacity can lead to a decrease in sediment oxygenation and carbon cycling and result in higher sedimentary OC concentrations [48]. On the other hand, oxygen consumption remains unaffected by trawling in our experiment which is in accordance with other studies suggesting that oxygen consumption is either unaffected [15,17] or decreased [18,19,21] by trawling due to the removal of the reactive surface sediment and the consequent reduction in carbon mineralization.

Besides the changes in environmental parameters, trawl fishing causes changes in species composition either through the direct removal of animals or by decreasing the settlement succession of species with pelagic larvae [49] from the control areas in the trawled areas and vice versa. This may result in alterations both in the functional effects and response traits of a community, that in turn may have broad implications for the overall ecosystem functioning [50]. In our study, trawling affected the links between species traits and biogeochemistry. Specifically, the untrawled sites presented more variable associations between specific traits and biogeochemical processes than the trawled sites. This indicated that in the undisturbed area, biogeochemistry was mainly controlled by macrofaunal community. In contrast, trawled sites appeared to have less associations between macrofaunal traits and biogeochemical variables. This could be explained either by lower functional redundancy within the disturbed sites due to fewer species overall or by the stimulation of microbial activity against the macrofaunal-induced metabolism in the disturbed sites [18,24,51]. It is also worth noting that more associations between species functional traits and biochemical processes were recorded in winter than in summer. It was anticipated that higher temperatures in summer would have favored macrofaunal activity [52] and as a result, an increased number of significant associations between functional traits and specific biogeochemical processes. Nevertheless, there is not always a straightforward relation between single species and biogeochemical cycling processes but rather complex interactions of species competition for space and food and habitat characteristics since macrofaunal activities have different effects across different environments [53].

In the undisturbed sites, many traits were found to be related to benthic biogeochemistry indicating a balanced and diverse community. Bioturbation, burrowing activities and feeding mode are some of the traits influencing sediment geochemistry. Specifically, bioturbation and burrowing affect nutrient fluxes across the sediment water interface, organic carbon concentrations, chlorophyll burial and decomposition, oxygen consumption and mediation of nitrogen [22,24,54–58]. In addition, sessile species that live in tubes influence the stability and accumulation rates of sediment, which are key drivers of organic carbon burial and storage [28,59]. Burrowing fauna can increase the stability and accumulation rate of sediment if there is an increase in biogenic materials such as tubes or mucus production [60].

In contrast to the undisturbed sites, long-lived bioturbators, sessile and burrowing fauna that are particularly vulnerable to damage from mobile demersal fishing, in the trawled sites showed reductions due to the resuspension and the following deposition of the surface sediments caused by trawling [8,18,61]. Subsequently, small opportunistic deposit feeders with low bioturbation ability may take advantage of the aforementioned reduction and appeared to be linked with the biogeochemical parameters in the trawled sites. Thus, under trawling disturbance, the shift towards opportunistic lifestyles can decouple the macrofauna-sediment relations that facilitate nutrient cycling and lead to a microbially driven metabolism [17,51,60].

## 5. Conclusions

Trawling had an impact on the links between macrofaunal functional traits and biogeochemical processes. Nevertheless, this impact was not adequately reflected in all the biogeochemical variables studied. As a result, in the trawled sites, either there was an uncoupling of species-sediment relations towards a microbially induced metabolism or the benthic community, and specifically ephemeral deposit feeders, took advantage of the organic matter availability due to sediment resuspension and preserved benthic biogeochemistry. The differences recorded in the associations of species functional traits and biogeochemical processes in both trawled and untrawled sites underlined the importance of unique functional traits on ecosystem functioning. In addition, in such an experimental design, quantitative effects of trawling are likely to be underestimated due to reduced suitable settling area for species also within the undisturbed area due to its spatial vicinity to the trawled sites.

Finding the links and relating species functional traits to important specific ecosystem processes will help scientists and policymakers to better predict and communicate the impact of fishing disturbance on benthic ecosystem functioning and set appropriate thresholds for adverse effects. Exceeding these thresholds should trigger management response actions. To this end, successful management measures could include confining the trawling footprint within historically trawled areas and/or allocating a longer closed period for trawling to restore the area [62].

**Supplementary Materials:** The following supporting information can be downloaded at: https://www.mdpi.com/article/10.3390/biology11101378/s1, Table S1: Species trait scores, Table S2: Species abundance per sampling core, Table S3: ANOVA results.

**Author Contributions:** I.T.: Conceptualization, Methodology, Formal analysis, Investigation, Writing—Original draft, Visualization. C.J.S.: Methodology, Formal analysis, Investigation, Resources, Visualization, Supervision, Funding acquisition, Writing—review and editing. K.N.P.: Investigation, Methodology, Formal analysis, Writing—review and editing, Funding acquisition. M.C.A.: Investigation, Writing—review and editing, Funding acquisition. All authors have read and agreed to the published version of the manuscript.

**Funding:** The work was carried out under the EU-funded project Cost-Impact: Costing the impact of demersal fishing on marine ecosystem processes and biodiversity. FP5, Quality of Life: Q5RS-2001-00993. Co-financing was provided by the General Secretariat for Research and Technology, Ministry of Development, Greece. The authors acknowledge the support by the MarBEF Network of Excellence 'Marine Biodiversity and Ecosystem Functioning' which is funded by the Sustainable Development, Global Change and Ecosystems Programme of the European Community's Sixth Framework Programme (contract no. GOCE-CT-2003-505446).

**Institutional Review Board Statement:** Not applicable.

**Informed Consent Statement:** Not applicable.

**Data Availability Statement:** Data on the benthic community can be found in the Supplementary Materials.

**Acknowledgments:** The authors would like to thank their colleagues at HCMR for help with sampling, construction of the incubation equipment and analyses, the Captain and crew of the RV Philia, and the partners of the Cost-Impact project for their inputs into the work.

**Conflicts of Interest:** The authors declare no conflict of interest.

## References

1. Bolam, S.G.; Garcia, C.; Eggleton, J.; Kenny, A.J.; Buhl-Mortensen, L.; Gonzalez-Mirelis, G.; van Kooten, T.; Dinesen, G.; Hansen, J.; Hiddink, J.G.; et al. Differences in Biological Traits Composition of Benthic Assemblages between Unimpacted Habitats. *Mar. Environ. Res.* **2017**, *126*, 1–13. [CrossRef]
2. Bremner, J. Species' Traits and Ecological Functioning in Marine Conservation and Management. *J. Exp. Mar. Biol. Ecol.* **2008**, *366*, 37–47. [CrossRef]
3. Amoroso, R.O.; Pitcher, C.R.; Rijnsdorp, A.D.; McConnaughey, R.A.; Parma, A.M.; Suuronen, P.; Eigaard, O.R.; Bastardie, F.; Hintzen, N.T.; Althaus, F.; et al. Bottom Trawl Fishing Footprints on the World's Continental Shelves. *Proc. Natl. Acad. Sci. USA* **2018**, *115*, E10275–E10282. [CrossRef] [PubMed]
4. Eigaard, O.R.; Bastardie, F.; Hintzen, N.T.; Buhl-Mortensen, L.; Buhl-Mortensen, P.; Catarino, R.; Dinesen, G.E.; Egekvist, J.; Fock, H.O.; Geitner, K.; et al. The Footprint of Bottom Trawling in European Waters: Distribution, Intensity, and Seabed Integrity. *ICES J. Mar. Sci.* **2017**, *74*, 847–865. [CrossRef]
5. Eigaard, O.R.; Bastardie, F.; Breen, M.; Dinesen, G.E.; Hintzen, N.T.; Laffargue, P.; Mortensen, L.O.; Nielsen, J.R.; Nilsson, H.C.; O'Neill, F.G.; et al. Estimating Seabed Pressure from Demersal Trawls, Seines, and Dredges Based on Gear Design and Dimensions. *ICES J. Mar. Sci. J. Cons.* **2016**, *73*, i27–i43. [CrossRef]
6. Smith, C.J.; Papadopoulou, K.N.; Diliberto, S. Impact of Otter Trawling on an Eastern Mediterranean Commercial Trawl Fishing Ground. *ICES J. Mar. Sci.* **2000**, *57*, 1340–1351. [CrossRef]
7. Hinz, H.; Prieto, V.; Kaiser, M.J. Trawl Disturbance on Benthic Communities: Chronic Effects and Experimental Predications. *Ecol. Appl.* **2009**, *19*, 761–773. [CrossRef]
8. Tillin, H.M.; Hiddink, J.G.; Jennings, S.; Kaiser, M.J. Chronic Bottom Trawling Alters the Functional Composition of Benthic Invertebrate Communities on a Sea-Basin Scale. *Mar. Ecol. Prog. Ser.* **2006**, *318*, 31–45. [CrossRef]
9. Jennings, S.; Kaiser, M.J. The Effects of Fishing on Marine Ecosystems. In *Advances in Marine Biology*; Blaxter, J.H.S., Southward, A.J., Tyler, P.A., Eds.; Academic Press: London, UK, 1998; Volume 34, pp. 1–475.
10. Kaiser, M.J.; Clarke, K.R.; Hinz, H.; Austen, M.C.V.; Somerfield, P.J.; Karakassis, I. Global Analysis of Response and Recovery of Benthic Biota to Fishing. *Mar. Ecol. Prog. Ser.* **2006**, *311*, 1–14. [CrossRef]
11. Queirós, A.M.; Hiddink, J.G.; Kaiser, M.J.; Hinz, H. Effects of Chronic Bottom Trawling Disturbance on Benthic Biomass, Production and Size Spectra in Different Habitats. *J. Exp. Mar. Biol. Ecol.* **2006**, *335*, 91–103. [CrossRef]
12. de Juan, S.; Thrush, S.F.; Demestre, M. Functional Changes as Indicators of Trawling Disturbance on a Benthic Community Located in a Fishing Ground (NW Mediterranean Sea). *Mar. Ecol. Prog. Ser.* **2007**, *334*, 117–129. [CrossRef]
13. Olsgard, F.; Schaanning, M.T.; Widdicombe, S.; Kendall, M.A.; Austen, M.C. Effects of Bottom Trawling on Ecosystem Functioning. *J. Exp. Mar. Biol. Ecol.* **2008**, *366*, 123–133. [CrossRef]
14. Van Denderen, P.D.; Bolam, S.G.; Hiddink, J.G.; Jennings, S.; Kenny, A.; Rijnsdorp, A.D.; Van Kooten, T. Similar Effects of Bottom Trawling and Natural Disturbance on Composition and Function of Benthic Communities across Habitats. *Mar. Ecol. Prog. Ser.* **2015**, *541*, 31–43. [CrossRef]
15. Trimmer, M.; Petersen, J.; Sivyer, D.B.; Mills, C.; Young, E.; Parker, E.R. Impact of Long-Term Benthic Trawl Disturbance on Sediment Sorting and Biogeochemistry in the Southern North Sea. *Mar. Ecol. Prog. Ser.* **2005**, *298*, 79–94. [CrossRef]
16. Mayer, L.M.; Schick, D.F.; Findlay, R.H.; Rice, D.L. Effects of Commercial Dragging on Sedimentary Organic Matter. *Mar. Environ. Res.* **1991**, *31*, 249–261. [CrossRef]
17. Sciberras, M.; Parker, R.; Powell, C.; Robertson, C.; Kröger, S.; Bolam, S.; Geert Hiddink, J. Impacts of Bottom Fishing on the Sediment Infaunal Community and Biogeochemistry of Cohesive and Non-Cohesive Sediments. *Limnol. Oceanogr.* **2016**, *61*, 2076–2089. [CrossRef]
18. Morys, C.; Brüchert, V.; Bradshaw, C. Impacts of Bottom Trawling on Benthic Biogeochemistry in Muddy Sediments: Removal of Surface Sediment Using an Experimental Field Study. *Mar. Environ. Res.* **2021**, *169*, 105384. [CrossRef]
19. Tiano, J.C.; Witbaard, R.; Bergman, M.J.N.; Van Rijswijk, P.; Tramper, A.; van Oevelen, D.; Soetaert, K. Acute Impacts of Bottom Trawl Gears on Benthic Metabolism and Nutrient Cycling. *ICES J. Mar. Sci.* **2019**, *76*, 1917–1930. [CrossRef]
20. Bradshaw, C.; Jakobsson, M.; Brüchert, V.; Bonaglia, S.; Mörth, C.M.; Muchowski, J.; Stranne, C.; Sköld, M. Physical Disturbance by Bottom Trawling Suspends Particulate Matter and Alters Biogeochemical Processes on and Near the Seafloor. *Front. Mar. Sci.* **2021**, *8*, 1127. [CrossRef]
21. De Borger, E.; Tiano, J.; Braeckman, U.; Rijnsdorp, A.D.; Soetaert, K. Impact of Bottom Trawling on Sediment Biogeochemistry: A Modelling Approach. *Biogeosciences* **2021**, *18*, 2539–2557. [CrossRef]
22. Wrede, A.; Beermann, J.; Dannheim, J.; Gutow, L.; Brey, T. Organism Functional Traits and Ecosystem Supporting Services—A Novel Approach to Predict Bioirrigation. *Ecol. Indic.* **2018**, *91*, 737–743. [CrossRef]

23. Tsikopoulou, I.; Lampa, M.; Tsiola, A.; Pitta, P.; Tsapakis, M.; Karakassis, I. Functional Adaptations of Benthic Communities to Organic Matter Enrichment at the Edge of an Allowable Zone of Effect (AZE). *Estuar. Coast. Shelf Sci.* **2021**, *262*, 107596. [CrossRef]
24. Sciberras, M.; Tait, K.; Brochain, G.; Hiddink, J.G.; Hale, R.; Godbold, J.A.; Solan, M. Mediation of Nitrogen by Post-Disturbance Shelf Communities Experiencing Organic Matter Enrichment. *Biogeochemistry* **2017**, *135*, 135–153. [CrossRef]
25. Pilskaln, C.H.; Churchill, J.H.; Mayer, L.M. Resuspension of Sediment by Bottom Trawling in the Gulf of Maine and Potential Geochemical Consequences. *Conserv. Biol.* **1998**, *12*, 1223–1229. [CrossRef]
26. Duplisea, D.E.; Jennings, S.; Malcolm, S.J.; Parker, R.; Sivyer, D.B. Modelling Potential Impacts of Bottom Trawl Fisheries on Soft Sediment Biogeochemistry in the North Sea. *Geochem. Trans.* **2001**, *2*, 112–117. [CrossRef] [PubMed]
27. Widdicombe, S.; Austen, M.C.; Kendall, M.A.; Olsgard, F.; Schaanning, M.T.; Dashfield, S.L.; Needham, H.R. Importance of Bioturbators for Biodiversity Maintenance: Indirect Effects of Fishing Disturbance. *Mar. Ecol. Prog. Ser.* **2004**, *275*, 1–10. [CrossRef]
28. LaRowe, D.E.; Arndt, S.; Bradley, J.A.; Estes, E.R.; Hoarfrost, A.; Lang, S.Q.; Lloyd, K.G.; Mahmoudi, N.; Orsi, W.D.; Shah Walter, S.R.; et al. The Fate of Organic Carbon in Marine Sediments—New Insights from Recent Data and Analysis. *Earth Sci. Rev.* **2020**, *204*, 103146. [CrossRef]
29. Dounas, C.; Davies, I.; Triantafyllou, G.; Koulouri, P.; Petihakis, G.; Arvanitidis, C.; Sourlatzis, G.; Eleftheriou, A. Large-Scale Impacts of Bottom Trawling on Shelf Primary Productivity. Cont. *Shelf Res.* **2007**, *27*, 2198–2210. [CrossRef]
30. Petihakis, G.; Smith, C.J.; Triantafyllou, G.; Sourlantzis, G.; Papadopoulou, K.-N.; Pollani, A.; Korres, G. Scenario Testing of Fisheries Management Strategies Using a High Resolution ERSEM–POM Ecosystem Model. *ICES J. Mar. Sci.* **2007**, *64*, 1627–1640. [CrossRef]
31. Tsikopoulou, I.; Smith, C.J.; Papadopoulou, N.K.; Eleftheriadou, E.; Karakassis, I. A Fishing Ground Benthic Ecosystem Improved during the Economic Crisis. *ICES J. Mar. Sci.* **2019**, *76*, 402–409. [CrossRef]
32. Smith, C.J.; Rumohr, H.; Karakassis, I.; Papadopoulou, K.N. Analysing the Impact of Bottom Trawls on Sedimentary Seabeds with Sediment Profile Imagery. *J. Exp. Mar. Biol. Ecol.* **2003**, *285–286*, 479–496. [CrossRef]
33. Yentsch, C.S.; Menzel, D.W. A Method for the Determination of Phytoplankton Chlorophyll and Phaeophytin by Fluorescence. Deep Sea Res. *Oceanogr. Abstr.* **1963**, *10*, 221–231. [CrossRef]
34. Parsons, T.R.; Maita, Y.; Lalli, C.M. *A Manual of Chemical & Biological Methods for Seawater Analysis*; Pergamon Press: Oxford, UK, 1984.
35. Walkley, A.; Black, I.A. An Examination of the Degtjareff Method for Determining Soil Organic Matter, and a Proposed Modification of the Chromic Acid Titration Method. *Soil Sci.* **1934**, *37*, 29–38. [CrossRef]
36. Strickland, J.D.H.; Parsons, T.R. A Practical Handbook of Seawater Analysis. In *Bulletin/Fisheries Research Board of Canada*, 2nd ed.; Fisheries and Marine Service: Ottawa, ON, Canada, 1972; p. 167.
37. Bremner, J.; Rogers, S.; Frid, C. Assessing Functional Diversity in Marine Benthic Ecosystems: A Comparison of Approaches. *Mar. Ecol. Prog. Ser.* **2003**, *254*, 11–25. [CrossRef]
38. Bolam, S.G.; McIlwaine, P.; Garcia, C. Marine Macrofaunal Traits Responses to Dredged Material Disposal. *Mar. Pollut. Bull.* **2021**, *168*, 112412. [CrossRef] [PubMed]
39. Jumars, P.A.; Dorgan, K.M.; Lindsay, S.M. Diet of Worms Emended: An Update of Polychaete Feeding Guilds. *Ann. Rev. Mar. Sci.* **2015**, *7*, 497–520. [CrossRef]
40. Queirós, A.M.; Birchenough, S.N.R.; Bremner, J.; Godbold, J.A.; Parker, R.E.; Romero-Ramirez, A.; Reiss, H.; Solan, M.; Somerfield, P.J.; Van Colen, C.; et al. A Bioturbation Classification of European Marine Infaunal Invertebrates. *Ecol. Evol.* **2013**, *3*, 3958–3985. [CrossRef]
41. Renz, J.R.; Powilleit, M.; Gogina, M.; Zettler, M.L.; Morys, C.; Forster, S. Community Bioirrigation Potential (BIPc), an Index to Quantify the Potential for Solute Exchange at the Sediment-Water Interface. *Mar. Environ. Res.* **2018**, *141*, 214–224. [CrossRef]
42. Dray, S.; Legendre, P. Testing the Species Traits Environment Relationships: The Fourth-Corner Problem Revisited. *Ecology* **2008**, *89*, 3400–3412. [CrossRef]
43. Dray, S.; Choler, P.; Dolédec, S.; Peres-Neto, P.R.; Thuiller, W.; Pavoine, S.; Ter Braak, C.J.F. Combining the Fourth-Corner and the RLQ Methods for Assessing Trait Responses to Environmental Variation. *Ecology* **2014**, *95*, 14–21. [CrossRef]
44. Dray, S.; Dufour, A.-B.; Chessel, D. The Ade4 Package—II: Two-Table and K-Table Methods. *R News* **2007**, *7*, 47–52.
45. De Madron, X.D.; Ferré, B.; Le Corre, G.; Grenz, C.; Conan, P.; Pujo-Pay, M.; Buscail, R.; Bodiot, O. Trawling-Induced Resuspension and Dispersal of Muddy Sediments and Dissolved Elements in the Gulf of Lion (NW Mediterranean). *Cont. Shelf Res.* **2005**, *25*, 2387–2409. [CrossRef]
46. Paradis, S.; Pusceddu, A.; Masqué, P.; Puig, P.; Moccia, D.; Russo, T.; Lo Iacono, C. Organic Matter Contents and Degradation in a Highly Trawled Area during Fresh Particle Inputs (Gulf of Castellammare, Southwestern Mediterranean). *Biogeosciences* **2019**, *16*, 4307–4320. [CrossRef]
47. Pusceddu, A.; Fiordelmondo, C.; Polymenakou, P.; Polychronaki, T.; Tselepides, A.; Danovaro, R. Effects of Bottom Trawling on the Quantity and Biochemical Composition of Organic Matter in Coastal Marine Sediments (Thermaikos Gulf, Northwestern Aegean Sea). *Cont. Shelf Res.* **2005**, *25*, 2491–2505. [CrossRef]
48. Solan, M.; Scott, F.; Dulvy, N.K.; Godbold, J.A.; Parker, R. Incorporating Extinction Risk and Realistic Biodiversity Futures: Implementation of Trait-Based Extinction Scenarios. In *Marine Biodiversity and Ecosystem Functioning*; Oxford University Press: Oxford, UK, 2012; Volume 15, pp. 127–148.
49. McLaverty, C.; Dinesen, G.; Gislason, H.; Brooks, M.; Eigaard, O. Biological Traits of Benthic Macrofauna Show Size-Based Differences in Response to Bottom Trawling Intensity. *Mar. Ecol. Prog. Ser.* **2021**, *671*, 1–19. [CrossRef]

50. Hooper, D.U.; Chapin, F.S.; Ewel, J.J.; Hector, A.; Inchausti, P.; Lavorel, S.; Lawton, J.H.; Lodge, D.M.; Loreau, M.; Naeem, S.; et al. Effects of Biodiversity on Ecosystem Functioning: A Consensus of Current Knowledge. *Ecol. Monogr.* **2005**, *75*, 3–35. [CrossRef]
51. Hale, R.; Godbold, J.A.; Sciberras, M.; Dwight, J.; Wood, C.; Hiddink, J.G.; Solan, M. Mediation of Macronutrients and Carbon by Post-Disturbance Shelf Sea Sediment Communities. *Biogeochemistry* **2017**, *135*, 121–133. [CrossRef]
52. Braeckman, U.; Provoost, P.; Gribsholt, B.; Van Gansbeke, D.; Middelburg, J.J.; Soetaert, K.; Vincx, M.; Vanaverbeke, J. Role of Macrofauna Functional Traits and Density in Biogeochemical Fluxes and Bioturbation. *Mar. Ecol. Prog. Ser.* **2010**, *399*, 173–186. [CrossRef]
53. Toussaint, E.; De Borger, E.; Braeckman, U.; De Backer, A.; Soetaert, K.; Vanaverbeke, J. Faunal and Environmental Drivers of Carbon and Nitrogen Cycling along a Permeability Gradient in Shallow North Sea Sediments. *Sci. Total Environ.* **2021**, *767*, 144994. [CrossRef]
54. Birchenough, S.N.R.; Parker, R.E.; McManus, E.; Barry, J. Combining Bioturbation and Redox Metrics: Potential Tools for Assessing Seabed Function. *Ecol. Indic.* **2012**, *12*, 8–16. [CrossRef]
55. Gogina, M.; Morys, C.; Forster, S.; Gräwe, U.; Friedland, R.; Zettler, M.L. Towards Benthic Ecosystem Functioning Maps: Quantifying Bioturbation Potential in the German Part of the Baltic Sea. *Ecol. Indic.* **2017**, *73*, 574–588. [CrossRef]
56. Gogina, M.; Zettler, M.L.; Vanaverbeke, J.; Dannheim, J.; Van Hoey, G.; Desroy, N.; Wrede, A.; Reiss, H.; Degraer, S.; Van Lancker, V.; et al. Interregional Comparison of Benthic Ecosystem Functioning: Community Bioturbation Potential in Four Regions along the NE Atlantic Shelf. *Ecol. Indic.* **2020**, *110*, 105945. [CrossRef]
57. Braeckman, U.; Foshtomi, M.Y.; Van Gansbeke, D.; Meysman, F.; Soetaert, K.; Vincx, M.; Vanaverbeke, J. Variable Importance of Macrofaunal Functional Biodiversity for Biogeochemical Cycling in Temperate Coastal Sediments. *Ecosystems* **2014**, *17*, 720–737. [CrossRef]
58. Josefson, A.B.; Norkko, J.; Norkko, A. Burial and Decomposition of Plant Pigments in Surface Sediments of the Baltic Sea: Role of Oxygen and Benthic Fauna. *Mar. Ecol. Prog. Ser.* **2012**, *455*, 33–49. [CrossRef]
59. Middelburg, J.J. Reviews and Syntheses: To the Bottom of Carbon Processing at the Seafloor. *Biogeosciences* **2018**, *15*, 413–427. [CrossRef]
60. Epstein, G.; Middelburg, J.J.; Hawkins, J.P.; Norris, C.R.; Roberts, C.M. The Impact of Mobile Demersal Fishing on Carbon Storage in Seabed Sediments. *Glob. Chang. Biol.* **2022**, *28*, 2875–2894. [CrossRef]
61. Rijnsdorp, A.D.; Bolam, S.G.; Garcia, C.; Hiddink, J.G.; Hintzen, N.T.; van Denderen, P.D.; van Kooten, T. Estimating Sensitivity of Seabed Habitats to Disturbance by Bottom Trawling Based on the Longevity of Benthic Fauna. *Ecol. Appl.* **2018**, *28*, 1302–1312. [CrossRef]
62. McConnaughey, R.A.; Hiddink, J.G.; Jennings, S.; Pitcher, C.R.; Kaiser, M.J.; Suuronen, P.; Sciberras, M.; Rijnsdorp, A.D.; Collie, J.S.; Mazor, T.; et al. Choosing Best Practices for Managing Impacts of Trawl Fishing on Seabed Habitats and Biota. *Fish Fish.* **2020**, *21*, 319–337. [CrossRef]

biology

MDPI

*Article*

# Estimating Effects of Sea Level Rise on Benthic Biodiversity and Ecosystem Functioning in a Large Meso-Tidal Coastal Lagoon

**Olivia Dixon [1,†], Johanna Gammal [1,*,†], Dana Clark [2], Joanne I. Ellis [3] and Conrad A. Pilditch [1]**

1 School of Science, The University of Waikato, Hamilton 3240, New Zealand
2 Cawthron Institute, Nelson 7010, New Zealand
3 School of Science, The University of Waikato, Tauranga 3110, New Zealand
* Correspondence: johanna.gammal@waikato.ac.nz
† These authors contributed equally to this work.

**Simple Summary:** Estuaries are among the world's most productive ecosystems, but due to their location between land and open sea, they are affected by many anthropogenic pressures, including the consequences of climate change. A rising sea level is one major consequence, which will affect both humans and ecosystems, especially in estuaries with extensive intertidal habitats. There is, however, a lack of knowledge regarding the ecological implications of losing intertidal habitats. Therefore, we investigated how seafloor macrofauna communities and their contribution to ecosystem functioning may change due to rising sea levels. Based on a spatially extensive dataset on macrofauna and environmental variables, we identified three main community groups representing intertidal, shallow subtidal, and deep subtidal habitats. Functional trait analysis indicated low functional redundancy for a key intertidal suspension-feeding bivalve (*Austrovenus stutchburyi*) and the lack of a shallow subtidal functional replacement should intertidal habitats become inundated (i.e., become shallow subtidal habitats). These findings thus strongly suggest that sea level rise and the associated environmental changes will modify the seafloor macrofauna communities in estuaries, and subsequently, the ecosystem functions that they influence will be altered.

**Abstract:** Estuaries are among the world's most productive ecosystems, but due to their geographic location, they are at the forefront of anthropogenic pressures. Sea level rise (SLR) is one major consequence of climate change that poses a threat to estuaries with extensive intertidal habitats. The ecological implications of intertidal habitat loss have been largely overlooked despite their likely significance. We aimed to address this knowledge gap by investigating how benthic macroinvertebrate communities and their contributions to ecosystem function are likely to respond to SLR. Based on a spatially extensive dataset (119 sites) from a large coastal lagoon, depth, sediment chlorophyll concentrations, mud content, and average current speed were identified as the main drivers of community compositional turnover. Shifts in benthic community structure and associated functional implications were then evaluated using depth as a proxy for SLR. Three main macrofaunal groups representing intertidal, shallow subtidal, and deep subtidal habitats were identified. Functional trait analysis indicated low functional redundancy for a key intertidal suspension-feeding bivalve (*Austrovenus stutchburyi*) and the lack of a shallow subtidal functional replacement should intertidal habitats become inundated. These findings strongly suggest SLR and the associated environmental changes will alter estuarine macroinvertebrate communities, with implications for future ecosystem function and resilience.

**Keywords:** sea-level rise; estuaries; intertidal area loss; benthic macrofauna; functional groups; Aotearoa New Zealand

## 1. Introduction

Estuaries comprise some of the world's most productive and widespread ecosystems and deliver vital ecosystem services used by humans around the globe [1,2]. The ecosystem services provided by these dynamic and complex environments include food provision, nutrient and carbon processing, coastal protection, and recreational activities [3–5]. Furthermore, the production from estuaries is fueling the wider coastal food webs [6], contributing to valuable nursery grounds for fish [7], and important foraging habitats for sea birds [8]. This high level of functionality is underpinned by soft-sediment benthic communities and the processes they regulate, e.g., [9,10]. Thus, understanding how these ecosystems respond to changing environmental conditions is critical for understanding broader-scale changes within coastal ecosystems.

Estuarine habitats, situated at the interface between the land and the sea, are at the forefront of localized (within catchment) anthropogenic stressors such as sedimentation, eutrophication, and pollution, often resulting from excess inputs of terrestrial sediment, nutrients, and contaminants [11,12]. There is a rich body of literature describing the impacts of these stressors on benthic communities and associated ecosystem functioning [13–17]. However, these ecosystems are also vulnerable to global scale climate change, in particular sea level rise (SLR, amongst others), of which the ecological impacts have received considerably less attention (but see, e.g., [18–22]).

Coastal barrier lagoons are a common type of estuary globally, and they are characterized by deeper, permanently submerged channels and extensive intertidal flats [23]. Such systems are particularly vulnerable to SLR because even a small increase in water depth can result in large reductions in intertidal areas. As an example, using a one-meter SLR scenario, Mangan et al. [18] estimated up to an 80% loss of intertidal area within 12 Aotearoa New Zealand estuaries. Under current predictions, the global mean sea level is estimated to rise between 0.3 m and 2 m by 2100 (compared to 2000 levels), following lowest and highest global greenhouse gas emission pathways, respectively [24,25]. While extensive research has focused on physical impacts such as changes to coastal geomorphology [26–28], including quantifying the loss of intertidal areas [8,19,29], the ecological impacts on soft sediment ecosystems have received much less attention.

The ecological changes (i.e., shifts in biodiversity and ecosystem function) that are likely to arise in response to an altered coastal environment (i.e., deeper water column, steeper slopes, and changed sedimentary environment) are not well documented but see e.g., [18–22]. Given the important role macro-benthic communities play in regulating ecosystem functions, understanding how communities might change with SLR will provide some insight into how ecosystem functions and services might be impacted. Here we make use of an extensive data set from a large shallow coastal lagoon to explore how macrobenthic community composition may be altered with SLR. By considering the functional traits of the macrofaunal species, we also explore whether shifts in community composition translate to potential shifts in function.

We focus on macrofaunal communities because their significance in sustaining valued ecosystem services is well recognized. Through bioturbation and feeding activities, these organisms enhance ecosystem functionality contributing to primary and secondary production, nutrient cycling/processing, sediment stabilization, habitat formation, and carbon sequestration, e.g., [30–35]. However, not all species contribute equally, with some making a disproportionate contribution. For example, evidence indicates that larger individuals play a greater role in facilitating solute fluxes (e.g., nitrogen and oxygen) and maintaining community structure than smaller ones, directly due to their body mass and indirectly through the larger impact of their bioturbation or generation of habitat [36]. Taxonomic groups distinguished by certain functional traits (e.g., bioturbating bivalves), therefore, often hold unique roles in ecosystem functionality. This suggests that the functional group diversity response to altered environmental conditions should be considered, in addition to changes in species diversity.

Grouping species based on their functional traits versus focusing on the roles held by individual species is increasingly widespread in community ecology [37]. In marine sediments, loss of functional diversity can impact important biogeochemical processes, including oxygen and nutrient fluxes [38], thus having flow-on effects on overall ecosystem function [39]. Functional redundancy is usually determined by the number and abundance of species sharing similar traits and, therefore, carrying out similar functions [40]. Exploring functional trait diversity allows us to gauge the resilience (determined by the degree of functional redundancy) associated with functional groups and, therefore, the functions regulating important processes such as primary production and nutrient cycling [41]. For example, functional groups that possess a high level of resilience include those with a greater number of species that can persist under varying stressors, such that a loss of an individual species will not necessarily mean loss of the key functions and the respective ecosystem services they contribute to [40,42,43]. Thus, reinforcing the importance of exploring how both species and functional group diversity may shift with rising anthropogenic stressors.

The central objective of this study was to explore the potential implications of SLR on macroinvertebrate community structure and ecosystem functioning within an estuarine setting. Due to the long-time scales associated with climate change, we used space as a proxy for time [44]. We aimed to reveal the response of community structure to shifts in water column depth and the influence of additional environmental factors that are also expected to shift with SLR. Additionally, using Gradient Forest analysis, we investigated if there was evidence of thresholds along environmental gradients where disproportionately greater shifts in community structure (represented by community turnover) occurred. To further understand the ecosystem-level consequences of changes in community structure, we also assessed the response of functional group community structure to shifts in water column depth (and the respective environmental characteristics). Analysis was based on a comprehensive dataset entailing both biological and environmental data (including water column depth) collected in a large barrier-enclosed coastal lagoon.

## 2. Materials and Methods

### 2.1. Study Area

Data used for this research were collected within Tauranga Harbor (37°40′ S, 176°10′ E; Figure 1) on the north-eastern coast of Aotearoa, New Zealand's North Island. Tauranga Harbor is characterized as a large (~200 $km^2$), shallow (<10 m depth, mean depth ~3 m), barrier-enclosed estuarine lagoon [45,46]. The harbor has an extensive intertidal area constituting approximately 66% of the estuary [45] and experiences a semi-diurnal tidal cycle with a tidal range of up to 2 m [47]. The harbor catchment is extensive (~1300 $km^2$) and includes horticultural, agricultural, and urban land, where water runs from these landscapes into the large estuary. The estuary is well-flushed and vertically well-mixed (tidal and wind mixing), but there is some spatial variation in salinity within the estuary ranging from 28 to 34 [48]. The relative coastal sea level is estimated to have risen at a rate of $2.3 \pm 0.26$ mm $yr^{-1}$ measured outside Tauranga Harbor over the period of 1974-2020 [49]. The current sea level rise median projections for 2100 according to IPCC SSP1-2.6 and SSP5-8.5 scenarios within the Tauranga Harbor areas vary between 0.3 and 1 m when also taking into account the vertical land movement [50].

### 2.2. Data Acquisition

Data were acquired and combined from two ecological surveys. The first focused primarily on intertidal habitats and was conducted between December 2011 and February 2012 (austral summer) and spanned 75 sites throughout the harbor [16,51]. The second was a subtidal survey carried out between March and May 2016 (late austral summer/autumn) and included 44 sites [52]. Sampling locations were selected to ensure a broad range of environmental gradients were represented, aiming to cover the full spatial extent and depth range (up to 9 m corrected to chart datum) of the harbor. Although these data sets were collected in different years, we believe they can be combined because the large number of sites included in the analysis means that the spatial variation in community structure (i.e., entire harbor) is likely to be greater than any between-year variation.

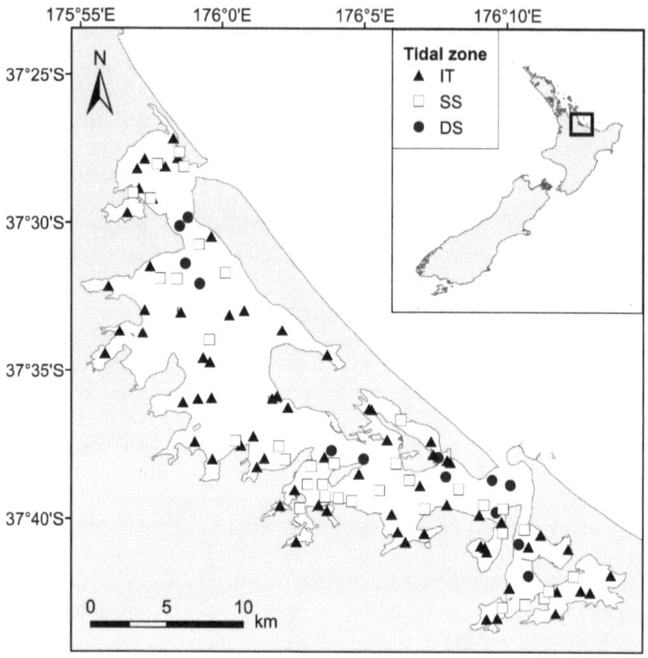


**Figure 1.** Location of Tauranga Harbor on the northeast coast of New Zealand (insert) and sample site locations within the harbor. The symbols indicate each site's tidal zone assignation; intertidal (IT), shallow subtidal (SS), and deep subtidal (DS).

### 2.3. Environmental Variables

An array of the water column and sedimentary variables were obtained at each site, but only variables measured in both surveys were included in this analysis (Supplementary Table S1). The sampling design and methods were consistent with Aotearoa, New Zealand's standardized Estuary Monitoring Protocol [53]. At each site, sediment samples were collected with cores 20 mm dia. and 20 mm deep (n = 10 and 6 cores per site within the intertidal survey and the subtidal survey, respectively). For both surveys, the replicates were composited into a single sample, and the sediment was analyzed for grain size, chlorophyll *a* (Chl *a*), nutrient content (total phosphorus TP; total nitrogen TN), organic content (OM, measured by loss on ignition), and heavy metals (lead Pb; zinc Zn; copper Cu) (see Supplementary Table S2 for extraction methods and further details in Ellis et al. [54]). Current speeds were estimated for the coordinates of each site from the Estuary Transport Module [55], where average and maximum values were obtained for this study. Chart datum (CD) depths for each site were determined by subtracting 1.08 m from mean sea level (MSL) values (MSL to CD conversion published by LINZ [Land Information New Zealand]) obtained from a hydrodynamic model grid developed by de Ruiter et al. [56] that incorporates LiDAR data, multibeam survey measurements, and LINZ bathymetric data.

### 2.4. Macrofauna Data

Three replicate core (13 cm diameter, 15 cm deep) samples were taken at each site and sieved on a 0.5 mm mesh to obtain the macrofauna. The macrofauna was preserved in ethanol (70%), counted, and identified to the lowest attainable taxonomic resolution (usually species). Taxa identifications were performed by experts and based on relevant guides (e.g., New Zealand Coastal Marine Invertebrates [57], National Institute of Water and Atmospheric Research (NIWA) Invertebrate Collection [58]) and consultation with taxonomic experts at NIWA. The intertidal and subtidal benthic macrofauna survey datasets were combined, and the taxonomic resolution was standardized to be consistent across surveys. Where individual taxa counts were low (<10 individuals across the combined data set) and the taxonomic resolution was poor (e.g., higher than class), the taxa groups were removed from further analysis (in total, <10 individuals were removed). Counts of larvae and juveniles were also removed from the dataset to reduce any influence of recruitment events on the statistical models. Site averages for macrofauna abundance data (i.e., average abundance per core) were calculated and used for all analyses.

### 2.5. Functional Group Assignment

In order to assess the prospective implications of environmental change on ecosystem function, each taxon was assigned to one of the 26 functional groups developed by Greenfield et al. [59]. The functional groups considered a range of functional traits representing life history, physical morphology, and behavioral characteristics that influence ecosystem functioning and stability in estuarine ecosystems (Table 1). In this study, as the taxonomic resolution was not always to species level, in the case where taxa can exhibit many attributes of a trait (e.g., different species from the family Spionidae qualified for different functional groups), the most dominant attribute of a trait was assigned as the functional group.

**Table 1.** Summary of defining traits for each functional group (1–26) as described by Greenfield et al. [59]. An example species are given for each group. Taxonomic class indicated within brackets; T – Thecostraca, B = Bivalvia, G = Gastropoda, A = Anthozoa, P = Polychaeta, M = Malacostraca.

| Functional Group | Description of Traits | Example Species |
|---|---|---|
| 1 | Calcified, Suspension feeding, Attached | *Austrominius modestus* (T) |
| 2 | Calcified, Suspension feeding, Top 2 cm, Freely mobile | *Austrovenus stutchburyi* (B) |
| 3 | Calcified, Suspension feeding, Top 2 cm, Limited mobility | *Arthritica bifurca* (B) |
| 4 | Calcified, Suspension feeding, Top 2 cm, Sedentary | *Arcuatula senhousia* (B) |
| 5 | Calcified, Deposit/Pred.Scav/Grazer, Above surface, Freely mobile | *Zeacumantus subcarinatus* (G) |
| 6 | Calcified, Deposit feeding, Top 2 cm, Limited mobility | *Linucula hartvigiana* (B) |
| 7 | Calcified, Deposit feeding, Predator/Scavenger, Top 2 cm, Freely mobile | *Pisinna zosterophila* (G) |
| 8 | Calcified, Deposit feeding, Deep, Limited mobility, No habitat structure, Large | *Macomona Liliana* (B) |
| 9 | Soft-bodied, Suspension feeding, Attached | *Anthopleura aureoradiata* (A) |
| 10 | Soft-bodied, Suspension feeding, Tube structure | *Euchone* sp. (P) |
| 11 | Soft-bodied, Deposit feeding, Top 2 cm, Freely mobile | Spaerodoridae (P) |
| 12 | Soft-bodied, Deposit feeding, Below surface, Freely mobile | Spionidae (P) |
| 13 | Soft-bodied, Deposit feeding, Below surface, Limited mobility | *Heteromastus filiformis* (P) |
| 14 | Soft-bodied, Deposit feeding, Deep | *Hyboscolex longiseta* (P) |
| 15 | Soft-bodied, Below surface, Tube structure | Terebellidae (P) |
| 16 | Soft-bodied, Predator/Scavenger, Top 2 cm, Freely mobile | Sigalionidae (P) |
| 17 | Soft-bodied, Predator/Scavenger, Top 2 cm, Limited mobility | Syllidae (P) |
| 18 | Soft-bodied, Predator/Scavenger, Below surface + Deep, Freely mobile, No habitat structure | *Perinereis* sp. (P) |
| 19 | Soft-bodied, Predator/Scavenger, Below surface, Limited mobility | Oligochaeta |
| 20 | Soft-bodied, Above surface, Top 2 cm, Below surface, Deep, Sedentary, Tube structure | *Owenia petersenae* (P) |
| 21 | Rigid, Suspension feeding, Top 2 cm | Tanaidacea (M) |
| 22 | Rigid, Deposit feeding, Predator/Scavenger, Top 2 cm, Freely mobile, No habitat structure | Amphipoda (M) |
| 23 | Rigid, Above surface, Freely mobile | Cumacea (M) |
| 24 | Rigid, Above surface, Freely mobile, Large | Ophiuroidea |
| 25 | Rigid, Predator/Scavenger, Attached | No individuals identified |
| 26 | Rigid, Predator/Scavenger, Below surface, Freely mobile, Large burrow former | *Hemiplax hirtipes* (M) |

### 2.6. Statistical Analyses

#### 2.6.1. Determining Critical Points of Compositional Turnover along Key Environmental Gradients (Gradient Forest Modelling)

To reveal how macroinvertebrate community structure changed along environmental gradients, Gradient Forest (GF) modeling was employed [60,61]. GF identifies critical points along environmental gradients where large shifts in rates of benthic macroinvertebrate compositional turnover occur [60]. GF models allow for the identification of compositional turnover thresholds by aggregating regression-tree-based Random Forest (RF) models. Species considered rare ($\leq$3 occurrences across all 119 sites) were additionally excluded from GF analysis as models are constrained by limited data. The three sites where depth exceeded 6 m were also removed from GF models as there were not enough data to adequately model species turnover beyond this depth. Two key processes are undertaken for GF modeling. The first process uses an extension R package, "extendedForest" [62], which calls on the R package "randomForest" to fit an ensemble of RF models for the input species. These RF models describe the relationship between the species distribution and a set of environmental variables. The second process uses the R package "gradientForest" to aggregate all of the individual split points determined from these models, estimating the most important points of species turnover along each environmental gradient to provide a measure of compositional turnover that represents the entire community.

RF models [61] are a flexible and robust way of modeling non-linear predictor-response relationships. The RF models for individual species are built based on an ensemble of regression trees (in this study, 500) where observations are repeatedly partitioned based on the 'best' individual split. This split point is indicative of a measure of importance reflecting the magnitude of change in abundance. The predictive power of individual RF models ($R^2f$) is explained by the proportion of out-of-bag variance for each species [60] and the importance of each predictor variable ($R^2$; a dimensionless value representing cumulative importance). Model performance degradation was used to select variables included in the final model as each environmental predictor is randomly permuted [63]. Multicollinearity between predictor variables is accounted for by using a conditional approach, allowing RF models to be robust to highly correlated variables.

GF modeling aggregates split importance values across each environmental gradient that were determined by the RF models, where species models with positive fits ($R^2f > 0$) are collated to form distributions reflecting compositional turnover relative to each environmental predictor [60,63]. As the distribution is formed, individual RF models with higher predictive importance (i.e., high $R^2f$) have a greater influence on the turnover distribution than models with lower predictive importance (i.e., low $R^2f$). The shape of the distribution constructed for each environmental variable indicates the predicted rate of compositional change along the respective gradient, where increased slope steepness indicates an increased rate of community compositional turnover [60,63]. Each GF model was bootstrapped 100 times to gauge model performance and certainty. In each bootstrap iteration, a random subsample of the macroinvertebrate data was taken, and each measure of compositional turnover was integrated when constructing final GF models for each environmental predictor. All GF analyses were conducted in statistical software R version 4.1.0 [64].

#### 2.6.2. Benthic Macrofauna–Defining Tidal Zones

To examine if there were distinct shifts in macrofauna community structures with water column depth and the other environmental variables, a hierarchical cluster analysis with the SIMPROF test [65] was performed on square-root transformed macrofauna abundance data. We aimed to reveal if there were unidentified assemblages of sites that could group together based on community and environmental similarities, which would enable a closer examination of community changes with SLR (i.e., intertidal areas becoming subtidal). Three groups of clusters were essentially identified based on the macrofauna community structure, and together with the environmental data, they broadly represent different tidal zones (Supplementary Figure S1). The (dis)similarities in community structure between clusters were assessed with a similarity percentage analysis SIMPER; [65]. In order to ensure adequate sampling effort within the identified clusters, species accumulation curves (SAC) were produced by plotting the number of species against the number of sites surveyed (Supplementary Figure S2).

A distance-based redundancy analysis (dbRDA) ordination plot was used to illustrate the relationship between the set of the most influential environmental predictors (represented as vector overlays that indicate direction and strength), explaining the disparities in community structure. Collinearity between predictors was examined, but no action was required (all $r < |0.8|$). The multivariate analyses were conducted in PRIMER version 7.0.13 [65].

#### 2.6.3. Functional Group Analysis–Implications for Ecosystem Functioning

In order to investigate potential implications for ecosystem functioning associated with environmental conditions, shifts in macrofauna community structure and associated functional groups with SLR were examined. To confirm if the functional group community structure also differed between the initial cluster groups, a one-way PERMANOVA and PERMDISP were employed together with post-hoc pairwise tests using the functional group abundance data (square-root transformed). SIMPER analysis was performed using Bray-Curtis dissimilarities to identify the contributions of each functional group to the overall dissimilarity between the clustered groups. The analyses were conducted in PRIMER 7 with the PERMANOVA+ add-on [66].

## 3. Results

### 3.1. Relative Importance of Environmental Gradients for Predicting Compositional Turnover

Gradient Forest (GF) analysis was employed to investigate thresholds of community compositional turnover for environmental gradients known to influence community structure. GF effectively modeled taxa turnover for 85 of the 157 input taxa based on 100 bootstrapped model runs. All 12 environmental predictor variables included were considered important for predicting patterns of macroinvertebrate community compositional turnover, contributing to 48% combined cumulative importance. Depth was, however, revealed as the most important predictor (6.6% of the conditional importance), followed by sediment Chl *a* concentration (6.5% of the conditional importance). Other environmental gradients considered important predictors by GF were average current speed, gravel, copper (Cu), mud content, total phosphorous (TP), lead (Pb), total nitrogen (TN), organic content (OM), zinc (Zn) and sand content (3–5% of the conditional importance each).

Non-linear curves representing rates of macroinvertebrate compositional turnover were observed for all environmental gradients except for Pb and sand, which had comparatively linear relationships indicating a constant rate of compositional turnover for these predictors (Figure 2). Steep sections in the cumulative importance curves indicated large shifts in community structure (i.e., rapid compositional turnover), whereas plateaued sections of the curves indicated more comparable communities. For depth, relatively constant rates of compositional turnover were observed but with a few rapid changes around 1, 3, and 4.5 m (Figure 2). For Chl *a*, the turnover rates increased relatively constantly, but around 30,000 μg/kg, a rapid increase was indicated. However, the variability in mean predicted cumulative change (measured by the 95% prediction interval) was noticeably high due to few data points above this value. The compositional turnover along the gradient of average current speed indicated gradual rates of increase at low current speeds but larger change around 0.3, 0.5, and 0.7 m/s. Regarding grain size, there were low turnover rates until 3% mud content, followed by more rapid changes. Similar patterns were indicated for the nutrients (TP and TN), low turnover with low nutrient concentrations followed by steadily increasing turnover.

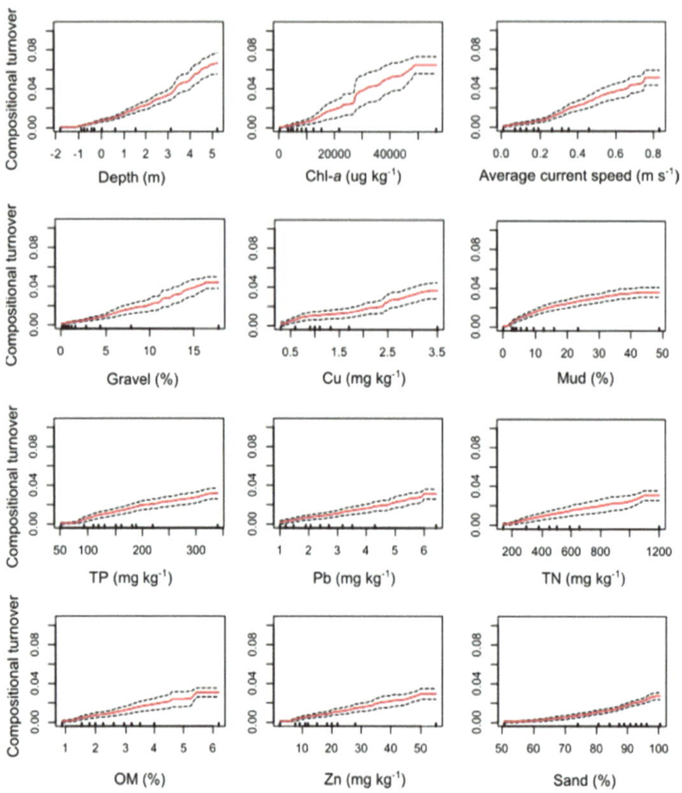


**Figure 2.** Cumulative importance curves (with 95% prediction intervals) visualizing the overall pattern of compositional turnover (in $R^2$-importance units) for all species across all environmental predictors included in gradient forest models. Rug plots along the x-axis represent deciles across each environmental gradient.

### *3.2. Definition of Tidal Zones Based on Community and Environmental Data*

Hierarchical cluster analysis performed on the complete species abundance dataset indicated six macrofaunal community clusters on a level of 37% similarity (Supplementary Figure S1). Based on a comparison of the inter-cluster characteristics, similar sites were combined into clusters representing three tidal zones (intertidal IT, shallow subtidal SS, and deep subtidal DS) for further analysis. The IT sites were shallowest, with an average water depth of −0.6 m (Table 2), compared to the SS and DS sites, with average depths of 1.5 m and 3.0 m, respectively. There was a small number of sites that overlapped in depth between groups. However, the average depths, environmental characteristics, and position within the harbor (Table 2, Figure 1) indicated that these clusters represented different tidal zones. All sites were generally sandy (>85% sand on average), but the mud and OM content varied as expected between the tidal zones, with the highest values at IT sites compared to SS and DS sites (Table 2). The average current speeds were accordingly lowest at the IT (0.15 m/s) sites, compared to SS (0.33 m/s) and DS (0.53 m/s) sites. The environmental variables explaining the variation in the macrofauna community composition at each site were illustrated by a dbRDA (Figure 3). Separation of the tidal zones occurred along the x-axis, aligning with the variables depth, chlorophyll *a* and average current speed, and along the y-axis due to sediment characteristics, mud, sand, and gravel content.

**Table 2.** Summary of average environmental and univariate macrofauna diversity measures (min-max) measured in Tauranga Harbor as a function of tidal zones; intertidal (IT, n = 70), shallow subtidal (SS, n = 36) and deep subtidal (DS, n = 13).

| | | IT | SS | DS |
|---|---|---|---|---|
| **Environmental Variables** | | | | |
| | Depth (m) | −0.6 (−2.0–3.0) | 1.5 (−1.0–7.9) | 3.0 (−0.2–9.0) |
| | Mud (%) | 13.6 (0.1–76.4) | 9.0 (2.6–25.4) | 3.0 (0.6–5.0) |
| | Sand (%) | 85 (24–100) | 87 (67–96) | 91 (78–99) |
| | Gravel (%) | 1.8 (0.1–14.6) | 4.7 (0.1–15.0) | 5.9 (0.1–17.8) |
| | OM (%) | 2.9 (0.9–10.0) | 2.8 (1.3–6.2) | 1.7 (1.0–3.0) |
| | Chl *a* (μg/kg) | 6107 (210–16,000) | 16,678 (5900–41,300) | 17,685 (2000–56,300) |
| | TP (mg/kg) | 168 (51–580) | 152 (79–340) | 121 (81–180) |
| | TN (mg/kg) | 484 (140–1900) | 548 (499–1200) | 452 (190–499) |
| | Cu (mg/kg) | 1.3 (1.0–6.1) | 1.1 (0.4–3.5) | 0.7 (0.3–1.0) |
| | Pb (mg/kg) | 2.7 (1.0–13.0) | 3.0 (1.6–6.4) | 2.0 (1.0–3.8) |
| | Zn (mg/kg) | 17.7 (2.5–55.0) | 17.7 (7.7–37.0) | 12.2 (6.4–25.0) |
| | Av. current speed (m/s) | 0.15 (0.01–0.52) | 0.33 (0.01–0.67) | 0.53 (0.23–0.83) |
| **Benthic community** | | | | |
| | S (taxa per core) | 19 (6–31) | 25 (18–37) | 15 (10–21) |
| | N (ind. per core) | 109 (27–329) | 234 (49–744) | 70 (22–183) |
| | Occurrence (% of sites taxa occurs at) | 23 (1–100) | 20 (3–100) | 23 (8–100) |
| | H′ (per core) | 1.92 (0.11–2.71) | 2.02 (0.76–2.74) | 1.72 (0.45–2.55) |
| | Most abundant taxa | Amphipoda (M)<br>Spionidae (P)<br>*Heteromastus filiformis* (P)<br>*Austrovenus stutchburyi* (B)<br>*Linucula hartvigiana* (B) | Spionidae (P)<br>Amphipoda (M)<br>Oligochaeta<br>*Aricidea* sp. (P)<br>*Heteromastus filiformis* (P) | *Paphies australis* (B)<br>Amphipoda (M)<br>Hesionidae (P)<br>Syllidae (P)<br>*Magelona* sp. (P) |

Chl *a* Chlorophyll *a*, TN total nitrogen, TP total phosphorus, Cu Copper, Pb Lead, Zn Zinc, OM organic content, Av. average current speed, S average number of taxa, N average abundance, H′ Shannon-Wiener diversity index. Class indicated for the most abundant taxa: M Malacostraca, P Polychaeta, B Bivalvia.

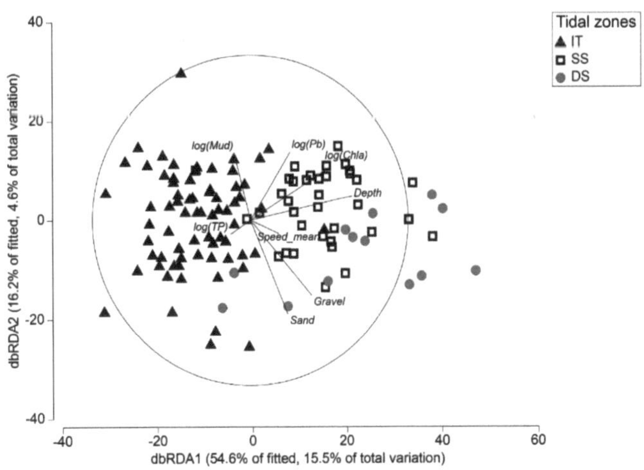


**Figure 3.** Distance-based redundancy analysis (dbRDA) plot visualizing the direction and influence of environmental predictors on shifts in macroinvertebrate community structure.

### *3.3. Differences in Macrofauna Communities between Tidal Zones*

Species accumulation curves (SAC) were generated to provide evidence that IT, SS, and DS clusters had been sampled adequately. For IT and SS sites, the SAC indicated that adequate sampling (i.e., a noticeable decrease in species accumulation rates with increasing sampling effort) occurred after 10 sites (30 cores) and 12 sites (36 cores), respectively (Supplementary Figure S2). For the DS cluster, species accumulation rates were less clear due to fewer samples, and results regarding this group need to be interpreted with care.

In this study, there were 157 different taxa identified across all sites. The highest number of taxa occurred in the SS tidal zone, with a total of 126 taxa recorded, compared to the IT and DS tidal zone, with a total of 83 and 66 taxa recorded, respectively. The abundance per core was also highest in the SS (234 ind./core) compared to the IT (109 ind./core) and DS (70 ind. per core; Table 2). Shannon-Wiener diversity indices for all three tidal zones were similarly varying between the tidal zones, with the highest at SS, then IT, and the lowest at DS (2.02, 1.92, and 1.72).

Of the top five most abundant taxa determined for each zone, IT had three taxa in common with SS (Amphipoda, Spionidae, and *Heteromastus filiformis*) and only one taxon (Amphipoda) shared with DS (Table 2). Similarly, there was only one taxon in common among the SS and DS top five most abundant (Amphipoda). A SIMPER analysis revealed that overall dissimilarity between IT and SS sites was 66%, and this was largely driven by differences in taxa abundance of Amphipoda (e.g., Caprellidae), Spionidae polychaetes (e.g., *Aonides trifida, Boccardia syrtis*), the polychaete *Aricidea* sp., oligochaete worms and the polychaete *Heteromastus filiformis* (Supplementary Table S3). The overall dissimilarity between IT and DS was 77%, where differences in community structure was primarily attributed to the polychaetes Spionidae (e.g., *A. trifida, B. syrtis*), Amphipoda (e.g., Caprellidae), the bivalve *Paphies australis* and polychaete *H. filiformis* (Supplementary Table S3). There was 72% dissimilarity between SS and DS community structures, largely attributed to Spionidae polychaetes (e.g., *A. trifida, B. syrtis*), oligochaete worms, polychaetes *H. filiformis*, *Aricidea* sp., Amphipoda (e.g., Caprellidae) and bivalve *P. australis*. The top taxa (except *P. australis*) generally had greater abundances at the IT and SS sites than at the DS sites (Supplementary Table S3).

### 3.4. Functional Group Analysis

The functional group structures between the different tidal zones were analysed to reveal if the shift in macrofauna communities potentially translates into a shift in the functionality of the benthic ecosystems. Significant differences in functional group structure between all tidal zones were indicated (PERMANOVA; Pseudo-F = 15.09; $p < 0.001$; Supplementary Table S4). Additionally, there were homogenous dispersions (PERMDISP $p > 0.05$) between the tidal zones except between SS and DS tidal zones (PERMDISP $p < 0.05$). Subsequently, the functional group differences between SS and DS should be interpreted with care.

A SIMPER analysis revealed that overall dissimilarity between the IT and SS functional group communities was 51% and was largely driven by differences in functional group abundance of FG13 (Soft-bodied, deposit-feeding, below the surface, limited mobility; e.g., polychaete *H. filiformis*), FG22 (Rigid, deposit-feeding, predator/scavenger, top 2 cm, mobile; e.g., Amphipoda), FG12 (Soft-bodied, deposit-feeding, below the surface, mobile; e.g., polychaete Spionidae), FG17 (Soft-bodied, predator/scavenger, top 2 cm, limited mobility; e.g., polychaetes Syllidae), FG19 (Soft-bodied predator/scavenger, below the surface, limited mobility; e.g., Oligochaeta) (Supplementary Table S5). In most instances, functional group abundance tended to be lower in IT compared to SS, with the exception of FG2 (Calcified, suspension-feeding, top 2 cm, mobile; e.g., bivalve *A. stutchburyi*) and FG6 (Calcified, deposit-feeding, top 2 cm, limited mobility; e.g., bivalve *L. hartvigiana*), where average abundance per core was greater for IT than SS (by factors of 1.8 and 2.8 respectively). The overall dissimilarity between IT and DS was 58%, mostly attributed to abundance differences of FG12 (Soft-bodied, deposit-feeding, below surface, mobile; e.g., polychaetes Spionidae), FG22 (Rigid, deposit-feeding, predator/scavenger, top 2 cm, mobile; e.g., Amphipoda), FG2 (Calcified, suspension feeding, top 2 cm, mobile; e.g., bivalve *A. stutchburyi*), FG13 (Soft-bodied, deposit-feeding, below the surface, limited mobility; e.g., polychaete *H. filiformis*), FG19 (Soft-bodied, predator/scavenger, below surface, limited mobility; e.g., Oligochaeta) (Supplementary Table S5). There was 56% dissimilarity between SS and DS, mostly driven by differences in FG12 (Soft-bodied, deposit-feeding, below surface, mobile; e.g., polychaete Spionidae), FG13 (Soft-bodied, deposit-feeding, below the surface, limited mobility; e.g., polychaete *H. filiformis*), FG22 (Rigid, deposit-feeding, predator/scavenger, top 2cm, mobile; e.g., Amphipoda), FG19 (Soft-bodied, predator/scavenger, below the surface, limited mobility; e.g., Oligochaeta), FG2 (Calcified, suspension-feeding, top 2 cm, mobile; e.g., bivalve *A. stutchburyi*).

## 4. Discussion

The aim of this study was to attempt to fill gaps in the scientific literature around the implications of sea level rise (SLR) on estuarine biodiversity and ecosystem functioning. To date, there has been little research addressing this aspect of coastal climate change ecology despite the growing relevance of diffuse climate change stressors. The findings indicated that there would be significant shifts in estuarine macroinvertebrate community structure with future SLR. Additionally, some species-specific shifts may trigger functional consequences. For example, the functionally important large cockle *Austrovenus stutchburyi*, in the intertidal zone, is unlikely to have a substitute in the shallow subtidal zone. The results thus demonstrate that localized gains and losses of individual species and functional traits within the community will likely have implications for the overall estuarine ecosystem functioning.

### *4.1. Environmental Drivers of Macroinvertebrate Community Structure and Compositional Turnover*

Water column depth was identified as the most important predictor of rates of community compositional turnover. The influence of depth on the spatial distribution of marine organisms has been well studied, e.g., [67–71]. However, links to SLR are generally ignored. In estuaries, increasing depths will be a key outcome of SLR [72]. Therefore, gaining an understanding of how macroinvertebrate communities shift with depth allows us to consider the prospective implications of SLR.

The GF modeling indicated constantly increasing compositional turnover rates of macroinvertebrate communities with increasing depth and rapid changes around 1, 3, and 4.5 m. Using SLR predictions with current global emission rates, we can expect a rise between 0.6 to 1.1 m by 2100 [25], and regional predictions based on different climate change scenarios, including local variations in the harbor estimate a range of 0.3–1 m [50]. Using depth as a proxy for SLR (assuming spatial and temporal variability is equal; [44]), the results indicated that the upper prediction reflects a threshold where a small increase in SLR at 1 m will drive a disproportionately greater change in macroinvertebrate community structure than that perceived for preceding SLR scenarios. This may be explained by the expected reductions suffered by intertidal species that are constrained by their optimal spatial distribution [73–75], impeding their ability to thrive in deeper submerged habitats. This shift observed in macroinvertebrate community structure would also align with that expected of the projected intertidal habitat loss under a 1.1 m SLR scenario (~85% reduction by the year 2100; [18]). Nonetheless, steady rates of compositional turnover were still observed approaching 1 m depth, providing an indication that even small changes in SLR will alter macroinvertebrate community structure, perhaps irreversibly, within Tauranga Harbor, as also earlier indicated by modeled distributions for a subset of species by Rullens and Mangan et al. [20].

These findings highlight the importance of depth as a predictor of species and community responses to SLR. However, it is unlikely that depth alone is driving observed responses, but instead acts as a surrogate for a combination of co-varying factors known to shape patterns of estuarine macroinvertebrate biodiversity. The high relative importance of depth may be owed to relationships with water column and sedimentary environment characteristics such as sea temperature, salinity, sediment grain size, and nutrient content, which are all known to also influence patterns of macroinvertebrate biodiversity [76–80].

This suggests that depth can represent a host of co-varying environmental parameters that will also shift with SLR.

The gradient forest analysis also indicated average current speed as an important factor for predicting patterns in macroinvertebrate community structure and compositional turnover (Figure 2). Flow rates are often highly variable throughout estuaries, largely owed to the complex bathymetry of the seafloor (e.g., channels) and bordering landforms (e.g., tombolos) that influence flow dynamics [81,82]. Here, average current speeds measured at each site varied from 0.01–0.83 m/s (Table 2), with a rapid increase in compositional turnover rate around current speeds of 0.3, 0.5, and 0.7 m/s (Figure 2). Average current speeds exceeding this value generally existed at deeper sites in the harbor, often in the center of channels, which is likely explained by the strong influence of tidal exchange on current speeds in main channels [83,84].

The flow dynamics associated with these channels often support increased delivery rates of particulate food, which is favorable to filter-feeding organisms [85]. This may explain why high densities of the filter-feeding bivalve *Paphies australis* were generally restricted to deep subtidal sites in this study, as also shown in earlier studies e.g., [73]. As estuary depth is expected to increase with future SLR, we can also anticipate altered current speeds (i.e., likely reduced in deep channels) due to the influence of basin geometry (e.g., degree of channel constriction) and depth on flow dynamics [56,86]. This indicates that although overall water column depth will increase with SLR, which could suggest that species like, for example, *P. australis* will extend their spatial distribution, their distribution is likely to be constrained if altered current speeds do not match those required to support high densities. From this, we can deduce that some species will not necessarily extend their spatial distribution to 'follow' their optimal depth range if other environmental factors are altered that may limit their distribution.

### 4.2. Comparisons across Tidal Zones and Implications of Reduced Intertidal Area

Estuarine benthic macroinvertebrate community structure differed across intertidal (IT), shallow subtidal (SS), and deep subtidal (DS) zones. The findings demonstrated that species richness and average abundance were highest at the SS sites (Table 2). An explanation for this is that SS represents a transitional zone comprising a mixture of species that occur in IT and DS habitats [87,88]. The lower species richness and average abundance at the IT sites were expected as many estuarine species lack unique adaptations (e.g., desiccation prevention) required to endure environmental circumstances typical of IT habitats, (e.g., air exposure during periods of tidal emergence) [89]. Thus, the subtidal habitat is preferable to more species. Although the general consensus within ecological studies is that increased diversity positively influences ecosystem function [90–92], this can be context dependent [93,94]. In estuarine ecosystems, certain species make a disproportionate contribution to ecosystem function (e.g., *Austrovenus stutchburyi*) [36] owed to key factors (e.g., abundance/dominance, functional traits) influencing important ecological processes and functions (e.g., sediment destabilization, primary production, ecosystem engineering) [30,34,35,41,95,96]. It is thus critical to recognize that greater species richness does not always reflect better ecosystem performance, particularly when functionally important or unique species are reduced or lost [94].

Under future SLR conditions, it is suggested that intertidal areas will essentially become subtidal as they become permanently inundated [72]. In this study, functional group community structure significantly differed between IT, SS, and DS habitats, indicating dominant IT functional groups may experience reductions whilst those of SS will become more widespread. Based on our results, such a shift would suggest a two-fold increase in the average abundance of soft-bodied deposit-feeders located below the sediment surface, such as polychaetes (FG13 and FG12), in areas where this habitat shift occurs. Species included in this functional group are considered important drivers of community structure, and many are key bioturbators that contribute to ecological processes and promote ecosystem function (e.g., nutrient cycling, sediment destabilization) [97,98], which at a glance suggests this shift could be desirable. Additionally, these functional groups were relatively abundant across all tidal zones and had a high degree of redundancy, indicating high ecological resilience to environmental change. However, the results also suggested that calcified suspension- and deposit-feeders at the sediment surface, such as the bivalves *Austrovenus stutchburyi* and *Linucula hartvigiana* (FG2 and FG6), will experience reductions to less than half of the average abundance in areas that shift from IT to SS, indicating this shift may have a large impact on the ecosystem. The dominant species of FG2, *A. stutchburyi*, plays an important role in intertidal habitats as an ecological engineer and positively influences ecological processes such as primary production, denitrification, and reworking of the sedimentary environment [31,36,99,100]. Additionally, the distribution of species contributing to these dominant IT functional groups is generally more spatially constrained and displays lower redundancy. Therefore the expected habitat shifts associated with SLR indicate implications for ecosystem function due to the predicted reductions suffered by these groups.

There is often a degree of functional redundancy within estuarine taxa where multiple species can offer similar contributions to ecosystem processes [59]. However, *A. stutchburyi* and *Paphies australis* were the only abundant species characterized by FG2 (suspension-feeding, mobile, top 2 cm), indicating low redundancy and resilience despite occurring in high abundances. Furthermore, as *P. australis* is generally restricted to subtidal regions, particularly where depth and current speeds are greater [73] (i.e., DS habitats in this study), it is unlikely to move into all newly submerged areas due to SLR (equivalent to SS habitats in this study). Thus, despite similar contributions to functionality, *P. australis* is unlikely to provide functional resilience should intertidal *A. stutchburyi* populations be reduced or lost following a shift to shallow subtidal habitats. This is concerning as SLR will reduce intertidal area [8,18,101], whilst shallow subtidal coverage is expected to increase. Furthermore, modeling studies in this estuary have indicated that SLR will cause the loss of high-density areas of *A. stutchburyi* and these locations coincide with areas that exhibit the highest potential for ecosystem services [20,102]. This highlights the vulnerability of *A. stutchburyi* and its functional role in intertidal habitat loss. Thus, we can anticipate significant implications on ecosystem functions and the ecosystem services they underpin due to SLR.

An important aspect to acknowledge is the possibility that time scales associated with geomorphic and ecological shifts due to SLR may differ. Generally, ecological shifts can occur very rapidly as changing environmental conditions can often have a direct impact on species distributions [103]. Changes to geomorphology, however, can take place over a much longer period of time [104]. This suggests that if intertidal habitats become flooded by SLR, the projected changes to the sedimentary environment (i.e., lower mud content/coarser sediments) may display a time lag, whereas the response of species distributions and their respective communities to SLR is expected to be much more immediate. Moreover, there is the possibility of intertidal habitat 'legacy effects' (i.e., residual qualities of the former habitat) [105] hindering the transition of 'muddy' sediments to 'sandy' sediments typical of the shallow subtidal habitats observed in this study. This suggests that species currently thriving in shallow subtidal habitats (e.g., some polychaetes) may not necessarily occur in the same densities in inundated intertidal areas if sediment composition limits their distribution. As we employed a space-for-time approach, key findings of this study heavily rely on the assumption that intertidal habitats will essentially become shallow subtidal as they become permanently inundated by SLR. Therefore, unknown legacy effects of intertidal habitats may influence macroinvertebrate community responses to SLR that have not been accounted for in this study.

## 5. Conclusions

SLR resulting from a warming planet will significantly modify coastal geomorphology, influencing tidal dynamics, currents, and the sedimentary environment. Reduced intertidal coverage will impact estuarine ecosystems and their communities at local and global scales, yet the ecological repercussions have been largely dismissed despite their prospective significance. This study demonstrated that macrofauna community structure differed significantly across tidal zones and that patterns in macrofauna biodiversity will shift in response to altered depth and concomitant changes to the water column and sedimentary environment. Thus, it highlights that SLR will significantly alter estuarine macroinvertebrate communities and subsequently result in repercussions for ecosystem function and resilience. The results of this study also indicated that the ecological impacts of species loss would be dependent on the species-specific contributions to ecosystem function. Many species-specific contributions are, however, unknown, which may mean that there are implications that are not yet fully recognized for ecosystem function if intertidal habitats are lost to SLR. We do, however, know that intertidal habitats hold a significant role in maintaining important ecological processes (e.g., primary production, and denitrification) [106,107], often exceeding that of the adjacent subtidal habitats [108]. Additionally, species abundance has been shown to strongly influence these processes, e.g., [32,109]. Therefore, it is reasonable to assume that highly abundant intertidal species will have a significant influence on ecological processes, although the nature of the effect will likely depend on species identity. To obtain an extensive understanding of the implications, we can expect to arise under future SLR conditions. We must understand the unique roles of all species and their functional roles that are vulnerable to expected habitat shifts. Management efforts targeting biodiversity in coastal environments should also recognize the expected shifts in community structure that will occur through habitat loss. This will be fundamental for ensuring management strategies are indeed effective for maintaining biodiversity, particularly for systems such as marine protected areas that often treat habitats as fixed in space over time. Well-informed management of biological communities and coastal environments will be critical for ensuring that the ecosystem functions and services valued by society are conserved for future generations.

**Supplementary Materials:** The following supporting information can be downloaded at: https://www.mdpi.com/article/10.3390/biology12010105/s1, Table S1: Environmental and biological variables at each site; Table S2: Summary of analysis methods and units of the measured environmental variables; Figure S1: Cluster analysis of the benthic communities for identification of tidal zones; Figure S2: Species accumulation curves for each tidal zone; Table S3: SIMPER analysis results based on species abundance data within tidal zones; Table S4: Results of PERMANOVA comparing functional group composition between tidal zones; Table S5: SIMPER analysis results based on functional group data within tidal zones.

**Author Contributions:** Conceptualization, C.A.P., J.G. and J.I.E.; Methodology, all authors; Formal Analysis, O.D., D.C. and J.G.; Data Curation, J.G. and O.D.; Writing—Original Draft Preparation, O.D. and J.G.; Writing—Review & Editing, all authors.; Visualization, O.D., D.C. and J.G.; Funding Acquisition, C.A.P. All authors have read and agreed to the published version of the manuscript.

**Funding:** This research was funded by the National Science Challenge Sustainable Seas, project no. C01X1901.

**Institutional Review Board Statement:** Not applicable.

**Informed Consent Statement:** Not applicable.

**Data Availability Statement:** Restrictions apply to the availability of these data. Data was obtained from the Research programs Manaaki Taha Moana (MTM) and Oranga Taiao Oranga Tangata (OTOT) and can be made available upon request to Dana Clark (Cawthron Institute) with the permission of the Research Programmes.

**Acknowledgments:** We are grateful for the opportunity to use the extensive data sets collected as part of the New Zealand Ministry of Business, Innovation and Employment funded projects Manaaki Taha Moana and Oranga Taiao Oranga Tangata, which were led by Murray Patterson from Massey University, and thank all those involved.

**Conflicts of Interest:** The authors declare no conflict of interest.

## References


1. Costanza, R.; de Groot, R.; Sutton, P.; van der Ploeg, S.; Anderson, S.J.; Kubiszewski, I.; Farber, S.; Turner, R.K. Changes in the global value of ecosystem services. *Glob. Environ. Chang.* **2014**, *26*, 152–158. [CrossRef]
2. Levin, L.A.; Boesch, D.F.; Covich, A.; Dahm, C.; Erséus, C.; Ewel, K.C.; Kneib, R.T.; Moldenke, A.; Palmer, M.A.; Snelgrove, P.; et al. The Function of Marine Critical Transition Zones and the Importance of Sediment Biodiversity. *Ecosystems* **2001**, *4*, 430–451. [CrossRef]
3. Barbier, E.B.; Hacker, S.D.; Kennedy, C.; Koch, E.W.; Stier, A.C.; Silliman, B.R. The value of estuarine and coastal ecosystem services. *Ecol. Monogr.* **2011**, *81*, 169–193. [CrossRef]
4. Duarte, C.M.; Losada, I.J.; Hendriks, I.E.; Mazarrasa, I.; Marbà, N. The role of coastal plant communities for climate change mitigation and adaptation. *Nat. Clim. Chang.* **2013**, *3*, 961–968. [CrossRef]
5. Shepard, C.C.; Crain, C.M.; Beck, M.W. The Protective Role of Coastal Marshes: A Systematic Review and Meta-analysis. *PLoS ONE* **2011**, *6*, e27374. [CrossRef]
6. Christianen, M.J.A.; Middelburg, J.J.; Holthuijsen, S.J.; Jouta, J.; Compton, T.J.; van der Heide, T.; Piersma, T.; Sinninghe Damsté, J.S.; van der Veer, H.W.; Schouten, S.; et al. Benthic primary producers are key to sustain the Wadden Sea food web: Stable carbon isotope analysis at landscape scale. *Ecology* **2017**, *98*, 1498–1512. [CrossRef]
7. Sheaves, M.; Baker, R.; Nagelkerken, I.; Connolly, R.M. True Value of Estuarine and Coastal Nurseries for Fish: Incorporating Complexity and Dynamics. *Estuaries Coasts* **2015**, *38*, 401–414. [CrossRef]
8. Galbraith, H.; Jones, R.; Park, R.; Clough, J.; Herrod-Julius, S.; Harrington, B.; Page, G. Global Climate Change and Sea Level Rise: Potential Losses of Intertidal Habitat for Shorebirds. *Waterbirds* **2002**, *25*, 173–183. [CrossRef]
9. Snelgrove, P.V.R.; Thrush, S.F.; Wall, D.H.; Norkko, A. Real world biodiversity–ecosystem functioning: A seafloor perspective. *Trends Ecol. Evol.* **2014**, *29*, 398–405. [CrossRef]
10. Jones, H.F.E.; Pilditch, C.A.; Hamilton, D.P.; Bryan, K.R. Impacts of a bivalve mass mortality event on an estuarine food web and bivalve grazing pressure. *N. Z. J. Mar. Freshw. Res.* **2017**, *51*, 370–392. [CrossRef]
11. Smith, V.H. Eutrophication of freshwater and coastal marine ecosystems a global problem. *Environ. Sci. Pollut. Res.* **2003**, *10*, 126–139. [CrossRef]
12. Jaffe, B.E.; Smith, R.E.; Foxgrover, A.C. Anthropogenic influence on sedimentation and intertidal mudflat change in San Pablo Bay, California: 1856–1983. *Estuar. Coast. Shelf Sci.* **2007**, *73*, 175–187. [CrossRef]
13. Bonsdorff, E.; Blomqvist, E.M.; Mattila, J.; Norkko, A. Coastal eutrophication: Causes, consequences and perspectives in the Archipelago areas of the northern Baltic Sea. *Estuar. Coast. Shelf Sci.* **1997**, *44*, 63–72. [CrossRef]
14. Thrush, S.; Hewitt, J.; Cummings, V.; Ellis, J.; Hatton, C.; Lohrer, A.; Norkko, A. Muddy waters: Elevating sediment input to coastal and estuarine habitats. *Front. Ecol. Environ.* **2004**, *2*, 299–306. [CrossRef]
15. Douglas, E.J.; Pilditch, C.A.; Kraan, C.; Schipper, L.A.; Lohrer, A.M.; Thrush, S.F. Macrofaunal Functional Diversity Provides Resilience to Nutrient Enrichment in Coastal Sediments. *Ecosystems* **2017**, *20*, 1324–1336. [CrossRef]
16. Ellis, J.I.; Clark, D.; Atalah, J.; Jiang, W.; Taiapa, C.; Patterson, M.; Sinner, J.; Hewitt, J. Multiple stressor effects on marine infauna: Responses of estuarine taxa and functional traits to sedimentation, nutrient and metal loading. *Sci. Rep.* **2017**, *7*, 12013. [CrossRef]
17. Bernardino, A.F.; Sanders, C.J.; Bissoli, L.B.; Gomes, L.E.d.O.; Kauffman, J.B.; Ferreira, T.O. Land use impacts on benthic bioturbation potential and carbon burial in Brazilian mangrove ecosystems. *Limnol. Oceanogr.* **2020**, *65*, 2366–2376. [CrossRef]
18. Mangan, S.; Bryan, K.R.; Thrush, S.F.; Gladstone-Gallagher, R.V.; Lohrer, A.M.; Pilditch, C.A. Shady business: The darkening of estuaries constrains benthic ecosystem function. *Mar. Ecol. Prog. Ser.* **2020**, *647*, 33–48. [CrossRef]
19. Fujii, T.; Raffaelli, D. Sea-level rise, expected environmental changes, and responses of intertidal benthic macrofauna in the Humber estuary, UK. *Mar. Ecol. Prog. Ser.* **2008**, *371*, 23–35. [CrossRef]
20. Rullens, V.; Mangan, S.; Stephenson, F.; Clark, D.E.; Bulmer, R.H.; Berthelsen, A.; Crawshaw, J.; Gladstone-Gallagher, R.V.; Thomas, S.; Ellis, J.I.; et al. Understanding the consequences of sea level rise: The ecological implications of losing intertidal habitat. *N. Z. J. Mar. Freshw. Res.* **2022**, 1–18. [CrossRef]
21. Beukema, J.J. Expected changes in the benthic fauna of Wadden Sea tidal flats as a result of sea-level rise or bottom subsidence. *J. Sea Res.* **2002**, *47*, 25–39. [CrossRef]
22. Yamanaka, T.; Raffaelli, D.; White, P.C.L. Non-Linear Interactions Determine the Impact of Sea-Level Rise on Estuarine Benthic Biodiversity and Ecosystem Processes. *PLoS ONE* **2013**, *8*, e68160. [CrossRef] [PubMed]
23. Flemming, B.W. Siliciclastic Back-Barrier Tidal Flats. In *Principles of Tidal Sedimentology*; Davis, R.A., Jr., Dalrymple, R.W., Eds.; Springer: Dordrecht, The Netherlands, 2012; pp. 231–267.
24. Sweet, W.V.; Hamlington, B.D.; Kopp, R.E.; Weaver, C.P.; Barnard, P.L.; Bekaert, D.; Brooks, W.; Craghan, M.; Dusek, G.; Frederikse, T.; et al. *Global and Regional Sea Level Rise Scenarios for the United States: Updated Mean Projections and Extreme Water Level Probabilities Along U.S. Coastlines*; NOAA Technical Report NOS 01; National Oceanic and Atmospheric Administration, National Ocean Service: Silver Spring, MD, USA, 2022; p. 111.
25. Oppenheimer, M.; Glavovic, B.; Hinkel, J.; van de Wal, R.; Magnan, A.K.; Abd-Elgawad, A.; Cai, R.; Cifuentes-Jara, M.; Deconto, R.M.; Ghosh, T.; et al. Sea Level Rise and Implications for Low Lying Islands, Coasts and Communities. In *IPCC Special Report on the Ocean and Cryosphere in a Changing Climate*; Portner, H.-O., Roberts, D.C., Masson-Delmotte, V., Zhai, P., Tignor, M., Poloczanska, E., Mintenbeck, K., Alegria, A., Nicolai, M., Okem, A., et al., Eds.; Cambridge University Press: Cambridge, UK; New York, NY, USA, 2019; pp. 321–445. [CrossRef]

26. Crooks, S. The effect of sea-level rise on coastal geomorphology. *Ibis* **2004**, *146*, 18–20. [CrossRef]
27. Pethick, J.S.; Crooks, S. Development of a coastal vulnerability index: A geomorphological perspective. *Environ. Conserv.* **2000**, *27*, 359–367. [CrossRef]
28. Alpar, B. Vulnerability of Turkish coasts to accelerated sea-level rise. *Geomorphology* **2009**, *107*, 58–63. [CrossRef]
29. Mazaris, A.D.; Matsinos, G.; Pantis, J.D. Evaluating the impacts of coastal squeeze on sea turtle nesting. *Ocean. Coast. Manag.* **2009**, *52*, 139–145. [CrossRef]
30. Welsh, D.T. It's a dirty job but someone has to do it: The role of marine benthic macrofauna in organic matter turnover and nutrient recycling to the water column. *Chem. Ecol.* **2003**, *19*, 321–342. [CrossRef]
31. Lohrer, A.M.; Townsend, M.; Hailes, S.F.; Rodil, I.F.; Cartner, K.; Pratt, D.R.; Hewitt, J.E. Influence of New Zealand cockles (Austrovenus stutchburyi) on primary productivity in sandflat-seagrass (*Zostera muelleri*) ecotones. *Estuar. Coast. Shelf Sci.* **2016**, *181*, 238–248. [CrossRef]
32. Schenone, S.; Thrush, S.F. Unraveling ecosystem functioning in intertidal soft sediments: The role of density-driven interactions. *Sci. Rep.* **2020**, *10*, 11909. [CrossRef]
33. Snelgrove, P.V.R. The biodiversity of macrofaunal organisms in marine sediments. *Biodivers. Conserv.* **1998**, *7*, 1123–1132. [CrossRef]
34. Lohrer, A.M.; Thrush, S.F.; Gibbs, M.M. Bioturbators enhance ecosystem function through complex biogeochemical interactions. *Nature* **2004**, *431*, 1092–1095. [CrossRef] [PubMed]
35. Lohrer, A.M.; Halliday, N.J.; Thrush, S.F.; Hewitt, J.E.; Rodil, I.F. Ecosystem functioning in a disturbance-recovery context: Contribution of macrofauna to primary production and nutrient release on intertidal sandflats. *J. Exp. Mar. Biol. Ecol.* **2010**, *390*, 6–13. [CrossRef]
36. Thrush, S.F.; Hewitt, J.E.; Gibbs, M.; Lundquist, C.; Norkko, A. Functional Role of Large Organisms in Intertidal Communities: Community Effects and Ecosystem Function. *Ecosystems* **2006**, *9*, 1029–1040. [CrossRef]
37. Cadotte, M.W.; Carscadden, K.; Mirotchnick, N. Beyond species: Functional diversity and the maintenance of ecological processes and services. *J. Appl. Ecol.* **2011**, *48*, 1079–1087. [CrossRef]
38. Waldbusser, G.G.; Marinelli, R.L.; Whitlatch, R.B.; Visscher, P.T. The effects of infaunal biodiversity on biogeochemistry of coastal marine sediments. *Limnol. Oceanogr.* **2004**, *49*, 1482–1492. [CrossRef]
39. Foster, S.Q.; Fulweiler, R.W. Estuarine Sediments Exhibit Dynamic and Variable Biogeochemical Responses to Hypoxia. *J. Geophys. Res. Biogeosciences* **2019**, *124*, 737–758. [CrossRef]
40. Rosenfeld, J.S. Functional redundancy in ecology and conservation. *Oikos* **2002**, *98*, 156–162. [CrossRef]
41. O'Gorman, E.J.; Yearsley, J.M.; Crowe, T.P.; Emmerson, M.C.; Jacob, U.; Petchey, O.L. Loss of functionally unique species may gradually undermine ecosystems. *Proc. R. Soc. B: Biol. Sci.* **2011**, *278*, 1886–1893. [CrossRef]
42. Ellingsen, K.E.; Hewitt, J.E.; Thrush, S.F. Rare species, habitat diversity and functional redundancy in marine benthos. *J. Sea Res.* **2007**, *58*, 291–301. [CrossRef]
43. Naeem, S. Species Redundancy and Ecosystem Reliability. *Conserv. Biol.* **1998**, *12*, 39–45. [CrossRef]
44. Pickett, S.T.A. Space-for-Time Substitution as an Alternative to Long-Term Studies. In *Long-Term Studies in Ecology: Approaches and Alternatives*; Likens, G.E., Ed.; Springer: New York, NY, USA, 1989; pp. 110–135.
45. Inglis, G.; Gust, N.; Fitridge, I.; Floerl, O.; Woods, C.; Kospartov, M.; Hayden, B.J.; Fenwick, G.D.; Staff, M.B.N.Z.; Directorate, M.B.N.Z.P.-b. *Port of Tauranga, Second Baseline Survey for Non-Indigenous Marine Species: (Research Project ZBS2000/04)*; New Zealand Government-Ministry of Agriculture & Forestry-MAF Biosecurity New Zealand: Wellington, New Zealand, 2008.
46. Hume, T.M.; Snelder, T.; Weatherhead, M.; Liefting, R. A controlling factor approach to estuary classification. *Ocean. Coast. Manag.* **2007**, *50*, 905–929. [CrossRef]
47. LINZ. Standard Port Tidal Levels. Available online: https://www.linz.govt.nz/guidance/marine-information/tide-prediction-guidance/standard-port-tidal-levels (accessed on 15 May 2022).
48. Tay, H.W.; Bryan, K.R.; de Lange, W.P.; Pilditch, C.A. The hydrodynamics of the southern basin of Tauranga Harbour. *N. Z. J. Mar. Freshw. Res.* **2013**, *47*, 249–274. [CrossRef]
49. Bell, R.; Hannah, J.; Andrews, C. *Update to 2020 of the Annual Mean Sea Level Series and Trends Around New Zealand. Prepared for Ministry for the Environment*; NIWA Client Report No: 2021236HN; NIWA: Newmarket, New Zealand, 2022. Available online: https://environment.govt.nz/publications/update-to-2020-of-the-annual-mean-sea-level-series-and-trends-around-new-zealand (accessed on 2 January 2023).
50. NZSeaRise. Sea-Level Rise and Vertical land Movement, Map Portal. Available online: https://www.searise.nz/maps-2 (accessed on 27 December 2022).
51. Ellis, J.; Clark, D.; Hewitt, J.E.; Taiapa, C.; Sinner, J.; Patterson, M.; Hardy, D.; Park, S.; Gardner, B.; Morrison, A.; et al. *Ecological Survey of Tauranga Harbour*; Prepared for Manaaki Taha Moana, Manaaki Taha Moana Research Report No. 13. Cawthron Report No. 2321; Cawthron Institute: Nelson, New Zealand, 2013.
52. Clark, D.; Taiapa, C.; Sinner, J.; Taikato, V.; Culliford, D.; Battershill, C.N.; Ellis, J.I.; Hewitt, J.E.; Gower, F.; Borges, H.; et al. *2016 Subtidal Ecological Survey of Tauranga Harbour and Development of Benthic Health Models*; OTOT Research Report No. 4; Massey University: Palmerston, New Zealand, 2018; pp. 1–72.

53. Robertson, B.M.; Gillespie, P.A.; Asher, R.; Frisk, S.; Keeley, N.B.; Hopkins, G.A.; Thompson, S.J.; Tuckey, B.J. *Estuarine Environmental Assessment and Monitoring: A National Protocol. Part A. Development, Part B. Appendices, and Part C. Application*; Prepared for supporting Councils and the Ministry for the Environment, Sustainable Management Fund Contract No. 5096; Cawthron Institute: Nelson, New Zealand, 2002; p. Part A. 93p. Part B. 159p. Part C. 140p plus field sheets.
54. Ellis, J.I.; Hewitt, J.E.; Clark, D.; Taiapa, C.; Patterson, M.; Sinner, J.; Hardy, D.; Thrush, S.F. Assessing ecological community health in coastal estuarine systems impacted by multiple stressors. *J. Exp. Mar. Biol. Ecol.* **2015**, *473*, 176–187. [CrossRef]
55. Knight, B.R. *Estuary Transport Module for Tauranga Harbour. Prepared for Oranga Taiao Oranga Tāngata Research Programme*; Cawthron Report No. 3381. 31 p. plus appendices; Cawthron Institute: Nelson, New Zealand, 2019.
56. de Ruiter, P.J.; Mullarney, J.C.; Bryan, K.R.; Winter, C. The links between entrance geometry, hypsometry and hydrodynamics in shallow tidally dominated basins. *Earth Surf. Process. Landf.* **2019**, *44*, 1957–1972. [CrossRef]
57. Cook, S.D.C. (Ed.) *New Zealand Coastal Marine Invertebrates*; Canterbury University Press: Christchurch, New Zealand, 2010.
58. NIWA. NIWA Invertebrate Collection. v1.1. The National Institute of Water and Atmospheric Research (NIWA). Dataset/Occurrence. 2018. Available online: https://nzobisipt.niwa.co.nz/resource?r=obisspecify&v=1.1. (accessed on 2 January 2023).
59. Greenfield, B.L.; Kraan, C.; Pilditch, C.A.; Thrush, S.F. Mapping functional groups can provide insight into ecosystem functioning and potential resilience of intertidal sandflats. *Mar. Ecol. Prog. Ser.* **2016**, *548*, 1–10. [CrossRef]
60. Ellis, N.; Smith, S.J.; Pitcher, C.R. Gradient forests: Calculating importance gradients on physical predictors. *Ecology* **2012**, *93*, 156–168. [CrossRef] [PubMed]
61. Breiman, L. Random Forests. *Mach. Learn.* **2001**, *45*, 5–32. [CrossRef]
62. Liaw, A.; Wiener, M. Classification and regression by randomForest. *R News* **2002**, *2*, 18–22.
63. Pitcher, R.C.; Lawton, P.; Ellis, N.; Smith, S.J.; Incze, L.S.; Wei, C.-L.; Greenlaw, M.E.; Wolff, N.H.; Sameoto, J.A.; Snelgrove, P.V.R. Exploring the role of environmental variables in shaping patterns of seabed biodiversity composition in regional-scale ecosystems. *J. Appl. Ecol.* **2012**, *49*, 670–679. [CrossRef] [PubMed]
64. R Core Team. *R: A Language and Environment for Statistical Computing*; R Foundation for Statistical Computing: Vienna, Austria, 2019.
65. Clarke, K.R.; Gorley, R.N.; Somerfield, P.J.; Warwick, R.M. *Change in Marine Communities: An Approach to Statistical Analysis and Interpretation*, 3rd ed.; PRIMER-E: Plymouth, UK, 2014.
66. Anderson, M.; Gorley, R.; Clarke, K. *PERMANOVA+ for PRIMER: Guide to Software and Statistical Methods*; PRIMER-E: Plymouth, UK, 2008.
67. Mindel, B.L.; Neat, F.C.; Trueman, C.N.; Webb, T.J.; Blanchard, J.L. Functional, size and taxonomic diversity of fish along a depth gradient in the deep sea. *PeerJ* **2016**, *4*, e2387. [CrossRef] [PubMed]
68. Kramer, N.; Tamir, R.; Eyal, G.; Loya, Y. Coral Morphology Portrays the Spatial Distribution and Population Size-Structure Along a 5–100 m Depth Gradient. *Front. Mar. Sci.* **2020**, *7*, 615. [CrossRef]
69. Villalobos, V.I.; Valdivia, N.; Försterra, G.; Ballyram, S.; Espinoza, J.P.; Wadham, J.L.; Burgos-Andrade, K.; Häussermann, V. Depth-Dependent Diversity Patterns of Rocky Subtidal Macrobenthic Communities Along a Temperate Fjord in Northern Chilean Patagonia. *Front. Mar. Sci.* **2021**, *8*, 635855. [CrossRef]
70. Gogina, M.; Zettler, M.L. Diversity and distribution of benthic macrofauna in the Baltic Sea: Data inventory and its use for species distribution modelling and prediction. *J. Sea Res.* **2010**, *64*, 313–321. [CrossRef]
71. Flach, E.; de Bruin, W. Diversity patterns in macrobenthos across a continental slope in the NE Atlantic. *J. Sea Res.* **1999**, *42*, 303–323. [CrossRef]
72. Holleman, R.C.; Stacey, M.T. Coupling of Sea Level Rise, Tidal Amplification, and Inundation. *J. Phys. Oceanogr.* **2014**, *44*, 1439–1455. [CrossRef]
73. Rullens, V.; Stephenson, F.; Lohrer, A.M.; Townsend, M.; Pilditch, C.A. Combined species occurrence and density predictions to improve marine spatial management. *Ocean. Coast. Manag.* **2021**, *209*, 105697. [CrossRef]
74. Kraan, C.; Aarts, G.; van der Meer, J.; Piersma, T. The role of environmental variables in structuring landscape-scale species distributions in seafloor habitats. *Ecology* **2010**, *91*, 1583–1590. [CrossRef]
75. Puls, W.; van Bernem, K.H.; Eppel, D.; Kapitza, H.; Pleskachevsky, A.; Riethmüller, R.; Vaessen, B. Prediction of benthic community structure from environmental variables in a soft-sediment tidal basin (North Sea). *Helgol. Mar. Res.* **2012**, *66*, 345–361. [CrossRef]
76. Ysebaert, T.; Herman, P.M.J.; Meire, P.; Craeymeersch, J.; Verbeek, H.; Heip, C.H.R. Large-scale spatial patterns in estuaries: Estuarine macrobenthic communities in the Schelde estuary, NW Europe. *Estuar. Coast. Shelf Sci.* **2003**, *57*, 335–355. [CrossRef]
77. Liao, Y.; Shou, L.; Jiang, Z.; Gao, A.; Zeng, J.; Chen, Q.; Yan, X. Benthic macrofaunal communities along an estuarine gradient in the Jiaojiang River estuary, China. *Aquat. Ecosyst. Health Manag.* **2016**, *19*, 314–325. [CrossRef]
78. Denis-Roy, L.; Ling, S.D.; Fraser, K.M.; Edgar, G.J. Relationships between invertebrate benthos, environmental drivers and pollutants at a subcontinental scale. *Mar. Pollut. Bull.* **2020**, *157*, 111316. [CrossRef]
79. Engle, V.D.; Summers, J.K. Latitudinal gradients in benthic community composition in Western Atlantic estuaries. *J. Biogeogr.* **1999**, *26*, 1007–1023. [CrossRef]

80. Snelgrove, P.V.R. Diversity of marine species. In *Encyclopedia of Ocean Sciences*; Steele, J.H., Ed.; Academic Press: Oxford, UK, 2001; pp. 748–757.
81. Lee, J.; Valle-Levinson, A. Influence of bathymetry on hydrography and circulation at the region between an estuary mouth and the adjacent continental shelf. *Cont. Shelf Res.* **2012**, *41*, 77–91. [CrossRef]
82. Conroy, T.; Sutherland, D.A.; Ralston, D.K. Estuarine Exchange Flow Variability in a Seasonal, Segmented Estuary. *J. Phys. Oceanogr.* **2020**, *50*, 595–613. [CrossRef]
83. Chadwick, D.B.; Largier, J.L. Tidal exchange at the bay-ocean boundary. *J. Geophys. Res. Ocean.* **1999**, *104*, 29901–29924. [CrossRef]
84. Bolaños, R.; Brown, J.M.; Amoudry, L.O.; Souza, A.J. Tidal, Riverine, and Wind Influences on the Circulation of a Macrotidal Estuary. *J. Phys. Oceanogr.* **2013**, *43*, 29–50. [CrossRef]
85. Norkko, J.; Hewitt, J.E.; Thrush, S.F. Effects of increased sedimentation on the physiology of two estuarine soft-sediment bivalves, Austrovenus stutchburyi and Paphies australis. *J. Exp. Mar. Biol. Ecol.* **2006**, *333*, 12–26. [CrossRef]
86. Zarzuelo, C.; López-Ruiz, A.; Díez-Minguito, M.; Ortega-Sánchez, M. Tidal and subtidal hydrodynamics and energetics in a constricted estuary. *Estuar. Coast. Shelf Sci.* **2017**, *185*, 55–68. [CrossRef]
87. Schilthuizen, M. Ecotone: Speciation-prone. *Trends Ecol. Evol.* **2000**, *15*, 130–131. [CrossRef] [PubMed]
88. Kark, S.; van Rensburg, B.J. Ecotones: Marginal or Central Areas of Transition? *Isr. J. Ecol. Evol.* **2006**, *52*, 29–53. [CrossRef]
89. Newell, R.C. One-Adaptations to intertidal life. In *Adaptation to Environment*; Newell, R.C., Ed.; Butterworth-Heinemann: Oxford UK, 1976; pp. 1–82.
90. Hooper, D.U.; Chapin III, F.S.; Ewel, J.J.; Hector, A.; Inchausti, P.; Lavorel, S.; Lawton, J.H.; Lodge, D.M.; Loreau, M.; Naeem, S.; et al. Effects of biodiversity on ecosystem functioning: A consensus of current knowledge. *Ecol. Monogr.* **2005**, *75*, 3–35. [CrossRef]
91. Lefcheck, J.S.; Byrnes, J.E.K.; Isbell, F.; Gamfeldt, L.; Griffin, J.N.; Eisenhauer, N.; Hensel, M.J.S.; Hector, A.; Cardinale, B.J.; Duffy, J.E. Biodiversity enhances ecosystem multifunctionality across trophic levels and habitats. *Nat. Commun.* **2015**, *6*, 6936. [CrossRef] [PubMed]
92. Danovaro, R.; Gambi, C.; Dell'Anno, A.; Corinaldesi, C.; Fraschetti, S.; Vanreusel, A.; Vincx, M.; Gooday, A.J. Exponential Decline of Deep-Sea Ecosystem Functioning Linked to Benthic Biodiversity Loss. *Curr. Biol.* **2008**, *18*, 1–8. [CrossRef] [PubMed]
93. Stachowicz, J.J.; Bruno, J.F.; Duffy, J.E. Understanding the Effects of Marine Biodiversity on Communities and Ecosystems. *Annu. Rev. Ecol. Evol. Syst.* **2007**, *38*, 739–766. [CrossRef]
94. Grman, E.; Lau, J.A.; Schoolmaster, D.R., Jr.; Gross, K.L. Mechanisms contributing to stability in ecosystem function depend on the environmental context. *Ecol. Lett.* **2010**, *13*, 1400–1410. [CrossRef] [PubMed]
95. Hillman, J.R.; Lundquist, C.J.; O'Meara, T.A.; Thrush, S.F. Loss of Large Animals Differentially Influences Nutrient Fluxes Across a Heterogeneous Marine Intertidal Soft-Sediment Ecosystem. *Ecosystems* **2020**, *24*, 272–283. [CrossRef]
96. Luck, G.W.; Harrington, R.; Harrison, P.A.; Kremen, C.; Berry, P.M.; Bugter, R.; Dawson, T.P.; de Bello, F.; Díaz, S.; Feld, C.K.; et al. Quantifying the Contribution of Organisms to the Provision of Ecosystem Services. *BioScience* **2009**, *59*, 223–235. [CrossRef]
97. van der Linden, P.; Marchini, A.; Smith, C.J.; Dolbeth, M.; Simone, L.R.L.; Marques, J.C.; Molozzi, J.; Medeiros, C.R.; Patrício, J. Functional changes in polychaete and mollusc communities in two tropical estuaries. *Estuar. Coast. Shelf Sci.* **2017**, *187*, 62–73. [CrossRef]
98. Wouters, J.M.; Gusmao, J.B.; Mattos, G.; Lana, P. Polychaete functional diversity in shallow habitats: Shelter from the storm. *J. Sea Res.* **2018**, *135*, 18–30. [CrossRef]
99. Jones, H.F.E.; Pilditch, C.A.; Bruesewitz, D.A.; Lohrer, A.M. Sedimentary Environment Influences the Effect of an Infaunal Suspension Feeding Bivalve on Estuarine Ecosystem Function. *PLoS ONE* **2011**, *6*, e27065. [CrossRef]
100. Woodin, S.A.; Volkenborn, N.; Pilditch, C.A.; Lohrer, A.M.; Wethey, D.S.; Hewitt, J.E.; Thrush, S.F. Same pattern, different mechanism: Locking onto the role of key species in seafloor ecosystem process. *Sci. Rep.* **2016**, *6*, 26678. [CrossRef]
101. Elmilady, H.; van der Wegen, M.; Roelvink, D.; Jaffe, B.E. Intertidal Area Disappears Under Sea Level Rise: 250 Years of Morphodynamic Modeling in San Pablo Bay, California. *J. Geophys. Res. Earth Surf.* **2019**, *124*, 38–59. [CrossRef]
102. Rullens, V.; Townsend, M.; Lohrer, A.M.; Stephenson, F.; Pilditch, C.A. Who is contributing where? Predicting ecosystem service multifunctionality for shellfish species through ecological principles. *Sci. Total Environ.* **2022**, *808*, 152147. [CrossRef]
103. Pollard, A.I.; Reed, T. Benthic invertebrate assemblage change following dam removal in a Wisconsin stream. *Hydrobiologia* **2004**, *513*, 51–58. [CrossRef]
104. Tullos, D.D.; Finn, D.S.; Walter, C. Geomorphic and Ecological Disturbance and Recovery from Two Small Dams and Their Removal. *PLoS ONE* **2014**, *9*, e108091. [CrossRef]
105. Bain, D.J.; Green, M.B.; Campbell, J.L.; Chamblee, J.F.; Chaoka, S.; Fraterrigo, J.M.; Kaushal, S.S.; Martin, S.L.; Jordan, T.E.; Parolari, A.J.; et al. Legacy Effects in Material Flux: Structural Catchment Changes Predate Long-Term Studies. *BioScience* **2012**, *62*, 575–584. [CrossRef]
106. Drylie, T.P.; Lohrer, A.M.; Needham, H.R.; Bulmer, R.H.; Pilditch, C.A. Benthic primary production in emerged intertidal habitats provides resilience to high water column turbidity. *J. Sea Res.* **2018**, *142*, 101–112. [CrossRef]
107. Sigleo, A.C. Denitrification Rates Across a Temperate North Pacific Estuary, Yaquina Bay, Oregon. *Estuaries Coasts* **2019**, *42*, 655–664. [CrossRef] [PubMed]

108. Piehler, M.F.; Smyth, A.R. Habitat-specific distinctions in estuarine denitrification affect both ecosystem function and services. *Ecosphere* **2011**, *2*, 12. [CrossRef]
109. Dangles, O.; Malmqvist, B. Species richness–decomposition relationships depend on species dominance. *Ecol. Lett.* **2004**, *7*, 395–402. [CrossRef]

biology

MDPI

*Article*

# Role of Macrofaunal Communities in the Vistula River Plume, the Baltic Sea—Bioturbation and Bioirrigation Potential

**Natalia Anna Miernik, Urszula Janas * and Halina Kendzierska**

Institute of Oceanography, University of Gdańsk, al. Marszałka Piłsudskiego 46, 81-378 Gdynia, Poland
* Correspondence: urszula.janas@ug.edu.pl; Tel.: +48-58-523-6867

**Simple Summary:** Coastal areas, especially river plumes, are very diverse and dynamic zones where numerous geological, chemical and biological processes take place. This is because fresh water from the river with all substances, including pollutants from the land, mixes with salt water from the sea, creating specific living conditions for the organisms that inhabit the area. These organisms, e.g., macroscopic invertebrates such as mussels or worms, live in the sediment where their movement and feeding activities cause the sediment to mix and allow water to flow through it—these activities are called bioturbation and bioirrigation. Our research aimed to investigate how the structure and functioning of benthic marine ecosystems change with distance from the river mouth. We found that coastal areas are very diverse and host a wide range of organisms that bioturbate and bioirrigate and support sediment transformations relatively deep (up to 15 cm) into the sediment. Farther away from the river mouth, organisms were very scarce and occurred only on the sediment surface and did not burrow into the sediment, so bioturbation and bioirrigation did not take place. The coastal zone is like a hotspot where ecosystem processes and services are intensively reflected, and this is especially important when deeper areas are not functioning properly, as in the Baltic Sea. For this reason, we should consider how we can support the protection and recovery of marine ecosystems.

**Abstract:** Macrozoobenthos plays a key role in the transformation of inputs from rivers to the sea, such as nutrients, organic matter, or pollutants, and influences biogeochemical processes in the sediments through bioturbation and bioirrigation activity. The purpose of our study was to determine the structure of benthic communities, their bioturbation ($BP_C$) and bioirrigation potential ($IP_C$), and the vertical distribution of macrofauna in the Gulf of Gdańsk. The study revealed changes in the structure of benthic communities and, consequently, in the bioturbation and bioirrigation potential in the study area. Despite the presence of diverse and rich communities in the coastal zone, $BP_C$ and $IP_C$ values, although high, were formed by a few species. Both indices were formed mainly by the clam *Macoma balthica* and polychaetes, although the proportion of polychaetes in $IP_C$ was higher than in $BP_C$. In the deepest zones, the communities became poorer until they eventually disappeared, along with all macrofaunal functions. Both indices changed similarly with distance from the Vistula River mouth, and there was a very strong correlation between them. We also demonstrated that the highest diversity of the macrofauna was observed in the upper first cm of the sediment, but the highest biomass was observed in deeper layers—at a depth of up to 6 cm, and single individuals occurred even below 10 cm.

**Keywords:** macrozoobenthos; marine biodiversity; bioturbation; bioirrigation; coastal zone; Baltic Sea; Gulf of Gdańsk; Vistula River plume

## 1. Introduction

Coastal zones provide a variety of benefits derived by humans from ecosystem functions and processes. These include nutrient regulation or waste treatment functions, where biota play an important role in storage, recycling or removal of nutrients and compounds [1]. All of these functions help maintain healthy and productive marine ecosystems. Coastal ecosystems with high biodiversity of habitats and benthic communities, especially lagoons, bays and estuaries, play a special role in marine regulatory processes [2,3]. Benthic organisms play a key role in the circulation of chemical elements and nutrients directly by physiological processes such as feeding, respiration and excretion, as well as indirectly by reworking the sediment matrix through bioturbation and bioirrigation [4–9]. These activities can be positive for the ecosystem in terms of sediment oxygenation and increasing the surface area available for microbial activity [10–12]. Intensive bioturbation or bioirrigation may also lead to the intensification of degradation, transformation or burial of organic matter and contaminants [13,14]. On the other hand, sediment reworking may cause a release of contaminants accumulated in the deeper parts of sediments [14,15]. Thus, bioturbation and bioirrigation play a crucial role in biochemical cycles and production at the seafloor and basin scale [16,17].

At the same time, the coastal zone is particularly exposed to land-based pollution from i.a. increased industrialization, urbanization, agricultural and aquacultural development as well as climate change [18]. Nutrients, organic matter and contaminants from land enter the seas and oceans mainly through surface runoff. Nowadays, river pollution in most populated areas is severe and according to high urbanization future scenarios, about 80% of the global human population is projected to live in sub-basins with multi-pollutant problems [19].

The Gulf of Gdańsk, located in the southern part of the Baltic Sea, is a coastal system with a mixture of fresh and brackish water. Salinity, but also other parameters such as nitrogenous compound and chlorophyll *a* concentrations, change both with distance from the river mouth and with depth [20,21]. Research by Łukawska-Matuszewska et al. [22] showed that sediment toxicity in the Gulf of Gdańsk increases with distance from land, which is associated with an increase in the content of fine sediment fractions, hydrogen sulfide and black carbon, with the latter suggesting anthropogenic contamination of the sediment. The area of the entire Gulf is strongly affected by the Vistula River. It is the longest river flowing into the Baltic Sea, passing through agricultural land, forests and several urban agglomerations [23]. The river has the second largest drainage basin of the rivers flowing into the Baltic Sea (194,000 $km^2$, covering 11% of the whole Baltic Sea catchment area). The Vistula River contributes about 90% of the total inflow to the Gulf of Gdańsk [24]. Along with the river's waters come nutrients, organic matter and various pollutants: heavy metals, organic pollutants, including pharmaceuticals and emerging contaminants [25–27]. In addition to the Vistula River, there are other sources of pollutants such as dozens of watercourses, ports, industry, wastewater treatment plants, atmospheric deposition or disturbed sediment [26]. All these compounds reaching the sea can affect the structure and functioning of the ecosystem, while at the same time the presence of organisms such as zoobenthos can help process these compounds. To understand the role of the benthic fauna in these processes, it is necessary to determine how benthic animals are distributed in the vicinity of the Vistula River and how they function.

There is a strong need for indices that demonstrate the decline in ecosystem functioning under anthropopressure and improvement during sustainable ecosystem-based marine management [28,29]. This is due, i.a. to the demand for measures to maintain and improve the ecological status of the marine environment in accordance with the Marine Strategy Framework Directive. Existing bioturbation and bioirrigation potential indices can be used as a proxy of ecosystem processes [30–33]. Basic benthic monitoring parameters (i.e., abundance and biomass), as well as research-based knowledge (or, in many cases, expert knowledge) of benthic fauna traits related to their behavior in the sediment, are used for the calculations. So far, these coefficients have been successfully used and combined with studies of biogeochemical cycles [34], solutes exchange between water and sediment [33,35], studies of anaerobic episodes [36] and apparent redox discontinuity layer (aRPD) [37]. Although these indices appear simple, they carry some limitations related to insufficient knowledge of the activity of individual species and how it changes under the influence of various factors. However, being aware of these limitations, these tools can be applied in both scientific research and environmental monitoring. According to Queirós and colleagues [38], the bioturbation potential index also has limitations, and knowing this can contribute to more informed use of the index as an indicator of benthic function.

A few studies on the role of macrofauna carried out in the Gulf of Gdańsk have addressed the bioturbation potential index ($BP_C$) or nutrient fluxes between water and sediments [39–42]. Studies on the functioning of marine ecosystems in the Gulf of Gdańsk in recent years have also considered the influence of organic matter on the structure and functioning of trophic networks [43] and how organic matter is transformed by organisms [44]. So far, no research has been carried out in the Gulf of Gdańsk on bioirrigation processes. There are also few published studies on the distribution of organisms in the sediment. They mostly contain information on the depth of occurrence of individual macrofauna and meiofauna taxa [45–48], but only single studies addressed entire benthic communities [40,41,49].

The objective of this study was to determine the structure of benthic fauna as well as the bioturbation and bioirrigation potential of macrofauna in the sediments of the Vistula plume area, in the Gulf of Gdańsk. Furthermore, it was determined quantitatively how this impact of benthic communities varies depending on the proximity of the Vistula River mouth, as well as which species are the most responsible for sediment matrix reworking in the area. In addition, we have made an attempt to investigate the vertical distribution of macrofauna taxa, detailing their maximal and typical depth of occurrence in the sediment.

The results presented in this paper will help to demonstrate the zones where, due to the presence of animals in the sediments and their activity, nutrients, organic matter and pollutants carried into the Gulf of Gdańsk by the Vistula River are processed. They will also provide knowledge of the vertical distribution of species in the sediments necessary, among other things, for indices of functionality to assess the functioning of the seafloor and basin. Determining the role of macrofauna will also provide arguments for the protection and proper management of marine areas in estuaries.

## 2. Materials and Methods

### 2.1. Sampling

Bottom water, sediment and fauna were collected during two cruises in the Vistula River plume area and along an offshore depth transect in the Gulf of Gdańsk, the Baltic Sea (Figure 1). Samples from 11 sites were collected in July 2014 from the deck of RV Elisabeth Mann-Borgese. In March 2016, three more sites were sampled (VE04, VE06, VE07) during a cruise aboard RV Alkor. Bottom water temperature, salinity and dissolved oxygen (DO) concentration were measured at all sites approximately 0.5 m above the sediment using a Seabird CTD-system with an oxygen SBE43 sensor.

For sediment and macrofauna analysis sediment cores (inner diam. 10 cm) were collected with a multicorer and subsamples of coarse-grained sands were collected from a Haps corer. At each site the upper 10 cm sediment sample for sediment parameters was frozen and prior to all analysis the sediment was dried and homogenized. The organic matter content of the dry sediments was measured as the percentage loss on ignition (LOI) after dry combustion for 8 h at 450 °C and for 5 h at 550 °C for samples collected in March 2016. For grain size analysis, samples were sieved using a shaker and a set of standard test sieves with mesh diameters of 2, 1, 0.5, 0.25, 0.125 and 0.063 mm [50]. Based on a percentage of each class in the total sample mass, sediments were classified by the Udden–Wentworth grain-size scale (after Wentworth [51]).

### 2.2. Macrofauna

For benthic fauna analysis, 3 to 5 replicates were collected at each site, with the exception of station VE49, where only 2 replicates could be collected. Sediment cores were divided into layers: 0–1 cm, 1–3 cm, 3–6 cm, 6–10 cm, 10–15 cm and >15 cm depth. We sifted all layers separately through a 1 mm sieve to separate the macrofauna from the sediment and preserved with 4% formaldehyde until analysis (stored for at least 3 months).

In the laboratory, the fauna was sorted and taxa, with the exception of Oligochaeta and *Marenzelleria* spp. polychaetes, were identified to the species level. Taxa were counted and weighed to determine their abundance and biomass (wet mass) per square meter.

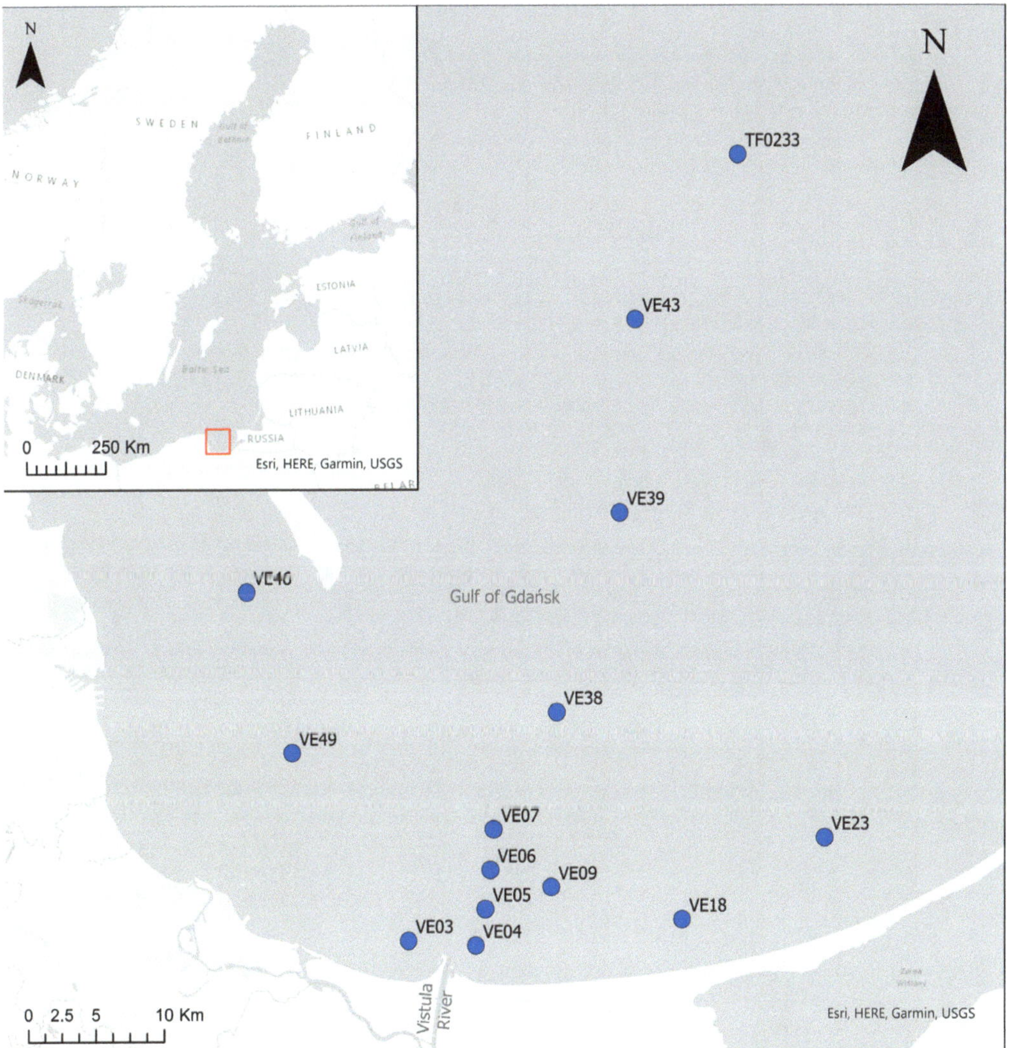


**Figure 1.** Study area with sampling sites. The red rectangle indicates the location of the study area on a map of the Baltic Sea.

### 2.3. Bioturbation Potential ($BP_C$) and Irrigation Potential ($IP_C$)

To calculate the bioturbation and bioirrigation potential, wet mass (WW) was converted to ash free dry mass (AFDW). The conversion was based on literature data; for bivalves, the coefficients were used for individuals with shells [52–55]. The Bioturbation

Potential Community Index ($BP_C$) at individual sites was calculated by summing the bioturbation potentials ($BP_i$) calculated for individual taxa [30,36].

$$BP_c = \sum BP_i \text{ where : } BP_i = \left(\frac{B_i}{A_i}\right)^{0.5} * A_i * M_i * R_i \quad (1)$$

where for taxon i: $B_i$ is biomass (in ash free dry mass $g \cdot m^{-2}$) and $A_i$ is abundance (ind.$\cdot m^{-2}$) at each sample, while $M_i$, mobility, and $R_i$, sediment reworking, are categorical scores assigned to each species (Table A1).

The Irrigation Potential Community Index ($IP_C$) at individual sites was calculated by summing the irrigation potentials ($IP_i$) calculated for individual taxa [56].

$$IP_c = \sum IP_i \text{ where : } IP_i = \left(\frac{B_i}{A_i}\right)^{0.75} * A_i * BT_i * FT_i * ID_i \quad (2)$$

where for taxon i: $B_i$ is biomass (in ash free dry mass $g \cdot m^{-2}$) and $A_i$ is abundance (ind.$\cdot m^{-2}$) at each sample, while feeding type ($FT_i$), burrow type ($BT_i$) and depth ($ID_i$) are scores for the trait categories assigned to each species (Table A1). Exponent 0.5 used in $BP_C$ emphasizes the importance of organisms with high density and relatively low biomass, while exponent 0.75 used in $IP_C$ emphasizes the activity of organisms with larger sizes but lower densities [33].

### *2.4. Vertical Distrbution of Macrofauna in Sediment*

The analysis of the vertical distribution of macrozoobenthos in the sediment to determine the maximum burial depth of each taxon and the entire community was performed for both the abundance and biomass of organisms from 51 cores. To present the vertical distribution, the benthic macrofauna abundance and biomass measured in separate sediment layers were recalculated per 1 $dm^3$ volume. The burial depth data were averaged for all cores in which a given taxon occurred. The percentage of individual taxa abundance and biomass (90%) in the studied layers was indicated to determine the typical depth of occurrence.

### *2.5. Data Analysis*

Principal Component Analysis (PCA) was carried out to determine the relationship between physicochemical conditions in bottom water and surface sediments, and the variability between the sites. A matrix with normalized data on bottom water temperature, salinity, dissolved oxygen concentration and organic matter content in surface sediments was used in statistical analysis. Environmental parameters were strongly correlated with the depth of the basin i.e., salinity (Pearson's $r = 0.95$), DO ($r = -0.79$), type of sediment ($r = 0.86$) and LOI (0.65).

Prior to biological data analysis, the biomass at each sampling site was averaged and square root transformed. Cluster (Bray–Curtis similarity) and SIMPROF analysis was used to determine the similarity of macrofauna samples. The SIMPER procedure was applied to identify species responsible for similarities/differences in macrozoobenthic communities between the analyzed sites [57]. Biota and Environment matching analysis (BEST BIO-ENV) was performed to determine the effects of temperature, salinity, DO concentration in bottom water and organic matter content in surface sediment on the formation of benthic fauna communities. Distance-based linear models (distLM) were used to examine the effects of environmental variables on biomass, maximum burrowing depth, $BP_C$ and $IP_C$ [58]. First, the relationships between the variables were examined and oxygen concentration was excluded from the analysis as being strongly correlated with salinity (Pearson's $r = -0.84$). The following three environmental variables were selected: temperature, salinity and LOI and log(x + 1) transformation was used before analysis. Stepwise selection and the AICc stopping criterion were used in distLM to investigate the role of environmental variables in predicting biological traits of macrozoobenthos. Resemblance matrices were based on the Euclidean distance similarities between the sites. The results of marginal tests indicate the proportion of the variation the predictor accounts for on its own, while the results from the sequential test indicate the proportion added by the predictor to the cumulative total proportion explained. The statistical analyses were computed in PRIMER v6 & PERMANOVA +. Maps with results were prepared using Arc GIS Pro 2.9.0, ESRI Inc., Redlands, California, the United States of America.

Data from individual cores were used to analyze the relationship between the biological parameters. The relationship between bioturbation and bioirrigation potential indices (calculated using WW and AFDW, and two different exponents in the case of $IP_C$) and the number of taxa, abundance, biomass and maximum burrowing depth were determined by Spearman's rank correlation test. In addition, $IP_C$ values obtained when considering the maximum burrowing depth of macrofaunal individuals in the sediment observed in this study were also compared with those assumed based on the literature and expert knowledge. Prior to the statistical analysis, the normality of the data was tested (Shapiro–Wilk test $p < 0.05$).

## 3. Results

### 3.1. Environmental Conditions

Bottom water temperature at the surveyed sites was relatively uniform (below 6.2 °C), except for sites VE03, VE05, and VE18, which were surveyed in the summer season, above thermocline (Table 1). Bottom water salinity was generally higher in the deeper parts and reached 12.7 in the Gdańsk Deep (site TF0233). The opposite situation was observed for dissolved oxygen concentrations in bottom water. Oxygen conditions were above 4.68 mL·dm$^{-3}$ at the shallow sites, but oxygen deficiency was observed in the deepest part—below 3.41 mL·dm$^{-3}$, and the two deepest sites (VE43 and TF0233) showed hypoxia (DO < 2 mL·dm$^{-3}$). Sediment variability was fairly typical for the coastal areas. The shallow sites were characterized by the presence of medium and fine-grained sands, while deeper sites were dominated by clay and silt.

**Table 1.** Values of sediment characteristics and environmental variables measured in the bottom waters at research sites.

| Station | Temperature [°C] | Salinity | Oxygen [mL·dm$^{-3}$] | Sediment Type | LOI [%] | Depth [m] |
|---|---|---|---|---|---|---|
| VE04 | 4.2 | 7.6 | 8.10 | Fine-grained sand | 1.25 | 15 |
| VE03 | 14.2 | 7.4 | 5.65 | Fine-grained sand | 4.49 | 16 |
| VE05 | 12.6 | 7.4 | 4.68 | Fine-grained sand | 4.03 | 24 |
| VE18 | 11.2 | 7.3 | 5.83 | Fine-grained sand | 0.91 | 24 |
| VE49 | 6.2 | 7.6 | 5.93 | Medium-grained sand | 0.89 | 25 |
| VE09 | 5.3 | 8.0 | 6.32 | Medium-grained sand | 0.81 | 32 |
| VE06 | 3.8 | 8.0 | 8.19 | Fine-grained sand | 0.88 | 38 |
| VE23 | 5.0 | 8.0 | 6.47 | Sandy silt | 4.24 | 48 |
| VE46 | 4.6 | 8.2 | 5.90 | Silt | 13.26 | 48 |
| VE07 | 3.7 | 8.0 | 8.33 | Fine-grained sand | 3.07 | 59 |
| VE38 | 4.5 | 9.1 | 3.41 | Silt | 4.40 | 67 |
| VE39 | 5.3 | 11.2 | 2.56 | Silty clay | 18.54 | 84 |
| VE43 | 5.8 | 12.3 | 1.52 | Silty clay | 3.18 | 94 |
| TF0233 | 5.6 | 12.7 | 1.59 | Silty clay | 15.49 | 105 |

PCA analysis was conducted to determine the effect of four physicochemical parameters on the variability between the sites (Figure 2). The first principal component explains 59.6% (eigenvalue 2.39), and together with the second principal component (eigenvalue 1.08) a total of 86.6% of the total variation (Table 2). Salinity with a coefficient of −0.612 has the largest contribution to the distribution along the PC1 axis. The distribution along the PC2 axis was most significantly affected by bottom water temperature (coefficient 0.939).

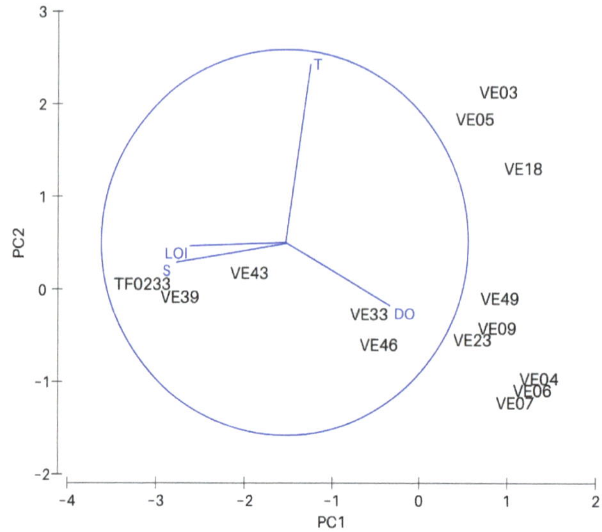


**Figure 2.** Results of principle component analysis (PCA). Variables included in the PCA are bottom water temperature (T), salinity (S) and oxygen concentration (DO), and organic matter content in the surface sediments (LOI).

**Table 2.** Percentage of variation and coefficients in the linear combinations of variables forming PCs.

| Variable | PC1 | PC2 | PC3 |
|---|---|---|---|
| Variation [%] | 59.6 | 26.9 | 11.7 |
| Temperature (T) | 0.128 | 0.939 | 0.104 |
| Salinity (S) | −0.612 | −0.100 | −0.366 |
| Oxygen (DO) | 0.576 | −0.329 | 0.366 |
| LOI | −0.527 | −0.015 | 0.849 |

### 3.2. Macrofauna

The study revealed the presence of a total of 23 macrofaunal taxa in the Gulf of Gdańsk. Taxa with the highest frequency in the Vistula estuary (above 70%) were the bivalve *Macoma balthica*, Oligochaeta, the polychaetes *Bylgides sarsi*, *Marenzelleria* spp., *Pygospio elegans*, as well as the crustacean *Corophium volutator* and the gastropod *Peringia ulvae* (data not shown). The biodiversity of benthic organisms decreased with depth—from 16 taxa at site VE05 to no organisms in the Gdańsk Deep. The main factors determining the structure of macrozoobenthos biomass were salinity and oxygen concentration in the bottom water (BIOENV, r = 0.74).

Based on cluster and SIMPROF analysis, three groups of sites were distinguished with respect to the biomass of the identified macrofauna taxa (Figure 3). In both group 1 and group 2, *M. balthica* was the most dominant species in the biomass and significantly contributed to the similarity of biomass in both groups (contribution to the total biomass of 67% and 79%, respectively). In addition, species that contributed to the similarity in group 1 were *Hediste diversicolor* (11%), *P. ulvae* (9%) and *Mya arenaria* (8%). Other taxa that accounted for the similarity between sites in group 2, in addition to *M. balthica*, were *Marenzelleria* spp. (7%) and *Halicryptus spinulosus* (5%). In addition, group 3 comprised the deepest sites, where only polychaetes represented by the species *B. sarsi* were observed. The average dissimilarity between group 1 or group 2 and group 3 was >99%. In both cases, *M. balthica* accounted for the highest proportion of dissimilarity (>54%).

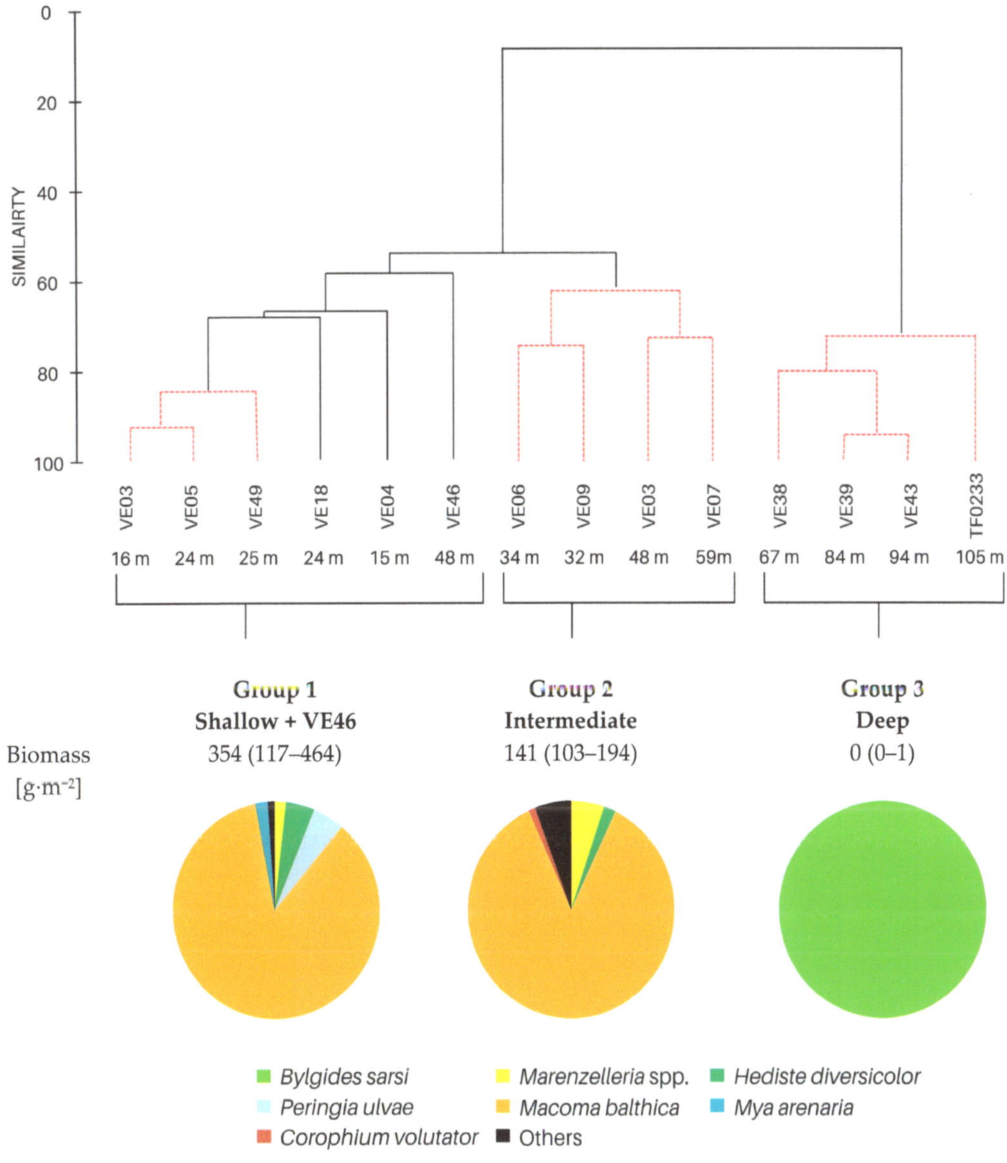


**Figure 3.** Groups of sites distinguished by cluster analysis on the basis of taxonomic composition and biomass (data transformation: $\sqrt{}$): **top**—cluster similarity of the study sites; **bottom**—contribution of taxa in macrofaunal biomass in three groups of sites.

All the biological parameters studied reached the highest values at the shallow sites and site VE46, and their values gradually decreased in subsequent groups with increasing depth (Table 3). Benthic communities at the shallow and intermediate sites were characterized by high taxonomic diversity of macrofauna. The highest values of density and biomass of macrofauna were observed at the shallow sites and decreased with depth. Similarly, the values of the $BP_C$ and $IP_C$ indices decreased, with the values of both indices being lower by half at the intermediate sites compared to the shallow sites.

**Table 3.** Number of taxa, maximum burial depth of macrofauna, mean values: abundance, $BP_C$ and $IP_C$ (min.–max), contribution of individual taxa to the formation of these parameters, and in each group of sites provided in Figure 3.

| | Group 1 | Group 2 | Group 3 |
|---|---|---|---|
| No. of taxa | 11 (6–16) | 8 (7–10) | 0 (0–1) |
| Max. burial depth [cm] | 14 (10–15) | 11 (6–15) | 0 (0–1) |
| | 11,030 (3628–30,557) | 3188 (2578–3851) | 63 (0–127) |
| Abundance [ind.$\cdot m^{-2}$] | 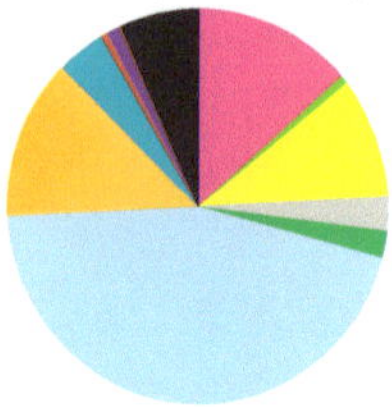 | 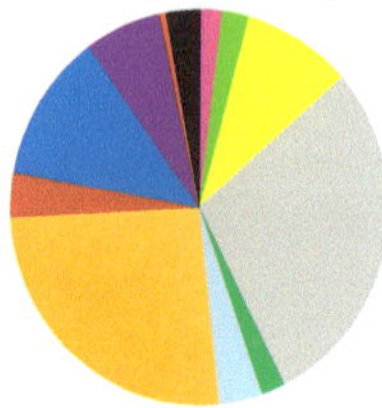 |  |
| | 3412 (1576–5000) | 1683 (1412–1890) | 5 (0–13) |
| $BP_C$ | 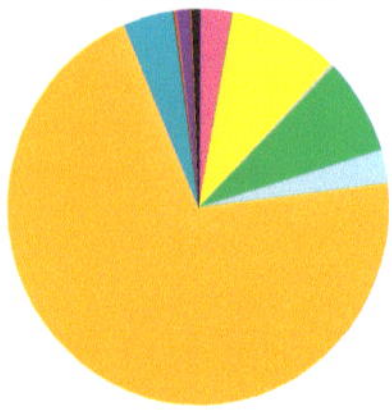 | 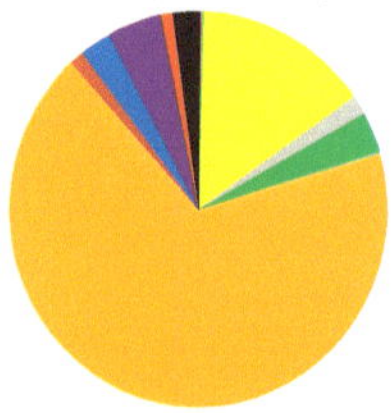 |  |
| | 1451 (581–1985) | 714 (632–785) | 1 (0–4) |
| $IP_C$ | 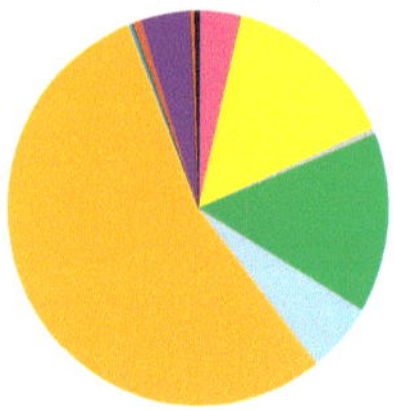 | 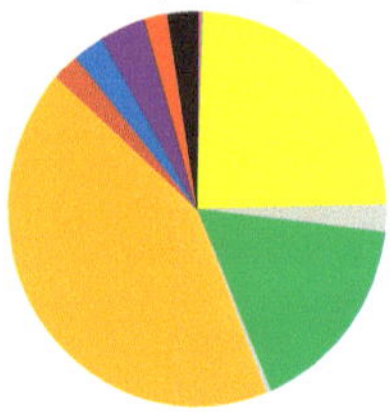 |  |

■ Oligochaeta ■ *Bylgides sarsi* ■ *Marenzelleria* spp. ■ *Pygospio elegans*
■ *Hediste diversicolor* ■ *Peringia ulvae* ■ *Macoma balthica* ■ *Mya arenaria*
■ *Corophium volutator* ■ *Monoporeia affinis* ■ *Pontoporeia femorata* ■ *Saduria entomon*
■ Others

The vertical distribution of organisms in each group differed in terms of both abundance and biomass (Figure 4). In all groups of sites, the largest number (>62%) of organisms was found in the shallowest layer of sediment. This was also the only layer in group 3 containing organisms. However, the maximum biomass of organisms was observed in the deeper sediment layers—as much as 42% of the biomass at the sites from group 1 was found in the 3–6 cm sediment layer, and in group 2, organisms were found in the shallower layers—almost 60% of the biomass was found in the 1–3 cm sediment layer. This is due to the dominance of *M. balthica* in the infaunal biomass.

*M. balthica* accounted for the largest proportion of biomass at all but the deepest sites (Figure 5) (Table A2). The biomass was also composed of *Marenzelleria* spp., *P. ulvae* and *H. diversicolor*. Only epifaunal *B. sarsi* was observed at the deepest sites. The coastal sites were characterized by the occurrence of taxa such as *Marenzelleria* spp. and *H. diversicolor*, which burrow to a depth of 15 cm. With the distance from the Vistula River, fewer taxa were observed burrowing deeper into the sediment.

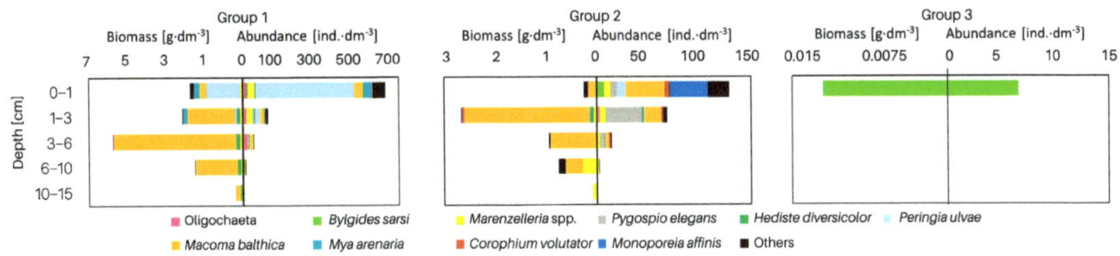


**Figure 4.** Vertical distribution of macrofaunal taxa deep into the sediment in each group of the sites shown in Figure 3. The scale of abundance and biomass differs for individual groups.

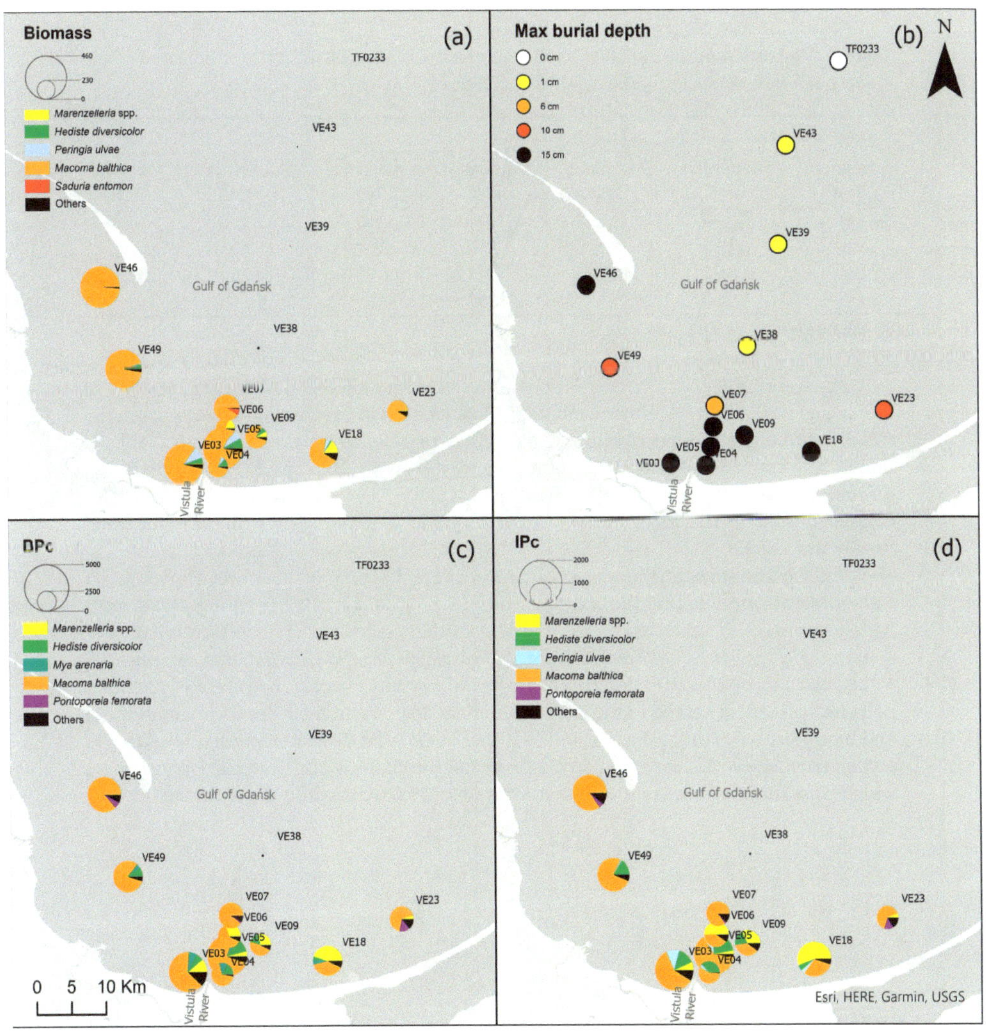


**Figure 5.** (**a**) Biomass [g·m$^{-2}$]; (**b**) maximum burial depth of organisms [cm]; (**c**) bioturbation potential index ($BP_C$) and (**d**) bioirrigation potential index ($IP_C$), and the proportion of taxa in the values of each parameter in the Gulf of Gdańsk.

Both the bioturbation potential index and the bioirrigation potential index followed the distribution of biomass, with the highest values in the shallow areas and in vicinity of the Vistula River mouth, and lower values in the deep area and no bioturbation activity in the Gdańsk Deep. $BP_C$ and $IP_C$ at all (except the deepest) sites were mainly formed by *M. balthica*. At the shallow sites, the polychaetes, *M. arenaria* and *Pontoporeia femorata* contributed relatively significantly to the formation of $BP_C$, while at VE18 it was mainly formed by *Marenzelleria* spp. In the formation of $IP_C$, *Marenzelleria* spp. contributed more than other taxa at several sites (VE18, VE06, VE09). The highest $BP_C$ (5001) and $IP_C$ (1958) values were recorded at site VE05.

Among environmental parameters, salinity (and highly correlated DO) was the most important predictor, explaining more than 44% of the variability in biomass, burial depth, $BP_C$ and $IP_C$ (Table 4). Salinity (and highly correlated DO) and temperature, and in the case of $BP_C$ also LOI, explained more than 80% of the data variation in $BP_C$ and $IP_C$.

**Table 4.** Proportion of the variables explaining the distLM model adjustment in marginal and sequential tests for biomass, burial depth, $BP_C$ and $IP_C$.

| | **Biomass** | | **Burial Depth** | | **$BP_C$** | | **$IP_C$** | |
|---|---|---|---|---|---|---|---|---|
| | **Marginal Test** | **Sequential Test** | **Marginal Test** | **Sequential Test** | **Marginal Test** | **Sequential Test** | **Marginal Test** | **Sequential Test** |
| Salinity | 0.442 ** | 0.442 * | 0.644 ** | 0.644 ** | 0.529 ** | 0.529 ** | 0.564 ** | 0.564 ** |
| Temperature | 0.302 * | 0.153 | 0.03 | | 0.425 * | 0.23 ** | 0.456 ** | 0.247 * |
| LOI | 0.01 | 0.1021 | 0.250 | | 0.027 | 0.076 * | 0.062 | |
| Total | | 0.697 | | 0.644 | | 0.835 | | 0.811 |

Significance levels * $p < 0.05$, ** $p < 0.01$.

There are strong positive correlations between bioturbation and bioirrigation potential indices, as well as between them and key characteristics of benthic communities, i.e., the number of taxa, abundance, biomass as well as maximum burrowing depth (Table 5). Similarly strong and significant relationships exist between $BP_C$ and $IP_C$ calculated from differently presented biomass data (WW and AFDW), as well as when comparing $IP_C$ calculated from benthic fauna burial depth data obtained in this study with the index using the literature data.

The examination of the macrofauna in different layers of the cores showed that the sediments are inhabited to a depth of 15 cm (Figure 6). All the studied taxa, with the exception of *H. spinulosus*, are observed in the shallowest layer of sediment. For some taxa (*Planaria torva*, *Ecrobia ventrosa*, *Potamopyrgus antipodarum*, *Saduria entomon*, *Mysis mixta* and *Neomysis integer*), this is the only layer of occurrence. Few—especially polychaetes—are observed in the deeper sediment layers, and their dominant abundance and biomass occurs in the 3–6 cm and 6–10 cm layers. The deepest recorded taxa are the polychaetes *Marenzelleria* spp. and *H. diversicolor*, the clams *M. balthica* and Oligochaetes, which were found in the layer up to a maximum of 15 cm deep into the sediment.

**Table 5.** Spearman's correlation coefficients of biological parameters from all sampling sites. Both indices were calculated using different exponents (0.5 or 0.75) and two types of animal biomass—wet formalin mass (WW) or ash free dry mass (AFDW). Significance level for all correlations was $p < 0.000001$.

| | $BPc^{0.5}_{WW}$ | $BPc^{0.5}_{AFDW}$ | $BPc^{0.75}_{AFDW}$ | $IPc^{0.75}_{WW}$ | $IPc^{0.75}_{AFDW}$ | $IPc^{0.75}_{AFDW}$ IDi lit. | Max. Burrowing Depth | No. of Taxa | Abundance | Biomass WW | Biomass AFDW |
|---|---|---|---|---|---|---|---|---|---|---|---|
| $BPc^{0.5}_{WW}$ | | 0.999 | 0.991 | 0.978 | 0.982 | 0.982 | 0.672 | 0.756 | 0.934 | 0.968 | 0.970 |
| $BPc^{0.5}_{AFDW}$ | | | 0.990 | 0.990 | 0.986 | 0.985 | 0.678 | 0.766 | 0.938 | 0.965 | 0.968 |
| $BPc^{0.75}_{AFDW}$ | | | | 0.984 | 0.977 | 0.976 | 0.663 | 0.732 | 0.915 | 0.989 | 0.990 |
| $IPc^{0.75}_{WW}$ | | | | | 0.999 | 0.997 | 0.708 | 0.785 | 0.940 | 0.980 | 0.963 |
| $IPc^{0.75}_{AFDW}$ | | | | | | 0.997 | 0.720 | 0.792 | 0.942 | 0.953 | 0.956 |
| $IPc^{0.75}_{AFDW}$ IDi lit. | | | | | | | 0.718 | 0.785 | 0.935 | 0.949 | 0.953 |
| Max. burrowing depth | | | | | | | | 0.730 | 0.717 | 0.637 | 0.637 |
| No. of taxa | | | | | | | | | 0.836 | 0.706 | 0.706 |
| Abundance | | | | | | | | | | 0.880 | 0.883 |
| Biomass WW | | | | | | | | | | | 0.999 |
| Biomass AFDW | | | | | | | | | | | |

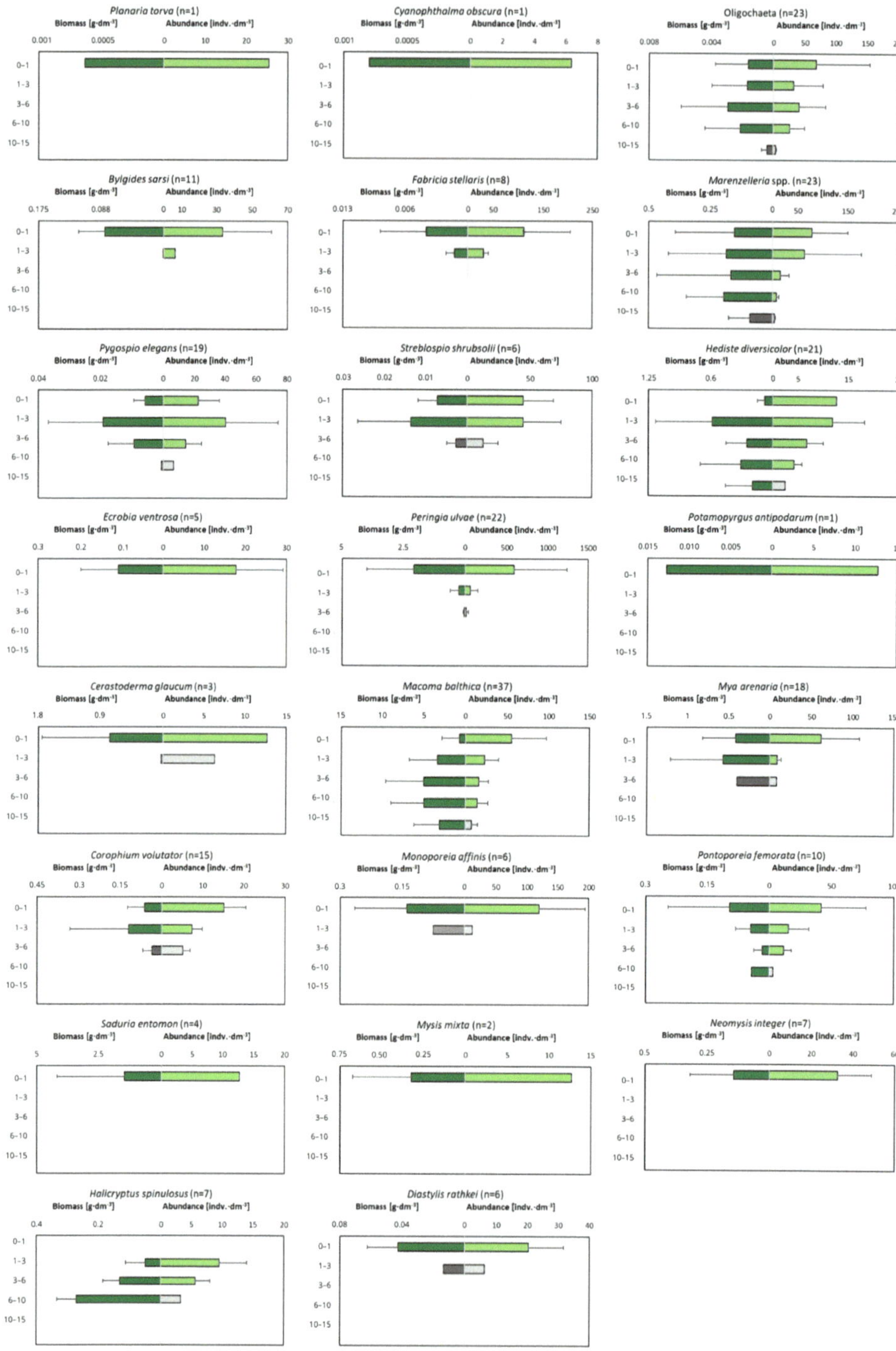


**Figure 6.** Depth of occurrence of individual macrofaunal taxa in the sediment. Green color indicates to what depth 90% of all organisms are observed. n is number of cores in which particular taxa was observed.

## 4. Discussion

### 4.1. Conditions of Bottom Water and Sediments and Their Impact on Macrozoobenthos

Coastal areas, such as estuaries, lagoons and bays, are dynamic environments with gradients of freshwater and seawater flows, representing transition zones between land and sea [3,59,60]. Gradients in physicochemical parameters of bottom water and surface sediments are typical for the Gulf of Gdańsk [41,42,61] [this research]. As the depth of the basin increases, salinity increases, DO decreases, while the proportion of the finest fraction, organic matter content and hydrogen sulfide concentration increases. The area of the Vistula outflow is characterized by the presence of increased amounts of organic matter and nutrients supplied with river runoff [62–65]. Organic matter and nutrients, as well as contaminants on their way from land to open sea are transformed, retained or removed by biota or moved unchanged to the offshore areas of the Baltic Sea [26,44,66–68].

The conditions prevailing in the bottom water and sediments affect the distribution and species composition of macrozoobenthos. In the case of benthic communities inhabiting the seabed of the Gulf of Gdańsk, the factors that had the greatest impact on biomass structure, macrofauna burial depth and indices of bioturbation and bioirrigation potential were conditions such as salinity and oxygen concentration in the water above the bottom, and factors strongly related to these, such as sediment conditions. It is known that as oxygen conditions in the water above the seabed deteriorate, the concentration of toxic hydrogen sulfide in the sediments increases [41,69–71].

### 4.2. Macrozoobenthos

The present study revealed the presence of 23 taxa of the benthic macrofauna in the study area. The results were similar to those obtained during other macrozoobenthos studies conducted in the Vistula estuary [39,41,72,73]. The greatest diversity of benthic organisms was observed in the coastal zone, where the density was dominated by *P. ulvae*, a gastropod species typical of the coastal zone in the Baltic Sea, while in the deeper zones the species composition of the benthic community shifted and the abundance was dominated by *P. elegans* and *M. balthica*, species also common in the Baltic Sea. The biomass in all but the deepest zones was dominated by *M. balthica*.

The highest number of taxa was observed at some distance from the Vistula estuary (at a depth of 16–24 m). Relatively few taxa were found at the shallowest site (15 m depth), due to the fact that the estuary is highly dynamic and the material carried by the river forms an unstable and easily eroded substrate, unfavourable to macrozoobenthos development [72–75]. Although organic matter carried with river runoff constitutes food resources for macrofauna [44], it can also cause benthic organisms to become covered and buried, leading in extreme cases to the complete disappearance of benthic macrofauna in a given area [76]. As the depth of the water body increases, both the taxonomic diversity and the biomass of the macrofauna decreases. At the deepest sites, the macrofauna is either absent or represented by single individuals of the surface-living, semi-pelagic polychaete *B. sarsi*. The reason for this is the decomposition of large amounts of organic matter accumulating on the bottom and stable stratification in the deeper area, which leads to oxygen deficiency or anoxia at the bottom and occurrence of hydrogen sulfide in the surface sediments [69]. These conditions adversely affect the behaviour, physiological processes, fitness of the benthic fauna, and consequently lead to a loss of functions performed by the benthic fauna [36,77,78]. Such a loss of biodiversity can result in reduced resistance of the environment to stress [79].

### 4.3. Bioturbation and Bioirrigation

The research carried out has shown that while the zoobenthos biomass in the Vistula estuary is completely dominated by *M. balthica*, the use of bioturbation and bioirrigation potential indices reveals the role of other species, i.e., those whose biomass is not large but it is known from experimental studies that their activity can significantly affect biogeochemical processes [80,81]. The benthic communities described in this study are characterized by their high bioturbation and bioirrigation potential in the coastal region. This is where their impact on various compounds is most likely to be greatest. *M. balthica*, whose intensive bioturbation and bioirrigation activity is relatively well studied, had the largest contribution to the indices [80,81]. Polychaetes of the genus *Marenzelleria* also contributed relatively significantly to the bioirrigation potential index. Experimental studies have shown that this species is an extremely effective bioirrigator and bioturbator [6,48,82,83]. In situ experiments in the Vistula plume showed a significant increase in nutrient fluxes from sediments inhabited by macrofauna, with the greatest impact observed in the presence of polychaetes [39]. In previous studies, a comparison between bioturbation and bioirrigation potential indices maps showed a very similar pattern, but also some differences [56,84]. For example, differences on a spatial scale were found in the German Bight, with higher $IP_C$ scores in areas where sessile or semi-sessile species (i.e., *Lanice conchilega* and *Notomastus latericeus*) were particularly abundant [56]. In the Vistula estuary, such a difference is apparent only for one site (VE18), where higher $IP_C$ values compared to $BP_C$ are due to the abundance of *Marenzelleria* spp.

The present study demonstrated a strong positive relationship between the two indices and their strong correlation with both the maximum burrowing depth of macrozoobenthos, the number of taxa, as well as the abundance and total biomass of macrozoobenthos. Interestingly, there was virtually no difference in these relationships regardless of how the calculations were made (i.e., wet or ash free dry mass). In an earlier study conducted in another region of the Baltic Sea, the authors found no relationship between the bioirrigation index and the number of taxa [84]. According to Queirós et al. [38], $BP_C$ was found to be a good predictor of bioturbation distance (average distance travelled by a sediment particle). However, it was found unsuitable for determining other attributes of infauna, such as bioturbation activity, bioturbation depth or diffusion transport. In addition, the index also appears to be a better predictor of community-level estimates, rather than those for individual species. Statistical models using experimental results showed that $BP_C$ explained a considerable amount of variance in oxic processes, i.a. oxic mineralization, total N mineralization, and nitrification [85]. Few studies have also been conducted to determine the correlation between bioirrigation potential index values and actual bioirrigation. However, the results of these studies are inconclusive and require further research. A study by De Borger et al. [86] showed that $IP_C$ correlates more strongly with burrow ventilation depth than with ventilation rate. The correlation between $IP_C$ and irrigation rate was not confirmed by Toussaint et al. [85].

The present study did not use the bioirrigation index ($BIP_C$) proposed by Renz et al. [32], the scoring system of which additionally takes into account the distinction between the advection and diffusion system performance. The use of this index would result in higher values for free living species and species living in burrows as well as facultative deposit/suspension feeders in advective sediments. Furthermore, it would result in even higher values of bioirrigation potential in the coastal zone, where the advection system dominates, and an even higher proportion of *M. balthica* or polychaetes *H. diversicolor* and *Marenzelleria* spp. in the index for this zone. At the deeper sites where diffusive sediments occur, bioirrigation potential would be much lower than in the coastal zones. The system by Renz and co-workers [32] would emphasize the variability of the bioirrigation index in the Gulf of Gdańsk and the gradual loss of this function in the environment with increasing depth of the water body. Both approaches to the determination of the bioirrigation potential index are certainly worth testing in further studies, especially those combining studies of benthic assemblages, including functional indices, with experimental studies of the impact of macrofauna on biogeochemical processes, or measurements of animal activity.

The indices used provide only a simplified approximation of the potential capabilities of benthic communities. Bioturbation and bioirrigation are dynamic and complex activities performed by these organisms. They are determined by a number of factors that affect the biological functions of these animals. $BP_C$ was observed to follow the seasonal pattern in seawater temperature, with the highest values in summer and autumn [38]. However, it should be kept in mind that temperature and food availability have the potential to impact bioturbation and bioirrigation intensity, as these factors affect physiological processes of benthic species. Studies conducted on the polychaete *Alitta virens* showed that sediment reworking processes could be affected by both low and high temperature, with the lowest bioturbation intensity under low temperature [87]. Oxygen depletion may also change the activity of animals in the sediment, thus affecting bioturbation and bioirrigation. Depending on the oxygen concentration and exposure time, these conditions can result in, for example, an increase in burrow ventilation, a decrease in animal activity or no activity at all [88–90].

### 4.4. Burrowing Depth

The burrowing depth of organisms provides, among other things, an indication of the depth to which they can affect the conditions and processes in the sediments. In the present study, most of the organisms (>62% of all individuals) inhabited the shallowest layer of sediment (0–1 cm). The maximum biomass of organisms can be found in the deeper layers of sediment—deeper layers (3–6 cm) at the shallowest sites and slightly shallower layers at the intermediate sites (1–3 cm). A similar distribution of organisms deep into the sediment was observed in earlier studies conducted in the Gulf of Gdańsk—the highest abundance of organisms was found in the shallowest layer of the sediment and it decreased with depth [40,41]. In contrast to the abundance, the biomass of organisms in the shallow water zone did not decrease with depth and its distribution was more varied—the highest biomass was usually observed in deeper layers, i.e., up to 6 cm into the sediment [40].

However, even organisms living on the sediment surface can play an extremely important role by being active in disrupting the diffusive boundary layer, which improves the oxygen conditions of the sediment [91]. Few organisms, i.e., bivalves and polychaetes, burrow naturally into the sediment and are rarely present on its surface [83,92], and their typical burrowing depth is 3–10 cm. The maximum depth of occurrence of a given taxon depends on the ability of the organism to contact the sediment surface, for example, the burial depth of *M. balthica* depends on the length of the clam's siphon, which is often also related to the size and age of the organism [46]. Our research showed the occurrence of *M. balthica* below a depth of 10 cm, which is also the maximum depth at which the bivalves bioturbate and bioirrigate the sediments. Other deep burrowing species—from the genus *Marenzelleria*—were found up to a depth of 15 cm, but some scientists indicate that these species can burrow as deep as 35 cm [93]. These deep burrowing organisms, such as polychaetes, form burrows that enable water transport in the sediment and aerobic chemical reactions in the deeper layers, as well as affect nutrient cycling [83,94,95].

The vertical distribution of organisms is determined by environmental factors. Organisms change the depth of their occurrence seasonally [46], e.g., *M. balthica* has been shown to burrow deepest in winter and remain shallowly buried in the sediment during the summer season. Oxygen deficiencies and hydrogen sulfide cause the animals to move to the sediment surface or they become periodically inactive [92,96,97]. While animals are present in the sediment, their functions may be temporarily impaired.

## 5. Conclusions

Our research has shown changes in the structure and functioning of benthic communities with increasing distance from the Vistula River mouth. Coastal zones are characterized by relatively high biodiversity and great burrowing depth of macrofauna, as well as high bioturbation and bioirrigation potential of benthic communities. However, this activity disappears in deep zones with the absence of benthic organisms. The lack of bioturbation and bioirrigation means there is no support for biogeochemical transformation by the macrofauna in the deep zones. In the study area, only a few species drive bioturbation and bioirrigation—the bivalve *M. balthica* and the polychaetes *H. diversicolor* and *Marenzelleria* spp. Other taxa had a marginal impact. Such a strong dominance of single taxa in performing bioturbation and bioirrigation could lead to instability in ecosystem functioning in the case that these organisms were to disappear as a result of an ecological disaster, environmental degradation or disease. At the same time, these large organisms were the only taxa burrowing deep into the sediment (below 10 cm), and thus the only ones supporting geochemical processes deep in the sediment. To summarize, the coastal zone, unlike the offshore zone, proved to be a hotspot for bioturbation- and bioirrigation-driven processes, which are responsible for the proper functioning of the seafloor and basin. However, very poor functional diversity of the benthic macrofauna in the deepest zones means that we should appreciate and protect coastal zones more efficiently.

**Author Contributions:** Conceptualization, N.A.M., U.J. and H.K.; methodology, N.A.M., U.J. and H.K.; formal analysis, N.A.M. and U.J.; investigation, N.A.M. and H.K.; data curation, N.A.M. and H.K.; writing—original draft preparation, N.A.M., U.J. and H.K.; writing—review and editing, N.A.M., U.J. and H.K.; visualization, N.A.M. and U.J.; project administration, U.J.; funding acquisition, U.J. and H.K. All authors have read and agreed to the published version of the manuscript.

**Funding:** This study was funded by the BONUS COCOA project, which was supported by BONUS (Art 185), jointly funded by the EU and the National Center for Research and Development (NCBiR, Poland) and Norway Grant (2020/37/K/NZ8/02782 "Benthic coastal buffers against climatic and eutrophication extremes, BUFFER").

**Institutional Review Board Statement:** Not applicable.

**Informed Consent Statement:** Not applicable.

**Data Availability Statement:** Not applicable.

**Acknowledgments:** We would like to thank Maren Voss for her excellent cruise leadership, scientific and operational crews and captains of the ships, especially Michael Pötzsch, Anna Borecka, Heather Reader, Natalia Kozak and Monika Lengier for their help during sampling. We also thank Ines Bartl for providing data on sediment parameters from two sites in 2016.

**Conflicts of Interest:** The authors declare no conflict of interest.

## Appendix A

**Table A1.** Categorical scores assigned to each taxon for $BP_c$ and $IP_c$ indices calculations according to Solan et al. [30], Villnäs et al. [36] Queirós et al. [31] and Wrede et al. [98] (modified), where: Mobility ($M_i$): 1, feeding on the sediment surface; limited movement on the sediment surface; sessile; 2, limited movement; 3, slow free movement; 4, free to movement. Reworking type ($R_i$): 1, epifauna; 2, surficial modifiers; 3, upward or downward conveyor; 4, biodiffusors. Burrow type ($BT_i$): 1, epifauna or internal irrigation (i.e., siphons); 2, open irrigation (i.e., Y- or U-shaped burrow); 3, blind ended burrow. Feeding type ($FT_i$): 1, surface filter feeder; 2, predator; 3, deposit feeder; 4, sub-surface filter feeder. Irrigation depth ($ID_i$): 1, 0–1 cm; 2, 1–3 cm; 3, 3–6 cm; 4, 6–10 cm; 5, 10–15 cm.

| Taxa | $BP_c$ | | $IP_c$ | | |
|---|---|---|---|---|---|
| | $M_i$ | $R_i$ | $BT_i$ | $FT_i$ | $ID_i$ |
| *Planaria torva* | 1 | 1 | 1 | 2 | 1 |
| *Cyanophthalma obscura* | 3 | 1 | 3 | 2 | 2 |
| Oligochaeta | 3 | 2 | 3 | 3 | 4 |
| *Bylgides sarsi* | 3 | 1 | 1 | 2 | 2 |
| *Fabricia stellaris* | 2 | 1 | 3 | 1 | 2 |
| *Marenzelleria* spp. | 4 | 4 | 3 | 3 | 5 |
| *Pygospio elegans* | 2 | 2 | 3 | 3 | 3 |
| *Streblospio shrubsolii* | 2 | 2 | 3 | 3 | 2 |
| *Hediste diversicolor* | 4 | 3 | 2 | 3 | 5 |
| *Ecrobia ventrosa* | 1 | 1 | 1 | 3 | 1 |
| *Peringia ulvae* | 1 | 1 | 1 | 3 | 2 |
| *Potamopyrgus antipodarum* | 1 | 1 | 1 | 3 | 1 |
| *Cerastoderma glaucum* | 3 | 2 | 1 | 1 | 1 |
| *Macoma balthica* | 3 | 4 | 1 | 3 | 4 |

**Table A1.** *Cont.*

| Taxa | BP$_c$ | | IP$_c$ | | |
|---|---|---|---|---|---|
| | M$_i$ | R$_i$ | BT$_i$ | FT$_i$ | ID$_i$ |
| *Mya arenaria* | 3 | 4 | 1 | 1 | 2 |
| *Corophium volutator* | 2 | 2 | 2 | 3 | 3 |
| *Monoporeia affinis* | 4 | 2 | 3 | 3 | 2 |
| *Pontoporeia femorata* | 4 | 4 | 3 | 3 | 4 |
| *Diastylis rathkei* | 3 | 2 | 3 | 3 | 2 |
| *Saduria entomon* | 4 | 2 | 3 | 3 | 1 |
| *Mysis mixta* | 4 | 1 | 3 | 3 | 1 |
| *Neomysis integer* | 4 | 1 | 3 | 3 | 1 |
| *Halicryptus spinulosus* | 3 | 4 | 3 | 3 | 4 |

**Table A2.** Average (±SD) number of taxa and biomass (g. m$^{-2}$) at the sampling sites of the dominant taxa present ($n$ = 2 for sites VE49, $n$ = 3 for TF0233, VE09, VE38, VE39, VE43; $n$ = 4 for VE04, VE06, VE07, VE18, VE23, VE46 and $n$ = 5 for VE03 and VE05).

| Site | No. of Taxa | *Marenzelleria* spp. | *Hediste diversicolor* | *Peringia ulvae* | *Macoma balthica* | *Saduria entomon* | Others * |
|---|---|---|---|---|---|---|---|
| VE03 | 8 ± 4 | 1.9 ± 2.0 | 21.7 ± 22.1 | 42.5 ± 26.1 | 370.7 ± 227.2 | 0.0 ± 0.0 | 14.6 ± 10.1 |
| VE04 | 5 ± 0 | 0.0 ± 0.0 | 16.2 ± 13.0 | 7.6 ± 8.4 | 90.1 ± 97.4 | 0.0 ± 0.0 | 3.4 ± 2 |
| VE05 | 8 ± 3 | 1.1 ± 2.1 | 35.4 ± 19.3 | 39.3 ± 8.6 | 375.5 ± 117.5 | 0.0 ± 0.0 | 10.8 ± 11.7 |
| VE06 | 5 ± 0 | 18.3 ± 6.8 | 0.0 ± 0.0 | 0.1 ± 0.2 | 81.8 ± 57.9 | 0.0 ± 0.0 | 3.2 ± 2.6 |
| VE07 | 4 ± 1 | 0.0 ± 0.0 | 0.0 ± 0.0 | 0.0 ± 0.0 | 177.2 ± 77.5 | 14.0 ± 27.9 | 3 ± 3.7 |
| VE09 | 5 ± 4 | 10.0 ± 8.7 | 10.6 ± 17 | 1.1 ± 1.9 | 109.1 ± 113.5 | 0.0 ± 0.0 | 6.1 ± 7.3 |
| VE18 | 9 ± 2 | 34.1 ± 30.3 | 4.6 ± 6.5 | 12.0 ± 7.4 | 158.7 ± 59.8 | 0.0 ± 0.0 | 18.2 ± 16.4 |
| VE23 | 6 ± 1 | 1.1 ± 0.9 | 0.0 ± 0.0 | 0.0 ± 0.0 | 118.0 ± 150.6 | 0.1 ± 0.1 | 10.3 ± 7.7 |
| VE38 | 0 ± 0 | 0.0 ± 0.0 | 0.0 ± 0.0 | 0.0 ± 0.0 | 0.0 ± 0.0 | 0.0 ± 0.0 | 1.8 ± 3.1 |
| VE39 | 0 ± 0 | 0.0 ± 0.0 | 0.0 ± 0.0 | 0.0 ± 0.0 | 0.0 ± 0.0 | 0.0 ± 0.0 | 0.2 ± 0.2 |
| VE43 | 0 ± 0 | 0.0 ± 0.0 | 0.0 ± 0.0 | 0.0 ± 0.0 | 0.0 ± 0.0 | 0.0 ± 0.0 | 0.4 ± 0.7 |
| VE46 | 5 ± 0 | 0.1 ± 0.1 | 0.0 ± 0.0 | 0.0 ± 0.0 | 458.1 ± 82.1 | 0.7 ± 1.4 | 5.8 ± 3.3 |
| VE49 | 7 ± 0 | 0.0 ± 0.0 | 15.9 ± 6.1 | 3.8 ± 0.8 | 374.7 ± 183.2 | 0.0 ± 0.0 | 10.2 ± 7.1 |
| TF0233 | 0 ± 0 | 0.0 ± 0.0 | 0.0 ± 0.0 | 0.0 ± 0.0 | 0.0 ± 0.0 | 0.0 ± 0.0 | 0.0 ± 0.0 |

* Species covered by the category "Others" include the taxa listed in Table A1.

## References


1. de Groot, R.S.; Wilson, M.A.; Boumans, R.M.J. A typology for the classification, description and valuation of ecosystem functions, goods and services. *Ecol. Econ.* **2002**, *41*, 393–408. [CrossRef]
2. McGlathery, K.; Sundbäck, K.; Anderson, I. Eutrophication in shallow coastal bays and lagoons: The role of plants in the coastal filter. *Mar. Ecol. Prog. Ser.* **2007**, *348*, 1–18. [CrossRef]
3. Carstensen, J.; Conley, D.J.; Almroth-Rosell, E.; Asmala, E.; Bonsdorff, E.; Fleming-Lehtinen, V.; Gustafsson, B.G.; Gustafsson, C.; Heiskanen, A.-S.; Janas, U.; et al. Factors regulating the coastal nutrient filter in the Baltic Sea. *AMBIO* **2019**, *49*, 1194–1210. [CrossRef] [PubMed]
4. Biles, C.L.; Paterson, D.M.; Ford, R.B.; Solan, M.; Raffaelli, D.G. Bioturbation, ecosystem functioning and community structure. *Hydrol. Earth Syst. Sci.* **2002**, *6*, 999–1005. [CrossRef]
5. Braeckman, U.; Provoost, P.; Gribsholt, B.; Van Gansbeke, D.; Middelburg, J.; Soetaert, K.; Vincx, M.; Vanaverbeke, J. Role of macrofauna functional traits and density in biogeochemical fluxes and bioturbation. *Mar. Ecol. Prog. Ser.* **2010**, *399*, 173–186. [CrossRef]
6. Kauppi, L.; Bernard, G.; Bastrop, R.; Norkko, A.; Norkko, J. Increasing densities of an invasive polychaete enhance bioturbation with variable effects on solute fluxes. *Sci. Rep.* **2018**, *8*, 1–12. [CrossRef]
7. Benelli, S.; Bartoli, M.; Ribaudo, C.; Fano, E.A. Contrasting Effects of an Alien Worm on Benthic N Cycling in Muddy and Sandy Sediments. *Water* **2019**, *11*, 465. [CrossRef]

8. Bartoli, M.; Benelli, S.; Magri, M.; Ribaudo, C.; Moraes, P.; Castaldelli, G. Contrasting Effects of Bioturbation Studied in Intact and Reconstructed Estuarine Sediments. *Water* **2020**, *12*, 3125. [CrossRef]
9. Borawska, Z.; Szymczycha, B.; Silberberger, M.J.; Koziorowska-Makuch, K.; Szczepanek, M.; Kędra, M. Benthic fluxes of dissolved silica are an important component of the marine Si cycle in the coastal zone. *Estuar. Coast. Shelf Sci.* **2022**, *273*, 107880. [CrossRef]
10. Politi, T.; Zilius, M.; Castaldelli, G.; Bartoli, M.; Daunys, D. Estuarine Macrofauna Affects Benthic Biogeochemistry in a Hypertrophic Lagoon. *Water* **2019**, *11*, 1186. [CrossRef]
11. Politi, T.; Barisevičiūte, R.; Bartoli, M.; Bonaglia, S.; Cardini, U.; Castaldelli, G.; Kančauskaitė, A.; Marzocchi, U.; Petkuviene, J.; Samuiloviene, A.; et al. A bioturbator, a holobiont, and a vector: The multifaceted role of *Chironomus plumosus* in shaping N-cycling. *Freshw. Biol.* **2021**, *66*, 1036–1048. [CrossRef]
12. Samuiloviene, A.; Bartoli, M.; Bonaglia, S.; Cardini, U.; Vybernaite-Lubiene, I.; Marzocchi, U.; Petkuviene, J.; Politi, T.; Zaiko, A.; Zilius, M. The Effect of Chironomid Larvae on Nitrogen Cycling and Microbial Communities in Soft Sediments. *Water* **2019**, *11*, 1931. [CrossRef]
13. Kristensen, E.; Mikkelsen, O.L. Impact of the burrow-dwelling polychaete *Nereis diversicolor* on the degradation of fresh and aged macroalgal detritus in a coastal marine sediment. *Mar. Ecol. Prog. Ser.* **2003**, *265*, 141–153. [CrossRef]
14. Gilbert, F.; Rivet, L.; Bertrand, J.-C. The in vitro influence of the burrowing polychaete *Nereis diversicolor* on the fate of petroleum hydrocarbons in marine sediments. *Chemosphere* **1994**, *29*, 1–12. [CrossRef]
15. Granberg, M.E.; Gunnarsson, J.S.; Hedman, J.E.; Rosenberg, R.; Jonsson, P. Bioturbation-Driven Release of Organic Contaminants from Baltic Sea Sediments Mediated by the Invading Polychaete *Marenzelleria neglecta*. *Environ. Sci. Technol.* **2008**, *42*, 1058–1065. [CrossRef] [PubMed]
16. Lohrer, A.M.; Thrush, S.F.; Gibbs, M.M. Bioturbators enhance ecosystem function through complex biogeochemical interactions. *Nature* **2004**, *431*, 1092–1095. [CrossRef] [PubMed]
17. Snelgrove, P.V.; Soetaert, K.; Solan, M.; Thrush, S.; Wei, C.-L.; Danovaro, R.; Fulweiler, R.W.; Kitazato, H.; Ingole, B.; Norkko, A.; et al. Global Carbon Cycling on a Heterogeneous Seafloor. *Trends Ecol. Evol.* **2018**, *33*, 96–105. [CrossRef] [PubMed]
18. Elliott, M.; Day, J.; Ramachandran, R.; Wolanski, E.; Fang, Q.; Sheehan, M.; Seen, A.; Ellison, J. A Synthesis: What Is the Future for Coasts, Estuaries, Deltas and Other Transitional Habitats in 2050 and Beyond? In *Coasts and Estuaries: The Future*; Wolanski, E., Day, J.W., Elliott, M., Ramachandran, R., Eds.; Elsevier: Oxford, UK, 2019; pp. 1–28. ISBN 9780128140031.
19. Strokal, M.; Bai, Z.; Franssen, W.; Hofstra, N.; Koelmans, A.A.; Ludwig, F.; Ma, L.; van Puijenbroek, P.; Spanier, J.E.; Vermeulen, L.C.; et al. Urbanization: An increasing source of multiple pollutants to rivers in the 21st century. *NPJ Urban Sustain.* **2021**, *1*, 1–13. [CrossRef]
20. Kruk-Dowgiałło, L.; Szaniawska, A. Gulf of Gdańsk and Puck Bay. In *Ecology of Baltic Coastal Waters. Ecological Studies*; Schiewer, U., Ed.; Springer: Berlin/Heidelberg, Germany, 2008; Volume 197.
21. Bartl, I.; Hellemann, D.; Rabouille, C.; Schulz, K.; Tallberg, P.; Hietanen, S.; Voss, M. Particulate organic matter controls benthic microbial N retention and N removal in contrasting estuaries of the Baltic Sea. *Biogeosciences* **2019**, *16*, 3543–3564. [CrossRef]
22. Łukawska-Matuszewska, K.; Burska, D.; Niemirycz, E. Toxicity Assessment by Microtox® in Sediments, Pore Waters and Sediment Saline Elutriates in the Gulf of Gdańsk (Baltic Sea). *CLEAN–Soil Air Water* **2009**, *37*, 592–598. [CrossRef]
23. HELCOM. *Input of Nutrients by the Seven Biggest Rivers in the Baltic Sea Region*; Baltic Sea Environment Proceedings; Helsinki Commission: Helsinki, Finland, 2018; No.161.
24. Witek, Z.; Humborg, C.; Savchuk, O.; Grelowski, A.; Łysiak-Pastuszak, E. Nitrogen and phosphorus budgets of the Gulf of Gdańsk (Baltic Sea). *Estuar. Coast. Shelf Sci.* **2003**, *57*, 239–248. [CrossRef]
25. Staniszewska, M.; Koniecko, I.; Falkowska, L.; Krzymyk, E. Occurrence and distribution of bisphenol A and alkylphenols in the water of the gulf of Gdansk (Southern Baltic). *Mar. Pollut. Bull.* **2015**, *91*, 372–379. [CrossRef]
26. Zaborska, A.; Siedlewicz, G.; Szymczycha, B.; Dzierzbicka-Głowacka, L.; Pazdro, K. Legacy and emerging pollutants in the Gulf of Gdańsk (southern Baltic Sea)–loads and distribution revisited. *Mar. Pollut. Bull.* **2019**, *139*, 238–255. [CrossRef] [PubMed]
27. Saniewska, D.; Bełdowska, M.; Szymczak, E.; Kuliński, K.; Bełdowski, J.; Voss, M.; Pryputniewicz-Flis, D.; Burska, D. Processes affecting the transformation of mercury in the coastal zone in the vicinity of two river mouths in the southern Baltic Sea. *Mar. Chem.* **2022**, *238*, 104065. [CrossRef]
28. Van Hoey, G.; Permuy, D.C.; Vandendriessche, S.; Vincx, M.; Hostens, K. An ecological quality status assessment procedure for soft-sediment benthic habitats: Weighing alternative approaches. *Ecol. Indic.* **2013**, *25*, 266–278. [CrossRef]
29. Rees, W.E. Ecological Footprints and Appropriated Carrying Capacity: What Urban Economics Leaves Out. *Environ. Urban.* **1992**, *4*, 121–130. [CrossRef]
30. Solan, M.; Cardinale, B.J.; Downing, A.L.; Engelhardt, K.A.M.; Ruesink, J.L.; Srivastava, D.S. Extinction and Ecosystem Function in the Marine Benthos. *Science* **2004**, *306*, 1177–1180. [CrossRef]
31. Queirós, A.M.; Birchenough, S.N.R.; Bremner, J.; Godbold, J.A.; Parker, R.E.; Romero-Ramirez, A.; Reiss, H.; Solan, M.; Somerfield, P.J.; Van Colen, C.; et al. A bioturbation classification of European marine infaunal invertebrates. *Ecol. Evol.* **2013**, *3*, 3958–3985. [CrossRef]
32. Renz, J.; Powilleit, M.; Gogina, M.; Zettler, M.; Morys, C.; Forster, S. Community bioirrigation potential (BIPc), an index to quantify the potential for solute exchange at the sediment-water interface. *Mar. Environ. Res.* **2018**, *141*, 214–224. [CrossRef]
33. Wrede, A.; Holstein, J.; Brey, T. Choosing the lesser evil—A case study on quantification of sediment reworking rates in multi corer type sediment cores. *J. Exp. Mar. Biol. Ecol.* **2018**, *514–515*, 1–9. [CrossRef]

34. Nasi, F.; Ferrante, L.; Alvisi, F.; Bonsdorff, E.; Auriemma, R.; Cibic, T. Macrofaunal bioturbation attributes in relation to riverine influence: What can we learn from the Po River lagoonal system (Adriatic Sea)? *Estuar. Coast. Shelf Sci.* **2020**, *232*, 106405. [CrossRef]
35. Braeckman, U.; Foshtomi, M.Y.; Van Gansbeke, D.; Meysman, F.; Soetaert, K.; Vincx, M.; Vanaverbeke, J. Variable Importance of Macrofaunal Functional Biodiversity for Biogeochemical Cycling in Temperate Coastal Sediments. *Ecosystems* **2014**, *17*, 720–737. [CrossRef]
36. Villnäs, A.; Norkko, J.; Lukkari, K.; Hewitt, J.; Norkko, A. Consequences of Increasing Hypoxic Disturbance on Benthic Communities and Ecosystem Functioning. *PLoS ONE* **2012**, *7*, e44920. [CrossRef]
37. Birchenough, S.; Parker, R.; McManus, E.; Barry, J. Combining bioturbation and redox metrics: Potential tools for assessing seabed function. *Ecol. Indic.* **2012**, *12*, 8–16. [CrossRef]
38. Queirós, A.M.; Stephens, N.; Cook, R.; Ravaglioli, C.; Nunes, J.; Dashfield, S.; Harris, C.; Tilstone, G.H.; Fishwick, J.; Braeckman, U.; et al. Can benthic community structure be used to predict the process of bioturbation in real ecosystems? *Prog. Oceanogr.* **2015**, *137*, 559–569. [CrossRef]
39. Thoms, F.; Burmeister, C.; Dippner, J.W.; Gogina, M.; Janas, U.; Kendzierska, H.; Liskow, I.; Voss, M. Impact of Macrofaunal Communities on the Coastal Filter Function in the Bay of Gdansk, Baltic Sea. *Front. Mar. Sci.* **2018**, *5*, 1–19. [CrossRef]
40. Janas, U.; Burska, D.; Kendzierska, H.; Pryputniewicz-Flis, D.; Łukawska-Matuszewska, K. Importance of benthic macrofauna and coastal biotopes for ecosystem functioning–Oxygen and nutrient fluxes in the coastal zone. *Estuar. Coast. Shelf Sci.* **2019**, *225*, 106238. [CrossRef]
41. Kendzierska, H.; Łukawska-Matuszewska, K.; Burska, D.; Janas, U. Benthic fluxes of oxygen and nutrients under the influence of macrobenthic fauna on the periphery of the intermittently hypoxic zone in the Baltic Sea. *J. Exp. Mar. Biol. Ecol.* **2020**, *530–531*, 151439. [CrossRef]
42. Silberberger, M.J.; Koziorowska-Makuch, K.; Borawska, Z.; Szczepanek, M.; Kędra, M. Disentangling the Drivers of Benthic Oxygen and Dissolved Carbon Fluxes in the Coastal Zone of the Southern Baltic Sea. *Estuaries Coasts* **2022**, *45*, 2450–2471. [CrossRef]
43. Szczepanek, M.; Silberberger, M.J.; Koziorowska-Makuch, K.; Nobili, E.; Kędra, M. The response of coastal macrobenthic food-web structure to seasonal and regional variability in organic matter properties. *Ecol. Indic.* **2021**, *132*, 108326. [CrossRef]
44. Szczepanek, M.; Silberberger, M.J.; Koziorowska-Makuch, K.; Kędra, M. Utilization of riverine organic matter by macrobenthic communities in a temperate prodelta. *Front. Mar. Sci.* **2022**, *9*, 1–15. [CrossRef]
45. Dye, A.H. Seasonal Fluctuations in the Vertical Distribution of Meiofauna in Estuarine Sediments. *Zool. Afr.* **1978**, *13*, 207–212. [CrossRef]
46. Reading, C.; McGrorty, S. Seasonal variations in the burying depth of *Macoma balthica* (L.) and its accessibility to wading birds. *Estuar. Coast. Mar. Sci.* **1978**, *6*, 135–144. [CrossRef]
47. Hill, C.; Elmgren, R. Vertical Distribution in the Sediment in the Co-Occurring Benthic Amphipods Pontoporeia affinis and *P. femorata*. *Oikos* **1987**, *49*, 221. [CrossRef]
48. Renz, J.; Forster, S. Effects of bioirrigation by the three sibling species of *Marenzelleria* spp. on solute fluxes and porewater nutrient profiles. *Mar. Ecol. Prog. Ser.* **2014**, *505*, 145–159. [CrossRef]
49. Haque, A.; Szymelfenig, M.; Węsławski, J.M. Small-scale vertical distribution of zoobenthos in the sandy littoral of the Gulf of Gdańsk. *Oceanologia* **1997**, *39*, 433–446.
50. Łęczyński, L.; Szymczak, E. Własności fizyczne osadów dennych. In *Fizyczne, Biologiczne i Chemiczne Badania Morskich Osadów Dennych*; Wydawnictwo UG: Gdańsk, Poland, 2010.
51. Wentworth, C.K. A Scale of Grade and Class Terms for Clastic Sediments. *J. Geol.* **1922**, *30*, 377–392. [CrossRef]
52. Salonen, K.; Sarvala, J.; Hakala, I.; Viljanen, M.-L. The relation of energy and organic carbon in aquatic invertebrates1. *Limnol. Oceanogr.* **1976**, *21*, 724–730. [CrossRef]
53. Rumohr, H.; Brey, T.; Ankar, S. A Compilation of Biometric Conversion Factors for Benthic Invertebrates of the Baltic Sea. *Balt. Mar. Biol.* **1987**, *9*, 62.
54. Zettler, M.L. Population dynamics, growth and production of the neozoon *Marenzelleria* cf. *viridis* (Verrill, 1873) (Polychaeta: Spionidae) in a coastal water of the southern Baltic Sea. *Aquat. Ecol.* **1997**, *13*, 177–186. [CrossRef]
55. Ricciardi, A.; Bourget, E. Weight-to-weight conversion factors for marine benthic macroinvertebrates. *Mar. Ecol. Prog. Ser.* **1998**, *163*, 245–251. [CrossRef]
56. Wrede, A.; Beermann, J.; Dannheim, J.; Gutow, L.; Brey, T. Organism functional traits and ecosystem supporting services—A novel approach to predict bioirrigation. *Ecol. Indic.* **2017**, *91*, 737–743. [CrossRef]
57. Clarke, K.R. Non-parametric multivariate analyses of changes in community structure. *Aust. J. Ecol.* **1993**, *18*, 117–143. [CrossRef]
58. Anderson, M.; Gorley, R.; Clarke, K.P. *For PRIMER: Guide to Software and Statistical Methods*; PRIMER-E: Plymouth, UK, 2008.
59. Elliott, M.; Whitfield, A. Challenging paradigms in estuarine ecology and management. *Estuar. Coast. Shelf Sci.* **2011**, *94*, 306–314. [CrossRef]
60. Ducrotoy, J.P.; Michael, E.; Cutts, N.; Franco, A.; Little, S.; Mazik, K.; Wilkinson, M. Temperate Estuaries: Their Ecology Under Future Environmental Changes. In *Coasts and Estuaries: The Future*; Wolanski, E., Day, J.W., Elliott, M., Ramachandran, R., Eds.; Elsevier: Oxford, UK, 2019; pp. 577–594. ISBN 9780128140031.

61. Janas, U.; Szaniawska, A. The influence of hydrogen sulphide on macrofaunal biodiversity in the Gulf of Gdańsk. *Oceanologia* **1996**, *38*, 127–142.
62. Pempkowiak, J.; Kupryszewski, G. The input of organic matter to the Baltic from the Vistula River. *Oceanology* **1980**, *12*, 79–98.
63. Voss, M.; Liskow, I.; Pastuszak, M.; Rüβ, D.; Schulte, U.; Dippner, J. Riverine discharge into a coastal bay: A stable isotope study in the Gulf of Gdańsk, Baltic Sea. *J. Mar. Syst.* **2005**, *57*, 127–145. [CrossRef]
64. Voss, M.; Asmala, E.; Bartl, I.; Carstensen, J.; Conley, D.J.; Dippner, J.W.; Humborg, C.; Lukkari, K.; Petkuviene, J.; Reader, H.; et al. Origin and fate of dissolved organic matter in four shallow Baltic Sea estuaries. *Biogeochemistry* **2020**, *154*, 385–403. [CrossRef]
65. Reader, H.E.; Thoms, F.; Voss, M.; Stedmon, C.A. The Influence of Sediment-Derived Dissolved Organic Matter in the Vistula River Estuary/Gulf of Gdansk. *J. Geophys. Res. Biogeosci.* **2019**, *124*, 115–126. [CrossRef]
66. Asmala, E.; Carstensen, J.; Conley, D.J.; Slomp, C.P.; Stadmark, J.; Voss, M. Efficiency of the coastal filter: Nitrogen and phosphorus removal in the Baltic Sea. *Limnol. Oceanogr.* **2017**, *62*, S222–S238. [CrossRef]
67. Nybom, I.; Horlitz, G.; Gilbert, D.; Berrojalbiz, N.; Martens, J.; Arp, H.P.H.; Sobek, A. Effects of Organic Carbon Origin on Hydrophobic Organic Contaminant Fate in the Baltic Sea. *Environ. Sci. Technol.* **2021**, *55*, 13061–13071. [CrossRef]
68. Kuliński, K.; Rehder, G.; Asmala, E.; Bartosova, A.; Carstensen, J.; Gustafsson, B.; Hall, P.O.J.; Humborg, C.; Jilbert, T.; Jürgens, K.; et al. Biogeochemical functioning of the Baltic Sea. *Earth Syst. Dyn.* **2022**, *13*, 633–685. [CrossRef]
69. Janas, U.; Wocial, J.; Szaniawska, A. Seasonal and annual changes in the macrozoobenthic populations of the Gulf of Gdańsk with respect to hypoxia and hydrogen sulphide. *Oceanologia* **2004**, *46*, 85–102. Available online: http://www.iopan.gda.pl/oceanologia/86 (accessed on 15 November 2022).
70. Lukawska-Matuszewska, K.; Kielczewska, J. Effects of near-bottom water oxygen concentration on biogeochemical cycling of C, N and S in sediments of the Gulf of Gdansk (southern Baltic). *Cont. Shelf Res.* **2016**, *117*, 30–42. [CrossRef]
71. Łukawska-Matuszewska, K.; Graca, B. Pore water alkalinity below the permanent halocline in the Gdańsk Deep (Baltic Sea)-Concentration variability and benthic fluxes. *Mar. Chem.* **2018**, *204*, 49–61. [CrossRef]
72. Kotwicki, L. Macrozoobenthos of the sandy littoral zone of the Gulf of Gdańsk. *Oceanologia* **1997**, *39*, 447–460.
73. Włodarska-Kowalczuk, M.; Mazurkiewicz, M.; Jankowska, E.; Kotwicki, L.; Damrat, M.; Zajączkowski, M. Effects of fluvial discharges on meiobenthic and macrobenthic variability in the Vistula River prodelta (Baltic Sea). *J. Mar. Syst.* **2016**, *157*, 135–146. [CrossRef]
74. Zajaczkowski, M.; Włodarska-Kowalczuk, M. Dynamic sedimentary environments of an Arctic glacier-fed river estuary (Adventfjorden, Svalbard). I. Flux, deposition, and sediment dynamics. *Estuar. Coast. Shelf Sci.* **2007**, *74*, 285–296. [CrossRef]
75. Wodarska-Kowalczuk, M.; Kedra, M. Surrogacy in natural patterns of benthic distribution and diversity: Selected taxa versus lower taxonomic resolution. *Mar. Ecol. Prog. Ser.* **2007**, *351*, 53–63. [CrossRef]
76. Rhoads, D.C.; Boesch, D.F.; Zhican, T.; Fengshan, X.; Liqiang, H.; Nilsen, K.J. Macrobenthos and sedimentary facies on the Changjiang delta platform and adjacent continental shelf, East China Sea. *Cont. Shelf Res.* **1985**, *4*, 189–213. [CrossRef]
77. Diaz, R.J.; Rosenberg, R. Marine benthic hypoxia: A review of its ecological effects and the behavioural responses of benthic macrofauna. *Oceanogr. Mar. Biol. Annu. Rev.* **1995**, *33*, 245–303.
78. Levin, L.A.; Ekau, W.; Gooday, A.J.; Jorissen, F.; Middelburg, J.J.; Naqvi, S.W.A.; Neira, C.; Rabalais, N.N.; Zhang, J. Effects of natural and human-induced hypoxia on coastal benthos. *Biogeosciences* **2009**, *6*, 2063–2098. [CrossRef]
79. Villnäs, A.; Norkko, J.; Hietanen, S.; Josefson, A.B.; Lukkari, K.; Norkko, A. The role of recurrent disturbances for ecosystem multifunctionality. *Ecology* **2013**, *94*, 2275–2287. [CrossRef] [PubMed]
80. Michaud, E.; Desrosiers, G.; Mermillod-Blondin, F.; Sundby, B.; Stora, G. The functional group approach to bioturbation: The effects of biodiffusers and gallery-diffusers of the *Macoma balthica* community on sediment oxygen uptake. *J. Exp. Mar. Biol. Ecol.* **2005**, *326*, 77–88. [CrossRef]
81. Michaud, E.; Desrosiers, G.; Mermillod-Blondin, F.; Sundby, B.; Stora, G. The functional group approach to bioturbation: II. The effects of the *Macoma balthica* community on fluxes of nutrients and dissolved organic carbon across the sediment–water interface. *J. Exp. Mar. Biol. Ecol.* **2006**, *337*, 178–189. [CrossRef]
82. Quintana, C.O.; Tang, M.; Kristensen, E. Simultaneous study of particle reworking, irrigation transport and reaction rates in sediment bioturbated by the polychaetes *Heteromastus* and *Marenzelleria*. *J. Exp. Mar. Biol. Ecol.* **2007**, *352*, 392–406. [CrossRef]
83. Kristensen, E.; Hansen, T.; Delefosse, M.; Banta, G.T.; Quintana, C.O. Contrasting effects of the polychaetes *Marenzelleria viridis* and Nereis diversicolor on benthic metabolism and solute transport in sandy coastal sediment. *Mar. Ecol. Prog. Ser.* **2011**, *425*, 125–139. [CrossRef]
84. Gogina, M.; Renz, J.R.; Forster, S.; Zettler, M.L. Benthic Macrofauna Community Bioirrigation Potential (BIPc): Regional Map and Utility Validation for the South-Western Baltic Sea. *Biology* **2022**, *11*, 1085. [CrossRef]
85. Toussaint, E.; De Borger, E.; Braeckman, U.; De Backer, A.; Soetaert, K.; Vanaverbeke, J. Faunal and environmental drivers of carbon and nitrogen cycling along a permeability gradient in shallow North Sea sediments. *Sci. Total. Environ.* **2021**, *767*, 144994. [CrossRef]
86. De Borger, E.; Tiano, J.; Braeckman, U.; Ysebaert, T.; Soetaert, K. Biological and biogeochemical methods for estimating bioirrigation: A case study in the Oosterschelde estuary. *Biogeosciences* **2020**, *17*, 1701–1715. [CrossRef]
87. Ouellette, D., Desrosiers, G., Gagné, J.; Gilbert, F.; Poggiale, J.; Blier, P.; Stora, G. Effects of temperature on in vitro sediment reworking processes by a gallery biodiffusor, the polychaete *Neanthes virens*. *Mar. Ecol. Prog. Ser.* **2004**, *266*, 185–193. [CrossRef]

88. Wohlgemuth, S.; Taylor, A.; Grieshaber, M. Ventilatory and metabolic responses to hypoxia and sulphide in the lugworm *Arenicola marina* (L.). *J. Exp. Biol.* **2000**, *203*, 3177–3188. [CrossRef] [PubMed]
89. Hagerman, L. Physiological flexibility; a necessity for life in anoxic and sulphidic habitats. *Hydrobiologia* **1998**, *375*, 241–254. [CrossRef]
90. Haselmair, A.; Stachowitsch, M.; Zuschin, M.; Riedel, B. Behaviour and mortality of benthic crustaceans in response to experimentally induced hypoxia and anoxia in situ. *Mar. Ecol. Prog. Ser.* **2010**, *414*, 195–208. [CrossRef]
91. Lindström, M.; Sandberg-Kilpi, E. Breaking the boundary—The key to bottom recovery? The role of mysid crustaceans in oxygenizing bottom sediments. *J. Exp. Mar. Biol. Ecol.* **2008**, *354*, 161–168. [CrossRef]
92. Tallqvist, M. Burrowing behaviour of the Baltic clam *Macoma balthica*: Effects of sediment type, hypoxia and predator presence. *Mar. Ecol. Prog. Ser.* **2001**, *212*, 183–191. [CrossRef]
93. Zettler, M.; Bick, A.; Arch, R.B. Fish. Mar. Res, and Undefined 1995, Distribution and population dynamics of *Marenzelleria viridis* (Polychaeta, Spionidae) in a coastal water of the southern Baltic. *Arch. Fish. Mar. Res.* **2005**, *42*, 209–224. Available online: https://www.academia.edu/download/45531524/Distribution_and_population_dynamics_of_20160511-21785-periqn.pdf (accessed on 7 November 2022).
94. Bonaglia, S.; Bartoli, M.; Gunnarsson, J.; Rahm, L.; Raymond, C.; Svensson, O.; Yekta, S.S.; Brüchert, V. Effect of reoxygenation and *Marenzelleria* spp. bioturbation on Baltic Sea sediment metabolism. *Mar. Ecol. Prog. Ser.* **2014**, *482*, 43–55. [CrossRef]
95. Maximov, A.; Bonsdorff, E.; Eremina, T.; Kauppi, L.; Norkko, A.; Norkko, J. Context-dependent consequences of *Marenzelleria* spp. (Spionidae: Polychaeta) invasion for nutrient cycling in the Northern Baltic Sea. *Oceanologia* **2015**, *57*, 342–348. [CrossRef]
96. Johansson, B. Behavioural response to gradually declining oxygen concentration by Baltic Sea macrobenthic crustaceans. *Mar. Biol.* **1997**, *129*, 71–78. [CrossRef]
97. Villnäs, A.; Norkko, A.; Lehtonen, K.K. Multi-level responses of *Macoma balthica* to recurring hypoxic disturbance. *J. Exp. Mar. Biol. Ecol.* **2019**, *510*, 64–72. [CrossRef]
98. Wrede, A.; Dannheim, J.; Gutow, L.; Brey, T. Who really matters: Influence of German Bight key bioturbators on biogeochemical cycling and sediment turnover. *J. Exp. Mar. Biol. Ecol.* **2017**, *488*, 92–101. [CrossRef]

biology MDPI

Article

# Population Genetics of Manila Clam (*Ruditapes philippinarum*) in China Inferred from Microsatellite Markers

**Sichen Zheng [1,2], Tianshi Zhang [1,3], Kang Tu [4], Li Li [5], Zhihong Liu [1,3], Biao Wu [1,3], Liqing Zhou [1,3] and Xiujun Sun [1,3,*]**

1 National Key Laboratory of Mariculture Biobreeding and Sustainable Goods, Yellow Sea Fisheries Research Institute, Chinese Academy of Fishery Sciences, Qingdao 266071, China
2 College of Fisheries and Life Science, Shanghai Ocean University, Shanghai 201306, China
3 Laboratory for Marine Fisheries Science and Food Production Processes, Pilot National Laboratory for Marine Science and Technology, Qingdao 266237, China
4 Putian Institute of Aquaculture Science of Fujian Province, Putian 351100, China
5 National Oceanographic Center, Marine Science Research Institute of Shandong Province, Qingdao 266104, China
* Correspondence: xjsun@ysfri.ac.cn

**Simple Summary:** The Manila clam (*Ruditapes philippinarum*) is one of the most commercially important bivalves along the coast of China. The increasing expanding of clam culture may result in some serious problems. In this paper, we investigated the genetic diversity and differentiation of *R. philippinarum* populations and tested the hypothesis that clam population differentiation is influenced by the southern breeding and northern culture. The present findings will provide useful information for natural resource conservation and genetic breeding of the Manila clam in China.

**Abstract:** The Manila clam (*Ruditapes philippinarum*) is one of the most commercially important bivalves along the coast of China. With the continuous expansion of clam farming scale, it may lead to some serious problems, including loss of genetic variation, inbreeding depression, and reduced effective population size ($N_e$). In the present study, eleven microsatellite markers were used to investigate the genetic diversity and differentiation among 13 clam populations along the coast of China. As a result, 150 alleles were detected according to the genotyping results of eleven microsatellite loci. The observed heterozygosity ($H_o$) was estimated to be ranging from 0.437 to 0.678, while the expected heterozygosity ($H_e$) was calculated to be varying from 0.587 to 0.700. $F_{st}$ values between populations ranged from 0.0046-0.1983. In particular, the Laizhou population had the highest genetic variability, which was significantly different from the others (all $F_{st}$ values > 0.1). For all the clam populations, there was no significant linear regression between genetic and geographic distance, indicating that these populations do not follow a pattern of isolation by distance (IBD). Genetic structure was estimated according to NJ, principal coordinates (PCoA), and structure-based clustering. Estimates of effective population size range from dozens to thousands among different populations, based on linkage-disequilibrium and molecular coancestry methods. The results reveal the genetic diversity of clams and verify the hypothesis that clam population differentiation may be influenced by the mode of southern breeding and northern culture, providing guiding information for natural resource conservation and genetic breeding of clams.

**Keywords:** *Ruditapes philippinarum*; SSR; genetic diversity; genetic differentiation; effective population size

## 1. Introduction

The Manila clam (*Ruditapes philippinarum*) is an important marine bivalve living in the intertidal zone and has the second largest production among bivalve mollusks [1]. In China, it is widely distributed in the coastal areas from Liaoning in the north to Hainan in the south [2]. It has become one of the most commercially important bivalves in the shellfish industry, with an annual production of more than three million tons [3]. In recent years, more than 60% of adult clams are produced in Liaoning and Shandong provinces [3]. In contrast, clam seeds for culture in northern China are mainly purchased from the artificial breeding in southern China. The mode of southern breeding and northern culture may have some negative impacts in local populations such as loss of genetic variation, inbreeding depression, and reduced effective population size [2,4]. Artificial breeding with a small number of parents may increase the probability of cross-generation inbreeding depression, possibly decreasing their ability to adapt to new and challenging environments. [5]. However, the current genetic structure in a wide range of clam populations remains largely unknown. Therefore, it is essential to investigate the genetic diversity and differentiation of clam populations along the coast of China.

Genetic variation can affect the ability of aquatic animals to adapt to environmental changes [6]. Examination of genetic variation is critically important for the suitable management and conservation of natural and cultured populations in aquatic animals [7]. Molecular genetic markers are powerful tools to detect genetic variation among populations in fisheries [8]. Among the available molecular markers, microsatellite or simple sequence repeat (SSR) markers have been widely accepted as the popular molecular tools in population genetics and parentage analyses due to their high polymorphism and codominance [9]. For instance, the application of SSR in population genetics has been reported in a variety of aquatic animals, such as pearl mussel (*Hyriopsis cumingii*), ridgetail white prawn (*Exopalaemon carinicauda*), Silond catfish (*Silonia silondia*), Pacific abalone (*Haliotis Discus hannai*), blood clam (*Barbatia virescens*), and crab (*Portunus trituberculatus*) [10–15]. Despite this, most of these microsatellite studies are relying on the traditional silver staining of DNA in polyacrylamide gels, which may cause some typical sources of scoring errors capable of biasing biological conclusions, such as stuttering and null alleles [16]. SSRs are also limited by the relatively low-throughput genotyping because of their difficulties for automation and data management compared with SNPs. Despite this, SSRs can be accomplished through co-amplification of multiple microsatellites in a single PCR cocktail by multiplexing, which has been improved by decreasing genotyping costs and increasing throughput, e.g., using labelled M13-tails [17–19]. However, the current practices of multiplexing microsatellites in population genetics are lagging, especially in mollusks.

In this study, the new multiplex SSR method has been performed by using labelled M13-tails, providing a cost-effective method for SSR genotyping in clams. Eleven polymorphic microsatellite markers were selected to analyze the genetic diversity and differentiation of the Manila clam (*R. philippinarum*) along the coast of China. The examination of population genetic structure and differentiation of the clams aims to verify the hypothesis that clam population differentiation may be affected by the mode of southern breeding and northern culture. The present findings will not only provide useful information for genetic structure in a wide range of populations, but also help to promote natural conservation and genetic breeding of clam *R. philippinarum*.

## 2. Materials and Methods

### 2.1. Sample Collection and DNA Extraction

A total of 406 clams (*R. philippinarum*) were collected from the northern and southern coast of China (Figure 1). The sampling time, locations, and quantities for the clam samples are summarized in Table 1. Six populations were collected from the southern coast, including Chaozhou (CZ), Lianjiang (LJ), Ningbo (NB), Sanya (SY), Zhangzhou (ZZ), and Beihai (BH). Meanwhile, six populations were collected from the northern coast, including Laizhou (LZ), Rizhao (RZ), Qingdao (QD), Haiyang (HY), Donggang (DG), and Zhuanghe (ZH). In addition, the sample of XY was collected from the selected clam population for rapid growth. For each sample, the foot muscle of clams was dissected and preserved in 100% ethanol. The traditional phenol chloroform method was used for DNA extraction from the foot muscle. After DNA extraction, the quality of DNA was assessed by 1.5%

agarose gel electrophoresis. The DNA concentration was measure by the Nanodrop Lite ultra-micro spectrophotometer. All the DNA samples were diluted into 50 ng/μL and stored at −20 °C.

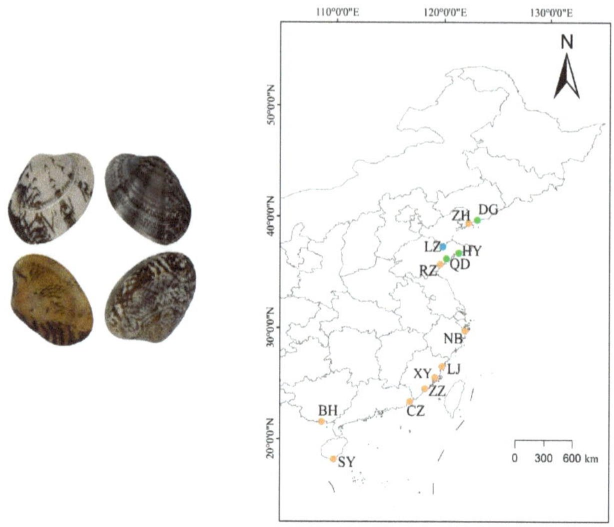


**Figure 1.** The sampled location for the Manila clam (*Ruditapes philippinarum*) along the coast of China. The map of the People's Republic of China is downloaded from the website of http://bzdt.ch.mnr.gov.cn/, accessed on 7 December 2021. The color dots representing the three groups revealed by the PCoA analysis: blue (LZ), green (DG, HY, QD), and orange (ZH, RZ, NB, LJ, XY, ZZ, CZ, BH, and SY).

**Table 1.** Sample code, location, collection date, and sample sizes for all populations of *R. philippinarum*.

| Sample Code | Name | Location | Collection Date | Sample Size |
|---|---|---|---|---|
| CZ | Chaozhou | Chaozhou, Guangdong Province | July 2020 | 32 |
| LZ | Laizhou | Laizhou, Shandong Province | August 2020 | 32 |
| LJ | Lianjiang | Lianjiang, Fujian Province | July 2020 | 32 |
| NB | Ningbo | Ningbo, Zhejiang Province | July 2020 | 32 |
| SY | Sanya | Sanya, Hainan Province | June 2020 | 32 |
| RZ | Rizhao | Rizhao, Shandong Province | August 2020 | 32 |
| ZZ | Zhangzhou | Zhangzhou, Fujian Province | July 2020 | 32 |
| BH | Beihai | Beihai, Guangxi Province | June 2020 | 32 |
| QD | Qingdao | Qingdao, Shandong Province | August 2020 | 29 |
| HY | Haiyang | Haiyang, Shandong Province | August 2020 | 28 |
| DG | Donggang | Donggang, Liaoning Province | August 2020 | 24 |
| ZH | Zhuanghe | Zhuanghe, Liaoning Province | August 2020 | 29 |
| XY | Selected population | Putian, Fujian Province | August 2020 | 40 |

### 2.2. Primer Screening and PCR Amplification

Eleven pairs of microsatellite markers with stable amplification were selected from the previous reports [20,21]. The basic information for primer sequences and PCR conditions is shown in Table 2. The fluorescent labeling for SSRs using M13 tails were performed according to the previous study with minor modifications [17]. Briefly, three primers were used for each PCR amplification: (1) the first one was a forward primer with M13 tails at the 5′ end; (2) the second one was an SSR reverse primer; (3) the third one was an M13 universal primer with a fluorescent label (the 5′ end labeled with 6-carboxy-fluorescine (Fam), hexachloro-6-carboxy-fluorescine (Hex), 6-carboxy-X-rhodamine (Rox), and tetramethylrhodamine (Tamra) fluorescent groups). The selected primer pairs were sorted according to the size ranges. The similar size fragments were labeled with different fluorescence, while different size fragments were labeled with the same fluorescence (Figure S1, Table 2). The PCR reaction system included template DNA 50 ng, 2× Taq plus Master

Mix II 8 μL (Nanjing Vazyme Biotechnology Co., Ltd., Nanjing, China), forward primer 0.04 μL (10 μmol/L), reverse primer 0.16 μL (10 μmol/L), and fluorescent labeled M13 primer 0.16 μL (10 μmol/L), plus dd $H_2O$ to 16 μL. The PCR reactions were performed as follows: 94 °C for 5 min; 30 cycles of 94 °C for 30 s, 53 °C for 45 s, and 72 °C for 45 s; 8 cycles of 94 °C 30 s, 53 °C 45 s, 72 °C 45 s; a final extension at 72 °C for 10 min. The quality of PCR products was detected by 1.5% agarose gel electrophoresis. Finally, 1 μL of PCR products was added to 22 μL formamide and 0.5 μL ROX standard and run on the ABI 3730XL (Shanghai Sangon Bioengineering Co., Ltd., Shanghai, China).

**Table 2.** Primer sequences and information about microsatellite loci from *Ruditapes philippinarum*.

| Sequence | Locus | Accession | Primer (5′-3′) | Repeat Unit | Tm (°C) | Fluorescent Labelling | Size Ranges/bp |
|---|---|---|---|---|---|---|---|
| 1 | Rpt23 | KC811247 | F: AGCGTGTTGCTGCTCTTC<br>R: ATTACTCCCACTGTTCGT | (AGC)6 | 48 | FAM | 81–117 |
| 2 | Rp-07 | AM874000 | F: TATGGCTGGTTTGGACTG<br>R: TCCCGTTACACTTACTTTCA | (AT)7 | 51 | TAM | 119–151 |
| 3 | Asari16 | AB257421 | F: GCTCGAGTCTGATTGGCTACTGAA<br>R: GGTATCTAGTCAGCTCTTGCAGTA | (CT)12 | 55 | ROX | 151–174 |
| 4 | Rp-03 | AM873616 | F: CCGCTGTGAGGAGACCAA<br>R: CCGCCTATGTGACAAAATGA | (TTG)6 | 58 | FAM | 170–213 |
| 5 | Rpt36 | KC811251 | F: TTGAGGCATCAATAACTTTC<br>R: ACTTCTGCATCTCGGCTA | (TTG)8 | 50 | TAM | 230–268 |
| 6 | Rpt100 | KC811260 | F: TCATTTCCAAGGCAGGTA<br>R: GAGGTGTTGAAGGAGCAG | (ATG)5 | 50 | ROX | 237–274 |
| 7 | Rpt106 | KC811263 | F: ACCTCAGTTCAAATGTCT<br>R: AATACTAACGCTGTGGAT | (AGT)6 | 48 | HEX | 373–409 |
| 8 | Rpt105 | KC811262 | F: GGTATGGTGGTAAATGGA<br>R: TCATAGGTAGGGTGGTTT | (GTT)5 | 46 | FAM | 375–411 |
| 9 | Rpt67 | KC811255 | F: GGGTTCTTCTGTAGTTGG<br>R: TGAGAAATCAGACCCAAT | (GAA)5 | 46 | TAM | 379–415 |
| 10 | Rpt32 | KC811249 | F: TCACTTTCTGCTCCTACA<br>R: AAAGGGAATCTCGTGGTG | (CAT)5 | 47 | ROX | 415–451 |
| 11 | Rpt83 | KC811257 | F: GGTCGCCTAATTTCGTAG<br>R: TAATAATTTTCCTGGAGCTCTGGCG | (TGT)7 | 46 | HEX | 429–472 |

## 2.3. Data Processing

The software MSAnalyzer 4.05 was used to calculate number of alleles (N), the observed heterozygosity ($H_o$), and the expected heterozygosity ($H_e$) [22]. The allelic richness ($A_r$) and inbreeding coefficient ($F_{is}$) were calculated through FSTAT 2.9.3 [23]. The significant positive $F_{is}$ values indicate inbreeding within populations (excess of homozygotes), whereas the significant negative $F_{is}$ values represent an excess of heterozygosity. The differences of allelic richness among different groups were compared by the Kruskal–Wallis test of SPSS 26. Furthermore, differences in the allelic richness for each population at each locus was tested using a Kruskal–Wallis rank sum analysis [24]. Hardy–Weinberg equilibrium test and genetic differentiation coefficient ($F_{st}$) were calculated by Genepop 4.0 [25]. For the STRUCTURE analysis, the optimal K value was calculated according to the procedure of Evanno [26], and then the Q value corresponding to the optimal K value was obtained through the repeated sampling analysis of the structure operation results by the CLUMPP software [27]. The genetic structure figure of 13 populations were finally constructed by the software distruct1.1. Genetic distance (Ds) was calculated based on POPULATION software, and then MEGA X was used to build NJ and ME evolutionary trees [28]. An analysis of molecular variance (AMOVA) was performed by the ARLEQUIN program ver. 3.0 to measure the components of variance among and within the populations [29]. A principal component analysis (PCoA) was performed based on the covariance matrix of allele frequencies using GenAlEx 6.3. Mantel test was also performed with GenAlEx 6.3. Linkage-disequilibrium (LD) and molecular coancestry (Cn) methods were used to estimate Ne by using LDNe and NeEstimator v2.0 [30,31].

## 3. Results

### *3.1. Genetic Diversity within Populations*

For the eleven microsatellite loci, the genotyping results of 408 individuals were derived from 13 clam populations with sample sizes ranging from 24 to 40. The descriptive genetic statistics (e.g., N, $A_r$, $G_D$, $H_o$, and $H_e$) were shown for each locus and population in Table S1. As a result, more than nine alleles were found in each of the eleven microsatellite loci, with the maximum alleles (18 alleles) detected in Rp-03. The mean allelic richness ($A_r$) varied from 3.2 (Rpt100 and Rp-07) to 5.5 (Rpt106). At the population level, the average of observed heterozygosity ($H_o$) was calculated to be ranging from 0.437 to 0.678, while the expected heterozygosity ($H_e$) was estimated to be varying from 0.587 to 0.700. Among the eleven loci, the highest $H_e$ value (0.700) was detected in the selected population of XY, while the lowest value (0.587) was found in the QD population. The number of alleles per locus in each population ranged from 2 to 11, and allelic richness per locus varied from 1.7 to 5.5. For all these populations, the LZ population had the largest number of alleles (6.9), as well as the maximum of allele richness (4.1). In contrast, the least number of alleles (5.1) and the minimum of allele richness (3.3) were found in the ZH population. Despite this, no significant difference in allelic richness was detected among these populations (Kruskal–Wallis test, $p > 0.05$). The positive values of $F_{is}$ were consistently found in all the populations, except for HY population (Table S1). A total of 69 (48.3%) of the 143 locus–population combinations were significantly deviant from Hardy–Weinberg equilibrium (HWE) after the Bonferroni correction ($p < 0.005$).

### *3.2. Genetic Differentiation among Populations*

Pairwise $F_{st}$ values among the 13 populations were shown in Table 3. Pairwise $F_{st}$ values across all samples were ranging from 0.0046 to 0.1983 (Table 3). The lowest genetic differentiation was detected between population LJ and NB ($F_{st}$ = 0.0046, $p < 0.01$), whereas the highest differentiation was found between the QD and LZ populations ($F_{st}$ = 0.1983, $p < 0.01$). The genetic differentiation between the LZ population and other populations is relatively high, varying from 0.1020 to 0.1983 ($p < 0.01$). The genetic distances ($D_S$) among populations were also displayed in Table 3. The lowest genetic distance (0.0446) was detected between CZ and LJ, while the largest value (0.4702) was found between LZ and DG. The genetic distances between LZ and the other 12 populations were ranging from 0.1141 to 0.4702. The NJ and ME clustered dendrograms were constructed based on the pairwise genetic distances (Figure 2). As illustrated, no obvious pattern of genetic differentiation was detected among the populations from the northern and southern coast. As displayed in Figure 2A, three northern populations (QD, HY, and RZ) and one southern population (SY) were clustered into one independent branch. In the meantime, two northern populations (ZH and DG) and the selected population (XY) were clustered into another branch. Subsequently, the two small branches were merged with some southern populations (NB, CZ, LJ, BH, and ZZ). The large branch was finally clustered with the LZ population. The clustering result of the ME tree is similar to the NJ tree.

**Table 3.** Pairwise $F_{st}$ (under diagonal) and Nei's genetic distance Ds (above diagonal) of *R. philippinarum*.

| Population | CZ | LZ | LJ | NB | SY | RZ | ZZ | BH | QD | HY | DG | ZH | XY |
|---|---|---|---|---|---|---|---|---|---|---|---|---|---|
| CZ | 0 | 0.3291 | 0.0446 | 0.0927 | 0.1573 | 0.1847 | 0.1353 | 0.0810 | 0.2450 | 0.2361 | 0.2355 | 0.1407 | 0.0973 |
| LZ | 0.1084 ** | 0 | 0.3005 | 0.3113 | 0.3484 | 0.3194 | 0.3144 | 0.2863 | 0.4629 | 0.4265 | 0.4702 | 0.4178 | 0.3777 |
| LJ | 0.0048 | 0.1020 ** | 0 | 0.0492 | 0.1161 | 0.1349 | 0.1117 | 0.0556 | 0.1863 | 0.2015 | 0.2198 | 0.1525 | 0.0956 |
| NB | 0.0270 ** | 0.1021 ** | 0.0046 | 0 | 0.1050 | 0.1422 | 0.1141 | 0.0639 | 0.1517 | 0.1835 | 0.2201 | 0.1461 | 0.0899 |
| SY | 0.0474 ** | 0.1089 ** | 0.0285 ** | 0.0227 ** | 0 | 0.0835 | 0.1874 | 0.1731 | 0.1091 | 0.1284 | 0.1954 | 0.1685 | 0.1320 |
| RZ | 0.0648 ** | 0.1065 ** | 0.0435 ** | 0.0434 ** | 0.0260 ** | 0 | 0.1324 | 0.1868 | 0.1898 | 0.1660 | 0.2514 | 0.1676 | 0.1701 |
| ZZ | 0.0561 ** | 0.1181 ** | 0.0423 ** | 0.0375 ** | 0.0593 ** | 0.0484 ** | 0 | 0.0979 | 0.2433 | 0.2337 | 0.2641 | 0.1674 | 0.1423 |
| BH | 0.0425 ** | 0.1084 ** | 0.0252 ** | 0.0145 ** | 0.0348 ** | 0.0654 ** | 0.0413 ** | 0 | 0.1775 | 0.1881 | 0.2136 | 0.1410 | 0.1035 |
| QD | 0.1349 ** | 0.1983 ** | 0.1157 ** | 0.0857 ** | 0.0809 ** | 0.1048 ** | 0.1224 ** | 0.0794 ** | 0 | 0.1417 | 0.2242 | 0.2310 | 0.1688 |
| HY | 0.0982 ** | 0.1453 ** | 0.0864 ** | 0.0725 ** | 0.0488 ** | 0.0773 ** | 0.0904 ** | 0.0478 ** | 0.0679 ** | 0 | 0.2373 | 0.1507 | 0.1284 |
| DG | 0.1058 ** | 0.1653 ** | 0.0947 ** | 0.0827 ** | 0.0716 ** | 0.0977 ** | 0.1060 ** | 0.0716 ** | 0.0742 ** | 0.0559 ** | 0 | 0.2263 | 0.1750 |
| ZH | 0.0610 ** | 0.1444 ** | 0.0612 ** | 0.0510 ** | 0.0519 ** | 0.0599 ** | 0.0679 ** | 0.0580 ** | 0.0933 ** | 0.0508 ** | 0.0801 ** | 0 | 0.1510 |
| XY | 0.0415 ** | 0.1096 ** | 0.0405 ** | 0.0290 ** | 0.0410 ** | 0.0561 ** | 0.0557 ** | 0.0332 ** | 0.0602 ** | 0.0427 ** | 0.0510 ** | 0.0490 ** | 0 |

Note: "**" indicates the $F_{st}$ reaches significant level at $p < 0.01$.

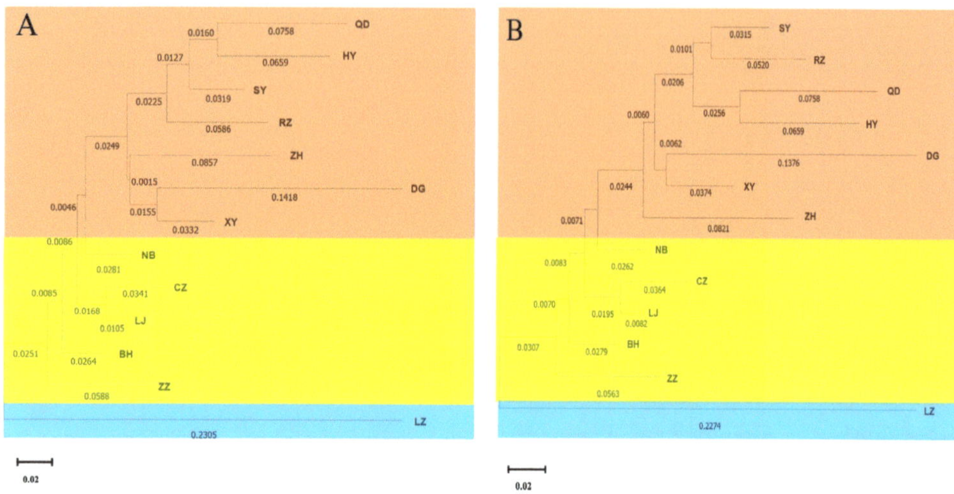


**Figure 2.** Cluster analysis of 13 populations of *R. philippinarum* by NJ (the neighbor-joining) and ME (the minimum evolution) methods. (**A**) NJ tree; (**B**) ME tree. The different colors are used to differentiate the clustered clades among the clam populations. The orange color represents the independent branch of the clam populations, including QD, HY, SY, RZ, ZH, DG, and XY. The orange clade clusters with the yellow branches (NB, CZ, LJ, BH, and ZZ), forming into a higher-level clade. The blue color represents the independent branch for the LZ population different from the large clade.

For each locus, the $F_{st}$ value was ranging from 0.0357 to 0.1729 ($p < 0.01$), with an average of 0.0663 (Table S2). The $N_m$ value of gene flow was varying from 1.1958 to 6.6869, with an average of 4.5900. The $F_{is}$ value for each locus was calculated to be ranging from −0.1021 to 0.6411, with an average of 0.1855. The STRUCTURE analysis revealed K = 3 was the most probable number of populations to explain the observed genotypes (Figure 3). As indicated by STRUCTURE analysis, all the individuals can be divided into three subgroups (Blue, Green, and Red; Figure 3). Consistently, the individuals from each population were also classified into the three genetic clusters, suggesting the high gene flow of these clam populations. According to AMOVA analysis, the greatest number of variances occurred within individuals (67.33%), compared to 25.7% among individuals and 6.97% among the populations (Table 4).

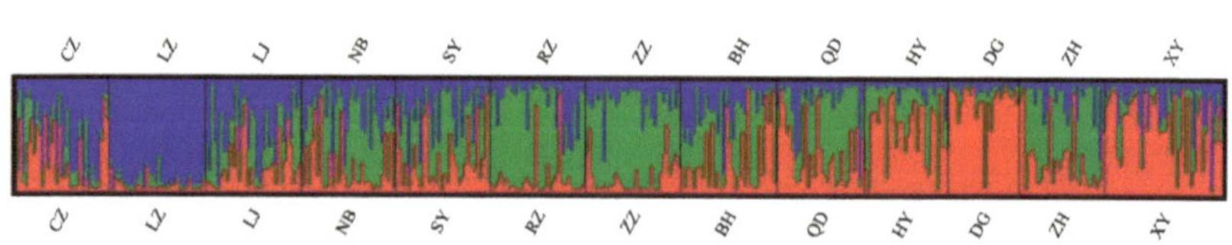


**Figure 3.** Estimated genetic clusters of thirteen *R. philippinarum* populations. The graph is based on the proportion of individuals per population in the inferred clusters according to STRUCTURE. Each of the three colors represents a different genetic cluster, and black lines separate the populations.

**Table 4.** Analysis of molecular variance (AMOVA) in thirteen populations of *R. philippinarum*.

| Source of Variation | d.f. | Sum of Squares | MS | Est. Var. | Percentage Variation |
|---|---|---|---|---|---|
| Among Populations | 12 | 260.237 | 21.686 | 0.273 | 6.97% |
| Among Individuals | 393 | 1828.026 | 4.651 | 1.007 | 25.70% |
| Within Individuals | 406 | 1071.000 | 2.638 | 2.638 | 67.33% |
| Total | 811 | 3159.262 | | 3.918 | 100.00% |

Degree of freedom (d.f.), mean square (MS), variance component (Est. Var.).

The visual representation of genetic distances among the 13 populations revealed by PCoA analysis was displayed in Figure 4. In accordance with STRUCTURE results, PCoA analysis indicated that these 13 populations were mainly formed into three main groups: group I (LZ), group II (HY, DG, and QD), and group III (XY, SY, ZH, BH, RZ, NB, ZZ, LJ, and CZ) (Figure 4). A plot of the first and second principal coordinates is presented, accounting for 37.63% and 19.25% of the total variation, respectively. Samples from group I were well-differentiated from others on the first and second axes, while samples from group II were mainly separated on the first axis. Although four northern sites (Group I and Group II) seem to be different from other sites (Figure 4), the genetic difference within northern populations (Group I and Group II) are much greater than the difference between Group II (northern populations) and Group III (comprises both northern and southern populations). Based on the Mantel tests in GenAlex6.51, no significant linear relationship was detected between genetic distance and geographic distance in the clam samples ($Y = -5.41 \times 10^{-6}X + 0.1959$, $R^2 = 0.0018$, $p > 0.05$; Figure 5). The results indicate that the clam populations do not follow a pattern of isolation by distance (IBD; Figure 5), and this is evidenced by high gene flow among populations within the large geographic scales (Group III, Figure 4). For instance, ZH and BH are thousands of kilometers apart (>2500 km), but they have a relatively low genetic differentiation level, $F_{st} = 0.058$.

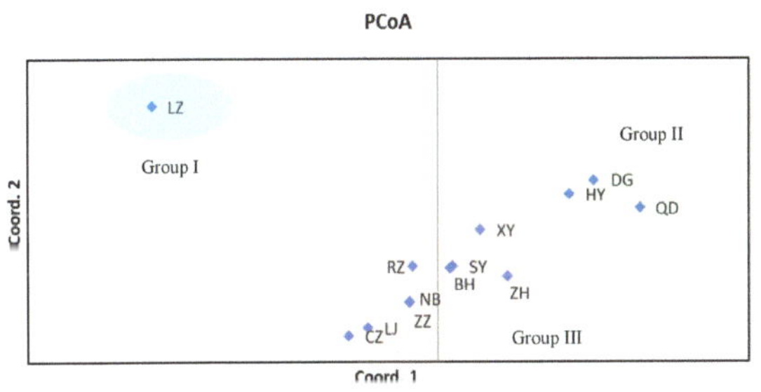


**Figure 4.** Principal coordinate analysis of genetic similarity among 13 clam populations.

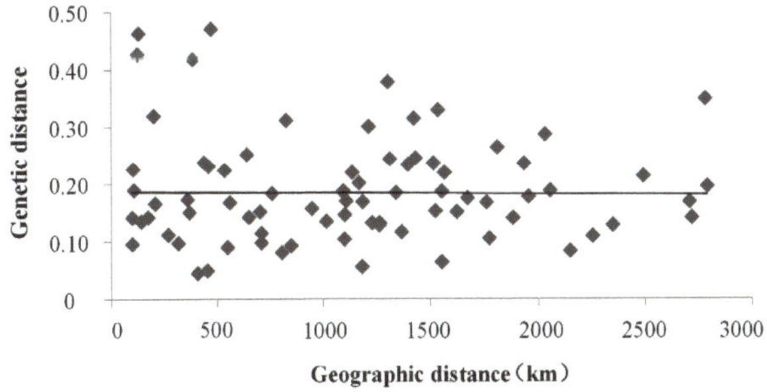


**Figure 5.** The non-significant linear regression between genetic distance and geographic distance based on eleven microsatellite loci in the clam samples ($Y = -5.41 \times 10^{-6}X + 0.1959$, $R^2 = 0.0018$, 10,000 permutations).

### 3.3. Estimation of Effective Population Size ($N_e$)

Two single-sample methods were used to estimate $N_e$ for all 13 samples collected in Table 5. The LDNe method yielded part negative $N_e$ estimates (Table 5). According to the LD and Cn methods, the $N_e$ values of most populations were low except for the XY population ($N_e = 375.4$). The lowest values of $N_e$ were found in Chaozhou and Donggang, having extremely low $N_e$ of less than the critical value ($N_e = 50$). Generally, the $N_e$ values estimated from the Cn method were relatively lower than those from the LD method.

**Table 5.** Effective population sizes ($N_e$) for *R. philippinarum* populations estimated by the linkage disequilibrium (LD) and molecular coancestry (Cn) methods.

| | LD | | | | Cn | | |
|---|---|---|---|---|---|---|---|
| **Pop** | **n** | $: \hat{r}^2$ | $E (: \hat{r}^2)$ | $N_e$ **(95% CI)** | **Pop** | **n** | $N_e$ **(95% CI)** |
| CZ | 28.5 | 0.0483 | 0.0391 | 31.7 (20.3–58.8) | CZ | 30.1 | 7.2 (3.4–12.5) |
| LZ | 31.0 | 0.0384 | 0.0355 | 114.9 (50.1–Infinite) | LZ | 31.5 | Infinite (Infinite–Infinite) |
| LJ | 29.5 | 0.0373 | 0.0377 | −876.7 (92.3–Infinite) | LJ | 30.8 | Infinite (Infinite–Infinite) |
| NB | 28.8 | 0.0375 | 0.0387 | −258.6 (129.9–Infinite) | NB | 30.4 | 32.8 (0–164.9) |
| SY | 28.6 | 0.0434 | 0.0390 | 68.2 (33.3–510.2) | SY | 29.8 | 20.7 (1.5–64.4) |
| RZ | 29.0 | 0.0418 | 0.0384 | 88.3 (36.7–Infinite) | RZ | 30.5 | 18.9 (0–94.8) |
| ZZ | 29.2 | 0.0374 | 0.0380 | −513.7 (74.7–Infinite) | ZZ | 30.3 | Infinite (Infinite–Infinite) |
| BH | 24.7 | 0.0419 | 0.0458 | −79.9 (290.4–Infinite) | BH | 27.9 | 54.3 (0.1–272.7) |
| QD | 17.7 | 0.0526 | 0.0672 | 22.7 (-39.9–Infinite) | QD | 19.0 | 11.9 (2–30.6) |
| HY | 18.2 | 0.0526 | 0.0649 | −26.5 (-47.9–Infinite) | HY | 22.4 | 11.1 (2.7–25.4) |
| DG | 14.9 | 0.0776 | 0.0827 | −62.1 (58.3–Infinite) | DG | 19.5 | 6.2 (4.2–8.5) |
| ZH | 22.9 | 0.0497 | 0.0498 | −4158.7 (47.7–Infinite) | ZH | 26.5 | 17.0 (2.8–43.7) |
| XY | 30.3 | 0.0374 | 0.0365 | 375.4 (73.5–Infinite) | XY | 34.8 | Infinite (Infinite–Infinite) |

Mean sample sizes per locus (n), mean squared correlation of allelic frequencies over ($: \hat{r}^2$) the expectation of $: \hat{r}^2$ based on mean sample size ($E (: \hat{r}^2)$).

## 4. Discussion

### *4.1. Genetic Diversity of Manila Clams in Different Populations from North to South*

High levels of genetic diversity appear to be a common feature of marine bivalves [32]. In this study, microsatellite analysis of *R. philippinarum* populations revealed a relatively higher level of genetic diversity ($H_e$ = 0.636) than those estimates from allozymic analysis and other DNA-based analyses, such as mtDNAs, AFLP, and RAPD [4,33,34]. Consistent with our study, high levels of genetic diversity estimated from microsatellite markers were also observed in other bivalves, such as *Crassostrea gasar* ($H_e$ = 0.843 [35], *Barbatia virescens* ($H_e$ = 0.790 [9]), and *Crassostrea ariakensis* ($H_e$ = 0.805 [36]). Large population sizes and high nucleotide mutation rates are likely to be the major contributors to the high levels of genetic diversity estimated from microsatellites [37,38].

Departures from Hardy–Weinberg equilibrium (HWE) were measured through the significance of permutation tests for the null hypothesis, $F_{is} = 0$ [39]. In the present study, the significant heterozygote deficiency was detected in clam populations according to these genotyped microsatellite loci ($F_{is}$ = 0.1855; $p < 0.05$). In addition to clams, multi-locus heterozygosity deficiencies have been previously widely reported in many other bivalves [40,41]. Early explanation for the departure from HWE in bivalves mainly involved null alleles, natural selection, inbreeding, and Wahlund effects [42]. However, the recent hypothesis of genetic load shows more compelling evidence for this phenomenon, indicating the large genetic load of partially dominant or additive detrimental mutations in wild adult populations [43,44]. It is therefore suggested that the high genetic load is largely responsible for heterozygote deficits in wild populations and segregation distortion in pair crosses, resulting in substantial genetic sterility [44]. Further studies will be needed to elucidate the genetic load by pair crosses of clams.

### *4.2. The Genetic Differentiation among Clam Populations*

The overall genetic differentiation among these populations was moderate but highly significant (global $F_{st}$ = 0.066, $p < 0.001$),indicating the existence of the genetic heterogeneity among populations. As the wild population in Laizhou Bay (Shandong province, North China), the LZ population remains the population with the highest level of genetic variability, showing great differentiation with other populations. This is consistent with the previous studies, supporting the natural status of clam populations with high levels of genetic variability [2,45]. As reported, natural selection continuously removes neutral diversity linked to either beneficial or deleterious variants [46]. In contrast to the LZ populations, other clam populations are likely to have low differentiation and high glow flow according to cluster and PCoA analysis. In the present study, the low differentiation among different populations supports the hypothesis that the genetic structure of clams may be influenced by the mode of southern breeding and northern culture.

In this study, the clam populations do not follow a pattern of isolation by distance, and this contrasts with the reported IBD pattern caused by larval dispersal in other coastal bivalve species [41]. For the clam populations, high gene flow among populations so far apart seems unlikely to be caused by the larval dispersal. The more reasonable explanation for this is probably due to seeds' transplantation by local farmers among different culture regions. In recent decades, there is considerable translocation of clam seeds cultivated in Fujian province (south) to culture sites in Shandong and Liaoning provinces (north) [4,20,21]. Therefore, the artificial breeding and culture of clams may increase the gene flow of clams, resulting in the low genetic differentiation between northern and southern populations, as evidenced by our present results. The low genetic differentiation between northern and southern populations has also been detected in the previous studies [2,47]. Therefore, the present findings do not support the typical pattern of genetic differentiation between northern and southern populations due to geographic isolation. The translocation of clam seeds may be served as one of the major factors influencing the population genetic structure of the clams. Adapted conservation measures for wild populations are required to maintain high levels of genetic diversity of clams on the coast of China. In order to protect the wild clam populations, it is necessary to take measures to prevent excessive harvesting and formulate laws and regulations to limit the number and time for clam harvesting. It is also important to ensure that natural habitats of clams have not been occupied or damaged by environmental pollution. We recommend the use of responsible conservation aquaculture protocols, such as large numbers of local adult clams for bloodstocks and new techniques reducing hatchery selection to facilitate the management of genetic variability [46]. However, simply increasing the number of breeders does not necessarily increase the effective breeding numbers in shellfish hatcheries. Therefore, the development of breeding strategies and optimization of production is also important in the maintaining of genetic diversity, such as pedigree monitoring by genetic markers and performing controlled spawning [17–19]. Recently, the rapid development of high-throughput sequencing methods have facilitated the incorporation of genomic tools in clam breeding programs by control parental contribution [49]. Overall, these strategies are recommended for the retention of high genetic variability in clam *R. philippinarum*, especially for the wild population in Laizhou Bay.

### *4.3. Estimation of Effective Population Sizes in Clam Populations*

The effective population size ($N_e$), a key parameter in evolutionary biology, determines the rates of genetic drift and loss of genetic variability and modulates the effectiveness of selection [50]. For wild populations, the supplement with artificially breeding individuals can lead to the $N_e$ reduction, known as the Ryman–Laikre effect [51]. As reported, the reduction of $N_e$ would lead to a collapse of local genetic adaptation, which could expose local populations to adverse effects [52–54].

The previous studies have indicated that $N_e$ of shellfish bloodstocks should be large enough to produce the first generation with relatively medium or high genetic diversity [55]. The small $N_e$ population will lead to the depletion of rare alleles, increasing of the random drift of the original population, and thus threatening of the sustainability of populations [56]. As a rule-of-thumb in populations, $N_e$ in the short term should not be less than 50, and in the long term should not be less than 500 [57]. The estimates of $N_e$ thresholds for avoiding inbreeding depression ($N_e$ = 50) and retention of genetic variation for future adaptations ($N_e$ = 500) can be used as a guiding principle to indicate the short- and long-term genetic viability of populations [55,56].

In the present study, small $N_e$ values (less than 50) have been obtained in several populations (e.g., QD and CZ) according to LD and Cn methods. The small $N_e$ values may be caused by inadvertent selection of the best offspring produced by a few parents and asymmetric reproduction [58]. Despite this, the accumulation of inbreeding might have some negative effects on survival rates of clams in these populations with small $N_e$ values [21,22]. Thus, it is essential to recover the local populations by the conservation programs (e.g., broodstock management and controlled spawning) to maintain a minimum viable population to maintain the evolutionary potential [59,60]. Surprisingly, negative $N_e$ values from the LD method have been detected in multiple populations of clams, probably due to the linkage disequilibrium generated by the sampling process and inadequate correction [61]. If $N_e$ is very large or limited data are available, by chance $^{r2}$ (mean squared correlation of allelic frequencies) can be smaller than the sample size correction, resulting in the negative estimates of $N_e$ [58,62]. Therefore, the negative estimates may occur when genetic results can be explained entirely by sampling error without invoking any genetic drift, interpreted as the infinite $N_e$ [58]. This is also supported by the computer simulations, indicating that the LD method is biased when the sample size is small (<100) and below the true $N_e$ [61]. Despite this, the lower bound of CIs in this study can provide some useful information for the plausible limits of these negative $N_e$ values. The future estimation of $N_e$ needs an extensive evaluation in larger sample sizes using increased numbers of loci and alleles. Despite uncertainties related to the small sampling size, $N_e$ estimates obtained by the two applied methods can provide useful complementary information for conservation programs to prevent inbreeding depression and loss of genetic variation. According to the present findings, the small $N_e$ values, as well as the low differentiation, may be caused by few broodstock used in southern hatcheries, with offspring transferred to the northern coast for culture at the mode of southern breeding and northern culture.

## 5. Conclusions

In this study, genetic diversity and differentiation were investigated by 11 microsatellite loci for *R. philippinarum* (Manila clam) populations from the coastal areas of China. The multiplex PCR using the labelled M13-tails was shown to be a cost-effective method for SSR genotyping in clams and mollusks, provided that the sufficient sampling size is ensured. The present findings support that the genetic population structure of clams may be influenced by the mode of southern breeding and northern culture. The assessment of the genetic diversity of *R. philippinarum* populations is of considerable importance for the optimal development of programs aimed at the conservation of cultivated and wild genotypes in the ecosystems. The present findings will provide guiding information on natural resource conservation and genetic breeding of the Manila clam in China. The highest level of genetic variability and greatest differentiation with other populations was confirmed for the wild Laizhou population. It was suggested that multi-locus heterozygote deficiency and segregation distortion in such populations may be caused by high genetic load. No relation was found between genetic and geographic distance, implying clam aquaculture may be served as one of the major factors influencing clam population genetic structure. Despite uncertainties related to the small sampling size, $N_e$ estimates obtained by the applied methods can provide useful complementary information for conservation programs to warn about inbreeding depression and loss of genetic variation, thereby serving the needs of natural resource conservation.

**Supplementary Materials:** The following supporting information can be downloaded at: https://www.mdpi.com/article/10.3390/biology12040557/s1, Table S1: Genetic diversity parameters of thirteen *R. philippinarum* populations. Table S2: Genetic differentiations and gene flow of thirteen *R. philippinarum* populations at different loci. Figure S1: SSR profiles generated on a capillary sequencer for a single sample. Green color, Hex fluorescence; Red color, Rox fluorescence; Blue color, Fam fluorescence; Black color, Tamra fluorescence.

**Author Contributions:** Conceptualization, X.S.; methodology, S.Z.; software, Z.L.; sample collection, K.T. and B.W.; data curation, L.Z.; writing—original draft preparation, S.Z.; writing—review and editing, X.S.; resources and visualization, L.L. and L.Z.; project administration, T.Z.; funding acquisition, Z.L. All authors have read and agreed to the published version of the manuscript.

**Funding:** This study was supported by research grants from the Science and Technology Innovation Program of the Laoshan Laboratory (LSKJ202203803) and National Natural Science Foundation of China (32273107). This work was also supported by Central Public—Interest Scientific Institution

Basal Research Fund, YSFRI, CAFS (20603022022001), Qingdao Shinan District Science and Technology Plan Project (2022-2-026-ZH), and Putian City Science and Technology Plan Project (2021NJJ002).

**Institutional Review Board Statement:** Not applicable.

**Informed Consent Statement:** Not applicable.

**Data Availability Statement:** Not applicable.

**Conflicts of Interest:** The authors declare that they have no competing interest.

## References

1. De Montaudouin, X.; Lucia, M.; Binias, C.; Lassudrie, M.; Baudrimont, M.; Legeay, A.; Raymond, N.; Jude-Lemeilleur, F.; Lambert, C.; Le Goïc, N. Why is Asari (=Manila) clam *Ruditapes philippinarum* fitness poor in Arcachon Bay: A meta-analysis to answer? *Estuar. Coast. Shelf Sci.* **2016**, *179*, 226–235. [CrossRef]
2. Tan, Y.; Fang, L.; Qiu, M.; Huo, Z.; Yan, X. Population genetics of the Manila clam (*Ruditapes philippinarum*) in East Asia. *Sci. Rep.* **2020**, *10*, 21890. [CrossRef]
3. DOF. *China Fisheries Statistic Yearbook;* China Agriculture Press: Beijing, China, 2021.
4. Liu, X.; Zhenmin, B.; Jingjie, H.; Shi, W.; Aibin, Z.; Hui, L.; Jianguang, F.; Rucai, W. AFLP analysis revealed differences in genetic diversity of four natural populations of Manila clam (*Ruditapes philippinarum*) in China. *Acta Oceanol. Sin.* **2007**, *1*, 150–158.
5. Kristensen, T.N.; Hoffmann, A.A.; Pertoldi, C.; Stronen, A.V. What can livestock breeders learn from conservation genetics and vice versa? *Front. Genet.* **2015**, *6*, 38. [CrossRef]
6. Conover, D.; Clarke, L.; Munch, S.; Wagner, G. Spatial and temporal scales of adaptive divergence in marine fishes and the implications for conservation. *J. Fish Biol.* **2006**, *69*, 21–47. [CrossRef]
7. Okumuş, İ.; Çiftci, Y. Fish population genetics and molecular markers: II-molecular markers and their applications in fisheries and aquaculture. *Turk. J. Fish. Aquat. Sci.* **2003**, *3*, 51–79.
8. Askari, G.; Shabani, A.; Kolangi Miandare, H. Application of molecular markers in fisheries and aquaculture. *Sci. J. Anim. Sci.* **2013**, *2*, 82–88.
9. Nie, H.; Niu, H.; Zhao, L.; Yang, F.; Yan, X.; Zhang, G. Genetic diversity and structure of Manila clam (*Ruditapes philippinarum*) populations from Liaodong peninsula revealed by SSR markers. *Biochem. Syst. Ecol.* **2015**, *59*, 116–123. [CrossRef]
10. An, H.; Lee, J.; Park, J. Population genetics of the Pacific abalone (*Haliotis discus hannai*) in Korea inferred from microsatellite marker analysis. *Genet. Mol. Res.* **2012**, *11*, 3904–3922. [CrossRef]
11. Bai, Z.; Zheng, H.; Lin, J.; Wang, G.; Li, J. Comparative analysis of the transcriptome in tissues secreting purple and white nacre in the pearl mussel *Hyriopsis cumingii. PLoS ONE* **2013**, *8*, e53617. [CrossRef]
12. Mandal, S.; Jena, J.; Singh, R.K.; Mohindra, V.; Lakra, W.; Deshmukhe, G.; Pathak, A.; Lal, K.K. De novo development and characterization of polymorphic microsatellite markers in a schilbid catfish, *Silonia silondia* (Hamilton, 1822) and their validation for population genetic studies. *Mol. Biol. Rep.* **2016**, *43*, 91–98. [CrossRef] [PubMed]
13. Liu, Q.; Cui, F.; Hu, P.; Yi, G.; Ge, Y.; Liu, W.; Yan, H.; Wang, L.; Liu, H.; Song, J. Using of microsatellite DNA profiling to identify hatchery-reared seed and assess potential genetic risks associated with large-scale release of swimming crab *Portunus trituberculatus* in Panjin, China. *Fish. Res.* **2018**, *207*, 187–196. [CrossRef]
14. Wang, L.; Yu, H.; Li, Q. Development of microsatellite markers and analysis of genetic diversity of *Barbatia virescens* in the southern coasts of China. *Genes Genom.* **2019**, *41*, 407–416. [CrossRef]
15. Li, J.; Li, J.; Chen, P.; Liu, P.; He, Y. Transcriptome analysis of eyestalk and hemocytes in the ridgetail white prawn *Exopalaemon carinicauda*: Assembly, annotation and marker discovery. *Mol. Biol. Rep.* **2015**, *42*, 135–147. [CrossRef]
16. DeWoody, J.; Nason, J.D.; Hipkins, V.D. Mitigating scoring errors in microsatellite data from wild populations. *Mol. Ecol. Notes* **2006**, *6*, 951–957. [CrossRef]
17. Schuelke, M. An economic method for the fluorescent labeling of PCR fragments. *Nat. Biotechnol.* **2000**, *18*, 233–234. [CrossRef]
18. Hayden, M.J.; Nguyen, T.M.; Waterman, A.; Chalmers, K.J. Multiplex-ready PCR: A new method for multiplexed SSR and SNP genotyping. *BMC Genom.* **2008**, *9*, 80. [CrossRef]
19. Guichoux, E.; Lagache, L.; Wagner, S.; Chaumeil, P.; Léger, P.; Lepais, O.; Lepoittevin, C.; Malausa, T.; Revardel, E.; Salin, F. Current trends in microsatellite genotyping. *Mol. Ecol. Resour.* **2011**, *11*, 591–611. [CrossRef] [PubMed]
20. Nie, H.L.J.H.; Zh; Guo, W.; Yan, X. Analysis of genetic variability in selected lines and a wild population of R*uditapes philippinarum* using microsatellite markers. *J. Fish. Sci. China* **2016**, *23*, 538–546.
21. Yu, Z.; Yan, X.; Yang, F.; Wang, J.; Zhang, Y.; Yang, F.; Zhang, G. Genetic diversity of different generations of the Dalian population of Manila clam *Ruditapes philippinarum* through selective breeding. *Acta Ecol. Sin.* **2011**, *31*, 4199–4206.
22. Dieringer, D.; Schlötterer, C. Microsatellite analyser (MSA): A platform independent analysis tool for large microsatellite data sets. *Mol. Ecol. Notes* **2003**, *3*, 167–169. [CrossRef]
23. Goudet, J. FSTAT, a Program to Estimate and Test Gene Diversities and Fixation Indices (Version 2.9.3). 2001. Available online: https://www2.unil.ch/popgen/softwares/fstat.htm (accessed on 19 October 2021.).
24. Sokal, R.R.; Rohlf, F.J.; Rohlf, J.F. *Biometry;* Macmillan: New York, NY, USA, 1995.
25. Raymond, M.; Rousset, F. An exact test for population differentiation. *Evolution* **1995**, *49*, 1280–1283. [CrossRef]

26. Evanno, G.; Regnaut, S.; Goudet, J. Detecting the number of clusters of individuals using the software STRUCTURE: A simulation study. *Mol. Ecol.* **2005**, *14*, 2611–2620. [CrossRef] [PubMed]
27. Jakobsson, M.; Rosenberg, N.A. CLUMPP: A cluster matching and permutation program for dealing with label switching and multimodality in analysis of population structure. *Bioinformatics* **2007**, *23*, 1801–1806. [CrossRef]
28. Kumar, S.; Stecher, G.; Li, M.; Knyaz, C.; Tamura, K. MEGA X: Molecular evolutionary genetics analysis across computing platforms. *Mol. Biol. Evol.* **2018**, *35*, 1547. [CrossRef]
29. Excoffier, L.; Laval, G.; Schneider, S. Arlequin (Version 3.0): An integrated software package for population genetics data analysis. *Evol. Bioinform.* **2005**, *1*, 47–50. [CrossRef]
30. Waples, R.S.; Do, C. LDNE: A program for estimating effective population size from data on linkage disequilibrium. *Mol. Ecol. Resour.* **2008**, *8*, 753–756. [CrossRef] [PubMed]
31. Do, C.; Waples, R.S.; Peel, D.; Macbeth, G.; Tillett, B.J.; Ovenden, J.R. NeEstimator v2: Re-implementation of software for the estimation of contemporary effective population size ($N_e$) from genetic data. *Mol. Ecol. Resour.* **2014**, *14*, 209–214. [CrossRef]
32. An, H.S.; Park, K.J.; Cho, K.C.; Han, H.S.; Myeong, J.-I. Genetic structure of Korean populations of the clam *Ruditapes philippinarum* inferred from microsatellite marker analysis. *Biochem. Syst. Ecol.* **2012**, *44*, 186–195. [CrossRef]
33. Vargas, K.; Asakura, Y.; Ikeda, M.; Taniguchi, N.; Obata, Y.; Hamasaki, K.; Tsuchiya, K.; Kitada, S. Allozyme variation of littleneck clam *Ruditapes philippinarum* and genetic mixture analysis of foreign clams in Ariake Sea and Shiranui Sea off Kyushu Island, Japan. *Fish. Sci.* **2008**, *74*, 533–543. [CrossRef]
34. Sekine, Y.; Yamakawa, H.; Takazawa, S.; Lin, Y.; Toba, M. Geographic variation of the COX1 gene of the short-neck clam *Ruditapes philippinarum* in coastal regions of Japan and China. *Venus* **2006**, *65*, 229–240.
35. Melo, M.A.D.; da Silva, A.R.B.; Varela, E.S.; Sampaio, I.; Tagliaro, C.H. Development and characterization of ten microsatellite markers for population studies of the native Brazilian oyster *Crassostrea gasar*. *Conserv. Genet. Resour.* **2012**, *4*, 583–586. [CrossRef]
36. Xiao, J.; Cordes, J.F.; Wang, H.; Guo, X.; Reece, K.S. Population genetics of *Crassostrea ariakensis* in Asia inferred from microsatellite markers. *Mar. Biol.* **2010**, *157*, 1767–1781. [CrossRef]
37. Hoshino, A.A.; Bravo, J.P.; Nobile, P.M.; Morelli, K.A. Microsatellites as Tools for Genetic Diversity Analysis. 2012. Available online: https://www.researchgate.net/profile/Andrea-Hoshino/publication/221925456_Microsatellites_as_Tools_for_Genetic_Diversity_Analysis/links/0d1c84f492b24ee605000000/Microsatellites-as-Tools-for-Genetic-Diversity-Analysis.pdf (accessed on 26 October 2021).
38. Hedgecock, D.; Li, G.; Hubert, S.; Bucklin, K.; Ribes, V. Widespread null alleles and poor cross-species amplification of microsatellite DNA loci cloned from the Pacific oyster, *Crassostrea gigas*. *J. Shellfish Res.* **2004**, *23*, 379–386.
39. Hedgecock, D.; Launey, S.; Pudovkin, A.; Naciri, Y.; Lapegue, S.; Bonhomme, F. Small effective number of parents (Nb) inferred for a naturally spawned cohort of juvenile European flat oysters *Ostrea edulis*. *Mar. Biol.* **2007**, *150*, 1173–1182. [CrossRef]
40. VanTassel, N.M.; Morris, T.J.; Wilson, C.G.; Zanatta, D.T. Genetic diversity maintained in comparison of captive-propagated and wild populations of *Lampsilis fasciola* and *Ptychobranchus fasciolaris* (Bivalvia: Unionidae). *Can. J. Fish. Aquat.* **2021**, *78*, 1312–1320. [CrossRef]
41. Mathias, P.T.; Hoffman, J.R.; Wilson, C.C.; Zanatta, D.T. Signature of postglacial colonization on contemporary genetic structure and diversity of *Quadrula quadrula* (Bivalvia: Unionidae). *Hydrobiologia* **2018**, *810*, 207–225. [CrossRef]
42. Chacón, G.M.; Arias-Pérez, A.; Freire, R.; Martínez, L.; Ojea, J.; Insua, A. Genetic characterization of wild, broodstock and seed samples of Polititapes *rhomboides* (Bivalvia: Veneridae): Implications for hatchery seed production. *Aquac. Rep.* **2021**, *20*, 100658. [CrossRef]
43. Plough, L.V.; Hedgecock, D. Quantitative trait locus analysis of stage-specific inbreeding depression in the Pacific oyster *Crassostrea gigas*. *Genetics* **2011**, *189*, 1473–1486. [CrossRef] [PubMed]
44. Plough, L.; Shin, G.; Hedgecock, D. Genetic inviability is a major driver of type III survivorship in experimental families of a highly fecund marine bivalve. *Mol. Ecol.* **2016**, *25*, 895–910. [CrossRef]
45. Lallias, D.; Taris, N.; Boudry, P.; Bonhomme, F.; Lapegue, S. Variance in the reproductive success of flat oyster *Ostrea edulis* L. assessed by parentage analyses in natural and experimental conditions. *Genet. Res.* **2010**, *92*, 175–187. [CrossRef]
46. Mackintosh, A.; Laetsch, D.R.; Hayward, A.; Charlesworth, B.; Waterfall, M.; Vila, R.; Lohse, K. The determinants of genetic diversity in butterflies. *Nat. Commun.* **2019**, *10*, 3466. [CrossRef]
47. Cordero, D.; Delgado, M.; Liu, B.; Ruesink, J.; Saavedra, C. Population genetics of the Manila clam (*Ruditapes philippinarum*) introduced in North America and Europe. *Sci. Rep.* **2017**, *7*, 39745. [CrossRef] [PubMed]
48. Teixeira, J.C.; Huber, C.D. The inflated significance of neutral genetic diversity in conservation genetics. *Proc. Natl. Acad. Sci. USA* **2021**, *118*, e2015096118. [CrossRef] [PubMed]
49. Ridlon, A.D.; Wasson, K.; Waters, T.; Adams, J.; Donatuto, J.; Fleener, G.; Froehlich, H.; Govender, R.; Kornbluth, A.; Lorda, J. Conservation aquaculture as a tool for imperiled marine species: Evaluation of opportunities and risks for Olympia oysters, *Ostrea lurida*. *PLoS ONE* **2021**, *16*, e0252810. [CrossRef]
50. Wang, J.; Santiago, E.; Caballero, A. Prediction and estimation of effective population size. *Heredity* **2016**, *117*, 193–206. [CrossRef]
51. Waples, R.S.; Hindar, K.; Karlsson, S.; Hard, J.J. Evaluating the Ryman–Laikre effect for marine stock enhancement and aquaculture. *Curr. Zool.* **2016**, *62*, 617–627. [CrossRef] [PubMed]
52. Christie, M.R.; Marine, M.; French, R.; Waples, R.S.; Blouin, M. Effective size of a wild salmonid population is greatly reduced by hatchery supplementation. *Heredity* **2012**, *109*, 254–260. [CrossRef]

53. Morvezen, R.; Boudry, P.; Laroche, J.; Charrier, G. Stock enhancement or sea ranching? Insights from monitoring the genetic diversity, relatedness and effective population size in a seeded great scallop population (*Pecten maximus*). *Heredity* **2016**, *117*, 142–148. [CrossRef] [PubMed]
54. Baskett, M.L.; Burgess, S.C.; Waples, R.S. Assessing strategies to minimize unintended fitness consequences of aquaculture on wild populations. *Evol. Appl.* **2013**, *6*, 1090–1108. [CrossRef]
55. Hold, N.; Murray, L.G.; Kaiser, M.J.; Hinz, H.; Beaumont, A.R.; Taylor, M.I. Potential effects of stock enhancement with hatchery-reared seed on genetic diversity and effective population size. *Can. J. Fish. Aquat. Sci.* **2013**, *70*, 330–338. [CrossRef]
56. Frankham, R.; Ballou, J.; Briscoe, D. *Introduction to Conservation Genetics*; Cambridge University Press: Cambridge, UK, 2002.
57. Franklin, I.; Frankham, R. How large must populations be to retain evolutionary potential? *Anim. Conserv. Forum* **1998**, *1*, 69–70. [CrossRef]
58. Waples, R.S.; Do, C. Linkage disequilibrium estimates of contemporary Ne using highly variable genetic markers: A largely untapped resource for applied conservation and evolution. *Evol. Appl.* **2010**, *3*, 244–262. [CrossRef]
59. Lallias, D.; Boudry, P.; Lapegue, S.; King, J.W.; Beaumont, A.R. Strategies for the retention of high genetic variability in European flat oyster (*Ostrea edulis*) restoration programmes. *Conserv. Genet.* **2010**, *11*, 1899–1910. [CrossRef]
60. Hedgecock, D.; Pudovkin, A.I. Sweepstakes reproductive success in highly fecund marine fish and shellfish: A review and commentary. *Bull. Mar. Sci.* **2011**, *87*, 971–1002. [CrossRef]
61. England, P.R.; Cornuet, J.-M.; Berthier, P.; Tallmon, D.A.; Luikart, G. Estimating effective population size from linkage disequilibrium: Severe bias in small samples. *Conserv. Genet.* **2006**, *7*, 303–308. [CrossRef]
62. Waples, R.S. A bias correction for estimates of effective population size based on linkage disequilibrium at unlinked gene loci. *Conserv. Genet.* **2006**, *7*, 167–184. [CrossRef]

biology

MDPI

*Article*

# A Highly Diverse Olenekian Brachiopod Fauna from the Nanpanjiang Basin, South China, and Its Implications for the Early Triassic Biotic Recovery

**Huiting Wu [1,2], Yang Zhang [2,3,*], Anfeng Chen [3] and Thomas L. Stubbs [4]**

1 School of Geoscience and Surveying Engineering, China University of Mining and Technology (Beijing), Beijing 100083, China; ht_wu415@163.com
2 State Key Laboratory of Palaeobiology and Stratigraphy, Nanjing Institute of Geology and Palaeontology, CAS, Nanjing 210008, China
3 School of Earth Sciences and Resources, China University of Geosciences (Beijing), Beijing 100083, China; af_chen18@163.com
4 School of Earth Sciences, University of Bristol, Bristol BS1 5QD, UK; tom.stubbs@bristol.ac.uk
* Correspondence: zhangy@cugb.edu.cn

**Simple Summary:** Brachiopods have been thought to be in very low diversity in the Early Triassic for a long time. There are only several Olenekian brachiopod fauna reported worldwide, all of which are in very low diversity. This paper reports the most diverse Olenekian brachiopod fauna so far, containing 14 species with nine genera. Among them, three new species are proposed, and six genera are found in the Early Triassic for the first time. This diverse fauna indicates that the diversity of Olenekian brachiopod fauna has been underestimated. Based on precise age constrained by conodont biostratigraphy and quantitative data of brachiopod, it can be inferred that brachiopod recovery in the studied section occurred in the latest Spathian rather than the Smithian when the environment started to ameliorate. Global brachiopod data also indicates that the initial recovery of brachiopods happened in the Spathian.

**Abstract:** As one of the predominant benthic organisms in the Palaeozoic, brachiopod was largely eliminated in the Permian–Triassic boundary mass extinction, and then highly diversified in the Middle Triassic. Since fossil data from the Early Triassic are rarely reported, the recovery patterns of Early Triassic brachiopods remain unclear. This study documents a well-preserved fauna that is the most diverse Olenekian brachiopod fauna so far (age constrained by conodont biostratigraphy) from the Datuguan section of ramp facies in South China. This fauna is composed of 14 species within nine genera, including six genera (*Hirsutella*, *Sulcatinella*, *Paradoxothyris*, *Dioristella*, *Neoretzia* and *Isocrania*) found in the Early Triassic for the first time and three new species, including *Paradoxothyris flatus* sp. nov., *Hirsutella sulcata* sp. nov. and *Sulcatinella elongata* sp. nov. The Datuguan fauna indicates that the diversity of Olenekian brachiopod fauna has been underestimated, which can be caused by a combination of reduced habitats (in geographic size and sedimentary type) compared with the end-Permian, great bed thickness making it difficult to find fossils and most species in the fauna having low abundance. Based on the faunal change in the Datuguan section and environmental changes in South China, it can be inferred that brachiopod recovery in the studied section occurred in the latest Spathian rather than the Smithian when the environment started to ameliorate. Global brachiopod data also indicates that the initial recovery of brachiopods happened in the Spathian, and many genera that widely occurred in the Middle or Late Triassic had originated in the Olenekian.

**Keywords:** Early Triassic; brachiopod; biotic recovery; benthos; Nanpanjiang basin; Datuguan section

## 1. Introduction

The Permian–Triassic boundary mass extinction (PTBME) event greatly disrupted marine ecosystems, which transformed from 'Palaeozoic-type' to 'Mesozoic- and Cenozoic-type' fauna [1–7]. The hostile environmental conditions (e.g., deadly high temperature [8]; anoxia event [9]) caused by volcanism were proposed to trigger this major biological crisis, and then these persistent environmental disasters and significantly decreased diversity limited the biotic recovery process in the post-extinction interval. Brachiopods were largely eliminated [10], but this gave way to a new evolutionary stage, and brachiopods subsequently evolved modest taxonomic, morphological, functional and ecological diversity in the Mesozoic and Cenozoic [11,12]. New ecomorphologies appeared, with changes from reclining and anchoring to pedicle-fixing ecologies [11], and changes from brachial ridges and spiralia to loops, spiralia and crura in the mineralised lophophore supports [12]. Through the whole Triassic, the Athyridida, Spiriferinida, Terebatulida and Rhynchonellida, which are characterised by the pedicle fixing type and loops, spiralia or crura supports, are the dominant brachiopod orders.

In the post-extinction interval, Induan brachiopod fauna is mostly reported from South China, characterised by the transient Permian holdovers (e.g., *Paryphella, Fusichonetes, Prelissorhynchia, Paracrurithyris*) [13–15], lingulids [16,17] and a few newcomers (e.g., *Meishanorhynchia, Lichuanorelloides*) [18,19]. In the Olenekian, brachiopod fauna is rarely reported (Idaho, western USA [20,21]; Qilian Area, north-western China [22]; Primorye, Russia [23]; Romania [24]; and Tibet, China [25]), and have very low diversity in most of these areas. Up until now, glimpses into the evolutionary dynamics of Early Triassic brachiopods (excluding Permian holdovers) show very low diversity and few occurrences. If this is true, what constraints (such as global warming, anoxia, and biotic interaction) brachiopod diversity and recovery patterns in the Early Triassic?

To investigate the early evolution of Triassic brachiopods, this study reports an Olenekian brachiopod fauna from the Datuguan section, Nanpanjiang Basin, southern Guizhou Province, South China. It is currently the most diverse fauna when compared to global contemporaries. This paper also provides several brachiopod fossil data models to outline how the main drivers (e.g., extinction event, environmental factors, sampling bias) influenced brachiopod diversity and recovery in the Early Triassic.

## 2. Geological Settings and Age

South China (especially the Yangtze Block part) is one of the few regions in the world yielding successive Lower to Middle Triassic strata, spanning a continuum of depositional environments from the nearshore clastic shelf, carbonate platform, offshore clastic shelf, ramp, isolated carbonate platform and basin [26]. Abundant trace fossils [27], and abundant and diversified marine organisms have been reported from these strata (e.g., gastropods [28–30]; brachiopods [18,31]; ammonoids [32,33]; ostracods [34]; bivalves [35]; and foraminifers [36]).

The Datuguan section is located 120 km south of Guiyang City and 5 km north of Luodian County. The section occurs on the southern ramp (below the storm wave base) of the Great Guizhou Bank from the Changhsingian (Late Permian) to Middle Triassic (Figure 1). At the Datuguan section, the upper Changhsingian strata belong to the Linghao Formation, which mainly contains dark grey thin-bedded siliceous mudstone, yellow-green thin-bedded calcareous mudstone and grey thick-bedded micritic limestone. The Induan and Olenekian strata are represented by the Luolou Formation, which is characterised by fawn medium-bedded calcareous mudstone and greyish-green medium-bedded siltstone, intercalated with dark grey medium-bedded micritic limestone. The Anisian strata are represented by the Xuman Formation, which mainly includes greyish-green tuff (only at the bottom), greyish-green medium-bedded siltstone, fawn medium-bedded calcareous mudstone, intercalated with dark grey medium-bedded micritic limestone (Figure 2).

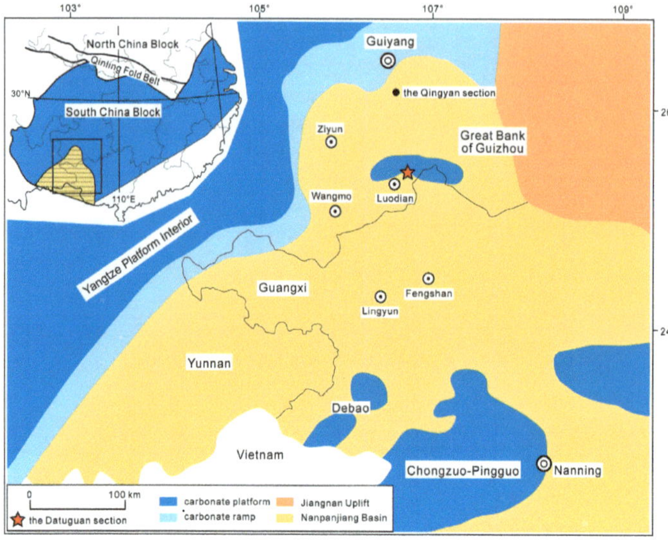


**Figure 1.** Early Triassic palaeogeographical map of Nanpanjiang Basin, South China (modified after [37]), showing the location of the Datuguan section and Qingyan section.

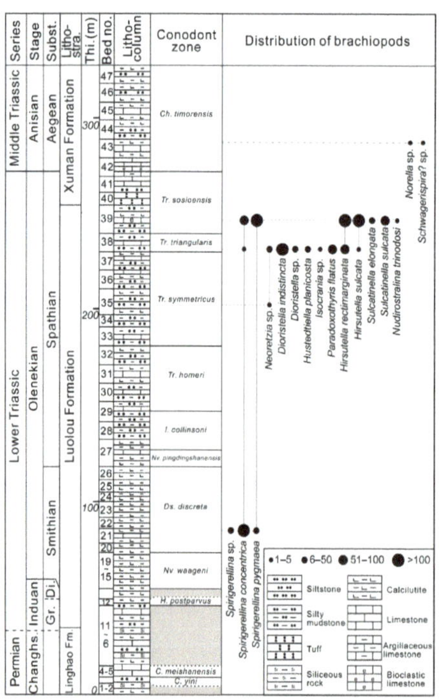


**Figure 2.** Distribution of brachiopods and zonations of conodont from the Linghao, Luolou and Xuman formations in the Datuguan section. Conodont data are from [38]. Subst., Substage; Lithostra., Lithostratigraphy; Thi., Thickness; Changhs., Changhsingian; Gr., Griesbachian; Di., Dienerian; Fm., Formation; *C.*, *Clarkina*; *H.*, *Hindeodus*; *Nv.*, *Novispathodus*; *Ds.*, *Discretella*; *I.*, *Icriospathodus*; *Tr.*, *Triassospathodus*, *Ch.*, *Chiosella*.

Strata of the Datuguan section from the Changhsingian (uppermost Permian) to the Anisian (lower Middle Triassic) are precisely defined by successive conodont biostratigraphy [38]. Based on the first occurrence of *Novispathodus waageni*, *Nv. pingdingshanensis* and *Chiosella timorensis*, the Induan–Olenekian boundary, the Smithian–Spathian boundary and the Olenekian–Anisian boundary are placed at the bottom of Bed 16, Bed 27 and the middle of Bed 42, namely the bottom of the Luolou Formation, the middle of the Luolou Formation and the bottom of the Xuman Formation (Figure 2).

## 3. Materials and Methods

In total, 1583 complete brachiopod specimens were collected from the Luolou Formation (Beds 21, 35, 38 and 39) and basal part of the Xuman Formation (Bed 46). All the specimens (accessible upon request from the corresponding author) are and will be permanently deposited in the Laboratory of Palaeontology, College of Geoscience and Surveying Engineering, China University of Mining and Technology, Beijing, China, with the prefixes LD.

To estimate completeness of sampling, a rarefaction analysis was applied and conducted using PAST (Palaeontological Statistics [39]). In order to investigate the changes to craniformean and rhychonelliformean brachiopod diversity and abundance from the Wuchiapingian (Late Permian) to the Anisian (Middle Triassic), brachiopod genera and occurrence data were collected from the Paleobiology Database (PBDB) (http://paleobiodb.org, up to 30 June 2022) and Treatise on Invertebrate Palaeontology Part H: Brachiopoda, Volume 2–6 [40–43]. When counting the occurrence frequency of a genus, specimens occurred in the same section were counted as one occurrence. Since this paper mainly focused on the 'Mesozoic-type' brachiopod, the occurrence data of Lingulida were not collected. The data downloaded from the PBDB used the following parameters: time intervals = Changhsingian and Rhaetian, and Taxon = brachiopoda. All brachiopod data have been checked and revised according to the most recently accepted classifications, and records with uncertainty were not included herein.

## 4. Results

A total of 16 species in 11 genera are recognised in the Datuguan brachiopod fauna (Figure 2), including three newly proposed species, *Hirsutella sulcata* sp. nov., *Paradoxothyris flatus* sp. nov. and *Sulcatinella elongata* sp. nov. Among the 11 genera of the Datuguan fauna, six of them are found in the Early Triassic for the first time (*Hirsutella*, *Sulcatinella*, *Paradoxothyris*, *Dioristella*, *Neoretzia* and *Isocrania*), and occupied nearly a quarter of brachiopod genera reported in the Olenekian (Figure 3).

There are only several Olenekian brachiopod fauna found worldwide so far (Table 1), mainly including those reported from western North America [20,21], the Balkan region [24], the Far East of Russia [23], north-western China [22], and Tibet [25]. Compared with those fauna, the Datuguan brachiopod fauna shows the highest richness (14 species in nine genera in the Olenekian) (Table 1, Figures 4–8).

**Table 1.** Main Olenekian brachiopod fauna worldwide, fauna with only one or two species not included herein.

| Location | Genus/ Species | Order | Age | Reference |
|---|---|---|---|---|
| Primorye, Russia | 6/6 | Rhynchonellida, Terebratulida, Spiriferinida, Athyridida | Olenekian | [23] |
| Qilian Area, China | 4/10 | Rhynchonellida, Terebratulida, Athyridida | Olenekian | [22] |
| Tibet, China | 3/3 | Rhynchonellida, Terebratulida, Athyridida | Smithian | [25] |
| Idaho, USA | 4/5 | Rhynchonellida, Terebratulida, Spiriferinida | Spathian | [20,21] |
| Dobrogea, Romania | 3/3 | Rhynchonellida | Spathian | [24] |
| Guizhou, China | 9/14 | Rhynchonellida, Terebratulida, Spiriferinida, Athyridida, Craniida | Olenekian | This study |

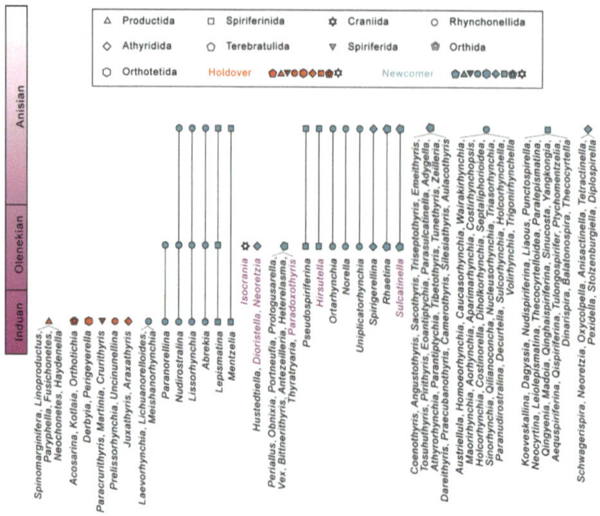


**Figure 3.** Occurrences of brachiopod genera (Lingulida excluded) worldwide during the Early and Middle Triassic based on data from the Datuguan section and PBDB. Genera shown in purple are described from the Olenekian for the first time in this paper.

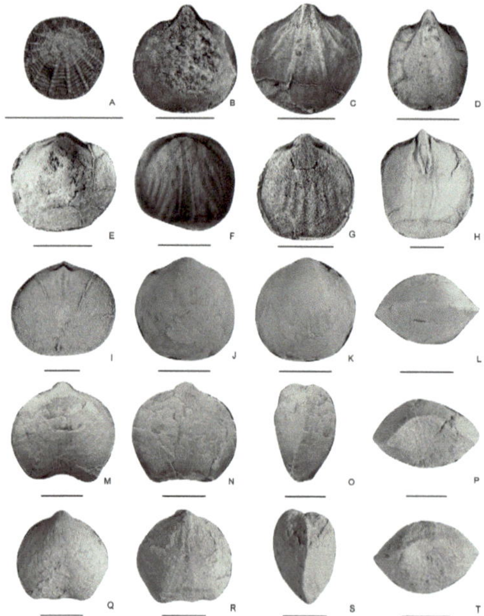

**Figure 4.** (**A**), *Isocrania* sp., ventral valve, LD380288. (**B–E**), *Dioristella indistincta*, (**B–D**), internal moulds of ventral valve, LD380119, LD380219, LD380396; I, an internal mould of a dorsal valve, LD380413. (**F,G**), *Dioristella* sp., internal moulds of dorsal valves, LD380200, LD380141. (**H–L**), *Spirigerellina concentrica*, (**H**), an internal mould of a ventral valve, LD385533; (**I**), an internal mould of a dorsal valve, LD385605; (**J–L**), ventral, dorsal and anterior views of a conjoined shell, LD385332. (**M–T**), *Spirigerellina pygmaea*, ventral, dorsal, lateral and anterior views of two conjoined shells, LD385207, LD385260. In (**A–L**), scale bar = 2 mm, in (**M–T**), scale bar = 5 mm.

**Figure 5.** (**A**,**B**), *Spirigerellina* sp., (**A**), an internal mould of a ventral valve, LD215553; (**B**), an internal mould of a dorsal valve, LD215554. (**C**,**D**), *Hustedtiella planicosta*, (**C**), an internal mould of a dorsal valve, LD380056; (**D**), an internal mould of a ventral valve, LD380272. (**E**), *Schwagerispira*? sp., a ventral valve, LD435623. (**F**,**G**), *Neoretzia* sp., (**F**), an external mould of a ventral valve, LD380267, (**G**), an internal mould of a dorsal valve, LD380172. (**H**), *Norella* sp., a ventral valve, LD435517. (**I–L**), *Nudirostralina trinodosi*, ventral, dorsal, anterior and lateral views of a conjoined shell, LD381185. Scale bar = 2 mm

## 5. Systematic Palaeontology

Order Spiriferinida Ivanova, 1972a [44]
Suborder Cyrtinidina Carter and Johnson in Carter, et al., 1994 [45]
Superfamily Suessioidea Waagen, 1883 [46]
Family Bittnerulidae Schuchert, 1929 [47]
Subfamily Hirsutellinae Xu and Liu in Yang, et al., 1983 [22]
Genus *Hirsutella* Cooper and Muir-Wood, 1951 [48]
*Hirsutella sulcata* sp. nov. (Figure 6F–L)

Derivation of name. In reference to the prominent sulcus in the ventral valve.

Diagnosis. Rounded subpentagonal to semicircular outline, distinct ventral sulcus with an extended median tongue.

Type specimen. Holotype, a ventral valve (LD384281, Figure 6K); paratype, a ventral valve (LD383719, Figure 6F).

Other material. A ventral valve (LD384237), a dorsal valve (LD383526).

Description. Shell medium in size for genus, 7.68–13.5 mm in length and 9.12–16.6 mm in width (12 specimens measured), rounded subpentagonal to semicircular in outline; maximum width at about midvalve; hinge slightly shorter than shell width; cardinal extremities rounded. Ventral valve moderately convex; umbo highly elevated and inflated, moderately incurved; beak narrow and pointed; sulcus beginning from umbo or midvalve, widening and deepening anteriorly, with a median tongue distinctly extended. Dorsal valve moderately convex, umbo slightly over hinge; lateral slopes strongly inclined, fold wide and elevated, not well demarcated from lateral slopes. External surface covered with costae, obscure at umbonal region, and even invisible when the surface layer of the shell is peeled off.

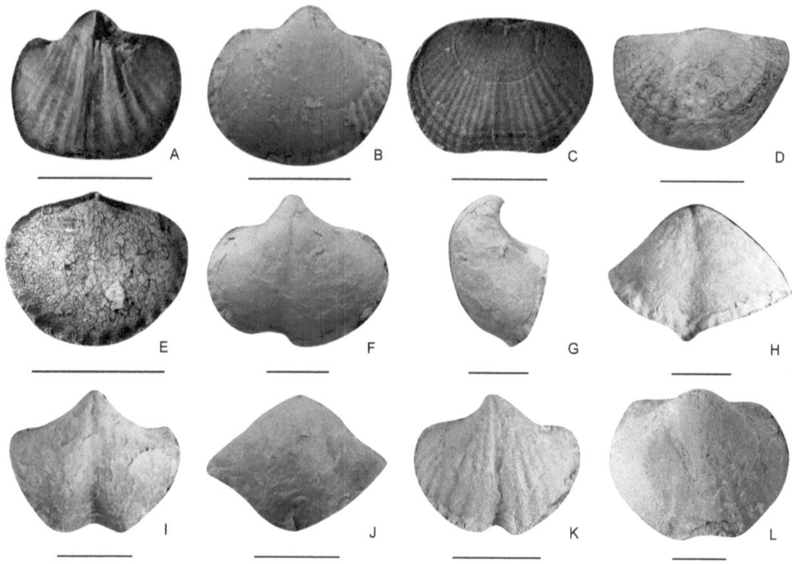


**Figure 6.** (**A–E**), *Hirsutella rectimarginata*, (**A**), an internal mould of a ventral valve, LD380429; (**B**), a ventral valve, LD383507; (**C**), an external mould of a dorsal valve, LD380440; (**D**), a dorsal valve, LD384020; (**E**), an internal mould of a dorsal valve, LD380253. (**F–L**), *Hirsutella sulcata* sp. nov., (**F–K**), ventral valves, (**F**), LD383719, (**I**), LD384237, (**K**), LD384281; (**G,H**), lateral and anterior views of (**F**); (**J**), anterior view of (**I**); (**L**), a dorsal valve, LD383526. Scale bar = 5 mm.

Remarks. The present species is similar to *Hirsutella extraruga* (Yang and Yin in Yang, et al., 1962) [49] in shell outline and lateral profile, but it has a more distinct sulcus and an extended median tongue occasionally developed in the ventral valve. *Hirsutella hirsuta* (Alberti, 1864) [50] is similar to the present species in shell outline and development of ventral sulcus, but differs by having a more elevated ventral beak. The current species is similar to *Sinucosta bifucata* Sun and Shi, 1985 [51] from the upper Triassic of Yunnan, China, in the rounded subpentagonal outline and moderately developed costae, but the former has a wider hinge, coarser costae and more distinct fold and sulcus. The Datuguan specimens resemble *Mentzelia subspherica* Yang and Xu, 1966 [52] from the Anisian of Guizhou, south China, in semicircular outline and sulcus beginning from beak, but the latter has more costae than most of the Datuguan specimens and developed spondylium. It is similar to *Dagyssia multicostata* (Yang and Xu, 1966) [52] from Qinghai, China, in the similar outline and feebly developed costae, but the latter has less developed sulcus and fold and more number of costae.

Distribution. Olenekian; China.

Order Terebratulida Waagen, 1883 [46]
Suborder Terebratulidina Waagen, 1883 [46]
Superfamily Dielasmatoidea Schuchert, 1913 [53]
Family Angustothyrididae Dagys, 1972b [54]
Genus *Paradoxothyris* Xu, 1978 [55]
*Paradoxothyris flatus* sp. nov. (Figure 7)

Derivation of name. In reference to the low convexity of both valves.

Diagnosis. Rounded lateral and anterior margins, variably developed median ridges on both valves, both valves slightly convex or nearly flattened, sometimes with regularly distributed costellae.

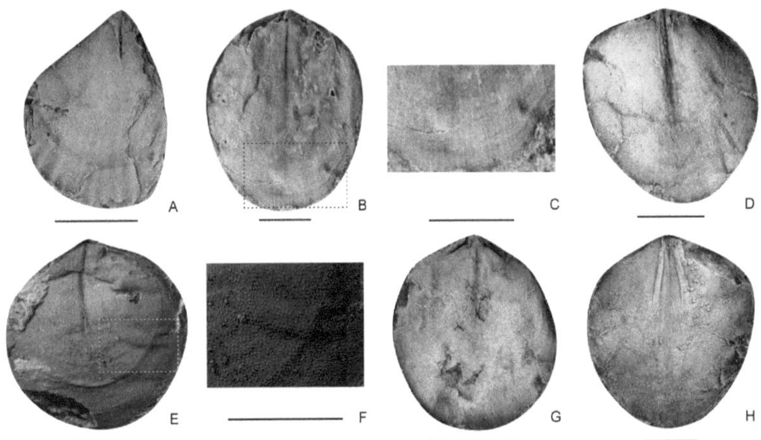


**Figure 7.** *Paradoxothyris flatus* sp. nov., (**A**,**B**), internal moulds of ventral valves, LD380016, LD380303, (**C**), enlarged area of (**B**); (**D**–**H**), internal moulds of dorsal valves, LD380100, LD380419, LD380125, LD380204, (**F**), enlarged area of (**E**). Scale bar = 2 mm.

Type specimen. Holotype, an internal mould of a ventral valve (LD380303, Figure 7B); paratype, an internal mould of a dorsal valve (LD380419, Figure 7E).

Other material. An internal mould of ventral valve (LD380016), and three internal moulds of dorsal valves (LD380100, LD380125, LD380204).

Description. Shell small to medium in size for genus, 2.92–9.44 mm in length and 2.25–8.39 mm in width (16 specimens measured), elongated suboval in outline; maximum width at middle to the anterior part of the shell. Ventral valve slightly convex to nearly flat; maximum convexity at umbo; posterior margin V-shaped, lateral and anterior margins very rounded; sulcus absent; interior with a weak median ridge beginning from beak and extending to one-fifth to a half of shell length, and absent in some specimens. Dorsal valve nearly flat; margins curved; fold absent; interior with a median ridge beginning from beak, and extending to about one-sixth to one-half of shell length; sockets long and narrow, and inner socket ridges thin and diverging at an angle of about 105°. Shell punctate; external surface sometimes ornamented with fine and dense costellae at the middle to the anterior part of the shell.

Remarks. The present species is similar to *Paradoxothyris cyclis* Xu, 1978 [55], *Paradoxothyris sangkaensis* (Jin, et al., 1979) [56] and *Paradoxothyris pentagona* (Jin, et al., 1979) [56] in the absence of a fold and sulcus, but differs in having a much less convex ventral valve and almost flat dorsal valve.

Distribution. Olenekian; China.

Family Dielasmatidae Schuchert, 1913 [53]
Subfamily Dielasmatinae Schuchert, 1913 [53]
Genus *Sulcatinella* Dagys, 1974 [57]
*Sulcatinella elongata* sp. nov. (Figure 8I–T and Figure 9)

Derivation of name. In reference to the elongated outline.

Diagnosis. Elongated rhombic to subpentagonal outline, distinctly inclined lateral slopes of ventral valve, strong unisulcate anterior commissure.

Type specimen. Holotype, a conjoined shell (LD384867, Figure 8Q–T), paratype, a conjoined shell (LD384907, Figure 8M–P).

Other material. A conjoined shell (LD384755).

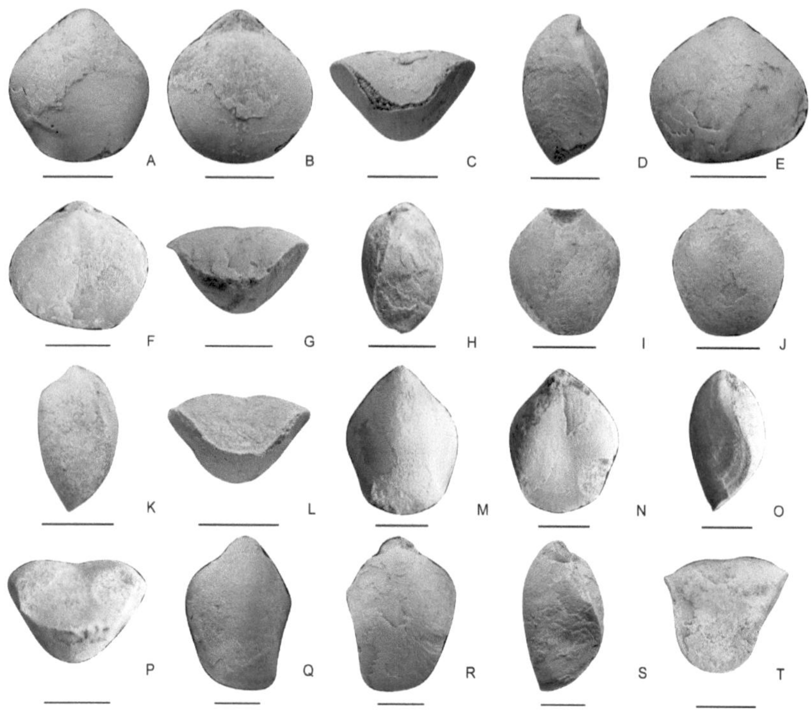


**Figure 8.** (**A–H**), *Sulcatinella sulcata*, ventral, dorsal, lateral and anterior views of two conjoined shells, LD384756, LD384786. (**I–T**), *Sulcatinella elongata* sp. nov., ventral, dorsal, lateral and anterior views of a conjoined shell, LD384755, LD384907, LD384867. Scale bar = 5 mm.

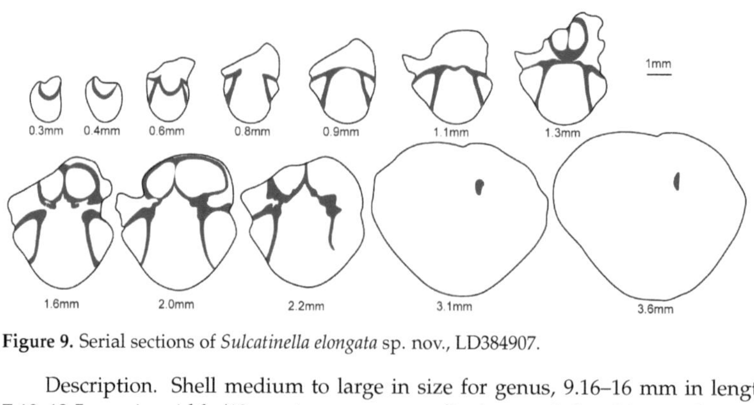


**Figure 9.** Serial sections of *Sulcatinella elongata* sp. nov., LD384907.

Description. Shell medium to large in size for genus, 9.16–16 mm in length and 7.19–12.5 mm in width (12 specimens measured), elongated rhombic to subpentagonal in outline; lateral commissure moderately to strongly incurved towards the dorsal side, anterior commissure strong unisulcate. Ventral valve moderately convex; beak slightly curved; posterior margin V-shaped, lateral and anterior margins straight to slightly curved; fold elevated, beginning from the umbonal region and widening anteriorly; lateral slopes flattened to slightly convex, distinctly inclined towards dorsal valve; interior with distinct and short pedicle collar; dental plates slightly diverging at an angle of about 30° (Figure 9). Dorsal valve slightly to moderately convex; sulcus strong, originating from umbo, distinctly widening and deepening anteriorly, strongly bending towards ventral valve at anterior part; interior with distinct and large crural bases, inner hinge plates converging at an angle of about 65° to form a V-shaped septalium, connected with median septum, septalium and septum disappear at about the same time (Figure 9).

Remarks. Shell length, width and thickness of ventral valve of *S. sulcata* and *S. elongata* specimens from the studied section are measured. The length-to-width ratio is adopted to represent the shell outline, and the thickness of ventral valve-to-width ratio is used to represent the inclination of lateral slopes of ventral valve. It is shown that the present species differs from *S. sulcata* by having a more elongated outline and more strongly inclined lateral slopes of ventral valve (Figure 10). It is similar to *S. incrassata* by Grădinaru and Gaetani [24] in the elongated subpentagonal outline and shell convexity, but differs by having a much more curved lateral commissure and wider dorsal sulcus. The Datuguan specimens resemble *Angustothyris qingyanensis* Guo et al., 2020 [31] from the Anisian in having an elongated outline, unisulcate anterior commissure and smooth shells, but differ by having distinctly developed dental plates and strongly declined lateral slopes of ventral valve.

Distribution. Olenekian; China.

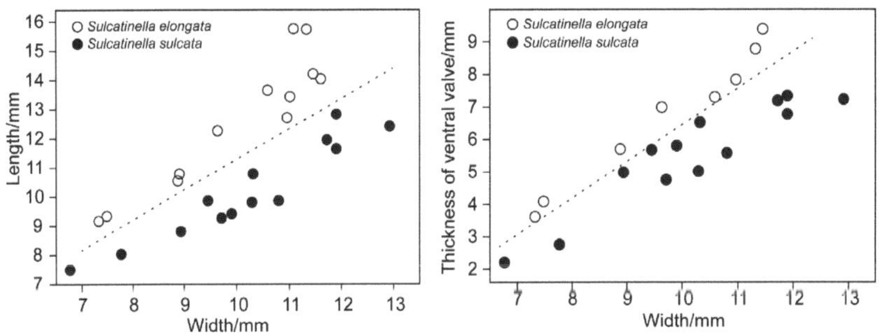


**Figure 10.** Graph of shell length to shell width and thickness of ventral valve to width of *S. sulcata* and *S. elongata* from the studied section.

## 6. Discussion

### 6.1. The Hidden Diversity in the Early Triassic

Olenekian brachiopod fauna was thought to be in very low diversity for a long time (as is shown in Table 1, wherein the species richness index was chosen to measure diversity); however, this newly discovered Luodian fauna shows a very high diversity (14 species in nine genera) and thus indicates a very likely hidden brachiopod diversity in the Early Triassic. Pietsch, et al., [58] referred to the hidden echinoid diversity of the Early Triassic. Massive diversity losses during the extinction event, coupled with hidden diversity in the recovery, result in evolutionary bottlenecks. If we examine the diversity changes of craniformean and rhynchonelliformean brachiopods from the Lopingian to Triassic, an evolutionary bottleneck existed in all the brachiopod orders Rhynchonellida, Spiriferinida, Terebratulida and Athyridida, which all have their lowest diversity in the Early Triassic [12,40–43] (Figures 3 and 11). Evident evolutionary bottlenecks widely occurred in marine organisms during the Early Triassic (e.g., radiolarians, foraminiferous, ammonoids [6]), and are a typical evolutionary pattern for the transitional interval between the mass extinction and subsequent completed ecosystem recovery. Generally, there are several parameters which can result in the phenomenon of hidden diversity, including the reduced habitat in geographic size and sedimentary type, taxa abundance and great bed thickness.

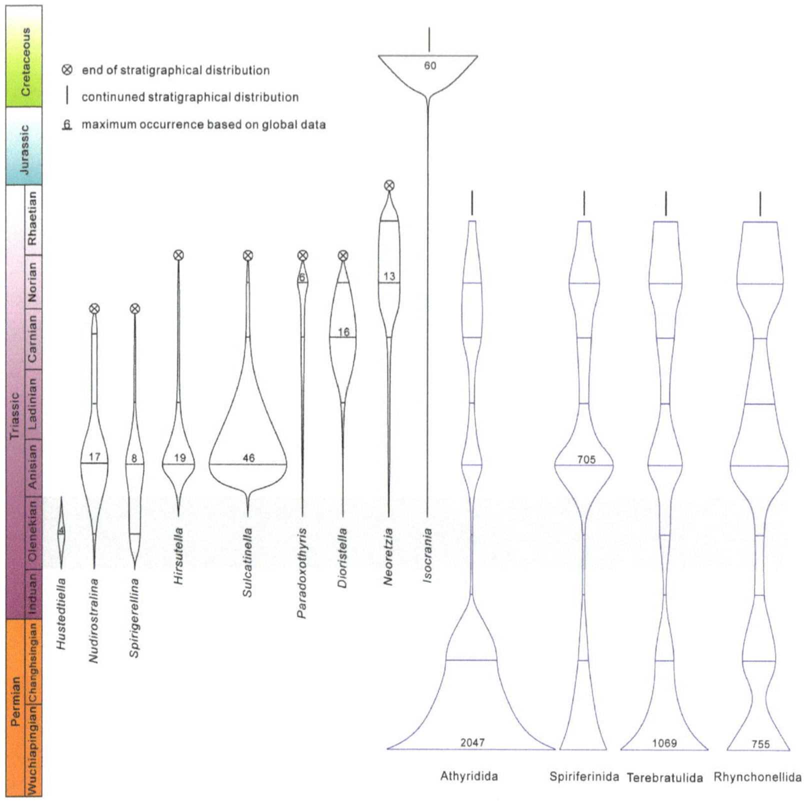


**Figure 11.** Stratigraphic ranges and occurrences of the Datuguan brachiopod genera (shown by black lines) and four brachiopod orders (shown by purple lines). Numbers above lines represent the highest occurrence frequency of genera.

During the Early Triassic, persistently deteriorated ocean environments made habitats hostile for marine organisms [8,9,59,60], especially for the benthos, and there were only a few 'habitable zones' in some specific environments [61–63]. In this case, the habitats of craniformean and rhynchonelliformean brachiopods were significantly reduced in the Early Triassic. The largely shrunken habitats in the Early Triassic oceanic environment would clearly reduce brachiopod abundance and the probability of fossil preservation and discovery, and could lead to considerable underestimation of Early Triassic brachiopod abundance and diversity.

How does taxa abundance affect diversity? We chose the Datuguan brachiopod fauna (Spathian, this study) and Jianzishan brachiopod fauna (Dienerian [19]) as examples to demonstrate the abundance model. As shown in Figure 12A,B, the brachiopod communities of Beds 38 and 39 are both characterised by one or two dominant species, and more than half of the species have very low abundance (less than 20 in Bed 38). The Jianzishan brachiopod fauna, which contains *Lichuanorelloides lichuanensis* (212 specimens), *Lissorhynchia* sp. (86 specimens) and *Crurithyris* sp. (eight specimens) [19], is a typical Induan-type fauna with an absolute dominant taxon. In this case, Early Triassic brachiopod fauna, which contains a dominant species and many low-abundance species, is very likely to underestimate diversity due to inadequate sampling

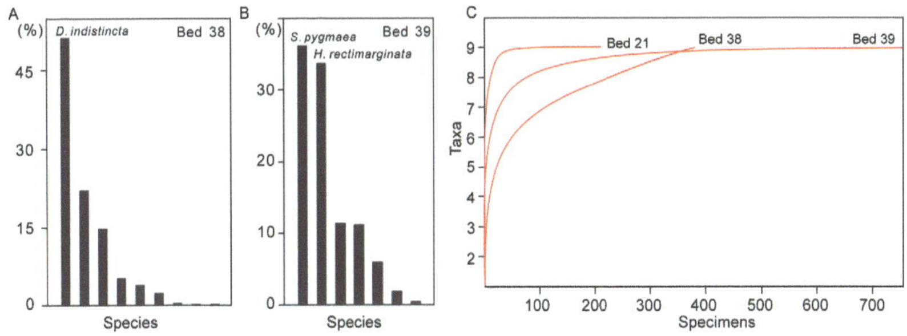


**Figure 12.** (**A**,**B**), frequency distribution of brachiopods from bed 38 (nine species) and bed 39 (seven species) in the Datuguan section; (**C**), results of rarefaction analysis of brachiopod data from three main beds yielding brachiopods in the Datuguan section. *D*: *Dioristella*; *S*: *Spirigerellina*; *H*: *Hirsutella*.

As for bed thickness, it is reasonable to speculate that fossil sampling is more difficult in thicker strata of the same duration. In South China, the Early Triassic strata are much thicker than the end-Permian strata. Most of the Early–Middle Triassic brachiopod fauna in South China are reported from ramp environments (the Yinkeng Formation [18]; the Xinyuan Formation [64]; the Daye Formation [19]; and the Qingyan Formation [30]), where much thinner strata are yielded than in shallow water settings. Weak hydrodynamic conditions in the deeper water environments could help to preserve fossils, and thinner strata would clearly increase the chance of fossil discovery.

To summarise, Early Triassic marine ecosystems, which existed between the collapse of Palaeozoic-type ecosystems and the final reconstruction of Mesozoic type ecosystems, are characterised by high dominance and low evenness. The great loss of diversity (extinction event) [65,66] and hostile environments [8,9] led to the high dominance of certain taxa (disasters, opportunists, newcomers) within these fauna, which further limited the abundance of other species. This pattern is one of the most important features of Early Triassic marine ecosystems, and could be one of the main reasons for substantially underestimated diversity. We examined the occurrences of nine genera found in the Olenekian from the studied section, and discovered that they all have the fewest records in the Olenekian, and six of them were first reported in the Early Triassic (Figure 11). In this case, adequate sampling (1583 complete specimens, Figure 12C) and condensed strata (the Luolou Formation in the ramp) should minimise the effect of sampling biases on brachiopod faunal diversity.

### *6.2. Brachiopod Recovery Pattern in the Post-Extinction Interval*

In the post-extinction interval, the Induan (especially Griesbachian) brachiopod fauna was mainly characterised by holdovers (26 genera), and the newcomers, which included a few genera of Rhynchonellida (seven genera) and Spiriferinida (two genera) (Figure 3). If we exclude the holdovers in the earliest Induan fauna, the Smithian and most Induan brachiopod fauna generally have very low diversity [18,19,25,67], which should be categorised in the 'survival stage' of the overall recovery process.

Most of these previously reported Olenekian fauna are mainly composed of Rhynchonellida, Terebratulida, Spiriferinida and Athyridida, except for that from Dobrogea (Romania), which only contains Rhynchonellida [24]. The faunal composition at the order level is consistent with the statistical data of brachiopod genera in the Triassic [43], and represents the initial stage of brachiopod evolution in the post-extinction interval. The Datuguan brachiopod fauna only includes Athyridida (three species within one genus) in the Smithian, and is dominated by Athyridida, Spiriferinida and Terebatulida in both species (13 species within nine genera), and specimen counts in the Spathian (Figure 13). This suggests that the brachiopod recovery occurred in the Spathian rather than Smithian, which is consistent with the global biotic recovery event for this substage [68–70].

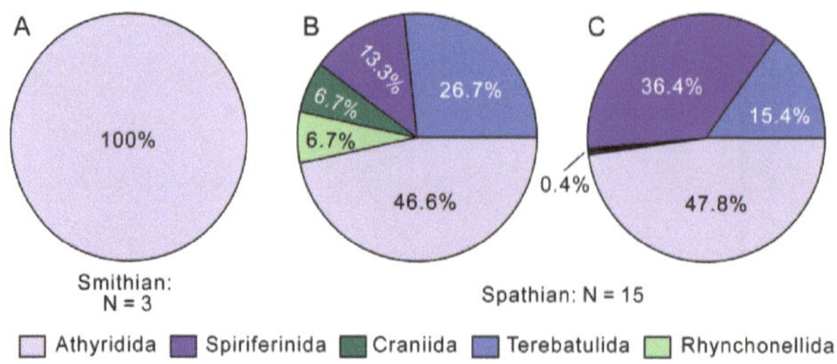


**Figure 13.** Composition of the Datuguan brachiopod fauna. (**A**,**B**) are based on species amount data, and (**C**) is based on specimen data. N, number of species.

The fossil horizons (Beds 38 and 39) of the Datuguan section yielding abundant brachiopods are restricted to the *Triassospathodus triangularis* and *Tr. sosioensis* conodont zones (Figure 2), which indicate a latest Spathian Age [71–73]. According to [8], the lethally hot temperatures started to fall, and a cooling event occurred in the latest Spathian (upper part of *Tr. homeri* Zone). Based on the evidence from pyrite framboids, the redox condition in ramp settings also started to improve from an anoxic–lower dysoxic to upper dysoxic–oxic environment in the latest Spathian of the Qingyan section (northern margin of Nanpanjiang Basin) (upper part of *Tr. homeri* Zone [9]) (Figure 1). As for the benthos, the Datuguan brachiopod fauna started to diversify in this improved habitat in the latest Spathian. The deteriorated environments (e.g., deadly temperatures, anoxia) in low latitudinal areas since the latest Permian evidently started to return to normal conditions after nearly five million years [8,9,59,60], and the reconstruction of the Triassic marine ecosystem truly began in the latest Spathian.

Unexpectedly, based on global data from online databases and published works of literature, the recovery rate of Olenekian (especially Spathian) brachiopod fauna (28 genera) has been substantially undervalued. Brachiopods showed high diversity in the late Olenekian, which is three times that in the Induan, and more than one-third of that in the Anisian (84 genera), and is therefore indicative of an initial recovery in the Spathian. It is noteworthy that some brachiopod genera, which have their maximum occurrences in the Middle or Late Triassic, already started to appear in the Olenekian age (Figure 11).

In addition, the fact that most contemporaneous brachiopod fauna are reported from the Balkan region, Primorye, the Qilianshan region, the Nanpanjiang basin and Idaho, indicates that the Palaeo-Tethys Sea region and the western margin of North America provided the most important habitats for brachiopods in the Spathian Age. These areas were actually the most hostile habitats during the Permian–Triassic boundary mass extinction event [59,74,75]. Environmental amelioration in tropical regions indicates an overall improvement of global oceanic environments, which might then have given rise to the subsequent overall recovery during the Middle Triassic.

## 7. Conclusions

1. A Olenekian brachiopod fauna which is the most diverse one so far, is reported in this study. It contains 14 species in nine genera, among which *Hirsutella*, *Sulcatinella*, *Paradoxothyris*, *Dioristella*, *Neoretzia* and *Isocrania* are found in the Early Triassic for the first time, and three species are newly proposed;
2. Brachiopod abundance and diversity data indicated that brachiopod recovery in the studied section happened in the latest Spathian when the environmental condition (deadly temperatures and anoxia) started to ameliorate;
3. One of the reasons that brachiopod was widely considered to be in very low abundance in the Early Triassic was the phenomenon of hidden diversity. It could be caused by the decrease of habitat, low taxa abundance and great thickness of strata.

**Author Contributions:** Conceptualization Y.Z. and H.W.; Data Curation H.W., Y.Z. and A.C.; Formal Analysis Y.Z., H.W. and T.L.S.; Investigation Y.Z.; Project Administration Y.Z. and H.W.; Supervision Y.Z.; Visualization Y.Z. and H.W.; Writing—Original Draft Preparation H.W., Y.Z. and T.L.S.; Writing—Review and Editing H.W., Y.Z. and T.L.S. All authors have read and agreed to the published version of the manuscript.

**Funding:** This research was funded by the National Natural Science Foundation of China (Grant No. 41902008), Fundamental Research Funds for the Central University (Grant Nos. 2652018131, 00/800015A302), and State Key Laboratory of Palaeobiology and Stratigraphy (Nanjing Institute of Geology and Palaeontology, CAS) (Nos. 223137, 223131).

**Institutional Review Board Statement:** Not applicable.

**Informed Consent Statement:** Not applicable.

**Data Availability Statement:** Data available on request from authors.

**Acknowledgments:** The authors thank Yuanlin Sun for help in preparing macrofossils, Jing Dou and Jingqi Liu for field assistance. We are very grateful for the helpful reviews by two anonymous reviewers.

**Conflicts of Interest:** The authors declare no conflict of interest.

## References

1. Vermeij, G.J. The Mesozoic marine revolution: Evidence from snails, predators and grazers. *Paleobiology* **1977**, *3*, 245–258. [CrossRef]
2. Gould, S.J.; Calloway, C.B. Clams and brachiopods—Ships that pass in the night. *Paleobiology* **1980**, *6*, 383–396. [CrossRef]
3. Sepkoski, J.J. A factor analytic description of the Phanerozoic marine fossil record. *Paleobiology* **1981**, *7*, 36–53. [CrossRef]
4. Raup, D.M.; Sepkoski, J.J., Jr. Mass extinctions in the marine fossil record. *Science* **1982**, *215*, 1501–1503. [CrossRef]
5. Bambach, R.K.; Knoll, A.H.; Wang, S.C. Origination, extinction, and mass depletions of marine diversity. *Paleobiology* **2004**, *30*, 522–542. [CrossRef]
6. Chen, Z.Q.; Benton, M.J. The timing and pattern of biotic recovery following the end-Permian mass extinction. *Nat. Geosci.* **2012**, *5*, 375–383. [CrossRef]
7. Brayard, A.; Krumenacker, L.J.; Botting, J.P.; Jenks, J.F.; Bylund, K.G.; Fara, E.; Vennin, E.; Olivier, N.; Goudemand, N.; Saucede, T.; et al. Unexpected Early Triassic marine ecosystem and the rise of the modern evolutionary fauna. *Sci. Adv.* **2017**, *3*, e1602159. [CrossRef]
8. Sun, Y.D.; Joachimski, M.M.; Wignall, P.B.; Yan, C.B.; Chen, Y.L.; Jiang, H.S.; Wang, L.N.; Lai, X.L. Lethally hot temperatures during the Early Triassic greenhouse. *Science* **2012**, *338*, 366–370. [CrossRef]
9. Huang, Y.G.; Chen, Z.Q.; Wignall, P.B.; Zhao, L.S. Latest Permian to Middle Triassic redox condition variations in ramp settings, South China: Pyrite framboid evidence. *GSA Bull.* **2017**, *129*, 229–243. [CrossRef]
10. Chen, J.; Chen, Z.Q.; Tong, J.N. Environmental determinants and ecologic selectivity of benthic faunas from nearshore to bathyal zones in the end-Permian mass extinction: Brachiopod evidence from South China. *Palaeogeogr. Palaeoclimatol. Palaeoecol.* **2011**, *308*, 84–97. [CrossRef]
11. Wang, Y.; Jin, Y.G.; Rong, J.Y. General conception on the evolutional trends of Brachiopoda. *Acta Palaeontol. Sin.* **1982**, *21*, 28–36. (In Chinese)
12. Carlson, S.J. The evolution of Brachiopoda. *Annu. Rev. Earth Planet. Sci.* **2016**, *44*, 409–438. [CrossRef]
13. Chen, Z.Q.; Kaiho, K.; George, A.D. Survival strategies of brachiopod faunas from the end-Permian mass extinction. *Palaeogeogr. Palaeoclimatol. Palaeoecol.* **2005**, *224*, 232–269. [CrossRef]
14. He, W.H.; Shi, G.R.; Yang, T.L.; Zhang, K.X.; Yue, M.L.; Xiao, Y.F.; Wu, H.T.; Chen, B.; Wu, S.B. Patterns of brachiopod faunal and body-size changes across the Permian–Triassic boundary: Evidence from the Daoduishan section in Meishan area, South China. *Palaeogeogr. Palaeoclimatol. Palaeoecol.* **2016**, *448*, 72–84. [CrossRef]
15. Wu, H.T.; He, W.H.; Weldon, E.A. Prelude of benthic community collapse during the end-Permian mass extinction in siliciclastic offshore sub-basin: Brachiopod evidence from South China. *Glob. Planet. Chang.* **2018**, *163*, 158–170. [CrossRef]
16. Peng, Y.Q.; Shi, G.R. New Early Triassic Lingulidae (Brachiopoda) genera and species from South China. *Alcheringa* **2008**, *32*, 149–170. [CrossRef]
17. Zhang, Y.; Shi, G.R.; Wu, H.T.; Yang, T.L.; He, W.H.; Yuan, A.Y.; Lei, Y. Community replacement, ecological shift and early warning signals prior to the end-Permian mass extinction: A case study from a nearshore clastic-shelf section in South China. *Palaeogeogr. Palaeoclimatol. Palaeoecol.* **2017**, *487*, 118–135. [CrossRef]
18. Chen, Z.Q.; Shi, G.R.; Kaiho, K. A new genus of rhynchonellid brachiopod from the Lower Triassic of South China and implications for timing the recovery of Brachiopoda after the end-Permian mass extinction. *Palaeontology* **2002**, *45*, 149–164. [CrossRef]
19. Wang, F.Y.; Chen, J.; Dai, X.; Song, H.J. A new Dienerian (Early Triassic) brachiopod fauna from South China and implications for biotic recovery after the Permian–Triassic extinction. *Pap. Palaeontol.* **2017**, *3*, 425–439. [CrossRef]

20. Girty, G.H. Descriptions of new species of Carboniferous and Triassic fossils (from southeastern Idaho). *United States Geol. Surv. Prof. Pap.* **1927**, *152*, 411–446.
21. Perry, D.G.; Chatterton, B.D.E. Late Early Triassic brachiopod and conodont fauna, Thaynes Formation, southeastern Idaho. *J. Paleontol.* **1979**, *53*, 307–319.
22. Yang, Z.Y.; Yin, H.F.; Xu, G.R.; Wu, S.B.; He, Y.L.; Liu, G.C.; Yin, J.R. *Triassic of the South Qilian MTS*; Geological Publishing House: Beijing, China, 1983; p. 224, (In Chinese with English Abstract).
23. Popov, A.M.; Zakharov, Y.D. Olenekian brachiopods from the Kamenushka River Basin, South Primorye: New data on the brachiopod recovery after the end-Permian mass extinction. *Paleontol. J.* **2017**, *51*, 735–745. [CrossRef]
24. Grădinaru, E.; Gaetani, M. Upper Spathian to Bithynian (Lower to Middle Triassic) brachiopods from North Dobrogea (Romania). *Riv. Ital. Paleontol. Stratigr.* **2019**, *125*, 91–123.
25. Wang, F.Y.; Chen, J.; Dai, X.; Song, H.J. A new Early Triassic brachiopod fauna from southern Tibet, China: Implications on brachiopod recovery and the late Smithian extinction in southern Tethys. *J. Paleontol.* **2022**, *96*, 1–32. [CrossRef]
26. Feng, Z.Z.; Yang, Y.Q.; Jin, Z.K.; Li, S.W.; Bao, Z.D. *Lithofacies Paleogeography of Permian of South China*; Petroleum University Press: Dongying, China, 1997; p. 242, (In Chinese with English Abstract).
27. Zhang, L.J.; Buatois, L.A.; Mángano, M.G.; Qi, Y.A.; Tai, C. Early Triassic trace fossils from South China marginal-marine settings: Implications for biotic recovery following the end-Permian mass extinction. *GSA Bull.* **2019**, *131*, 235–251. [CrossRef]
28. Yin, H.F.; Yochelson, E.L. Middle Triassic Gastropoda from Qingyan, Guizhou Province, China: 1. Pleurotomariacea and Murchisoniacea. *J. Paleontol.* **1983**, *57*, 162–187.
29. Yin, H.F.; Yochelson, E.L. Middle Triassic Gastropoda from Qingyan, Guizhou Province, China: 2. Trochacea and Neritacea. *J. Paleontol.* **1983**, *57*, 515–538.
30. Yin, H.F.; Yochelson, E.L. Middle Triassic Gastropoda from Qingyan, Guizhou Province, China: 3. Euomphalacea and Loxonematacea. *J. Paleontol.* **1983**, *57*, 1098–1127.
31. Guo, Z.; Chen, Z.Q.; Harper, D.A.T. The Anisian (Middle Triassic) brachiopod fauna from Qingyan, Guizhou, south-western China. *J. Syst. Palaeontol.* **2020**, *18*, 647–701. [CrossRef]
32. Brühwiler, T.; Brayard, A.; Bucher, H.; Kuang, G.D. Griesbachian and Dienerian (Early Triassic) ammonoid faunas from northwestern Guangxi and southern Guizhou (South China). *Palaeontology* **2008**, *51*, 1151–1180. [CrossRef]
33. Dai, X.; Song, H.J.; Brayard, A.; Ware, D. A new Griesbachian–Dienerian (Induan, Early Triassic) ammonoid fauna from Gujiao, South China. *J. Paleontol.* **2019**, *93*, 48–71. [CrossRef]
34. Qiu, X.C.; Tian, L.; Wu, K.; Benton, M.J.; Sun, D.Y.; Yang, H.; Tong, J.N. Diverse earliest Triassic ostracod fauna of the non-microbialite-bearing shallow marine carbonate of the Yangou section, South China. *Lethaia* **2019**, *52*, 583–596. [CrossRef]
35. Song, T.; Tong, J.N.; Tian, L.; Chu, D.L.; Huang, Y.F. Taxonomic and ecological variations of Permian–Triassic transitional bivalve communities from the littoral clastic facies in southwestern China. *Palaeogeogr. Palaeoclimatol. Palaeoecol.* **2019**, *519*, 108–123. [CrossRef]
36. Liu, X.K.; Song, H.J.; Bond, D.P.G.; Tong, J.N.; Benton, M.J. Migration controls extinction and survival patterns of foraminifers during the Permian–Triassic crisis in South China. *Earth-Sci. Rev.* **2020**, *209*, 103329. [CrossRef]
37. Lehrmann, D.J.; Stepchinski, L.; Altiner, D.; Orchard, M.J.; Montgomery, P.; Enos, P.; Ellwood, B.B.; Bowring, S.A.; Ramezani, J.; Wang, H.M.; et al. An integrated biostratigraphy (conodont and foraminifers) and chronostratigraphy (paleomagnetic reversals, magnetic susceptibility, elemental chemistry, carbon isotopes and geochronology) for the Permian–Upper Triassic strata of Guandao section, Nanpanjiang Basin, South China. *J. Asian Earth Sci.* **2015**, *108*, 117–135.
38. Chen, A.F.; Zhang, Y.; Golding, M.L.; Wu, H.T.; Liu, J.Q. Upper Changhsingian to lower Anisian conodont biostratigraphy of the Datuguan section, Nanpanjiang Basin, South China. *Palaeogeogr. Palaeoclimatol. Palaeoecol.* **2023**, *616*, 111470. [CrossRef]
39. Hammer; Harper, D.A.T.; Ryan, P.D. PAST: Paleontological statistics software package for education and data analysis. *Palaeontol. Electron.* **2001**, *4*, 1–9.
40. Kaesler, R.L. *Treatise on Invertebrate Paleontology. Part H. Brachiopoda Vols 2 & 3. Linguliformea, Craniiforemea, & Rhynchonelliformea (Part)*; Geological Society of America & University of Kansas: Lawrence, KS, USA, 2000; p. 919.
41. Kaesler, R.L. *Treatise on Invertebrate Paleontology. Part H. Brachiopoda Vol. 4. Rhynchonelliformea (Part) (Revised)*; Geological Society of America & University of Kansas: Lawrence, KS, USA, 2002; pp. 921–1688.
42. Kaesler, R.L. *Treatise on Invertebrate Paleontology. Part H. Brachiopoda Vol. 5. Rhynchonelliformea (Part) (Revised)*; Geological Society of America & University of Kansas: Lawrence, KS, USA, 2006; pp. 1689–2320.
43. Selden, P.A. *Treatise on Invertebrate Paleontology. Part H. Brachiopoda Revised Vol. 6*; Geological Society of America & University of Kansas: Lawrence, KS, USA, 2007; pp. 2321–3226.
44. Ivanova, E.A. Osnovnye zakonomernosti evoliutsii spiriferid (Brachiopoda) (Main features of spiriferid evolution (Brachiopoda)). *Paleontol. Zhurnal* **1972**, *3*, 28–42.
45. Carter, J.L.; Johnson, J.G.; Rémy, G.; Hou, H.F. A revised classification of the spiriferid brachiopods. *Ann. Carnegie Mus.* **1994**, *63*, 327–374. [CrossRef]
46. Waagen, W.H. Salt Range fossils: Productus limestone fossils, Part 4 (1–5): Brachiopoda. *Palaeontol. Indica* **1883**, *13*, 329–770.
47. Schuchert, C. Classification of brachiopod genera, fossil and recent. 10–15. In *Fossilium Catalogus*; Pompeckj, J.F., Ed., W. Junk. Berlin, Germany, 1929; Volume 1.
48. Cooper, G.A.; Muir-wood, H.M. Brachiopod homonyms. *J. Wash. Acad. Sci.* **1951**, *41*, 195–196.

49. Yang, Z.Y.; Ding, P.Z.; Yin, H.F.; Zhang, S.X.; Fan, J.S. The brachiopod fauna of Carboniferous, Permian and Triassic in the Qilianshan region. In *Monograph on the Geology of the Qilian Mountains*; Institute of Geology and Palaeontology, Chinese Academy of Sciences, Institute of Geology, Chinese Academy of Sciences, Beijing College of Geology, Eds.; Science Press: Beijing, China, 1962; Volume 4, pp. 1–134. (In Chinese)
50. von Alberti, F.A. *Überblick Über die Trias, Mit Berücksichtigung Ihres Vorkommens in Den Alpen*; J.G. Cottaschen Buchhandlung: Stuttgart, Germany, 1864; Volume 8, p. 7.
51. Sun, D.L.; Shi, Q. Late Paleozoic and Triassic brachiopods from the east of the Qinghai-Xizang Plateau. In *Stratigraphy and Paleontology of Western Sichuan and Eastern Xizang China Part 3*; Regional Geological Survey Team of Sichuan Geological Bureau, Nanjing Institute of Geology and Paleontology, Eds.; Sichuan Science and Technology Press: Chengdu, China, 1985; pp. 186–212. (In Chinese)
52. Yang, Z.Y.; Xu, G.R. *Triassic Brachiopods of Central Guizhou Province, China*; China Industry Publishing House: Beijing, China, 1966; p. 151, (In Chinese with English Summary).
53. Schuchert, C. Class 2. Brachiopoda. *Text-Book Palaeontol.* **1913**, *1*, 355–420.
54. DAGYS, A.S. Postembrional'noe razvitie brakhidiia pozdnepaleozoiskikh i rannemezozoiskikh Terebratulida. *Akad. Nauk. SSSR Sib. Otd. Inst. Geol. I Geofiz. Tr.* **1972**, *112*, 22–58.
55. XU, Q.J. Brachiopoda, Mesozoic. 267–314. In *Paleontological Atlas of Southwest China, Sichuan Province, Volume 2, Carboniferous–Mesozoic*; Southwest Geological Science and Technology Research Institute, Ed.; Geological Publishing House: Beijing, China, 1978. (In Chinese)
56. Jin, Y.G.; Ye, S.L.; Xu, H.K.; Sun, D.L. Brachiopoda. 60–217. In *Paleontological Atlas of North-Western China, Qinghai Province*; Institute of Geological Science of Qinghai Province and Nanjing Institute of Geology and Palaeonotology, Chinese Academy of Sciences, Eds.; Geological Publishing House: Beijing, China, 1979; Volume 1. (In Chinese)
57. Dagys, A.S. *Triasovye Brakhiopody (Morfologiia, Sistema, Filogeniia, Stratigraficheskoe Znachenie I Biogeografiia) [Triassic Brachiopods (Morphology, Classification, Phylogeny, Stratigraphical Significance and Biogeography)]*; Sibirskoe Otdelenie Izdatel'stvo "Nauka.": Novosibirsk, Russia, 1974; Volume 214, pp. 1–49.
58. Pietsch, C.; Ritterbush, K.A.; Thompson, J.R.; Petsios, E.; Bottjer, D.J. Evolutionary models in the Early Triassic marine realm. *Palaeogeogr. Palaeoclimatol. Palaeoecol.* **2019**, *513*, 65–85. [CrossRef]
59. Wignall, P.B.; Bond, D.P.G.; Sun, Y.D.; Grasby, S.E.; Beauchamp, B.; Joachimski, M.M.; Blomeier, D.P.G. Ultra-shallow-marine anoxia in an Early Triassic shallow-marine clastic ramp (Spitsbergen) and the suppression of benthic radiation. *Geol. Mag.* **2016**, *153*, 316–331. [CrossRef]
60. Chen, J.; Shen, S.Z.; Zhang, Y.C.; Angiolini, L.; Gorgij, M.N.; Crippa, G.; Wang, W.; Zhang, H.; Yuan, D.X.; Li, X.H.; et al. Abrupt warming in the latest Permian detected using high-resolution in situ oxygen isotopes of conodont apatite from Abadeh, central Iran. *Palaeogeogr. Palaeoclimatol. Palaeoecol.* **2020**, *560*, 109973. [CrossRef]
61. Beatty, T.W.; Zonneveld, J.P.; Henderson, C.M. Anomalously diverse Early Triassic ichnofossil assemblages in northwest Pangaea: A case for a shallow-marine habitable zone. *Geology* **2008**, *36*, 771–774. [CrossRef]
62. Song, H.J.; Wignall, P.B.; Chu, D.L.; Tong, J.N.; Sun, Y.D.; Song, H.Y.; He, W.H.; Tian, L. Anoxia/high temperature double whammy during the Permian–Triassic marine crisis and its aftermath. *Sci. Rep.* **2014**, *4*, 4231. [CrossRef]
63. Foster, W.J.; Danise, S.; Sedlacek, A.; Price, G.D.; Hips, K.; Twitchett, R.J. Environmental controls on the post-Permian recovery of benthic, tropical marine ecosystems in western Palaeotethys (Aggtelek Karst, Hungary). *Palaeogeogr. Palaeoclimatol. Palaeoecol.* **2015**, *440*, 374–394. [CrossRef]
64. He, W.H.; Zhang, K.X.; Chen, Z.Q.; Yan, J.X.; Yang, T.L.; Zhang, Y.; Gu, S.Z.; Wu, S.B. A new genus Liaous of early Anisian Stage (Middle Triassic) brachiopods from southwestern China: Systematics, reassessment of classification of the Spiriferinioidea, community paleoecology, and paleoenvironmental implications. *J. Paleontol.* **2015**, *89*, 966–979. [CrossRef]
65. Hautmann, M.; Bagherpour, B.; Brosse, M.; Frisk, A.; Hofmann, R.; Baud, A.; Nützel, A.; Goudemand, N.; Bucher, H. Competition in slow motion: The unusual case of benthic marine communities in the wake of the end-Permian mass extinction. *Palaeontology* **2015**, *58*, 871–901. [CrossRef]
66. Feng, X.Q.; Chen, Z.Q.; Benton, M.J.; Su, C.M.; Bottjer, D.J.; Cribb, A.T.; Li, Z.H.; Zhao, L.S.; Zhu, G.Y.; Huang, Y.G.; et al. Resilience of infaunal ecosystems during the Early Triassic greenhouse Earth. *Sci. Adv.* **2022**, *8*, eabo0597. [CrossRef] [PubMed]
67. Wu, H.T.; Zhang, Y.; Sun, Y.L. A brachiopod fauna from latest Permian to Induan of northern Guizhou, South China and its evolutionary pattern. *Geol. J.* **2021**, *56*, 6189–6198. [CrossRef]
68. Hautmann, M.; Smith, A.B.; Mcgowan, A.J.; Bucher, H. Bivalves from the Olenekian (Early Triassic) of south-western Utah: Systematics and evolutionary significance. *J. Syst. Palaeontol.* **2013**, *11*, 263–293. [CrossRef]
69. Foster, W.J.; Lehrmann, D.J.; Yu, M.; Ji, L.; Martindale, R.C. Persistent environmental stress delayed the recovery of marine communities in the aftermath of the latest Permian mass extinction. *Paleoceanogr. Paleoclimatol.* **2018**, *33*, 338–353. [CrossRef]
70. Woods, A.D.; Alms, P.D.; Monarrez, P.M.; Mata, S. The interaction of recovery and environmental conditions: An analysis of the outer shelf edge of western North America during the Early Triassic. *Palaeogeogr. Palaeoclimatol. Palaeoecol.* **2019**, *513*, 52–64. [CrossRef]
71. Kozur, H.; Krainer, K.; Mostler, H. Neospathodus sosioensis n. sp.; a new conodont species from the late Olenekian (uppermost Scythian) of western Sicily, Italy. In *Proceedings of the Fifth Radiolarian Symposium*; Special Volume; News of Osaka Micropaleontologists: Osaka, Japan, 1997; Volume 10, pp. 109–113.

72. Henderson, C.M.; Golding, M.L.; Orchard, M.J. Conodont sequence biostratigraphy of the Lower Triassic Montney Formation. *Bull. Can. Pet. Geol.* **2018**, *66*, 7–22.
73. Sun, Y.D.; Richoz, S.; Krystyn, L.; Grasby, S.E.; Chen, Y.L.; Banerjee, D.; Joachimski, M.M. Integrated bio-chemostratigraphy of Lower and Middle Triassic marine successions at Spiti in the Induan Himalaya: Implications for the Early Triassic nutrient crisis. *Glob. Planet. Chang.* **2021**, *196*, 103363. [CrossRef]
74. Pietsch, C.; Petsios, E.; Bottjer, D.J. Sudden and extreme hyperthermals, low-oxygen, and sediment influx drove community phase shifts following the end-Permian mass extinction. *Palaeogeogr. Palaeoclimatol. Palaeoecol.* **2016**, *451*, 183–196. [CrossRef]
75. Takahashi, S.; Hori, R.S.; Yamakita, S.; Aita, Y.; Takemura, A.; Ikehara, M.; Xiong, Y.J.; Poulton, S.W.; Wignall, P.B.; Itai, T.; et al. Progressive development of ocean anoxia in the end-Permian pelagic Panthalassa. *Glob. Planet. Chang.* **2021**, *207*, 103650. [CrossRef]

*biology* MDPI

*Article*

# Substrate Heterogeneity as a Trigger for Species Diversity in Marine Benthic Assemblages

**Katharina Romoth *,†, Alexander Darr †, Svenja Papenmeier, Michael L. Zettler and Mayya Gogina**

Leibniz Institute for Baltic Sea Research Warnemünde, Seestrasse 15, D-18119 Rostock, Germany; alexanderdarr@web.de (A.D.); svenja.papenmeier@io-warnemuende.de (S.P.); michael.zettler@io-warnemuende.de (M.L.Z.); mayya.gogina@io-warnemuende.de (M.G.)

* Correspondence: katharina.romoth@io-warnemuende.de

† These authors contributed equally to this work.

**Simple Summary:** An increasing number of different habitats leads to an increasing number of species and has been considered a key driver for biodiversity. However, there is no common understanding on how to measure habitat diversity. In this study, we tested a newly proposed measure of substrate heterogeneity by classifying changes on the seafloor with underwater video imaging. This analysis showed that the presence of small patches of different soft sediment types was associated with elevated species richness and a higher rate of occurrence of rare species.

**Abstract:** Many studies show that habitat complexity or habitat diversity plays a major role in biodiversity throughout different spatial scales: as structural heterogeneity increases, so does the number of available (micro-) habitats for the potential species inventory. The capability of housing species (even rare species) increases rapidly with increasing habitat heterogeneity. However, habitat complexity is not easy to measure in marine sublittoral sediments. In our study, we came up with a proposal to estimate sublittoral benthic habitat complexity using standard underwater video techniques. This tool was subsequently used to investigate the effect of habitat complexity on species richness in comparison to other environmental parameters in a marine protected area situated in the Fehmarn Belt, a narrow strait in the southwestern Baltic Sea. Our results show that species richness is significantly higher in heterogeneous substrates throughout all considered sediment types. Congruently, the presence of rare species increases with structural complexity. Our findings highlight the importance of the availability of microhabitats for benthic biodiversity as well as of the study area for regional ecosystem functioning.

**Keywords:** habitat complexity; macrozoobenthos; Baltic Sea; species richness; rare species

## 1. Introduction

Species composition and species richness of faunal communities are well known to depend on different environmental factors with respect to the considered spatial scale [1]. For example, diversity and species richness of endobenthic macrofauna assemblages in the Baltic Sea are mainly influenced by salinity at regional scale of hundreds of kilometers (i.e., the entire sea or its southwestern part, stretching through several sub-basins [2,3]). By contrast, on a sub-regional scale of tens of kilometers, substrate characteristics and other factors, often masked by water depth, become more important [4,5]. However, many terrestrial, limnic, and marine studies show that habitat complexity, or habitat diversity, plays a major role in biodiversity throughout different spatial scales (e.g., [6–8]): as structural heterogeneity increases, so does the number of available (micro-) habitats for the potential species inventory [9]. The effect of the available number of habitats might not only be additive in the sense that the species inventory of the different habitats is added in a small area, but also that highly specialized species might even be endemic in these areas [10]. Furthermore, ecosystems that are more complex show higher levels of multiple ecosystem functions than ecosystems with low habitat diversity [11]. The capability of housing species (even rare species) increases rapidly with increasing habitat heterogeneity. On the other hand, common and rare species potentially play an important role in ecosystem functioning, either by offering novel contributions to functional diversity or via functional redundancy [12]. In addition, in natural or anthropogenic stress phases, communities with an extensive set of functional traits have a higher probability of surviving and contribute to the stabilization of the system [13].

However, habitat complexity is not easy to measure in marine sublittoral sediments. On larger scales, seafloor morphology is often used as a proxy to capture habitat heterogeneity [14]. On smaller scales, heterogeneity in sediment characteristics can be a key factor for determining species diversity (e.g., [5]). Nevertheless, sediment heterogeneity is often not captured in standardized sampling with a low number of replicates at individual stations. Hence, sediment composition in heterogeneous areas can be determined by significantly increasing the number of repetitions in physical sampling, leading to a huge amount of additional effort [15]. Another challenge arises in connection with the amount of sediment that is taken to analyze sediment characteristics. Taking a small sub-sample for sediment analysis often does not represent the full range of the present grain size span. This is why sediment and infauna samples are often taken separately to guarantee enough sample material for both analyses [16]. However, taking separate samples to estimate substrate heterogeneity may lead to potentially significant spatial mismatches between biogenic and geological data and may restrict the ability to cover the full range of available sediment structures. In addition, potentially important geogenic and biogenic structures such as boulders, pebbles, macrophyte meadows, and bivalve shells are often overlooked. An alternative, efficient way to estimate habitat complexity of a patch is through the use of underwater video that often accompanies the physical sampling [17].

In our study, we came up with a proposal to estimate sublittoral benthic habitat complexity using standard underwater video techniques. Data derived using this tool were subsequently used to investigate the effect of habitat complexity on species richness in comparison to other environmental parameters in an area with steep environmental gradients at a relatively small spatial scale of a few tens of kilometers.

## 2. Materials and Methods

### 2.1. Study Area

The Natura 2000-site "Fehmarn Belt" (EU-code DE 1332-301, hereafter referred to as marine protected area, MPA) is located in the southwestern Baltic Sea and covers an essential part of a narrow strait between the Danish island Lolland and the German island Fehmarn (Figure 1). It covers an area of 280 km$^2$ and is characterized by a steep depth gradient. The Fehmarn Belt is part of the Baltic transition zone that is influenced by the inflow of saline water from the Atlantic and the outflow of brackish water from the Baltic Proper [18,19]. More than two-thirds of the water volume exchanged between the North Atlantic and the Baltic Sea passes by the Belt Sea and, hence, through Fehmarn Belt [20].

Seafloor morphology and surface sediment structure in this area are formed out of glacial and postglacial processes. While the eastern and the southwestern parts of the MPA mainly consist of wide areas of muddy sand and sandy mud, the central part of the area is characterized by a high grain size heterogeneity [21]. This part of the study area consists of a large abrasion platform with the lowest water depths of 10 m cut by a deep valley (Vinds Grav channel) from east to west. The highest water depth here is around 40 m [22], filled with fine-grained deposits. Coarse lag deposits dominate the abrasion platform north of the incision. Boulders, pebbles, shell gravel, and sand of different grain size form a highly patchy mosaic of microhabitats. Sediment classes can change within meters. Similar deposits can be found south of the valley but with increasing distance, closer to the coast of Fehmarn, sand partly covers the lag deposits. A remarkable geological feature is a field of sand ribbons and drowned dunes of a height up to 2 m [23]. The sand dunes generally consist of medium-to-coarse sand with finer grain sizes dominating in higher depths below 18 m. In addition, accumulating *Arctica*-shells and floating kelp also increase habitat variability in this ribbon field.

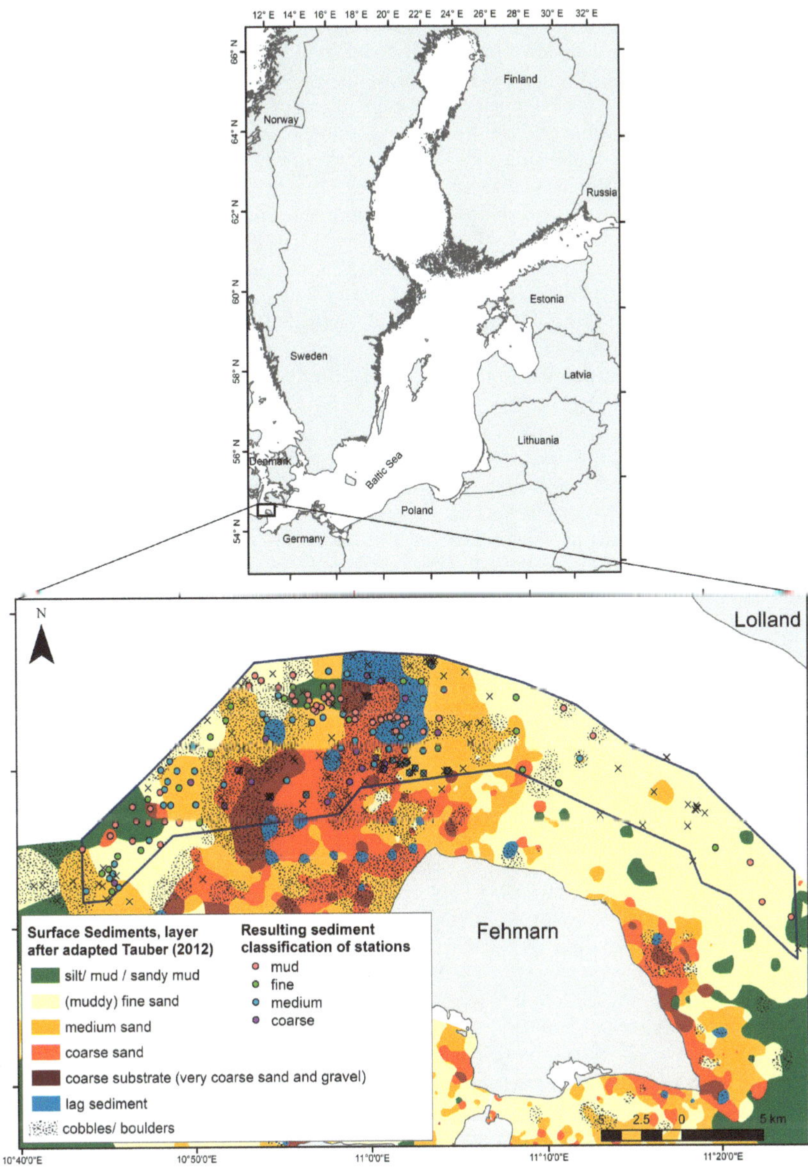


**Figure 1.** Map including the expected distribution of sediment types (after Tauber, 2012 [24]) and the location of sampling stations. Circles indicate 162 sampling locations (further referred to as stations) used for statistical analyses and the construction of general linear models, with circle colors referring to the results of the sediment type classification (crosses indicate samples excluded from the initial dataset ($n$ = 355) due to the criteria listed in Section 2.2). The blue line indicates the border of the Natura 2000-site and the white area indicates unmapped seabed in Danish waters.

### 2.2. Infauna and Sediment Data

Samples were collected during various projects at different spatial locations between 2012 and 2018. The locations (subsequently called stations) were selected in order to representatively cover the expected distribution of all major surface sediment types in the study area based on existing mapping, literature review, and the authors' experience (see Figure 1). Although the sampled stations included both monitoring stations and randomly placed mapping stations, the applied sampling method remained the same over the seven years. At each station, at least one sample for infauna and one for the analysis of sediment properties was collected using a Van Veen grab (sampling area 0.1 $m^2$), and one short video survey was conducted on the same day within a vicinity of 50–100 m.

Infauna samples were sieved using a 1 mm mesh sieve and fixated using a 4% buffered formaldehyde seawater solution. In the lab, samples were sorted using a binocular with a 10-fold magnification, and individuals were identified to the lowest possible taxonomical level (mainly species level) and counted. Taxonomy followed the World Register of Marine Species (WoRMS).

In contrast to the expected sediment types depicted in Figure 1, the final attribution of stations to the four sediment types was based on grain size distribution measurements using granulometric analyses. The considered sediment types were (i) muddy substrate (median grain size d50 < 63 µm), (ii) fine sand (d50 63–250 µm), (iii) medium sand (d50 250–500 µm), and (iv) coarse sand and gravel (d50 > 500 µm). The chosen number of classes was determined by the intention to have a sufficient number of stations within each class at the end and to adhere to the commonly accepted (coarse) sediment grain size classification. Sandy and gravelly sediments were dry-sieved automatically over a cascade of 10 sieves with differing mesh sizes ranging from 63 µm to 2 mm. Grain size distribution of muddy sediments was analyzed without chemical treatment by laser-diffraction particle size analyzer CILAS 1180L (3P Instruments GmbH & Co. KG, Odelzhausen, Germany). Parameters describing cumulative grain size distribution (namely median grain size, sorting, and skewness) were then calculated by using a skewed s-shape function, fitted to the cumulative grain size data with the least sum of squares method, applying a special fitting algorithm (the description is given in [25]). However, due to locally heterogeneous sediment conditions, additional information on sediment composition of the infauna sample was taken from the on-board optical sediment description. Samples were rejected from the analysis if a substantial mismatch occurred between the parameters of sediment sample and the on-board description of the sediment of infauna sample.

Depending on the particular aim of the project, one or three replicates were taken per infauna sampling event. However, only one randomly selected infauna sample per location and sampling event was included in the analysis to avoid spatial dependencies. To eliminate the overwhelming effects from epibenthic communities, samples with hard substrate (boulders exceeding approximately 5 cm in diameter) or kelp were also excluded from the analysis before randomly selecting one sample per site. Overall, 162 stations (unique sampling events at location) with infauna and sediment information were finally included in the analysis (Figure 1).

Unattached sessile epibenthic specimens as well as specimens that were not identified to species level were excluded from the following analysis. In addition, oligochaetes were excluded, as they were identified to species level only in some of the campaigns. In contrast, a few genera were included if they were never identified to species level and if the genus was known to be represented by a single species in the regional dataset (e.g., *Autolytus*, *Phoronis*, *Edwardsia*). Additionally, the frequency was determined, which represented the percentage of stations at which a species occurred. Rare species received special attention in the following analysis and were here defined as species occurring at fewer than four stations, corresponding to a frequency of <2%, and at none of the stations present in the abundance exceeding 3 individuals per sample (0.1 $m^2$).

Video transects were taken using a towed system with a SeaViewer underwater camera. Until 2014, an analogous camera was used that was subsequently replaced by an

HD camera of the same make. The video platform was equipped with additional light and towed over ground behind the floating vessel with the viewing direction ahead. The towing speed depended on the currents and wind speed and varied between 0.2 and 0.7 kn. The seabed was usually recorded for 5 minutes at approximately 0.5 m above the ground, depending on turbidity conditions. Only the first five minutes of the video were analyzed if the recording time exceeded this time span. Seafloor structures were categorized as follows: large boulders (hard substrates >50 cm), cobble/small boulders (hard substrates 5–50 cm), coarse gravel (2–5 cm), fine gravel (<2 cm), coarse or medium sand, fine sand, mud, bivalve shells (undestroyed or large pieces), and shell gravel. As the system was not equipped with laser pointers, no area calculation was possible. Consequently, the apparent occurrence of abiotic features was classified by estimation of the coverage using the following classes: absence of the feature, occasional occurrence (coverage <1% of the seafloor), frequent occurrence (1–10% coverage), dense occurrence (10–50%), and very dense occurrence (≥50%). For consistency in the video analysis and to avoid introducing observer-specific artefacts, the same person analyzed all the videos. As the video analysis only allows for a semi-quantitative approach, the substrate heterogeneity was described categorically. The four categories were defined as follows (see also Table 1 for schematic presentation):

- No heterogeneity (none): Other than the dominant (very dense) substrate class, at most one additional feature occurs occasionally;
- Low heterogeneity: Other than the dominant (very dense) substrate class, at most three additional features occur occasionally, or at most two additional features occur occasionally or frequently;
- Medium heterogeneity: Other than the dominant (very dense) substrate class, at most five additional features occur occasionally, or at most three additional features occur frequently of which one feature might occur densely;
- High heterogeneity: Any other combination, including at least four feature classes. Often, no single feature exceeds 50% coverage.

**Table 1.** Classification of substrate heterogeneity using a number of substrate features identified in short video transects.

<table>
<tr><th>Substrate</th><th colspan="4">Number of Features Occurring</th></tr>
<tr><th>Heterogeneity Class</th><th>Occasionally</th><th>Frequently</th><th>Densely</th><th>Very Densely</th></tr>
<tr><th></th><th>(≤1%)</th><th>(>1–10%)</th><th>(>10–50%)</th><th>(>50%)</th></tr>
<tr><td>none</td><td>≤1</td><td>0</td><td>0</td><td>1</td></tr>
<tr><td>low</td><td>≤3</td><td>0</td><td>0</td><td>1</td></tr>
<tr><td>or</td><td colspan="2">≤2</td><td>0</td><td>1</td></tr>
<tr><td>medium</td><td>≤ 5</td><td>0</td><td>0</td><td>1</td></tr>
<tr><td>or</td><td colspan="2">≤2</td><td>1</td><td>1</td></tr>
<tr><td>or</td><td></td><td>≤3</td><td></td><td>1</td></tr>
<tr><td>high</td><td>>5</td><td>0</td><td>0</td><td>1</td></tr>
<tr><td>or</td><td></td><td>>3</td><td></td><td>≤1</td></tr>
</table>

### 2.3. Analyses and Statistics

In this study, the species richness parameter was chosen to represent the species diversity. This metric is commonly used in studies addressing effects of habitat heterogeneity and complexity on biodiversity [26–28]. The Shannon–Wiener Index could alternatively be used [29,30], but equitability in distribution of species among a sample was outside the focus of this study.

All analyses were performed within the R environment [31]. Tests for normality in species richness were performed using a Shapiro–Wilk test [32]. Kruskal–Wallis and pairwise Mann–Whitney tests were used to initially evaluate the overall differences in species richness between the sediment types and substrate heterogeneity classes [33,34].

To compare diversity properties and account for possible sampling effort bias in estimating the expected number of observed species per sediment type, species-accumulation (rarefaction) curves were derived using the specaccum command of the R package vegan [35]. Default specaccum settings were used.

The dependency of species diversity on different environmental factors was tested using generalized linear models (GLM). GLM was chosen as the modelling method, as it was expected to have a higher power than linear models when analyzing count data [36]. The Shapiro–Wilk normality test suggested that species richness was not normally distributed. First, Poisson distribution was assumed for species richness (supported by the results of the Wilcoxon rank sum test) and correlations between numerical predictor variables were explored (see Supplementary Materials, Explanatory Text S1 for test results and Figure S3 for correlation graphs between numerical predictor variables). To reduce the complexity and find the best model, non-significant predictors were dropped, and backward selection based on the AIC information criterion [36] was carried out as the final step. However, the best-fitted Poisson model indicated overdispersion. To evaluate overdispersion, the DHARMa R package was used [37]. The variance was 3.8 times larger than the mean: plotted Pearson residuals considerably exceeded 1 (see Supplementary Materials, Explanatory Text S2). In order to address the detected overdispersion, we changed our distributional assumption to the negative binomial. To check if the distribution assumption could considerably influence our results, we also estimated the dispersion parameter within the model using the quasi-Poisson family. As there was no substantial difference in the interpretation, we focused on the outcome of the negative binomial model in the results, whereas the results of both dispersion-adjusted final models, side by side, are reported in the Supplementary Materials (Explanatory Text S2).

Overall, nine environmental parameters were tested in the initial model. The sediment variables loss on ignition, median grain size, skewness, and sorting were derived from the sediment analysis. Median grain size (in μm) indicated two outliers (values above 1500 μm): their influential effect was removed by transforming variable to phi units before entering the model [38]. Salinity and water depth were taken from measurements accompanying the sampling event. Slope and bathymetric position index (BPI) were derived from the bathymetry map by BSH [39], using the benthic terrain modeler extension (BTM, version 3.0) in ArcGIS [40]. Finally, substrate heterogeneity was estimated as described above and included as a 4-level categorical variable into the model (categories: none, low, medium, high). Sampling year and season (spring and summer) were included in GLM to test effects of temporal trends and seasonality.

Prior to entry into the model, numerical predictors were tested for collinearity using Spearman rank correlation (as mentioned above), and for the set including categorical predictors, the rule of Generalized Variance Inflation Factor GVIF $(1/(2 \times Df)) < 2.2$ (as equivalent to simple variance inflation value VIF < 5) was applied. Values of GVIF suitable for categorical predictors were adjusted to make them comparable across different numbers of parameters, as recommended by Fox and Monette [41]. Potentially important environmental parameters, such as oxygen depletion or the portion of particular grain size fraction in the sediment, were excluded from the analysis after testing for variable collinearity.

In order to obtain more insights on where the differences captured by the final model came from, a post hoc test was carried out for between-subject factors and interactions. For post hoc test, the emmeans R package [42] was used with the default settings of Tukey method for comparing estimates.

## 3. Results

### 3.1. Overall Species Inventory

Overall, 199 species were identified, with polychaetes (79 species), molluscs (54), and crustaceans (39) being the main contributors to species richness (Figure 2). Few species were present throughout the area, with *Scoloplos armiger* (147 records, frequency 90.7%), *Kurtiella bidentata* (140, 86.4%), *Diastylis rathkei* (131, 80.9%), *Ophiura albida* (122, 75.3%), and

*Abra alba* (122, 75.3%) being the most commonly occurring species. Overall, only 18 species were present in more than half of the stations. On the other hand, 25 species were identified in a singular sample and 48 species could be considered as rare in our dataset, following the definition given above (i.e., those occurring at frequency below 2% and with abundance at any station not exceeding 3 individuals per 0.1 m$^2$ sample). A complete list of species is provided in the Supplementary Materials—File S1. Of all the 199 species observed, 84 species were shared between all 4 sediment types considered, 9 were found only in mud, 2 were unique for fine sands, 14 for medium sands, and 15 for coarse sediments.

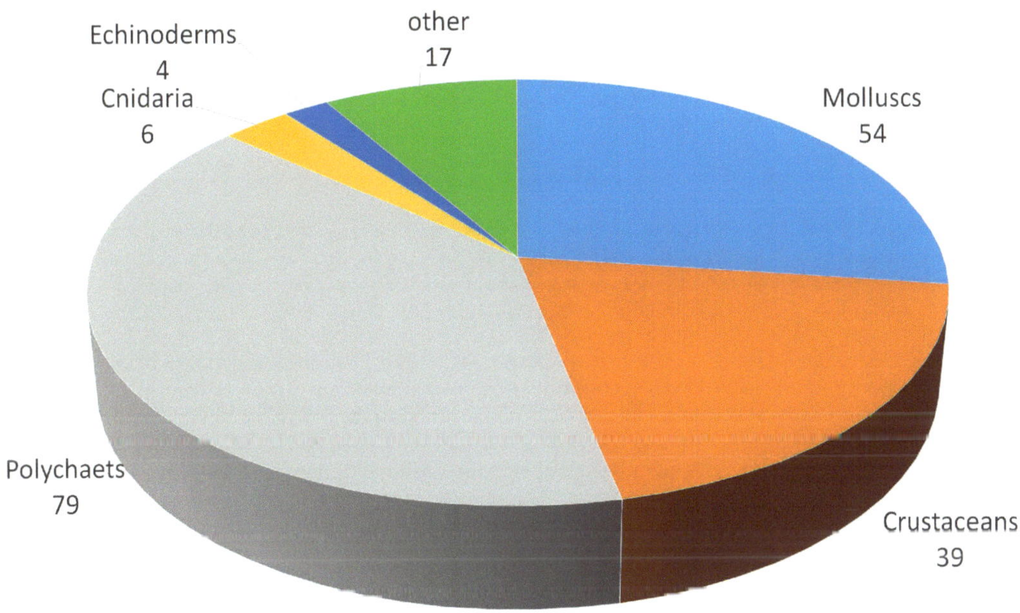


**Figure 2.** Composition of collected organisms. Polychaetes, molluscs, and crustaceans contributed most to species richness.

### 3.2. Species Richness in Different Sediment Types

Species richness varied between 6 and 70 species identified per 0.1 m$^2$, with a median of 27 species per sample. Median species richness per sediment type per sample varied between 17 taxa and 38 taxa per 0.1 m$^2$, with the lowest values in muddy substrate and the highest values in fine and medium sand (Figure 3). Although species richness in mud was significantly lower than in all other substrates ($p < 0.001$), no significant difference between the other sediment types were detected., Shapiro–Wilk tests for normality failed, indicating a non-normal distribution of species richness for all sediment types (Supplementary Materials—File S1).

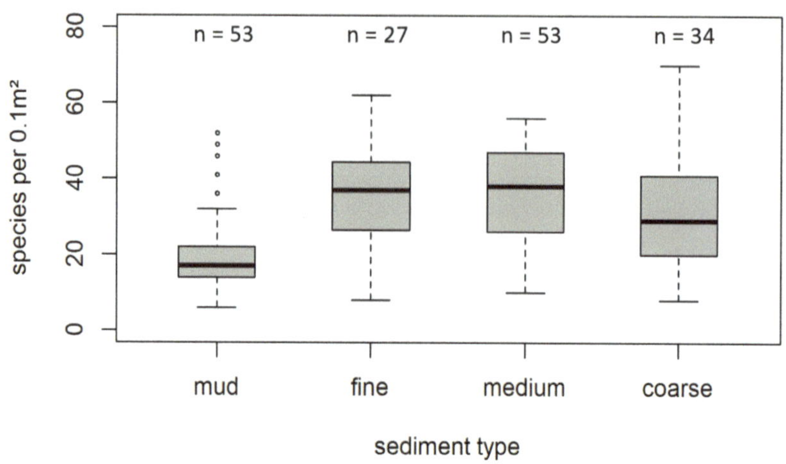


**Figure 3.** Boxplot showing species richness per grab and sediment type (mud, fine sand, medium sand, and coarse substrate). Boxes indicate the 25–75% interval, whiskers the 5–95% interval. Note that the black lines represent the median values that differ from the mean. Corresponding number of samples per category are given above each bar. Outliers are marked with circles.

Results from the species area curves (rarefaction analysis) for different sediment types consistently showed the lowest species richness in muddy substrates (Figure 4). The course of the species area curves for fine sand flattened earlier than the course for medium sand and coarse substrates. At an area of 1 $m^2$ (10 samples), 72 $\pm$ 10 species were identified in muddy substrates, whereas species richness exceeded 100 $m^{-2}$ in fine sand (104 $\pm$ 8), medium sand (109 $\pm$ 6), and coarse substrate (108 $\pm$ 11). In muddy substrates, a comparable number of species (105 $\pm$ 8) could only be found by aggregating 25 samples (corresponding to a cumulative sampled area of 2.5 $m^2$). At this spatial scale, the species richness in fine sand (129 $\pm$ 2) was also significantly lower than in medium sand (142 $\pm$ 5) and coarse substrate (146 $\pm$ 4).

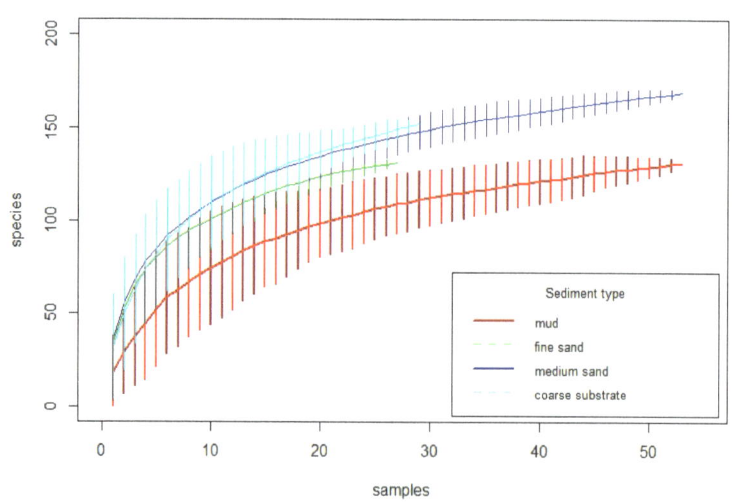


**Figure 4.** Species area curves for the four sediment types: mud ($n$ = 53), fine sand ($n$ = 27), medium sand ($n$ = 53), and coarse substrate ($n$ = 29), with vertical bars indicating the confidence interval at each step. A sample covers 0.1 $m^2$.

### 3.3. Testing the Relevance of Other Environmental Parameters

A negative test for normality indicated that other parameters in addition sediment type might influence the species richness of benthic communities in the investigation area. To explore the relative importance and explanatory power of both sediment type and substrate heterogeneity and to test for the influence of other parameters, a GLM was performed.

We included the years of sampling treated as a continuous variable in the GLM, in order to evaluate the presence of any temporal trend. Stations were sampled either in spring ($n = 36$) or in summer ($n = 126$). To test seasonality, we also included the season and its interaction with the sediment type. Mud results, in particular, suggested a significantly higher mean number of species in spring. However, this seasonal difference must be treated with caution, as it could be caused by the lower number of spring samples and be an artefact of an admittedly unbalanced sampling design, especially as the summer observations with the highest values of species richness in mud were statistically treated as outliers (see box-plot in Supplementary Materials). The post hoc analysis results (Supplementary Materials, Tables S3–S5) provided more insights on the significant between-classes differences of this interaction term: pairwise comparison of individual classes revealed only significantly lower number of species in spring observed in coarse sediment type compared to summer values in fine and medium sand sediment types in our dataset.

Median grain size (in phi units) had no significant effect on species richness when sediment type was included as a predictor and was dropped from the final model (the effect plot for this variable in the full model can be found in Supplementary Materials—File S1). The variable year was significant and had a negative estimate, suggesting some consistent reduction in species richness during the study period, particularly in "coarse" substrate and "fine" sand. Here, it is important to acknowledge the limits of this statistical inference due to possible temporal pseudoreplication. Sediment type and substrate heterogeneity both had significant effects on species richness. In particular, 'none' or 'low' heterogeneity showed the strongest linkage to a lower species number (Table 2 and Supplementary Materials—File S1). Common parameters describing seafloor topography (BPI and slope) where dropped from the final model for species richness.

**Table 2.** Results of the final GLM obtained using negative binomial distribution to explore the dependency of response variable species diversity (number of species) on different environmental factors of interest, appearance of multiannual trends, and seasonal differences. Substrate heterogeneity classes are abbreviated as "GeoClass" in the table.

| | Model (AIC: 1246) | | | | |
|---|---|---|---|---|---|
| | **Estimate** | **Std. Error** | **t-Value** | ***p*** | **Significance** |
| (Intercept) | 307.1 | 93.2 | 3.30 | 0.001 | *** |
| Factor (GeoClass)—low | −0.37 | 0.09 | −4.23 | 0.000 | *** |
| Factor (GeoClass)—medium | −0.05 | 0.09 | −0.60 | 0.550 | |
| Factor (GeoClass)—none | −0.62 | 0.15 | −4.04 | 0.000 | *** |
| Depth | −0.03 | 0.01 | −2.69 | 0.007 | ** |
| Salinity | 0.04 | 0.01 | 2.83 | 0.005 | ** |
| Year | −0.15 | 0.05 | −3.26 | 0.001 | ** |
| Factor (sediment)—fine | −144.7 | 174.5 | −0.83 | 0.407 | |
| Factor (sediment)—medium | −325.2 | 120.0 | −2.71 | 0.007 | ** |
| Factor (sediment)—mud | −374.7 | 139.6 | −2.68 | 0.007 | ** |
| Factor (Season)—summer | 0.29 | 0.17 | 1.74 | 0.082 | . |
| Year: (sediment)—fine | 0.07 | 0.09 | 0.83 | 0.405 | |
| Year: (sediment)—medium | 0.16 | 0.06 | 2.71 | 0.007 | ** |
| Year: (sediment)—mud | 0.19 | 0.07 | 2.69 | 0.007 | ** |
| Summer: (sediment)—fine | −0.26 | 0.28 | −0.92 | 0.356 | |
| Summer: (sediment)—medium | −0.15 | 0.22 | −0.67 | 0.502 | |
| Summer: (sediment)—mud | −0.55 | 0.23 | −2.44 | 0.015 | * |

* Significance codes: *** $p < 0.001$; ** $p < 0.01$; * $p < 0.05$; $p < 0.1$; Null deviance: 322.5 on 161 degrees of freedom; Residual deviance: 165.9 on 145 degrees of freedom.

The influence of substrate heterogeneity on species richness was illustrated using a boxplot (Figure 5). In homogeneous substrates, species richness barely exceeded 20 species per 0.1 $m^2$ (median: 14 species per 0.1 $m^2$). Species richness significantly increased by adding a few additional structural elements (substrate heterogeneity (GeoClass) 'low', median: 22 species per 0.1 $m^2$) and even more at medium and higher substrate heterogeneity (38 per 0.1 $m^2$ for substrate heterogeneity (GeoClass) 'medium' and 37 for substrate heterogeneity (GeoClass) 'high'). Looking separately at the four substrate classes described above revealed a similar pattern in all substrates (Figure 6). In all substrate classes, species richness was considerably lower in homogenous sediments or at low heterogeneity. However, due to the low number of samples in some combinations of substrate class and heterogeneity level, the significance of this pattern could not be verified. Results of the post hoc analyses (Supplementary Materials Figure S8) gives more detailed insights on species richness differences in independent categorical variables and interactions.

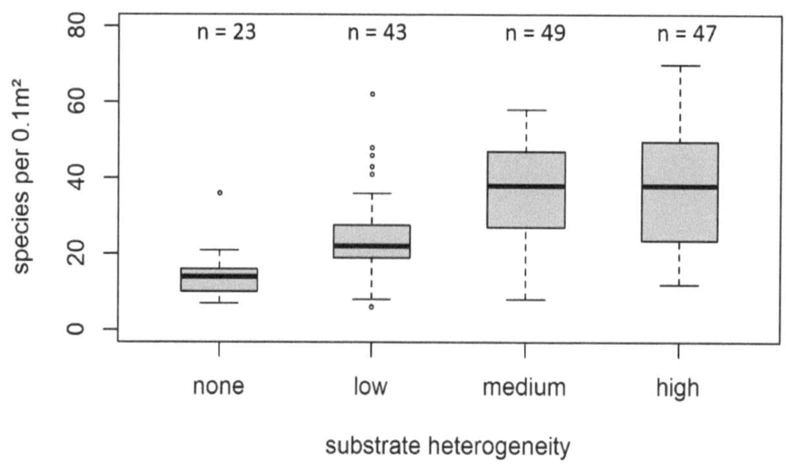


**Figure 5.** Boxplot showing species richness per grab and substrate heterogeneity class, as defined in Table 1. Boxes indicate the 25–75% interval, whiskers the 5–95% interval. Note that the black lines represent the median values that differ from the mean.

### *3.4. Occurrence of Rare Species*

Overall, 80 records of 48 rare species were identified. Based on the number of records and the number of samples, the rate of rare species detected per sample was calculated. The rate successively increased from homogeneous substrates (0.22 rare species per sample) to 0.83 rare species per sample in very heterogeneous substrates, when summarized across all sediment types. The occurrence of rare species differed between the sediment classes. The lowest probability of finding a rare species was discovered in fine sand (0.11 rare species per sample), whereas statistically more than one rare species could be identified per sample in coarse substrates (1.17). Moreover, the probability of finding a rare species was highest in highly heterogeneous coarse substrates if heterogeneity and sediment classes were considered separately (Table 3).

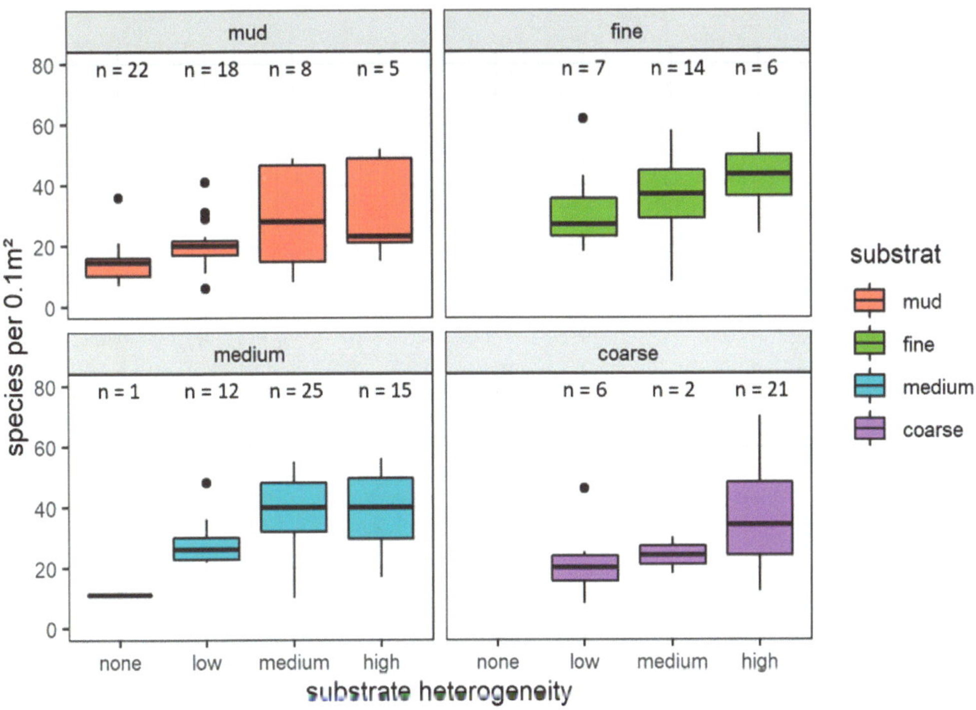


**Figure 6.** Boxplot showing species richness per grab separately for sediment types mud, fine sand (top line), medium sand, and coarse substrate (bottom) per substrate heterogeneity class. Boxes indicate the 25–75% interval, whiskers the 5–95% interval. Note that the black lines represent the median values that differ from the mean.

**Table 3.** Mean number of rare species per sample in different combinations of sediment classes and substrate heterogeneity. Numbers in brackets indicate the number of samples per combination. NA: combination not present.

| Substrate Heterogeneity | Mud | Fine Sand | Medium Sand | Coarse Substrate | Overall |
|---|---|---|---|---|---|
| None | 0.23 (22) | NA | 0 (1) | NA | 0.22 (23) |
| Low | 0.11 (18) | 0 (7) | 0.42 (12) | 0.83 (6) | 0.28 (43) |
| Medium | 0.25 (8) | 0.21 (14) | 0.72 (25) | 0.50 (2) | 0.49 (49) |
| High | 0.60 (5) | 0 (6) | 0.53 (15) | 1.33 (21) | 0.83 (47) |
| Overall | 0.23 (53) | 0.11 (27) | 0.58 (53) | 1.17 (29) | 0.49 (162) |

## 4. Discussion

In this study, we tested a newly proposed measure of substrate heterogeneity. It was derived from the frequency of morphological structures on the seafloor recorded with underwater video, and it was attributed to seafloor heterogeneity at a spatial scale somewhat larger than that of a standard grab sample (roughly 40 $m^2$ vs 0.1 $m^2$ [17]). Our results suggest that the sediment information value from a grab sample can be limited, especially when the sediment in the grab is homogenous but comes from an overall heterogeneous surrounding.

The data used in this study were not based on experiments, but rather on various projects that have been carried out within the study area over several years. Such an approach often carries the risk of an unrepresentative distribution of stations with regard to the relevant environmental gradients. We acknowledge that the results should be interpreted with caution due to a possible temporal pseudoreplication. Also, in this study, the data points were not evenly distributed along the considered substrate gradient and the substrate heterogeneity classes. However, the fact that the combinations of sediment type and heterogeneity class were not evenly distributed in the data mainly originates from the genesis and the amount of sediment supplied [43]. The coarse sediments were relicts of glacial deposits and were granulometrically poorly sorted by nature. Permanent hydrodynamic forces, winnowing the fine fraction that accumulated in low energetic areas (e.g., depressions, stone shadow), reinforced the heterogeneity of lag sediment-dominated areas. The coverage of the southern abrasion platform with mobile sands and, thus, a homogenous sediment distribution was related to the availability of large amounts of reworked nearshore sediments [23]. In this study, we also included only one replicate per sampling event in the analysis. This absence of replication may increase uncertainty in our results and cause limited reliability, due to unaccounted patchiness and existing fine-scale variability in benthic fauna distribution, which should be kept in mind.

Additionally, the comparatively large period of seven years and the fact that the data originate from different seasons increases the included natural variability in the biological data and, consequently, the associated uncertainty in the results. As the Fehmarn Belt is situated at the entrance of the Baltic Sea, inhabiting communities are frequently influenced by protruding saline waters from the Kattegat and Skagerrak. These water masses potentially carry pelagic larvae and also adult specimens with them, temporarily complementing the autochthonous species inventory. However, as both homogeneous and heterogeneous sediments have been sampled throughout the full time span, it is unlikely that this had significant impact on the overall pattern of the results.

In our study, we have focused on substrate characteristics and included comparatively few factors of water chemistry and physics that potentially may also affect species richness in the region. However, the included factors are known to be the most important for the distribution in the southwestern Baltic Sea and many other not included factors are known to be correlated with water depth; in particular, if the values describing them are derived from oceanographic models, this often remains the only option [44,45]. Additionally, other studies have shown that the available spatial resolution of such data (e.g., for drivers such as water currents or organic load) cannot act as a useful predictor on the considered scale of tens of kilometers [46]. Nevertheless, seasonal oxygen depletion mainly occurs in the deeper parts of the Fehmarn Belt and may (temporarily) reduce species richness in the predominantly homogenous muddy sediments. In addition, physical disturbance caused by anthropogenic activities, e.g., by demersal trawling, may have a negative impact on species richness (e.g., [47]). Bottom trawling mainly occurs on homogenous muddy and sandy sediments in the western and eastern parts of the study area [48], where species richness was detected to be comparably low. Nevertheless, due to limitations in our ability to adequately quantify the magnitude of this pressure in this area (c.f. [49]), its potential influence on species richness was not quantified here and needs to be addressed in future studies.

The way to estimate habitat complexity varies considerably between different studies dealing with marine benthic habitats [9,14,50]. The diversity in approaches is partly related to the particular considered spatial scale and the availability of data to describe the habitat complexity. However, no common understanding on how to measure habitat diversity is available and, consequently, the studies are often hardly comparable. In our study, we used a simple classification scheme of structures and substrates detected using underwater video. As one person analyzed all videos and the same approach was applied to all records, the approach can be considered as standardized within the study. However, the selection of the included features and their classification remained subjective. One potentially crucial issue is the handling of the surrounding boulders inhabited by their own epibenthic-dominated communities [46]. As the target of the study was to detect the influence of substrate heterogeneity on soft sediment communities, we tried to avoid including samples randomly taken on boulders or patches of dense stones by excluding all stations with a corresponding description of the substrate. Nevertheless, the presence of small stones in the samples could not be ruled out. As small stones are often populated by species-poor communities [46,51] that correspond to those often found on large bivalve shells, which are in turn considered as structuring elements in soft sediments, it was unfeasible to a priori deselect all sessile species. Consequently, a few sessile and many characteristic accessory species of hard-substrate communities were found in the sample and significantly contributed to overall species richness, observed especially in heterogeneous substrates. Large shells from *Arctica islandica* are the dominating biogenic hard substrate and can be found throughout the whole study area. They provide settling space for small epibenthic species, such as barnacles, tunicates, and epibenthic bivalves, and shelter for mobile or tube-building species, e.g., of polychaetes genera *Harmothoe* and *Flabelligera*. Likewise, the surrounding geogenic hard substrates such as boulders and cobbles add to the species inventory of the soft-sediment communities. This happens either by detached biogenic material, such as floating algae or pieces of sponge colonies, carrying specimens that inhabit them or by mobile species. However, not only the presence of geogenic hard substrate and its epibenthic community, but also the presence of different soft sediment types on small patches significantly raised species richness. The positive effect of habitat heterogeneity on biodiversity has been demonstrated for both hard-bottom and soft-bottom in previous studies on benthic systems [8,52–54]. Explanations for the mechanisms behind this effect include a greater number of niches due to increased microhabitat availability and, associated with greater surface area, a higher productivity and sampling effect [7]. High substrate heterogeneity may form greater variation in space sizes, providing habitable space to organisms with a wider variety of body sizes, thereby leading to higher species richness [50,55]. Furthermore, Kovalenko et al. [7] argue that increasing habitat complexity may decouple trophic interactions and subsequently increase ecosystem stability. Overall, our findings are in line with the results of other studies from marine and brackish waters (e.g., [8,14]). It could be shown that the variety of ecological niches in the heterogeneous areas in MPA Fehmarn Belt not only raise local biodiversity but additionally, and more importantly, provide habitats for rare species that were not found in homogeneous sediments. The role of these rare species in ecosystem function and stability is still not fully understood, but most studies concordantly highlight their potential role in functional redundancy and, consequently, in securing ecosystem resilience [11,13]. Consequently, the integrity of the heterogeneous areas and the inhabiting communities in the Fehmarn Belt can be of special interest, not only for nature conservation, but also for ecosystem function of the whole area.

## 5. Conclusions

To conclude, heterogeneous seabed forms structure habitat three-dimensionally, increase species richness, and buffer ecosystem functional diversity, thereby resisting fluctuating environmental factors. Areas with such a high multidimensional diversity are likely to be of outstanding importance in times of global overfishing, climate change, and exploration of offshore space and resources. The Fehmarn Belt is one these areas in the Baltic Sea, and its ecological development requires special attention to secure the future provision of related ecosystem services.

**Supplementary Materials:** The following supporting information can be downloaded at https://www.mdpi.com/article/10.3390/biology12060825/s1: Figure S1: Description of the occurring substrates at the study site; Explanatory Text S1: Testing the suitability of Poisson distribution; Figure S2: Poisson distribution; Figure S3: Checking predictors for collinearity; Table S1: Generalized Variance Inflation Factors; Figure S4: Effect plots for each predictor in the field model; Explanatory Text S2: Dispersion analysis and evaluation of how much the coefficient estimations are affected by overdispersion; Figure S5: Plot of estimated variance against the mean (Pearson residuals) for the best fitted Poisson model; Figure S6: DHARMa nonparametric dispersion test via sd of residuals fitted vs. simulated for the best fitted Poisson model: (dispersion = 3.8552, $p$-value < 0.0001) and plots of scaled residuals; Table S2: Results of GLM using the *Quasipoisson* family and alternatively used Negative Binomial instead of the Poisson model; Figure S7: DHARMa nonparametric dispersion test via sd of residuals fitted vs. simulated for negative binomial model (dispersion = 0.88992, *p*-value = 0.424) and plots of scaled residuals; Figure S8: Effect plots for each predictor in the final negative binomial model remained very similar; Tables S3–S5: Results of post-hoc tests for the final negative binomial GLM model. Upper triangle: *p* values adjust = "tukey"; diagonal: [Estimates] (emmean); lower triangle: Comparisons (estimate) earlier vs. later; Figure S9: Boxplot comparing the number of species in spring and summer; Figure S10: Positioning of stations sampled in mud; Figure S11: GLM results and effect plots for each predictor in the full model with the two influential points (outliers); Figure S12: Effect plots for each predictor in the full model with median grain size transformed in phi units; Table S6: List of species.

**Author Contributions:** Conceptualization, A.D. and M.L.Z.; methodology, A.D. and M.G.; validation, M.G. and K.R.; formal analysis, A.D., M.G., and K.R.; investigation, A.D. and K.R.; data curation, A.D. and K.R.; writing—original draft preparation, A.D. and K.R.; writing—review and editing, K.R., M.G., S.P. and M.L.Z.; visualization, A.D., M.G. and K.R.; project administration, K.R.; funding acquisition, A.D. and M.L.Z. All authors have read and agreed to the published version of the manuscript.

**Funding:** This research was funded by the Federal Agency for Nature Conservation (BfN), grant number 3 519 532 202. M.G. was partly supported by grants 03F0848A and 03F0937A (MGF Baltic Sea projects, funded by the German Federal Ministry of Education and Research). The publication of this article was funded by the Open Access Fund of the Leibniz Association.

**Institutional Review Board Statement:** Not applicable.

**Informed Consent Statement:** Not applicable.

**Data Availability Statement:** The data presented in this study are available on request from the corresponding author. The infauna data are going to be available via an online portal of the Federal Agency for Nature Conservation at the end of 2023 (contact: kathrin.heinicke@bfn.de).

**Acknowledgments:** We gratefully acknowledge the valuable work of all colleagues deployed in the sampling campaigns and laboratory analyses, including the staff of our research vessel Elisabeth Mann Borgese. We are also grateful for the support of the section physical and marine geology of the IOW for their help with the environmental data. We express our thanks to the anonymous reviewers for numerous constructive comments that greatly helped to improve the earlier version of the manuscript and to Irina Steinberg for correcting the English.

**Conflicts of Interest:** The authors declare no conflict of interest. The funders had no role in the design of the study; in the collection, analyses, or interpretation of data; in the writing of the manuscript; or in the decision to publish the results.

## References

1. Cochrane, S.K.J.; Andersen, J.H.; Berg, T.; Blanchet, H.; Borja, A.; Carstensen, J.; Elliott, M.; Hummel, H.; Niquil, N.; Renaud, P.E. What Is Marine Biodiversity? Towards Common Concepts and Their Implications for Assessing Biodiversity Status. *Front. Mar. Sci.* **2016**, *3*, 248. [CrossRef]
2. Gogina, M.; Nygård, H.; Blomqvist, M.; Daunys, D.; Josefson, A.B.; Kotta, J.; Maximov, A.; Warzocha, J.; Yermakov, V.; Gräwe, U.; et al. The Baltic Sea Scale Inventory of Benthic Faunal Communities. *ICES J. Mar. Sci.* **2016**, *73*, 1196–1213. [CrossRef]
3. Zettler, M.L.; Friedland, R.; Gogina, M.; Darr, A. Variation in Benthic Long-Term Data of Transitional Waters: Is Interpretation More than Speculation? *PLoS ONE* **2017**, *12*, e0175746. [CrossRef]
4. Glockzin, M.; Zettler, M.L. Spatial Macrozoobenthic Distribution Patterns in Relation to Major Environmental Factors—A Case Study from the Pomeranian Bay (Southern Baltic Sea). *J. Sea Res.* **2008**, *59*, 144–161. [CrossRef]
5. Martins, M.G.; Faria, J.; Rubal, M.; Neto, A.I. Linkages between Rocky Reefs and Soft-Bottom Habitats: Effects of Predation and Granulometry on Sandy Macrofaunal Assemblages. *J. Sea Res.* **2013**, *81*, 1–9. [CrossRef]
6. Buhl-Mortensen, L.; Vanreusel, A.; Gooday, A.J.; Levin, L.A.; Priede, I.G.; Buhl-Mortensen, P.; Gheerardyn, H.; King, N.J.; Raes, M. Biological Structures as a Source of Habitat Heterogeneity and Biodiversity on the Deep Ocean Margins. *Mar. Ecol.* **2010**, *31*, 21–50. [CrossRef]
7. Kovalenko, K.E.; Thomaz, S.M.; Warfe, D.M. Habitat Complexity: Approaches and Future Directions. *Hydrobiologia* **2012**, *685*, 1–17. [CrossRef]

8. Franz, M.; Ann Von Rönn, G.; Barboza, F.R.; Karez, R.; Reimers, H.-C.; Schwarzer, K.; Wahl, M. How Do Geological Structure and Biological Diversity Relate? Benthic Communities in Boulder Fields of the Southwestern Baltic Sea. *Estuaries Coasts* **2021**, *44*, 1994–2009. [CrossRef]
9. Burnett, M.R.; August, P.V.; Brown, J.H.; Killingbeck, K.T. The Influence of Geomorphological Heterogeneity on Biodiversity I. A Patch-Scale Perspective. *Conserv. Biol.* **1998**, *12*, 363–370. [CrossRef]
10. Burlakova, L.E.; Karatayev, A.Y.; Karatayev, V.A.; May, M.E.; Bennett, D.L.; Cook, M.J. Endemic Species: Contribution to Community Uniqueness, Effect of Habitat Alteration, and Conservation Priorities. *Biol. Conserv.* **2011**, *144*, 155–165. [CrossRef]
11. Alsterberg, C.; Roger, F.; Sundbäck, K.; Juhanson, J.; Hulth, S.; Hallin, S.; Gamfeldt, L. Habitat Diversity and Ecosystem Multifunctionality—The Importance of Direct and Indirect Effects. *Sci. Adv.* **2017**, *3*, e1601475. [CrossRef] [PubMed]
12. Jain, M.; Flynn, D.F.B.; Prager, C.M.; Hart, G.M.; Devan, C.M.; Ahrestani, F.S.; Palmer, M.I.; Bunker, D.E.; Knops, J.M.H.; Jouseau, C.F.; et al. The Importance of Rare Species: A Trait-Based Assessment of Rare Species Contributions to Functional Diversity and Possible Ecosystem Function in Tall-Grass Prairies. *Ecol. Evol.* **2014**, *4*, 104–112. [CrossRef] [PubMed]
13. Bolnick, D.I.; Amarasekare, P.; Araújo, M.S.; Bürger, R.; Levine, J.M.; Novak, M.; Rudolf, V.H.W.; Schreiber, S.J.; Urban, M.C.; Vasseur, D.A. Why Intraspecific Trait Variation Matters in Community Ecology. *Trends Ecol. Evol.* **2011**, *26*, 183–192. [CrossRef] [PubMed]
14. Zeppilli, D.; Pusceddu, A.; Trincardi, F.; Danovaro, R. Seafloor Heterogeneity Influences the Biodiversity-Ecosystem Functioning Relationships in the Deep Sea. *Sci. Rep.* **2016**, *6*, 26352. [CrossRef] [PubMed]
15. Mudroch, A.; McKnight, S.D. *Handbook of Techniques for Aquatic Sediments Sampling*, 2nd ed.; Taylor & Francis: London, UK, 2019. ISBN 978-0-367-44940-7.
16. Rumohr, H. *Soft-Bottom Macrofauna: Collection, Treatment, and Quality Assurance of Samples*; International Council for the Exploration of the Sea (ICES): Copenhagen, Denmark, 2009.
17. Beisiegel, K.; Darr, A.; Gogina, M.; Zettler, M.L. Benefits and Shortcomings of Non-Destructive Benthic Imagery for Monitoring Hard-Bottom Habitats. *Mar. Pollut. Bull.* **2017**, *121*, 5–15. [CrossRef]
18. Ojaveer, H.; Jaanus, A.; Mackenzie, B.R.; Martin, G.; Olenin, S.; Radziejewska, T.; Telesh, I.; Zettler, M.L.; Zaiko, A. Status of Biodiversity in the Baltic Sea. *PLoS ONE* **2010**, *5*, e12467. [CrossRef]
19. Matthäus, W. The History of Investigation of Salt Water Inflows into the Baltic Sea—From the Early Beginning to Recent Results. In *Meereswissenschaftliche Berichte No 65, Marine Science Reports*; Leibniz Institute for Baltic Sea Research Warnemünde: Rostock, Germany, 2006. [CrossRef]
20. Lass, H.U.; Schwabe, R.; Matthäus, W.; Francke, E. On the Dynamics of Water Exchange between Baltic and North Sea. *Beitr. Meereskd.* **1987**, *56*, 27–49.
21. Darr, A.; Heinicke, K.; Meier, F.; Papenmeier, S.; Richter, P.; Schwarzer, K.; Valerius, J.; Boedeker, D. Die Biotope Des Meeresbodens Im Naturschutzgebiet Fehmarnbelt. *BfN Schriften* **2022**, *636*. [CrossRef]
22. Schwarzer, K.; Diesing, M. *Erforschung der FFH-Lebensraumtypen Sandbank und Riff in der AWZ der Deutschen Nord-und Ostsee*; Report of the Institute for Geosciences at the University of Kiel on behalf of BfN; University of Kiel: Kiel, Germany, 2006; pp. 6–68.
23. Feldens, P.; Diesing, M.; Schwarzer, K.; Heinrich, C.; Schlenz, B. Occurrence of Flow Parallel and Flow Transverse Bedforms in Fehmarn Belt (SW Baltic Sea) Related to the Local Palaeomorphology. *Geomorphology* **2015**, *231*, 53–62. [CrossRef]
24. Tauber, F. *Meeresbodenrelief in der Deutschen Ostsee/Seabed Relief in the German Baltic Sea*; Bundesamt für Seeschifffahrt und Hydrographie: Hamburg, Germany, 2012.
25. Bobertz, B.; Harff, J.; Wolff, J.-O. Sediment Facies and Hydrodynamic Setting: A Study in the South Western Baltic Sea. *Ocean Dyn.* **2004**, *54*, 39–48. [CrossRef]
26. de Juan, S.; Ospina-Alvarez, A.; Hinz, H.; Moranta, J.; Barberá, C. The Continental Shelf Seascape: A Network of Species and Habitats. *Biodivers. Conserv.* **2023**, *32*, 1271–1290. [CrossRef]
27. Pitacco, V.; Mistri, M.; Granata, T.; Moruzzi, L.; Meloni, M.L.; Massara, F.; Sfriso, A.; Sfriso, A.A.; Munari, C. Habitat Heterogeneity: A Confounding Factor for the Effect of Pollutants on Macrobenthic Community in Coastal Waters. *Mar. Environ. Res.* **2021**, *172*, 105499. [CrossRef] [PubMed]
28. Sunday, J.M.; Fabricius, K.E.; Kroeker, K.J.; Anderson, K.M.; Brown, N.E.; Barry, J.P.; Connell, S.D.; Dupont, S.; Gaylord, B.; Hall-Spencer, J.M.; et al. Ocean Acidification Can Mediate Biodiversity Shifts by Changing Biogenic Habitat. *Nat. Clim. Chang.* **2017**, *7*, 81–85. [CrossRef]
29. Almond, P.M.; Linse, K.; Dreutter, S.; Grant, S.M.; Griffiths, H.J.; Whittle, R.J.; Mackenzie, M.; Reid, W.D.K. In-Situ Image Analysis of Habitat Heterogeneity and Benthic Biodiversity in the Prince Gustav Channel, Eastern Antarctic Peninsula. *Front. Mar. Sci.* **2021**, *8*, 614496. [CrossRef]
30. Sanvicente-Añorve, L.; Leprêtre, A.; Davoult, D. Diversity of Benthic Macrofauna in the Eastern English Channel: Comparison among and within Communities. *Biodivers. Conserv.* **2002**, *11*, 265–282. [CrossRef]
31. R Core Team. R: A Language and Environment for Statistical Computing 2019. Available online: https://www.r-project.org/ (accessed on 29 November 2019).
32. Shapiro, S.S.; Wilk, M.B. An Analysis of Variance Test for Normality (Complete Samples). *Biometrika* **1965**, *52*, 591–611. [CrossRef]
33. Kruskal, W.H.; Wallis, W.A. Use of Ranks in One-Criterion Variance Analysis. *J. Am. Stat. Assoc.* **1952**, *47*, 583–621. [CrossRef]

34. Mann, H.B.; Whitney, D.R. On a Test of Whether One of Two Random Variables Is Stochastically Larger than the Other. *Ann. Math. Stat.* **1947**, *18*, 50–60. [CrossRef]
35. Oksanen, J.; Blanchet, F.G.; Kindt, R.; Legendre, P.; Minchin, P.R.; O'Hara, R.B.; Simpson, G.L.; Solymos, P.; Stevens, H.; Wagner, H. Vegan: Community Ecology Package (Version 2.3-3) 2016. Available online: https://CRAN.R-project.org/package=vegan (accessed on 18 November 2022).
36. Akaike, H. Information Theory and an Extension of the Maximum Likelihood Principle. In Proceedings of the 2nd International Symposium on Information Theory, Tsahkadsor, Armenia, 2–8 September 1971; Petrov, B.N., Csák, F., Eds.; Akadémiai Kiadó: Budapest, Hungary, 1973; pp. 267–281.
37. Hartig, F.; Lohse, L. DHARMa: Residual Diagnostics for Hierarchical (Multi-Level/Mixed) Regression Models (R Package Version 0.4.6). 2022. Available online: https://CRAN.R-project.org/package=DHARMa (accessed on 26 May 2023).
38. Zuur, A.F.; Ieno, E.N.; Elphick, C.S. A Protocol for Data Exploration to Avoid Common Statistical Problems. *Methods Ecol. Evol.* **2010**, *1*, 3–14. [CrossRef]
39. BSH Bathymetry-Grid Coverage German EEZ Baltic Sea 2018. Available online: https://www.geoseaportal.de (accessed on 26 May 2023).
40. Walbridge, S.; Slocum, N.; Pobuda, M.; Esri, D.J.W. Unified Geomorphological Analysis Workflows with Benthic Terrain Modeler. *Geosciences* **2018**, *8*, 94. [CrossRef]
41. Fox, J.; Monette, G. Generalized Collinearity Diagnostics. *J. Am. Stat. Assoc.* **1992**, *87*, 178–183. [CrossRef]
42. Lenth, R. Emmeans: Estimated Marginal Means, Aka Least-Squares Means (R Package Version 1.7.2) 2022. Available online: https://CRAN.R-project.org/package=emmeans (accessed on 26 May 2023).
43. Feldens, P.; Schwarzer, K.; Hübscher, C.; Diesing, M. Genesis and Sediment Dynamics of a Subaqueous Dune Field in Fehmarn Belt (South-Western Baltic Sea). *Marbg. Geogr. Schr.* **2009**, *145*, 80–97.
44. Gogina, M.; Glockzin, M.; Zettler, M.L. Distribution of Benthic Macrofaunal Communities in the Western Baltic Sea with Regard to Near-Bottom Environmental Parameters. 1. Causal Analysis. *J. Mar. Syst.* **2010**, *79*, 112–123. [CrossRef]
45. Darr, A.; Gogina, M.; Zettler, M.L. Detecting Hot-Spots of Bivalve Biomass in the South-Western Baltic Sea. *J. Mar. Syst.* **2014**, *134*, 69–80. [CrossRef]
46. Beisiegel, K.; Darr, A.; Zettler, M.L.; Friedland, R.; Gräwe, U.; Gogina, M. Understanding the Spatial Distribution of Subtidal Reef Assemblages in the Southern Baltic Sea Using Towed Camera Platform Imagery. *Estuar. Coast. Shelf Sci.* **2018**, *207*, 82–92. [CrossRef]
47. Sciberras, M.; Parker, R.; Powell, C.; Robertson, C.; Kröger, S.; Bolam, S.; Geert Hiddink, J. Impacts of Bottom Fishing on the Sediment Infaunal Community and Biogeochemistry of Cohesive and Non-Cohesive Sediments. *Limnol. Oceanogr.* **2016**, *61*, 2076–2089. [CrossRef]
48. Transparency for a Sustainable Ocean. Available online: https://globalfishingwatch.org/ (accessed on 15 March 2023).
49. Schönke, M.; Clemens, D.; Feldens, P. Quantifying the Physical Impact of Bottom Trawling Based on High-Resolution Bathymetric Data. *Remote Sens.* **2022**, *14*, 2782. [CrossRef]
50. St. Pierre, J.I.; Kovalenko, K.E. Effect of Habitat Complexity Attributes on Species Richness. *Ecosphere* **2014**, *5*. [CrossRef]
51. Michaelis, R.; Hass, H.C.; Mielck, F.; Papenmeier, S.; Sander, L.; Gutow, L.; Wiltshire, K.H. Epibenthic Assemblages of Hard-Substrate Habitats in the German Bight (South-Eastern North Sea) Described Using Drift Videos. *Cont. Shelf Res.* **2019**, *175*, 30–41. [CrossRef]
52. Hewitt, J.E.; Thrush, S.F.; Halliday, J.; Duffy, C. The Importance of Small-Scale Habitat Structure for Maintaining Beta Diversity. *Ecology* **2005**, *86*, 1619–1626. [CrossRef]
53. Matias, M.G.; Underwood, A.J.; Hochuli, D.F.; Coleman, R.A. Independent Effects of Patch Size and Structural Complexity on Diversity of Benthic Macroinvertebrates. *Ecology* **2010**, *91*, 1908–1915. [CrossRef] [PubMed]
54. Menge, B.A.; Lubchenco, J.; Ashkenas, L.R. Diversity, Heterogeneity and Consumer Pressure in a Tropical Rocky Intertidal Community. *Oecologia* **1985**, *65*, 394–405. [CrossRef]
55. Tokeshi, M.; Arakaki, S. Habitat Complexity in Aquatic Systems: Fractals and Beyond. *Hydrobiologia* **2012**, *685*, 27–47. [CrossRef]

biology

MDPI

*Article*

# The Relationship between Lifespan of Marine Bivalves and Their Fatty Acids of Mitochondria Lipids

**Aleksandra Anatolyevna Istomina [1], Avianna Fayazovna Zhukovskaya [1], Andrey Nikolaevich Mazeika [2], Ekaterina Andreevna Barsova [2], Victor Pavlovich Chelomin [1], Marina Alexandrovna Mazur [3], Olesya Alexandrovna Elovskaya [1], Andrey Alexandrovich Mazur [1,*], Nadezda Vladimirovna Dovzhenko [1], Yuliya Vladimirovna Fedorets [1] and Alexander Alexandrovich Karpenko [3]**

1 Il'ichev Pacific Oceanological Institute, Far Eastern Branch, Russian Academy of Sciences, 690041 Vladivostok, Russia; s-istomina1@mail.ru (A.A.I.); avianna@poi.dvo.ru (A.F.Z.); chelomin@poi.dvo.ru (V.P.C.); olesya-sharova@mail.ru (O.A.E.); doreme_07@mail.ru (N.V.D.); zavertanova@poi.dvo.ru (Y.V.F.)

2 Institute of the World Ocean, Far Eastern Federal University, 690922 Vladivostok, Russia; mazeyka.an@dvfu.ru (A.N.M.); barsova.ea@dvfu.ru (E.A.B.)

3 A.V. Zhirmunsky National Scientific Center of Marine Biology, Far Eastern Branch, Russian Academy of Sciences, 690041 Vladivostok, Russia; proshinamarina94@gmail.com (M.A.M.); alex_karp@list.ru (A.A.K.)

* Correspondence: mazur.aa@poi.dvo.ru

**Simple Summary:** Determining the physiological and biochemical causes of aging in animals is important both because of the potential health utility for humans and because aging is related to growth, reproduction processes, and the response of organisms to environmental conditions and stress. It is assumed that the peculiarities of the fatty acid composition of mitochondrial membranes ("membrane-pacemaker" theory of aging) can influence the rate of oxidative damage in cells, as well as the rate of the aging process. This property, in turn, may be fundamental for all living organisms. In this study, the fatty acid composition of gill tissues' mitochondrial membranes, in marine bivalves, was determined and analyzed. The observed features in the lipid composition of mollusk mitochondria correlate well with the longevity of these animals.

**Abstract:** Marine bivalves belonging to the *Mytilidae* and *Pectinidae* Families were used in this research. The specific objectives of this study were: to determine the Fatty Acids (FAs) of mitochondrial gill membranes in bivalves with different lifespans, belonging to the same family, and to calculate their peroxidation index; to compare the levels of ROS generation, malondialdehyde (MDA), and protein carbonyls in the mitochondria of gills, in vitro, during the initiation of free-radical oxation; to investigate whether the FAs of mitochondria gill membranes affect the degree of their oxidative damage and the maximum lifespan of species (MLS). The qualitative membrane lipid composition was uniform in the studied marine bivalves, regardless of their MLS. In terms of the quantitative content of individual FAs, the mitochondrial lipids differed significantly. It is shown that lipid matrix membranes of the mitochondria of long-lived species are less sensitive to in vitro-initiated peroxidation compared with the medium and short-lived species. The differences in MLS are related to the peculiarities of FAs of mitochondrial membrane lipids.

**Keywords:** oxidative stress theory; peroxidation index; oxidative stress in vitro

## 1. Introduction

The process of biological aging is characterized by a progressive decline in the efficiency of physiological functions. The ability to maintain the homeostasis of basic cellular processes weakens with age, which ultimately leads to an increased risk of many diseases and increases the probability of death [1].

Currently, the most widely accepted explanation of the mechanisms of aging is the free radical theory proposed by Harman in 1956. According to this theory, reactive oxygen species (ROS) formed during metabolism exhibit high reactivity and inevitably damage important biological structures (including lipids, proteins, and nucleic acids). The accumulation of damages is accompanied by a decrease in physiological functions, and it ultimately leads to the aging and death of the organism [2]. However, the status of this theory is far from certain, as some studies have found a lack of correlation between oxidative damage and lifespan, and the genetic manipulation of antioxidant pathways in invertebrate models (e.g., the nematode worm *Caenorhabditis elegans* and the fruit fly *Drosophila melanogaster)* have yielded variable results on life span, whereas studies in higher animals (e.g., the naked mole-rat *Heterocephalus glaber)* have not, generally, supported a role for oxidative stress in modulating longevity [3–6].

Later, it became evident that there is a close link between ROS generation (mainly in the respiratory chain of mitochondria) and the aging process [7,8]. Most of the oxygen consumed by the cell is involved in mitochondrial oxidative phosphorylation. During this process, a stepwise one-electron reduction in an oxygen molecule occurs, with the generation of its active forms ($O_2^{\bullet-}$, $OH^{\bullet}$, $H_2O_2$) as intermediate products [9]. It has been shown that the rate of oxidative attacks, of ROS, on mitochondrial DNA is higher than on nuclear cell DNA [10]. At the same time, it turned out that the oxidative damage of mitochondrial DNA was inversely correlated with the lifespan of some birds and mammals since mutations, caused by damage in mitochondrial DNA, increased the aging process [10].

When explaining the mechanisms of aging, an attempt was made to combine the theory of oxidative stress ("free radical" theory) with the intensity of metabolism ("rate of living" theory). According to this viewpoint, organisms with a high metabolic rate are characterized by an increased production of oxyradicals that promote the rapid generation and accumulation of oxidative damage in the cell. On examples of certain representatives of mammals, birds, cephalopod mollusks, and the housefly (*Musca domestica*), by direct and indirect methods, it has been shown that ROS generation negatively correlates with lifespan [11–13].

However, as the experimental data were accumulated, this popular concept was considered to be erroneous. For example, it has been shown that spontaneous physical exercise and the associated increase in metabolic rate do not decrease the lifespans of mammals [1]. Another example is that birds and mammals have similar metabolic rates, but birds tend to live much longer than similarly sized mammals [14].

A relatively recent viewpoint has emerged, according to which the processes of aging and maximum lifespan (MLS) are closely connected to the susceptibility of a membrane lipid matrix to peroxidation, the key role in which is assigned to the nature of Fatty Acids (FAs). The composition of the FAs of membrane lipids subjected to oxidation correlates with the MLS of some birds and mammals, varies with body size [15,16], and is related to their MLS [17,18]. These studies played an important role in the development of "homeoviscous-longevity" theory and, later, the "membrane-pacemaker" theory of aging [19,20]. These theories assume that the characteristics of the FAs of mitochondrial membranes may influence the rate of oxidative damage in cells and the MLS of species.

It is known that the susceptibility of the same FAs to peroxidation increases exponentially with the number of double bonds of the carbon chain. Therefore, a single average value of susceptibility to peroxidation for any biological membrane, which has been named as peroxidation index (PI) [1,21], can be calculated from the profile of membrane FAs. The higher the value of the index, the more sensitive the lipid matrix is to oxidation.

The first indications of the relationship between the membrane composition and maximal lifespan were given by Pamplona and colleagues [22], who showed that the oxidation index PI of rat liver, guinea pig, and human mitochondria membranes correlated with their respective lifespan values. Later, it was shown that such a pattern is also typical for other tissues of humans and animals, including mammals, birds, and crustaceans [1].

Nevertheless, there are very few papers describing the applicability of this hypothesis to various invertebrate species. Despite the fact that there are extensive literature data on the FAs of lipids in invertebrate membranes, there are almost no papers linking it to the aging processes and lifespan of a particular species. For example, among invertebrates, the membrane lipid oxidation index has been calculated for individual representatives of bivalves [21,23]. The authors claim that there is a significant negative correlation between PI and the maximum lifespans (MLS) of these species. MLS is equivalent to the lifespan of the oldest observed specimen of a particular animal species, and it remains a frequently used trait in comparative biology [24].

Taking into account the fact that similar studies on phylogenetically similar species of marine mollusks have not been performed, we aimed to fill this gap and make some contribution to the development of this theory ("membrane-pacemaker" theory of aging). In addition, the specific objectives of this study were:

- to determine the FAs of mitochondrial gill membranes, in bivalves with different lifespans belonging to the same family, and to calculate their peroxidation index;
- to compare the levels of ROS generation, products of oxidative damage to lipids—malondialdehyde (MDA)—and protein carbonyls in the mitochondria of mollusk gills, in vitro, during the initiation of free-radical oxidation in the Fe-ascorbic acid model;
- to investigate whether the FAs of gill mitochondrial membranes affect the degree of their oxidative damage and the MLS of species.

Marine bivalves belonging to the *Mytilidae* Families (*Mytilus trossulus* Gould, 1850, *Modiolus kurilensis* Bernard, 1983, *Crenomytilus grayanus* Dunker, 1853) and *Pectinidae* Families (*Chlamys farreri* Jones and Preston, 1904, *Swiftopecten swiftii* Bernardi, 1858, *Mizuhopecten yessoensis* Jay, 1857) were used in this research. Bivalves are genetically intermediate to classical invertebrate models of aging (e.g., worms and flies) and mammals. This provides a better opportunity to understand the evolution of stress-response pathways and organismic aging [7]. Recent studies have shown that bivalves are excellent models for aging research [25,26]. First, some individuals can reach a significant age: for example, *Arctica islandica* (507 years) or *Crenomytilus grayanus* (150 years). At the same time, among them, there are also short-lived species, such as surf clams (Family *Donacidae*), with species of no more than a 1 year lifespan, as well as *Mytilus trossulus* (6 years). Second, it is possible to study the different-aged species living in the same environmental conditions and, accordingly, experiencing similar fluctuations in the environmental temperature during the year. It is likely that such species should have an approximately constant FA composition of membranes. Third, some bivalves are capable of maintaining their metabolism, at a basic level, under stressful conditions in the shelf zone. Among the mechanisms for maintaining such a state, one is the low susceptibility to membrane lipid peroxidation. In addition, the composition of mollusk membranes is very different from that of endothermic animals. Plasmalogens and non-methylene-interrupted FAs are found in significant amounts in the membrane lipids of all molluskan organs. It is assumed that they significantly affect the liquid crystal structure of the lipid matrix and act as retarders of the peroxidation processes in the membrane. Their presence also increases the antioxidant activity of lipids [8,27]. The variations in membrane FA composition may be an important missing link in the problem of explaining aging and the mechanisms that determine the maximum lifespan specific to each species. This is a testable hypothesis that requires further experiments.

## 2. Materials and Methods

### 2.1. Site of Bivalves Collection and Material

Mature mollusks were collected during the post-spawning period, in November 2021, in the waters of the Alekseev Bay and Stark Strait in the Sea of Japan (Figure 1).

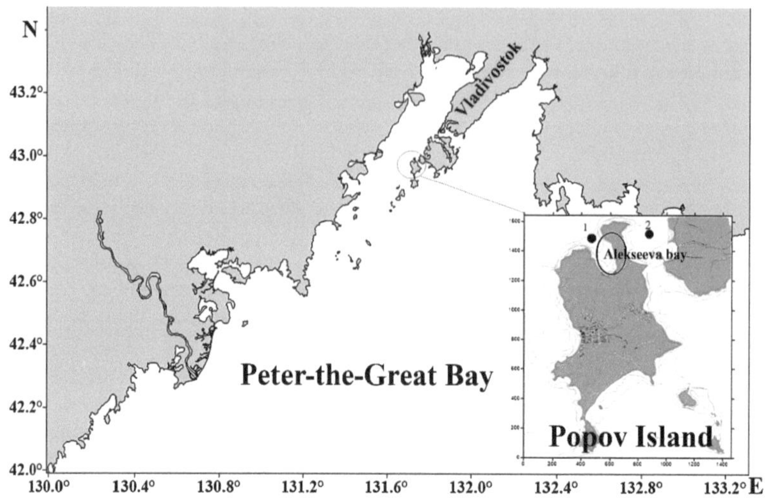


**Figure 1.** The location of the sampling sites in the Peter-the-Great Bay of the Japan Sea (Russia) (1—the collection site of mussels, 2—scallops).

The biological characteristics of bivalves are shown in Table 1.

**Table 1.** Biological characteristics of bivalves.

| Species | Length, mm | Approximate Age, Years | MLS, Years | Reference |
|---|---|---|---|---|
| *Mytilidae* | | | | |
| *Crenomytilus grayanus* | 116.9 ± 5.0 | 24 | 150 | [28,29] |
| *Modiolus kurilensis* | 111.7 ± 4.9 | 20 | 61 | [29,30] |
| *Mytilus trossulus* | 42.2 ± 3.9 | 4 | 6 | [31] |
| *Pectinidae* | | | | |
| *Mizuhopecten yessoensis* | 132.0 ± 11.2 | 5 | 22 | [32] |
| *Swiftopecten swiftii* | 83.9 ± 5.2 | 4 | 15 | [29,33] |
| *Chlamys farreri* | 92.3 ± 5.3 | 4 | 9 | [34] |

Note: MLS—maximum lifespan.

Mollusks were transported to the aquarium of the A.V. Zhirmunsky National Scientific Center of Marine Biology, where they were maintained at a constant temperature of 16 °C for 3 days to relieve the stress of transportation.

For FA analysis, 2 g of gills were obtained from one individual, for a total of 5 individuals for each species; for *M. trossulus*, tissue from 16 individuals was pooled for a total of 80 individuals. The mitochondria obtained for each mollusk species were separated into three samples ($n$ = 3). For ROS, MDA, and carbonyl analysis, mitochondria were obtained from gills weighing 0.6 g. For most mollusks, 1 sample was 1 individual, for a total of 6–8 samples ($n$ = 6–8). For *M. trossulus*, 1 sample was an assemblage of 3 individuals, for a total of 6–8 samples ($n$ = 6–8). The isolated gills were frozen in liquid nitrogen and stored for not more than 1 month before analysis. All procedures in the present work, as well as the mollusk disposal methods, were approved by the Commission on Bioethics at the V.I. Il'ichev Pacific Oceanological Institute, Far Eastern Branch of Russian Academy of Science (protocol №16 and date of approval 15 April 2021), Vladivostok, Russia.

The individual age of scallops and *M. trossulus* was estimated by growth retardation rings on the surface of the shell. The data were comparable with the growth curves obtained for these species by other authors (Table 1). The age of *C. grayanus* and *M. kurilensis* was determined by the curves of group linear growth (Table 1).

### 2.2. Mitochondria Isolation

Gills were homogenized on ice (1:5, weight/volume). Mitochondria were isolated in 0.5 M NaCl in a 0.05 M Tris-HCl (pH 7.5) medium containing 0.25 M sucrose, 1 mM EDTA, and 0.1 mM PMSF. The medium for homogenization was pre-blown with argon. The homogenate was centrifuged at 1000× $g$ for 12 min to remove large residual cells and nuclei. The resulting supernatant was centrifuged at 12,000× $g$ for 30 min. The mitochondria were washed from sucrose 3 times in 0.5 M NaCl in 0.05 M Tris-HCl (pH 7.5).

### 2.3. Biochemical Analysis

ROS levels were determined by the oxidation of DHR 123 (dihydrorhodamine 123) to fluorescent rhodamine 123 [35]. MDA content was determined by a color reaction with 2-thiobarbituric acid [36]. Protein carbonyl groups were determined by the alkaline method [37], and protein concentration was determined by the modified Lowry method [38].

### 2.4. Oxidative Stress In Vitro

The oxidative stress reaction was triggered by adding $Fe^{2+}$ and ascorbic acid (50 μM and 100 μM in the incubation medium, respectively) to mitochondria at 20 °C for one hour for MDA and carbonyl determination, as well as 15 min for ROS determination.

### 2.5. Determination of FAs

Lipid FAs were analyzed in the form of methyl esters using an Agilent 3700 chromatograph with a flame ionization detector. We used a Carbowax-20 M capillary column 25 m × 0.2 mm, a helium carrier gas, and a thermostat temperature of 200 °C [39]. FAs were identified by comparing the relative retention times of their methyl esters with the FA methyl esters of the standard mixture and the "carbon numbers" values [40]. The percentage of acids was calculated according to the method of Carrol [41]. FA methyl esters were obtained according to the method of Carreau and Dubacq [42].

### 2.6. Statistical Analysis

Statistical processing of the results was performed using Statistica 7. Breakdown and one-way ANOVA, as well as Statistics by Groups, Post-hoc were used to assess the reliability of parameter changes. Significance was established at $p < 0.05$.

## 3. Results

### 3.1. FAs in Mitochondrial Membranes of Mollusk Gill Cells

According to the results of the analysis presented in Table 2, the qualitative composition of the FAs of gill cell mitochondria lipids is uniform in all the studied marine bivalves, regardless of their MLS. However, in terms of the quantitative content of individual FAs, the mitochondrial lipids of mollusks differed significantly.

Despite significant variations (from 21.06 to 47.8%) in the content of total saturated fatty acids (SFAs), in all representatives of the *Mytilidae* Families and the *Pectinidae* Families, the palmitic and stearic acids (16:0 and 18:0) dominated. At the same time, the lowest content of SFAs was found in the mitochondrial lipids of the scallop *S. swiftii*, and the maximum was in the Pacific mussel *M. trossulus*. In general, it turned out that the SFAs in mitochondrial lipids in short-lived species was higher than in medium-lived and long-lived bivalves (Table 2).

In most mollusks, oleic acid (18:1 $n-7$) dominated among monounsaturated fatty acids (MUFAs), except for *C. grayanus* and *M. yessoensis*, in which eicosenoic acid (20:1 $n-9$) MUFAs predominated. The total content (MUFAs) in lipids also varied widely (from 11.4 to 20.5%) in the mollusks studied: the minimum amount was observed in the mitochondria of *M. trossulus*, and the maximum was in *C. grayanus* and *C. farreri*.

**Table 2.** Fatty acids (%) from gill mitochondria lipids. Values are mean ± SD, $n$ = 3.

| Fatty Acid | *Mytilidae* | | | *Pectinidae* | | |
|---|---|---|---|---|---|---|
| | *C. grayanus* | *M. kurilensis* | *M. trossulus* | *M. yessoensis* | *S. swiftii* | *C. farreri* |
| 12:0 | 1.0 ± 0.0 | 0.7 ± 0.2 | 1.1 ± 0.1 | 0.7 ± 0.0 | 0.7 ± 0.2 | 1.0 ± 0.7 |
| 14:0 ai | 0.2 ± 0.0 | 0.7 ± 0.1 | 0.8 ± 0.0 | 0.5 ± 0.0 | 0.3 ± 0.1 | 0.8 ± 0.2 |
| 14:0 | 0.6 ± 0.0 | 0.3 ± 0.1 | 0.7 ± 0.0 | 0.1 ± 0.0 | 0.4 ± 0.1 | 0.1 ± 0.0 |
| 15:1 n−7 | 1.0 ± 0.0 | 1.8 ± 0.0 | 1.6 ± 0.3 | 0.4 ± 0.0 | 0.4 ± 0.1 | 0.6 ± 0.0 |
| 16:0 | 17.9 ± 0.9 | 15.9 ± 0.3 | 19.0 ± 0.1 | 12.5 ± 0.6 | 12.7 ± 0.8 | 13.7 ± 0.7 |
| 16:1 $n$−9 | 2.1 ± 0.1 | 2.0 ± 0.5 | 2.7 ± 0.3 | 1.3 ± 0.0 | - | 2.2 ± 0.2 |
| 16:1 $n$−7 | 2.3 ± 0.1 | 2.3 ± 0.1 | 1.9 ± 0.0 | 1.4 ± 0.0 | 2.3 ± 0.5 | 1.2 ± 0.0 |
| 17:0 i | 0.6 ± 0.0 | 0.5 ± 0.0 | 0.5 ± 0.0 | 0.3 ± 0.0 | 0.9 ± 0.1 | 0.1 ± 0.0 |
| 17:0 ai | 1.1 ± 0.1 | 1.8 ± 0.1 | 1.8 ± 0.0 | 1.2 ± 0.0 | 0.3 ± 0.1 | 2.3 ± 0.0 |
| 17:0 | 0.8 ± 0.0 | 1.9 ± 0.1 | 1.7 ± 0.0 | 1.0 ± 0.0 | 0.9 ± 0.2 | - |
| 18:0 i | 2.8 ± 0.1 | 2.5 ± 0.2 | 4.0 ± 0.1 | 0.9 ± 0.0 | 1.2 ± 0.3 | 1.3 ± 0.1 |
| 18:0 | 9.6 ± 0.5 | 10.9 ± 0.3 | 19.8 ± 3.1 | 4.8 ± 0.2 | 3.3 ± 0.6 | 9.7 ± 0.1 |
| 18:1 $n$−9 | 0.7 ± 0.0 | - | - | - | 0.1 ± 0.1 | 1.0 ± 0.0 |
| 18:1 $n$−7 | 4.3 ± 0.2 | 5.1 ± 0.3 | 2.3 ± 1.6 | 4.3 ± 0.2 | 4.5 ± 0.4 | 5.4 ± 0.4 |
| 18:2 $n$−6 | 1.0 ± 0.0 | 2.4 ± 0.5 | 1.6 ± 0.3 | 2.0 ± 0.1 | 2.2 ± 0.2 | 2.5 ± 0.1 |
| 18:2 $n$−4 | 0.4 ± 0.0 | 1.5 ± 1.5 | 0.5 ± 0.2 | 0.6 ± 0.0 | 0.1 ± 0.0 | 0.3 ± 0.1 |
| 18:3 $n$−6 | - | - | 0.2 ± 0.0 | 0.2 ± 0.0 | 0.2 ± 0.0 | - |
| 18:3 $n$−3 | 0.8 ± 0.0 | 1.2 ± 0.0 | 0.3 ± 0.0 | 0.3 ± 0.0 | 0.2 ± 0.0 | 0.1 ± 0.0 |
| 20:0-i | 1.8 ± 0.1 | 1.7 ± 0.1 | 0.4 ± 0.1 | - | - | 0.8 ± 0.1 |
| 18:4 $n$−3 | 0.2 ± 0.0 | - | 0.4 ± 0.0 | - | 0.1 ± 0.0 | - |
| 20:1 $n$−13 | 2.2 ± 0.1 | 0.6 ± 0.1 | 0.8 ± 0.2 | 2.0 ± 0.1 | 1.8 ± 0.0 | 2.9 ± 0.1 |
| 20:1 $n$−9 | 5.8 ± 0.3 | 3.4 ± 0.1 | 1.9 ± 0.3 | 5.5 ± 0.3 | 4.2 ± 0.5 | 4.2 ± 0.2 |
| 20:1 $n$−7 | 2.2 ± 0.1 | 3.9 ± 0.0 | 1.3 ± 0.2 | 1.0 ± 0.1 | 0.6 ± 0.3 | 2.5± 0.1 |
| 20:2 (5,11) | 3.9 ± 0.2 | 1.9 ± 0.0 | 1.8 ± 0.1 | 5.4 ± 0.3 | 5.5 ± 0.6 | 0.9 ± 0.0 |
| 20:2 (5,13) | 1.6 ± 0.1 | 0.6 ± 0.1 | 1.1 ± 0.2 | 1.7 ± 0.1 | 1.0 ± 0.1 | 0.7 ± 0.1 |
| 20:4 $n$−6 | 5.5 ± 0.3 | 5.7 ± 0.5 | 3.5 ± 0.3 | 6.7 ± 0.3 | 5.8 ± 1.2 | 3.9 ± 0.2 |
| 20:5 $n$−3 | 4.2 ± 0.2 | 6.2 ± 0.5 | 6.0 ± 0.4 | 3.9 ± 0.2 | 4.7 ± 0.6 | 9.7 ± 0.3 |
| 22:2 | 8.2 ± 0.4 | 5.2 ± 0.1 | 4.1 ± 0.2 | 11.7 ± 0.6 | 9.0 ± 1.2 | 1.0 ± 0.0 |
| 22:6 $n$−3 | 8.6 ± 0.4 | 9.7 ± 0.3 | 9.3 ± 0.4 | 22.1 ± 1.1 | 25.8 ± 1.4 | 21.7 ± 1.3 |
| Total | 91.1 ± 4.6 | 90.4 ± 0.9 | 87.4 ± 4.0 | 93.6 ± 4.7 | 89.5 ± 7.2 | 92.0 ± 1.3 |
| SFAs | 36.3 ± 1.8 | 37.0 ± 0.0 | 47.8 ± 0.5 | 22.36 ± 1.1 | 21.0 ± 1.9 | 30.3 ± 1.4 |
| MUFAs | 20.5 ± 1.0 | 19.0 ± 1.0 | 11.4 ± 3.1 | 16.12 ± 0.8 | 13.4 ± 0.4 | 20.4 ± 0.6 |
| PUFAs | 34.3 ± 1.7 | 34.4 ± 0.1 | 28.2 ± 1.4 | 55.13 ± 2.8 | 55.0 ± 4.9 | 43.6 ± 5.7 |
| $\sum n-3$ | 13.6 ± 0.7 | 17.1 ± 0.8 | 15.4 ± 1.0 | 26.55 ± 1.3 | 30.8 ± 1.9 | 31.6 ± 1.6 |
| $\sum n-6$ | 6.5 ± 0.3 | 8.1 ± 1.0 | 5.2 ± 0.4 | 9.03 ± 0.5 | 8.3 ± 1.5 | 6.4 ± 0.4 |
| $n-3/n-6$ | 2.10 | 2.11 | 2.98 | 2.94 | 3.69 | 4.92 |
| SFAs/ PUFAs | 1.05 | 1.08 | 1.70 | 0.41 | 0.38 | 0.70 |
| ∑NMI FAs | 13.7 ± 0.01 | 7.75 ± 0.24 | 6.93 ± 0.15 | 18.9 ± 0.1 | 15.68 ± 1.63 | 2.76 ± 0.22 |
| $\sum 2n, 3n/\sum 4n, 6n$ | 1.11 | 0.84 | 0.73 | 0.77 | 0.58 | 0.22 |
| PI | 124.5 ± 6.2 | 146.7 ± 6.6 | 129.7 ± 7.9 | 238.2 ± 11.2 | 267.3 ± 19.5 | 252.7 ± 13.6 |
| MLS | 150 | 61 | 6 | 22 | 15 | 9 |

Note: MLS—maximum lifespan; PI—peroxidation index; SFAs—saturated fatty acids; MUFAs—monounsaturated fatty acids; PUFAs—polyunsaturated fatty acids; NMI FAs—non-methylene-interrupted fatty acids.

Non-methylene-interrupted fatty acids (NMI FAs), represented mainly by docosadienoic acid (22:2), were found in the FA composition of the mitochondria of marine mollusks. The greatest variation in the content of this acid was observed in pectinids: from 1% in the scallop *C. farreri* to 11.7% in the scallop *M. yessoensis*. The 22:2 level in representatives of the *Mytilidae* Family increased in the series: *M. trossulus—M. kurilensis—C. grayanus*; in representatives of the *Pectinidae* Family—in the series: *C. farreri—S. swiftii—M. yessoensis*. Regarding NMI FAs, both by the content of individual 22:2 and by the total level of NMI FAs, the total content of which varies from 2.76% (*C. farreri*) to 13.7% (*C. grayanus*), a direct connection with MLS is observed in species of the relevant family.

In representatives of the *Pectinidae* Family, polyunsaturated fatty acids (PUFAs) dominated in the FAs of gill mitochondrial cell membranes; their amount was greater than the total sum of SFAs and MUFAs. The *Mytilidae* Family showed a different pattern: the level of PUFAs did not exceed, and in some cases, it was lower than the total sum of saturated and monounsaturated acids (Table 2). Docosahexaenoic acid (22:6 $n-3$) significantly prevailed in the PUFA of mitochondria lipids, especially in pectinids. Among PUFAs, the ratio of $n-3/n-6$ acids varied from 2.10 to 4.92, with the minimum values observed in the *C. grayanus* and the *M. yessoensis*, and the maximum values were characteristic of the Pacific mussel *M. trossulus* and the Zhikong scallop *C. farreri* (Table 2).

### 3.2. PI of Mitochondrial Membranes

On the basis of the composition of FAs, according to the formula given in [21], the lipid peroxidation index was calculated, the values of which are shown in Table 2. From the analysis of these values, it follows that the propensity to oxidation of the FA lipids of mitochondrial membranes, in representatives of the *Pectinidae* Family, is higher than that in representatives of *Mytilidae* Family. At the same time, no correlation between the obtained values of PI and MLS of bivalves was revealed.

### 3.3. Constitutive Levels of ROS, MDA and Carbonyls

The basal levels of ROS generation and MDA content in mitochondria were highest in the long-lived Gray's mussel *C. grayanus* and the coastal scallop *M. yessoensis*, as compared with the medium and short-lived representatives of the respective families (Figure 2). In general, representatives of the *Mytilidae* Family differed from those of the *Pectinidae* Family (*S. swiftii, C. farreri*) in higher MDA content in the mitochondria of gill cells. At the same time, *C. grayanus* and *M. kurilensis* had a lower level of ROS generation compared to *M. yessoensis* and *C. farreri*, respectively. No interspecific differences in the content of protein carbonyls were found in any of the bivalves studied.

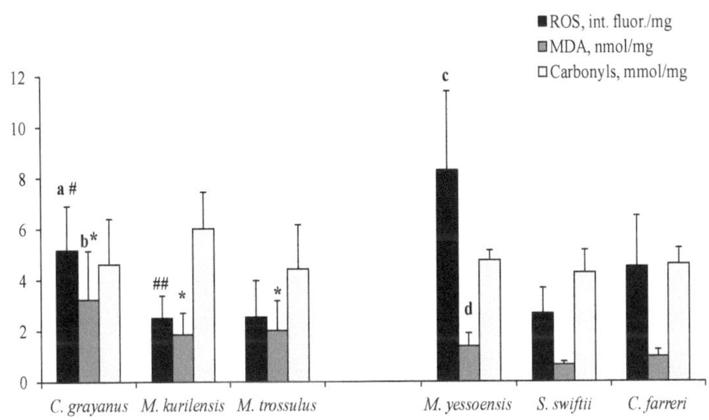


**Figure 2.** Constitutive levels of ROS (reactive oxygen species), MDA (malondialdehyde), and carbonyls in gill mitochondria. Significance of differences between: a—*C. grayanus* vs. *M. kurilensis* and

*M. trossulus*; b—*C. grayanus* vs. *M. kurilensis*; c—*M. yessoensis* vs. *S. swiftii* and *C. farreri*; d—*M. yessoensis* vs. *S. swiftii* and *C. farreri*; #—*C. grayanus* vs. *M. yessoensis*; * *C. grayanus, M. kurilensis* and *M. trossulus* vs. *S. swiftii* and *C. farreri*; ##—*M. kurilensis* vs. *C. farreri* ($n$ = 6–8; Post-hoc, $p < 0.05$).

### 3.4. Induction of Oxidative Stress In Vitro

The results of this series of experiments showed that, when free-radical processes were initiated using the Fenton reaction, the lowest level of oxygen radical generation was registered in the mitochondria of the long-lived mussel Gray's *C. grayanus* and the scallop *M. yessoensis* in contrast to the short-lived mussel *M. trossulus* and the scallop *C. farreri* (Figure 3). A similar pattern was observed in the formation of the main product of lipid oxidation—MDA.

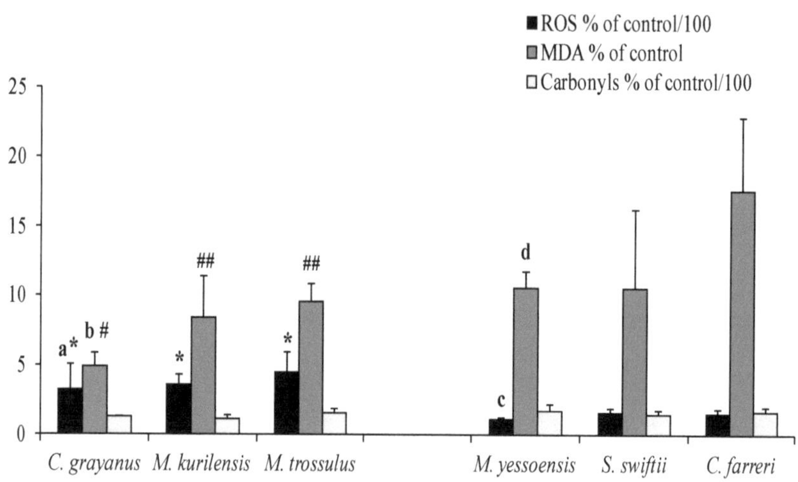


**Figure 3.** Response to induced oxidative stress in vitro. Significance of differences between: a—*C. grayanus* vs. *M. trossulus*; b—*C. grayanus* vs. *M. kurilensis* and *M. trossulus*; c—*M. yessoensis* vs. *S. swiftii* and *C. farreri*; d—*M. yessoensis* vs. *C. farreri*; *—*C. grayanus, M. kurilensis* and *M. trossulus* vs. *M. yessoensis, S. swiftii* and *C. farreri*; #—*C. grayanus* vs. *M. yessoensis, S. swiftii* and *C. farreri*; ##—*M. kurilensis* and *M. trossulus* vs. *C. farreri* ($n$ = 6–8; Post-hoc, $p < 0.05$).

Under these conditions of ROS generation initiation, the least amount of MDA accumulated in the mitochondrial lipids of long-lived mollusks (*C. grayanus* mussel and *M. yessoensis* scallop) compared with short-lived ones (*M. trossulus* and *C. farreri*). Mitochondrial membranes did not differ in protein carbonyl levels in bivalves from both families (Figure 3).

## 4. Discussion

### 4.1. Specific Features of FAs in Gill Mitochondria Lipids

Unsaturated fatty acids are easily subjected to oxidative damage in the cell, and the rate of oxidation increases with the number of double bonds. Therefore, unlike SFAs and MUFAs, which are relatively resistant to oxidation, PUFAs are easily and rapidly oxidized [8,43]. According to *homeoviscous theory*, the liquid crystalline state of the lipid matrix, necessary for the function of biological membranes, is maintained by regulating the degree of unsaturation of the acyl chains of phospholipids. In this respect, using mammalian and avian representatives as an example, it has been shown that the high unsaturation of membrane lipid FAs is associated with an increased level of oxidative lipid damage, but it negatively correlates with MLS [17,44]. Thus, in representatives of long-lived mammals, as compared to species with shorter lifespans (short-lived ones), a decrease in the ratio of acids with 4 or 6 double bonds and an increase in the level of FAs with 2 and 3 double bonds were found. At the same time, as noted by the authors, not only was a significant increase in lipid resistance to peroxidation observed but the corresponding fluidity of the lipid matrix was also maintained, and all the most important functions of membranes (receptor, ion transport, metabolite transport, etc.) were performed [8,44].

In fact, the results of this study of the FAs of mitochondrial lipids, in representatives of two families of marine bivalves, confirm this theory.

On the basis of the SFAs/PUFAs ratio, the authors showed that the amount of SFAs in gill membrane lipids was higher in representatives of the *Mytilidae* Family compared with the *Pectinidae* Family. It turned out that the long-lived *C. grayanus* and *M. yessoensis* had a lower SFAs/PUFAs ratio compared to the short-lived *M. trossulus* and *C. farreri* from the respective families. In addition, as in mammalian representatives, the proportion of FAs with 2 and 3 double bonds, in relation to FAs with 4 and 6 double bonds, was higher in long-lived species vs. short-lived species (Table 2).

It is known that the tendency of oxidation of acyl chains of lipids is determined not only by the degree of unsaturation but, also, by the position of double bonds. It was found that $n-3$ PUFAs are oxidized faster than $n-6$ PUFAs. Accordingly, membranes enriched with phospholipids with $n-6$ FAs are more stable in response to unfavorable environmental factors [45]. In addition, the ratio of these PUFAs ($\Sigma n-3/\Sigma n-6$) is an index characterizing the viscosity/liquidity of the lipid matrix of biological membranes. The lower is the ratio of $n-3/n-6$, the lower is the viscosity of the lipid matrix, but the higher is the resistance of lipids to oxidation, which is beneficial for the stability of membrane processes. In this respect, the paper of Valencak and Ruf [46] should be particularly noted. The authors revealed a negative correlation between the increase in the $n-3/n-6$ ratio in the skeletal muscle lipids of mammalian representatives and their lifespan. This interesting tendency is also clearly seen in marine bivalves. In this study, it was found that, in the mitochondrial lipids of long-lived mussel *C. grayanus* and scallop *M. yessoensis*, the ratio of $\Sigma n-3/\Sigma n-6$ is lower in comparison with the medium-lived and short-lived representatives of the respective families (Table 2). Based on this, there is every reason to believe that the lipid matrix of mitochondrial membranes of long-lived species is more stable in response to the effects of unfavorable environmental factors

In addition to the above characteristics of mitochondrial membrane lipids, the presence of NMI FAs draws attention, which can also have a significant influence on the structure and function of biological membranes. These unusual FAs can act as "structural antioxidants", slowing down the lipid matrix peroxidation processes [47]. The obtained results showed that the acids [Σ20:2 (5,11); 20:2 (5,13); 22:2] were present in far greater amounts in the lipids of long-lived *C. grayanus* and *M. yessoensis* mollusks than in the lipids of medium-lived and short-lived species from the respective families (Table 2). Therefore, it is logical to assume that the lifespans of the studied mollusks are related to the presence of these FAs in the lipids, which protect mitochondrial membranes from oxidative damage to a certain extent.

In general, the observed features in the lipid composition of mollusk mitochondria, through the presumed effect on lipid matrix oxidability, correlate well with the lifespans of these animals. Although the integral index (PI), calculated based on FA composition, demonstrated an increased sensitivity to the oxidative degradation of lipid membranes of representatives of the *Pectinidae* Family vs. representatives of the *Mytilidae* Family, it showed no relationship with the lifespans of bivalves. In this respect, the results of these studies and reasoning are consistent with those of Valencak and Ruf [46], who also found no correlation between the skeletal muscle lipid oxidation index (PI) and lifespan in 42 mammalian species.

The absence of such correlation calls into question the correctness of the calculation of this index, which does not consider additional factors influencing lipid oxidability. Among them, it should be emphasized that the high content of etheric lipids with alkyl and alkenyl fat radicals is characteristic of bivalves, whose contribution to lipid matrix oxidation of membranes is practically unstudied [48].

### 4.2. Constitutive Levels of ROS, MDA and Carbonyls in Gill Mitochondria

The "membrane-pacemaker" theory of aging suggests that lifespan can be related not only to lipid matrix oxidizability (based on PI) but, also, to the rate generation of ROS in the cell. The main source of ROS generation in the cell is the electron-transport chain localized in mitochondrial membranes. Taking into account their high reactivity, these ROS can initiate free-radical processes and cause the destruction of membrane lipids, proteins, and damage to mitochondrial DNA. In the latter case, there is strong evidence that the rate of aging is closely related to the frequency of mutations occurring in mitochondrial DNA [10]. In the lipid matrix of membranes, unsaturated fatty acids, especially PUFAs, become the preferred target for ROS. After the initiation of free-radical processes through a cascade of reactions, these FAs decompose to form highly reactive carbonyl compounds, such as malondialdehyde (MDA) and 4- hydroxynonenal (4-HN), which exhibit various cytotoxic and genotoxic properties [49]. It is likely that, through the regulation of ROS generation, mitochondria play a key role in preventing the formation and accumulation of various destructive damages affecting aging processes.

This opinion is based on the results of mammalian and bird studies in which it has been shown that, regardless of oxygen uptake rate, long-lived species show low rates of mitochondrial radical generation and contain lower constitutive levels of antioxidant activity [7,8,12,14]. Nevertheless, it has been shown that, in the long-lived (naked mole rat) *Heterocephalus glaber*, the endothelial and smooth muscle cells of carotid arteries and aorta produce comparable—or even higher—levels of ROS compared to short-lived mice [50].

The results of these studies also do not fit the general hypothesis. In long-lived *C. grayanus* and *M. yesonensis*, a relatively high baseline level of ROS generation and elevated MDA content in mitochondria were observed compared with other representatives of their families. Moreover, all this is realized against the background of a relatively low baseline level of antioxidant potential, including the activity of antioxidant enzymes and low molecular weight antioxidants [51].

Previously, in comparative studies of bivalves belonging to different families, it was shown that the isolated gill and heart mitochondria of long-lived *Arctica islandica* generated less ROS compared to short-lived *Mya arenaria*, *Spisula solidissima*, and *Mercenaria mercenaria* [27,51]. At the same time, the short-lived scallop *Argopecten irradians* and the long-lived *Tridacna derasa* did not significantly differ in ROS generation in gills, adductor muscles, and heart cells [52]. There was also no difference in the carbonyl content of the gills and adductor muscle in these species.

Comparing these results with the above examples, it is logical to assume that, in long-lived mollusks, against the background of low antioxidant protection, the hydrophobic component probably plays an important role in the mechanisms maintaining the oxidative stability of the lipid membrane matrix. These ideas, to a certain extent, were confirmed in the authors' experiments with the induction of mitochondrial lipid peroxidation initiated by the Fenton reaction. This approach makes it possible to not only estimate the potential ability of mitochondria to generate ROS but, also, to reveal the integral vulnerability of the hydrophobic matrix to oxidative degradation in case of oxidative stress.

### 4.3. Response to Induced Oxidative Stress In Vitro

The results showed that the mitochondria of long-lived *C. grayanus* and *M. yessoensis* produced lower levels of ROS and less MDA as compared to medium and short-lived representatives of the respective families, indicating greater resistance of their lipid matrix to in vitro-induced oxidative damage. As far as one can estimate from the published data, the presented results are not only characteristic for the study species. It was also found that exposure of the scallop *Argopecten irradians* to paraquat, rotenone, or organic hydroperoxide causing oxidative damage of mitochondria was accompanied by a faster death of these short lived mollusks compared to the long-lived *Mercenaria mercenaria*, *Arctica islandica*, and *Tridacna derasa* [53,54]. Moreover, there is evidence that experiments *in vitro*, fibroblasts, and lymphocytes of long-lived vertebrate species show increased resistance to induced oxidative stress [55].

When analyzing the results of mollusk mitochondrial resistance to the in vitro-induced oxidative damage of membrane lipids, the authors found two opposite trends. In representatives of the *Mytilidae* Family, in response to Fenton's reagents, mitochondria generated high levels of ROS, which nevertheless led to an insignificant accumulation of MDA. In similar experiments on representatives of the *Pectinidae* Family, a different picture was observed: against the background of an insignificant level of ROS generation, we detected a significant increase in the MDA content of mitochondrial membrane lipids. Although it is beyond the scope of this study to investigate the reasons for these peculiarities, the authors should admit that the mitochondrial membranes of these two families differ significantly not only in lipid matrix accessibility to peroxidation but, also, in the function of ROS generation centers.

As the experimental data accumulate, it becomes more and more evident that different mechanisms of stabilization of not only the lipid matrix but proteins can make a certain contribution to the processes ensuring cell resistance to stress and lifespan [51,56]. Nevertheless, according to the experimental data, the authors found no changes in the content of protein carbonyls at neither the baseline nor after ROS generation in bivalves with different lifespans. Apparently, the reparation processes of the damaged mitochondrial membrane proteins, with the participation of proteosomal and autophagic mechanisms, are stable and exhibit resistance to short-term exposure to oxidative stress in these representatives of marine mollusks. Regarding the poor study of this issue, the authors consider it necessary to perform further studies to identify the mechanisms maintaining the stability and integrity of membrane protein components with the involvement of representatives of other taxonomic groups with different lifespans.

## 5. Conclusions

The common features of the relationship between the FA composition of gill mitochondrial membranes and the MLS of species are revealed only in a comparative analysis of mollusks having a common origin within a family. The response to in vitro-induced oxidative stress also has a relationship with the MLS of species belonging to the same family.

The basal levels of ROS and MDA formation in gill mitochondria are higher in the long-lived *C. grayanus* and *M. yessoensis* vs. medium and short-lived representatives of the respective families, and the gill mitochondrial membranes of these species are more resistant to *in vitro*-induced oxidative stress (low levels of ROS and MDA).

It is likely that an important mechanism of lifespan maintenance in *C. grayanus* and *M. yessoensis* is a specific FA composition of mitochondrial membranes. It is characterized by a lower ratio of SFAs/PUFAs and $n{-}3/n{-}6$, a higher ratio of the sum of FAs with 2 and 3 bonds and the sum of FAs with 4 and 6 bonds, and higher content of the sum of NMI FAs vs. medium and short-lived species of the respective family.

**Author Contributions:** Conceptualization, A.A.I. and V.P.C.; methodology, A.F.Z., E.A.B. and N.V.D.; software, O.A.E. and A.A.K.; validation, A.A.K.; formal analysis, A.F.Z., A.N.M.; investigation, A.A.I., M.A.M. and Y.V.F.; resources, A.N.M., E.A.B., M.A.M., A.A.M. and N.V.D.; data curation, O.A.E.; writing—original draft preparation, A.A.I.; writing—review and editing, V.P.C.; visualization, A.A.M. and Y.V.F.; funding acquisition, V.P.C. All authors have read and agreed to the published version of the manuscript.

**Funding:** The research was supported by the State assignment for research work of V.I. Il'ichev Pacific Oceanological Institute, FEB RAS (no. 121021500052-9) at the marine experimental station "Popov Island" Il'ichev Pacific Oceanological Institute.

**Institutional Review Board Statement:** Not applicable.

**Informed Consent Statement:** Not applicable.

**Data Availability Statement:** Not applicable.

**Acknowledgments:** The authors are grateful to I.V. Popeko and A.A. Khudoleyev (NSCMB FEB RAS), who provided equipment for the aquarium and maintained the necessary conditions for animal housing.

**Conflicts of Interest:** The authors declare no conflict of interest.

## References

1. Hulbert, A.J.; Pamplona, R.; Buffenstein, R.; Buttemer, W.A. Life and death: Metabolic rate, membrane composition, and life span of animals. *Physiol. Rev.* **2007**, *87*, 1175–1213. [CrossRef] [PubMed]
2. Harman, D. Aging: A theory based on free radical and radiation chemistry. *J. Gerontol.* **1956**, *11*, 298–300. [CrossRef] [PubMed]
3. Blagosklonny, M.V. Aging: ROS or TOR. *Cell Cycle* **2008**, *7*, 3344–3354. [CrossRef]
4. Pérez, V.I.; Bokov, A.; Van Remmen, H.; Mele, J.; Ran, Q.; Ikeno, Y.; Richardson, A. Is the oxidative stress theory of aging dead? *Biochim. Biophys. Acta* **2009**, *1790*, 1005–1014. [CrossRef] [PubMed]
5. Salmon, A.B.; Richardson, A.; Pérez, V.I. Update on the oxidative stress theory of aging: Does oxidative stress play a role in aging or healthy aging? *Free Radic. Biol. Med.* **2010**, *48*, 642–655. [CrossRef]
6. Du, C.; Anderson, A.; Lortie, M.; Parsons, R.; Bodnar, A. Oxidative damage and cellular defense mechanisms in sea urchin models of aging. *Free Radic. Biol. Med.* **2013**, *63*, 254–263. [CrossRef]
7. Buttemer, W.A.; Abele, D.; Costantini, D. From bivalves to birds: Oxidative stress and longevity. *Funct. Ecol.* **2010**, *24*, 971–983. [CrossRef]
8. Pamplona, R.; Costantini, D. Molecular and structural antioxidant defenses against oxidative stress in animals. *Am. J. Physiol. Regul. Integr. Comp. Physiol.* **2011**, *301*, 843–863. [CrossRef]
9. Balaban, R.S.; Nemoto, S.; Finkel, T. Mitochondria, oxidants, and aging. *Cell* **2005**, *120*, 483–495. [CrossRef]
10. Skrha, J. Caloric restriction and oxidative stress. In *Oxidative Stress in Vertebrates and Invertebrates: Molecular Aspects of Cell Signaling*, 1st ed.; Blackwell: Hoboken, NJ, USA, 2012; p. 424.
11. Sohal, R.S.; Donato, H.; Biehl, E.R. Effect of age and metabolic rate on lipid peroxidation in the housefly, *Musca domestica* L. *Mech. Ageing. Dev.* **1981**, *16*, 159–167. [CrossRef]
12. Ku, H.H.; Sohal, R.S. Comparison of mitochondrial pro-oxidant generation and anti-oxidant defenses between rat and pigeon: Possible basis of variation in longevity and metabolic potential. *Mech. Ageing. Dev.* **1993**, *72*, 67–76. [CrossRef]
13. Zielinski, S.; Pörtner, H.O. Oxidative stress and antioxidative defense in cephalopods: A function of metabolic rate or age? *Comp. Biochem. Physiol. Part B Biochem. Mol. Biol.* **2000**, *125*, 147–160. [CrossRef]
14. Barja, G. Aging in vertebrates, and the effect of caloric restriction: A mitochondrial free radical production DNA damage mechanism? *Biol. Rev. Camb. Philos. Soc.* **2004**, *79*, 235–251. [CrossRef] [PubMed]
15. Couture, P.; Hulbert, A.J. Membrane fatty acid composition is related to body mass in mammals. *J. Membr. Biol.* **1995**, *148*, 27–39. [CrossRef] [PubMed]
16. Brand, M.D.; Turner, N.; Ocloo, A.; Else, P.L.; Hulbert, A.J. Proton conductance and fatty acyl composition of liver mitochondria correlates with body mass in birds. *Biochem. J.* **2003**, *376*, 741–748. [CrossRef] [PubMed]
17. Pamplona, R.; Portero-Otin, M.; Riba, D. Mitochondrial membrane peroxidizability index is inversely related to maximum lifespan in mammals. *J. Lipid. Res.* **1998**, *39*, 1989–1994. [CrossRef] [PubMed]
18. Barja, G. Rate of generation of oxidative stress-related damage and animal longevity. *Free Radic. Biol. Med.* **2002**, *33*, 1167–1172. [CrossRef]
19. Pamplona, R.; Barja, G.; Portero-Otin, M. Membrane fatty acid unsaturation, protection against oxidative stress, maximum life span: A homeoviscous-longevity adaptation? *Ann. NY Acad. Sci.* **2002**, *959*, 475–490. [CrossRef]
20. Hulbert, A.J.; Else, P.L. Membranes as possible pacemakers of metabolism. *J. Theor. Biol.* **1999**, *199*, 257–274. [CrossRef]
21. Munro, D.; Blier, P.U. The extreme longevity of *Arctica Islandica* is associated with increased peroxidation resistance in mitochondrial membranes. *Aging Cell* **2012**, *11*, 845–855. [CrossRef]
22. Pamplona, R. Low fatty acid unsaturation protects against lipid peroxidation in liver mitochondria from long-lived species: The pigeon and human case. *Mech. Ageing. Dev.* **1996**, *86*, 53–66. [CrossRef]
23. Rodríguez, E.; Dégletagne, C.; Hagen, T.M.; Abele, D.; Blier, P.U. Mitochondrial Traits Previously Associated with Species Maximum Lifespan Do Not Correlate with Longevity Across Populations of the Bivalve *Arctica islandica*. *Front. Physiol.* **2019**, *10*, 946. [CrossRef] [PubMed]
24. Speakman, J.R. Correlations between physiology and lifespan—Two widely ignored problems with comparative studies. *Aging Cell* **2005**, *4*, 167–175. [CrossRef] [PubMed]
25. Abele, D.; Brey, T.; Philipp, E. Bivalve models of aging and the determination of molluscan lifespans. *Exp. Gerontol.* **2009**, *44*, 307–315. [CrossRef] [PubMed]
26. Philipp, E.E.; Abele, D. Masters of longevity: Lessons from long-lived bivalves--a mini-review. *Gerontology* **2010**, *56*, 55–65. [CrossRef]
27. Zakhartsev, M.V.; Naumenko, N.V.; Chelomin, V.P. Non-methylene-interrupted fatty acids in phospholipids of the membranes of the mussel *Crenomytilus grayanus*. *Russ. J. Mar. Biol.* **1998**, *24*, 183–186.
28. Sadykhova, I.A. Gray mussel growth in the Peter-the-Great Bay (Sea of Japan). Biology of Gray mussel. *Nauka* **1983**, *1*, 62–68.

29. Zolotarev, V.N. Sklerohronologiya morskih dvustvorchatyh mollyuskov. Kiev. *Nauk. Dumka* **1989**, 112.
30. Selin, N.I. The composition and structure of a mixed population of *Crenomytilus grayanus* (Dunker, 1853) and *Modiolus kurilensis* (Bernard, 1983) (bivalvia: Mytilidae) in Peter the Great bay, Sea of Japan. *Russ. J. Mar. Bio.* **2018**, *449*, 363–372. [CrossRef]
31. Leskova, S.E. The growth of the Pacific mussel in the conditions of the Voevoda Bay (O. Russian). *Proc. Dalrybvtuz.* **2021**, *56*, 15–19.
32. Motavkin, P.A. Primorskiy Grebeshok. 1986. Available online: https://waves-vagues.dfo-mpo.gc.ca/library-bibliotheque/1162 22.pdf (accessed on 4 May 2023).
33. Ponurovsky, S.K.; Silina, A.V. Determination of the age and rate of linear growth of the Swift scallop. *Biol. Sea* **1983**, *1*, 20–24.
34. Karpenko, D.T. Growth rates of the Japanese scallop (*Chlamys farreri*) in the bays of the coastal zone of Russian Island (Peter-the-Great Bay, Sea of Japan). Complex research in the fisheries industry. In Proceedings of the V International Scientific and Technical Conference of Students, Postgraduates and Young Scientists Vladivostok, Vladivostok, Russia, 29 November 2020; pp. 14–18.
35. Viarengo, B.; Burlando, M.; Cavaletto, B.; Marchi, E.; Ponzano, J.; Blasco, J. Role of metallothionein against oxidative stress in the mussel *Mytilus galloprovincialis*. *Am. J. Physiol.* **1999**, *277*, 1612–1619. [CrossRef]
36. Buege, J.A.; Aust, S.D. Microsomal lipid peroxidation. *Meth. Enzymol.* **1978**, *52*, 302–310. [CrossRef]
37. Mesquita, C.S.; Oliveira, R.; Bento, F.; Geraldo, D.; Rodrigues, J.V.; Marcos, J.C. Simplified 2,4-dinitrophenylhydrazine spectrophotometric assay for quantification of carbonyls in oxidized proteins. *Anal. Biochem.* **2014**, *458*, 69–71. [CrossRef] [PubMed]
38. Markwell, M.; Haas, S.; Bieber, L.; Tolbert, N. A modification of the Lowry procedure to simplify protein determination in membrane and lipoprotein samples. *Anal. Biochem.* **1978**, *82*, 206–210. [CrossRef]
39. Kramer, J.K.G.; Fouchard, R.C.; Jenkins, K.J. Differences in chromatographic properties of fused silica capillary columns, coated, crosslnked, bonded, or crosslincked and bonded with polyethylene glycols (Carbowax 20M) using complex fatty acid methyl ester mixtures. *J. Chromatogr. Sci.* **1985**, *23*, 54–56. [CrossRef]
40. Christie, W.W. Equivalent chain length of methyl ester derivatives of fatty acids on gas chromatography. *J. Cromatogr.* **1988**, *447*, 305–314. [CrossRef]
41. Carrol, K.K. Quantitative estimation of peak areasin gas-liquid chromatography. *Nature* **1961**, *191*, 377–378. [CrossRef]
42. Carreau, J.P.; Dubacq, J.P. Adaptation of macro-scale method to the microscale for faty acid methyl trans esterification of biologicall lipid extracts. *J. Chromatogr. Sci.* **1985**, *23*, 54–56. [CrossRef]
43. Crockett, E.L. The cold but not hard fats in ectotherms: Consequences of lipid restructuring on susceptibility of biological membranes to peroxidation, a review. *J. Comp. Physiol. B* **2008**, *178*, 795–809. [CrossRef]
44. Naudí, A.; Jové, M.; Ayala, V.; Portero-Otín, M.; Barja, G.; Pamplona, R. Membrane lipid unsaturation as physiological adaptation to animal longevity. *Front. Physiol.* **2013**, *17*, 372. [CrossRef]
45. Hulbert, A.J. On the importance of fatty acid composition of membranes for aging. *J. Theor. Biol.* **2005**, *234*, 277–288. [CrossRef]
46. Valencak, T.G.; Ruf, T. N-3 polyunsaturated fatty acids impair lifespan but have no role for metabolism. *Aging Cell* **2007**, *6*, 15–25. [CrossRef]
47. Hanuš, L.O.; Levitsky, D.O.; Shkrob, I.; Dembitsky, V.M. Plasmalogens, fatty acids and alkyl glyceryl ethers of marine and freshwater clams and mussels. *Food Chem.* **2009**, *116*, 491–498. [CrossRef]
48. Yin, D. Biochemical basis of lipofuscin, ceroid, and age pigment-like fluorophores. *Free Radic. Biol. Med.* **1996**, *21*, 871–888. [CrossRef] [PubMed]
49. Labinskyy, N.; Csiszar, A.; Orosz, Z. Comparison of endothelial function, $O^{2}$ ·and $H_2O_2$ production, and vascular oxidative stress resistance between the longest-living rodent, the naked mole rat, and mice. *Am. J. Physiol. Heart Circ. Physiol.* **2006**, *291*, 2698–2704. [CrossRef] [PubMed]
50. Istomina, A.; Yelovskaya, O.; Chelomin, V.; Karpenko, A.; Zvyagintsev, A. Antioxidant activity of Far Eastern bivalves in their natural habitat. *Mar. Environ. Res.* **2021**, *169*, 105383. [CrossRef]
51. Ungvari, Z.; Ridgway, I.; Philipp, E.E.R. Extreme Longevity Is Associated with Increased Resistance to Oxidative Stress in *Arctica islandica*, the Longest-Living Non-Colonial Animal. *J. Gerontol. Ser. A Biomed. Sci. Med. Sci.* **2011**, *66*, 741–750. [CrossRef]
52. Munro, D.; Pichaud, N.; Paquin, F.; Kemeid, V.; Blier, P.U. Low hydrogen peroxide production in mitochondria of the long-lived *Arctica islandica*: Underlying mechanisms for slow aging. *Aging Cell* **2013**, *12*, 584–592. [CrossRef] [PubMed]
53. Ungvari, Z.; Csiszar, A.; Sosnowska, D. Testing Predictions of the Oxidative Stress Hypothesis of Aging Using a Novel Invertebrate Model of Longevity: The Giant Clam (*Tridacna derasa*). *J. Gerontol. Ser. A Biomed. Sci. Med. Sci.* **2012**, *68*, 359–367. [CrossRef]
54. Ungvari, Z.; Sosnowska, D.; Mason, J.B. Resistance to Genotoxic Stresses in *Arcticai slandica*, the Longest Living Noncolonial Animal: Is Extreme Longevity Associated with a Multistress Resistance Phenotype? *J. Gerontol. Ser. A Biomed. Sci. Med. Sci.* **2013**, *68*, 521–529. [CrossRef] [PubMed]
55. Kapahi, P.; Boulton, M.E.; Kirkwood, T.B. Positive correlation between mammalian life span and cellular resistance to stress. *Free Radic. Biol. Med.* **1999**, *26*, 495–500. [CrossRef] [PubMed]
56. Perez, V.I.; Buffenstein, R.; Masamsetti, V. Protein stability and resistance to oxidative stress are determinants of longevity in the longest-living rodent, the naked mole rat. *Proc. Natl. Acad. Sci. USA* **2009**, *106*, 3059–3064. [CrossRef] [PubMed]

biology

MDPI

Article

# Benthic Heterotrophic Protist Communities of the Southern Baltic Analyzed with the Help of Curated Metabarcoding Studies

**Maria Sachs, Manon Dünn and Hartmut Arndt ***

Institute of Zoology, General Ecology, Biocenter Cologne, University of Cologne, Zuelpicherstr. 47b, 51069 Cologne, Germany; msachs1@smail.uni-koeln.de (M.S.); manon.duenn@uni-koeln.de (M.D.)
* Correspondence: hartmut.arndt@uni-koeln.de

**Simple Summary:** Unicellular eukaryotes (organisms with a nucleus), or protists, are an extremely diverse group of organisms and inhabit almost all environments. In the world's oceans, they make up a large proportion of the overall diversity. Many heterotrophic protists feed on bacteria and, in this way, not only control bacterial abundance but also transport the bacterial-derived carbon to organisms at higher trophic levels in the food web. In recent years, many studies have focused on assessing the diversity of planktonic protists (organisms in the water column), but studies on seafloor dwelling (benthic) protists are much less frequent. So far, there are no extensive studies present that try to access the benthic protist communities in the Baltic Sea, one of Earth's largest brackish water environments. Within our study, we try to make a first assessment of this diversity using the molecular technique of metabarcoding, which allows the simultaneous identification of many organisms from one sample via the barcoding of nucleic acids, such as DNA and RNA. To obtain an overview of how certain environmental factors such as salinity and water depth of the sediment may influence the community structure, we chose two regions of the southern Baltic.

**Abstract:** Heterotrophic protists are key components of marine ecosystems. They act as controllers of bacterial and microphytobenthos production and contribute significantly to the carbon flux to higher trophic levels. Still, metabarcoding studies on benthic protist communities are much less frequent than for planktonic organisms. Especially in the Baltic Sea, representing the largest brackish water environment on earth, so far, no extensive metabarcoding studies have been conducted to assess the diversity of benthic protists in this unique and diverse habitat. This study aims to give first insights into the diversity of benthic protist communities in two different regions of the Baltic Sea, Fehmarnbelt, and Oderbank. Using amplicon sequencing of the 18S rDNA V9 region of over 100 individual sediment samples, we were able to show significant differences in the community composition between the two regions and to give insights into the vertical distribution of protists within the sediment (0–20 cm). The results indicate that the differences in community composition in the different regions might be explained by several abiotic factors such as salinity and water depth, but are also influenced by methodological aspects such as differences between DNA and RNA results.

**Keywords:** Baltic Sea; brackish; unicellular eukaryotes; diversity; amplicon sequencing; sediment

## 1. Introduction

The marine benthal represents the largest habitat on earth, yet most studies focusing on the marine environment target the planktonic community. This is particularly detrimental for protist research, as benthic protist communities exhibit key ecosystem functions as main controllers of the bacterial and microphytobenthos production and the transfer of organic carbon to higher trophic levels. Moreover, marine sediments serve as seedbanks for planktonic communities [1] and, in this way, always comprise a mixture of actually active organisms, dormant stages, and free DNA [2]. Protists in general make up a large proportion of the molecular (and hence also functional) diversity in marine ecosystems [3]. This large diversity is naturally associated with a spread over several trophic levels. While phototrophic protists (e.g., diatoms and some dinoflagellates) may act as important contributors to primary production in shallow waters, heterotrophic protists, which form the focus of the present study, are voracious consumers of prokaryotes attached to particles in the sediment or being suspended in the pore water [4]. In addition, they act as decomposers for marine detritus [5], and as parasites [6,7]. The biology of heterotrophic protists comprises a large variety of lifestyles that have a strong influence on the marine carbon cycle through multiple food web connections [4].

Their adaptations to several trophic conditions allow them to exist in oxic as well as anoxic [8] environments and can make them indicators for certain environmental factors in the benthic realm.

The enclosed Baltic Sea represents one of the largest brackish water environments on Earth [9]. Through the inflow of saline water from the North Sea on the one hand and the inflow of freshwater from different rivers on the other hand, the Baltic Sea waters are stratified and offer a variety of salinity changes vertically and horizontally. The Baltic Sea is relatively shallow, with a mean depth of 60 m and—from a geological point of view—with 10,000 to 15,000 years being rather young [10] (its ecological age being approx. 8000 years [9]). The Baltic Sea has a large catchment area with heavy exploitation by humans, such as fisheries, pollution, and nutrient inflow via riverine runoff [10].

Within the Baltic Sea, several studies have focused on planktonic protist diversity, using "classic" cultivation methods [11,12] and clone libraries [13] within suboxic and anoxic waters [14], but also through metabarcoding in estuary regions [15] or along a salinity gradient [16]. Regarding benthic protists, some studies assessed diversity over live counting and staining [17–19] for the small-scale vertical distribution of heterotrophic protists in the sediment. Larger studies, estimating a broader benthic protist community (e.g., over metabarcoding), are so far missing for the Baltic Sea. Thus, the state of knowledge about benthic protists communities that are most likely shaped by the various abiotic conditions described above is poor with regard to the Baltic Sea. Previous studies have shown that especially grain size [20], as well as salinity [21], are important factors influencing benthic protist communities.

Earlier studies [22] usually tried to asses biodiversity through live counting of morphotypes, a task that not only requires a deep knowledge of the morphological characteristics of certain species, but can also be biased by the occurrence of cryptic species [23]. The large amount of metabarcoding studies in the past 10 years has contributed to resolving these issues and also unveiled several new protist lineages [24,25]. Nevertheless, classic methods have not lost their power. Classic taxonomical work gives sequences a "face" and an ecological meaning and is the backbone of public databases, without which an assignment and interpretation of the myriad of sequences produced through extended metabarcoding studies would simply not be possible.

Based on previous studies regarding benthic protist communities in littoral sites [22], we aimed to assess the diversity of benthic protist communities of the Baltic Sea through metabarcoding of the V9 region of the 18S rDNA in two sublittoral regions of the southern Baltic Sea.

## 2. Materials and Methods

### 2.1. Sampling

The sampling for this study took place during two different cruises. The research vessel R/V Elisabeth Mann Borgese (EMB238) collected sediment samples from Fehmarnbelt at eight stations in 2020, four within the marine protected area (MPA) and four within a reference area. In 2021, sediment samples were taken at eight stations in the Oderbank region (EMB267), five stations from the MPA, and three from the reference area (Figure 1). For the metabarcoding studies, only a selection of samples was analyzed by the metabarcoding study (see Table 1). All sediment samples were taken with a Multicorer System (MUC).

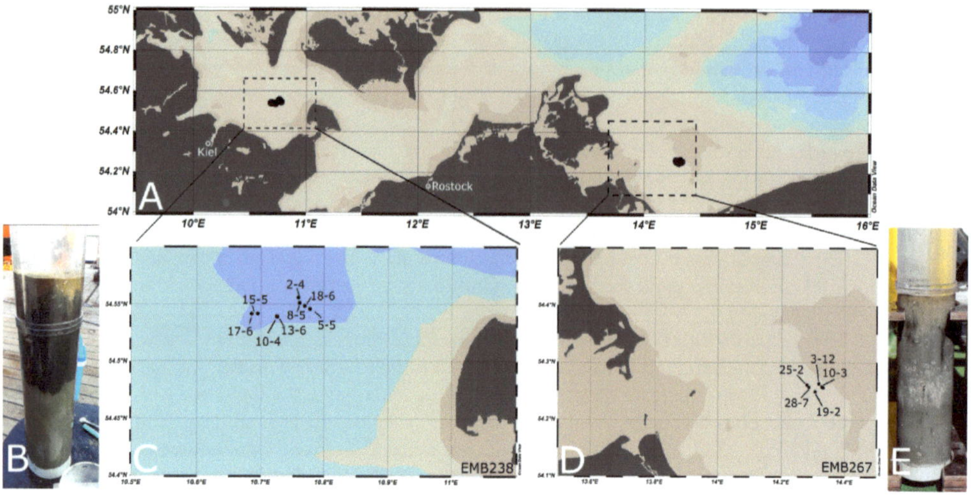


**Figure 1.** (**A**) Sampling regions of the two cruises in the western Baltic Sea, (**B**) MUC core taken from sediments of the Fehmarnbelt region, (**C**) close up of the sampling stations in the Fehmarnbelt region during cruise EMB238, (**D**) close up of the sampling stations in the Oderbank region during cruise EMB267, (**E**) MUC core taken from sediments of the Oderbank region. Maps were created using Ocean Data View [26].

**Table 1.** List of sampling stations relevant to this study. Sampling region (FB = Fehmarnbelt, OB = Oderbank), station/cast in the region, area (MPA = marine protected area, Ref. area = reference area), and the GPS position of the stations are given. The depth intervals at which MUC cores were cut are indicated (cmbsf = cm below seafloor), and cruise number and sediment type at the stations are added.

| Region | Station/Cast | Area | Longitude/Latitude | Depth Intervals [cmbsf] | Depth [m] | Cruise | Sediment Type |
|---|---|---|---|---|---|---|---|
| FB | 2-4 | MPA | 54°33.37′ 10°45.52′ | 0–1, 1–2, 2–4, 4–6, 6–10,10–15,15–20 | 23.5 | EMB238 | muddy |
| FB | 5-5 | MPA | 54°32.77′ 10°46.61′ | 0–1, 1–2, 2–4, 4–6, 6–10,10–15,15–20 | 23 | EMB238 | muddy |
| FB | 8-5 | MPA | 54°33.08′ 10°45.63′ | 0–1, 1–2, 2–4, 4–6, 6–10,10–15,15–20 | 23.9 | EMB238 | muddy |
| FB | 10-4 | Ref. area | 54°32.36′ 10°43.49′ | 0–1, 1–2, 2–4, 4–6, 6–10,10–15,15–20 | 22.8 | EMB238 | muddy |
| FB | 13-6 | Ref. area | 54°32.34′ 10°43.55′ | 0–1, 1–2, 2–4, 4–6, 6–10,10–15,15–20 | 23 | EMB238 | muddy |
| FB | 15-5 | Ref. area | 54°32.51′ 10°41.71′ | 0–1, 1–2, 2–4, 4–6, 6–10,10–15,15–20 | 23.2 | EMB238 | muddy |
| FB | 17-6 | Ref. area | 54°32.5′ 10°41.16′ | 0–1, 1–2, 2–4, 4–6, 6–10,10–15,15–20 | 23 | EMB238 | muddy |
| FB | 18-6 | MPA | 54°32.93′ 10°46.11′ | 0–1, 1–2, 2–4, 4–6, 6–10,10–15,15–20 | 24.4 | EMB238 | muddy |
| OB | 3-12 | MPA | 54°15.774′ 14°19.148′ | 0–1, 1–2, 2–3, 3–4, 6–7, 8–9, 10–11 | 15.3 | EMB267 | sandy |
| OB | 10-3 | MPA | 54°15.438′ 14°19.733′ | 0–1, 1–2, 2–3, 3–4, 6–7, 9–10, 10–15 | 14.9 | EMB267 | sandy |

**Table 1.** *Cont.*

| Region | Station/Cast | Area | Longitude/Latitude | Depth Intervals [cmbsf] | Depth [m] | Cruise | Sediment Type |
|---|---|---|---|---|---|---|---|
| OB | 19-2 | Ref. area | 54°14.934′ 14°18.435′ | 0–1, 1–2, 2–3, 3–4, 6–7, 9–10, 14–15 | 15.5 | EMB267 | sandy |
| OB | 25-2 | Ref. area | 54°15.655′ 14°16.873′ | 0–1, 1–2, 2–3, 3–4, 6–7, 9–10, 13.5–14.5 | 15.9 | EMB267 | sandy |
| OB | 28-7 | Ref. area | 54°15.406′ 14°17.241′ | 0–1, 1–2, 2–3, 3–4, 6–7, 9–10, 14–15 | 15.5 | EMB267 | sandy |

For each station, three cores per station were taken and sliced into seven layers. If cores were too short, the interval was adjusted (Table 1). The chosen sampling regions significantly differed in environmental conditions. At Fehmarnbelt, the sediment was fine, dense, and muddy with a median grain size of around 55 µm, and the salinity at the bottom ranged around 19 PSU; in the Oderbank region, the sediment was much coarser and sandy with a median grain size of around 178 µm, and the salinity at the bottom ranged around 8 PSU.

### 2.2. DNA and RNA Extraction and cDNA Synthesis

For the Fehmarnbelt stations 17-6 and 18-6, we extracted DNA and RNA of each of the three cores, each with seven depth intervals by using the ZymoBIOMICS DNA/RNA Miniprep Kit (Zymo Research, Freiburg, Germany) using 250 mg of sediment per sample, as recommended for soil samples by the manufacturer. In principle, the kit allows a simultaneous extraction of DNA and RNA from the same sample. When RNA concentration after extraction was not sufficient for downstream processing using this kit, RNA was re-isolated using the RNeasy PowerSoil Total RNA® Kit (Qiagen, Hilden, Germany) using 2 g of sediment. For the remaining stations, we extracted DNA from each of the three cores per station, but only for the upper two cm (0–1, 1–2 cmbsf = cm below seafloor), using the DNeasy Power Lyzer Power Soil® DNA Isolation Kit (Qiagen, Hilden, Germany) to extract whole genomic DNA, but added additional pre-washing steps with three different washing solutions to improve downstream applications through removing potential contaminants [27,28] and adding further heating steps after bead beating [27].

For sediment samples from the Oderbank, we extracted only RNA using the RNeasy PowerSoil Total RNA® Kit (Qiagen, Hilden, Germany). When RNA concentrations were too low after using 2 g of sediment per sample, we doubled the amount of sediment to 4 g, as recommended by the manufacturer.

For both sampling regions, RNA was synthesized to cDNA using the Thermo Scientific First strand cDNA Synthesis Kit (Thermo Fisher, Waltham, MA, USA) with an RNA template concentration of approx. 500 ng/µL per reaction using random hexamer primers included in the kit.

### 2.3. PCR Amplification and High-Throughput Sequencing

After quantification of total DNA and cDNA with a Quantus Fluorometer (Promega, Germany), the hypervariable V9 region of the 18S rDNA was amplified using the eukaryotic primer set 1389F (5′-TTG TAC ACA CCG CCC-3′) and 1510R (5′-CCT TCY GCA GGT TCA CCT AC-3′) [29] via PCR reaction. PCR mixtures contained 50 ng of total DNA/cDNA template, a final concentration of 0.35 µM for each primer, and VWR Red Taq DNA Polymerase Master Mix (VWR, Germany). The thermal program started with an initial denaturation step at 98 °C for 30 s followed by 25 cycles at 98 °C for 10 s, 57 °C for 30 s, 72 °C for 30 s, and completed with a final elongation step at 72 °C for 10 min. Chimera formation during PCR was reduced by a low number of cycles (25) [30]. To reduce intra-sample variability, PCR reactions were performed in triplicates. Because the results of metabarcoding data strongly depend on the targeted marker region, the hyper-variable V9 region was selected. While being much shorter than the hyper-variable V4 region and less present in public databases, the V9 region represents a good compromise to make a broad diversity of marine taxa visible, but also to recognize some rare species that are neglected when using V4 primers [31].

For subsequent quality measures during data analysis, we created an in vitro community, called a "mock community", comprising DNA of nine different protist cultures (Table 2) from the HFCC (Heterotrophic Flagellate Culture Collection Cologne). The species were chosen as representatives of the main protist supergroups. DNA of those cultures was isolated using the Quick g-DNA Miniprep kit (Zymo Research, Freiburg, Germany), amplified by PCR (V9 region of the 18S rDNA), purified, and quantified as described for the samples. PCR products of each member of the mock community were then pooled (50 ng of purified PCR product/strain) and added to each individual Next Generation Sequencing run. The Cologne Center for Genomics (CCG, University of Cologne) then performed a paired-end NovaSeq sequencing (2 × 150 bp) run of the amplified fragments.

**Table 2.** List of organisms used for the "mock community".

| HFCC No. | Species | Protist Group |
|---|---|---|
| 171 | Rhynchomonadidae undet. | Kinetoplastida |
| 175 | *Fabomonas tropica* | Ancyromonadida |
| 176 | *Massisteria marina* | Cercozoa |
| 178 | *Ministeria vibrans* | Opisthokonta |
| 203 | *Cafeteria burkhardae* | Stramenopiles |
| 744 | *Aristerostoma* sp. | Ciliophora |
| 766 | *Protocruzia* sp. | Ciliophora |
| 768 | *Halocafeteria* sp. | Stramenopiles |
| 828 | *Neobodo* sp. | Kinetoplastida |

### 2.4. Bioinformatic Processing

After sequencing, the raw reads were demultiplexed and processed as follows: barcode and primer sequences were clipped using cutadapt version 2.8 with parameters set to *no-indels*, $m = 30$, and $e = 0$ for the barcodes and $e = 0.2$ for the primer sequences [32]. The next steps were conducted using the dada2 package [33] in R version 4.1.2, starting with the *filter and trim* command and setting the parameters $maxEE = 1$, $truncQ = 11$, $truncLen = (125, 120)$, and $maxN = 0$ for quality filtering of the reads. The *errF* and *errR* functions were used to learn the error rates for the dataset. The *derepFastq* function was used for the dereplication of sequences and ASVs were inferred with the *dada* function. The *mergePairs* command merged paired reads with a minimum overlap of 12 nucleotides. As a last quality filtering step, chimeric sequences were removed using the *removeBimeraDenovo* function. By the addition of the V9 region of 150 protist strains from the Heterotrophic Flagellate Collection Cologne, we enlarged the existing PR$^2$ database and used it for taxonomic assignment of ASVs via the pairwise alignment function *usearch_global* (version v2.18.0; [34]). Retaining only heterotrophic protist sequences, Metazoa, fungi, autotrophic protists (determined on the basis of taxonomic assignment), as well as unassigned sequences were removed, keeping only ASVs with a pairwise identity of >80% to a reference sequence. As a last filtering step, we used the previously described mock community. Each library preparation was accompanied by one individual mock community, resulting in a total of 18 mock community datasets that were analyzed prior to sample analysis, as described above. For the main dataset of samples, we then chose individual minimum thresholds per sample according to the accompanying mock community on the respective sequencing lane. For calculation of these thresholds, we used the proportion of the lowest read number of an ASV in the mock community data set that could be assigned to the cultured species. ASVs in the sample data sets with a smaller read number than this calculated proportion were discarded. For the 18 accompanying mock communities, the calculated thresholds ranged between 0.02 and 0.07%.

### 2.5. Statistical Analyses

All statistical analyses as well as figures were conducted and plotted with RStudio v2023.03.0. To estimate sequence quality and depth, we calculated rarefaction curves as well as Shannon indices to compare the alpha diversity using the *vegan* package [35]. Non-metric multidimensional scaling (NMDS) analyses were performed to calculate the differences in protist communities between different sediment depths and sampling stations/regions. Therefore, the dissimilarity matrix was calculated based on the Jaccard distance. To compare if those differences were significant, we performed permutational multivariate analyses of variances (PermANOVA) using the *adonis* and *pairwise.adonis* functions. To visualize the proportion of shared and unique ASVs between stations and sediment layers, we used both the R package UpSetR [36] as well as the Treemapify package. To test whether abiotic factors such as salinity, grain size, and water depth had a significant impact on the community composition, we conducted a canonical correspondence analysis (CCA) using the *vegan* package [35] followed by a Monte Carlo permutation test.

The dataset used for the analysis of Fehmarnbelt consisted of six stations with three replicates for the upper 0–2 cm layer and two stations with three replicates for the seven-layer depth profile derived from DNA. Additionally, RNA extractions of samples from the vertical profile of these two stations were analyzed. The dataset for the Oderbank region is smaller, consisting of five stations with five-layer depth profiles (for the two deepest sediment layers, RNA yield was never sufficient for downstream analyses).

## 3. Results

### 3.1. Alpha Diversity of Benthic Protists in the Southern Baltic

After sequencing with NovaSeq, we received data for 129 sediment samples, resulting in a read number of 444,473,336 raw, demultiplexed reads, and 210,074 ASVs for the whole dataset. This results in an average of 3.4 ± 2.9 million reads per sample. Despite the high standard deviation (which was subsequently excluded from analysis), rarefaction curves of all but one sample reached saturation. Summed for sediment depth layers, all curves reached saturation (Figure 2). After the assembling and filtering steps, 293,254,105 reads could be assigned to a sequence from the V9 reference database with a pairwise identity of a minimum of 80%. After the exclusion of Metazoa, fungi, Streptophyta, and exclusively phototrophic taxa, 139,203,557 reads could be assigned to heterotrophic protists. After applying the read threshold derived from the mock community and after manual correction of ambiguous sequences, 78,023,157 reads were clustered into 1233 ASVs. From this dataset, only stations with complete depth profiles were used for further analyses.

In the Fehmarnbelt region, the uppermost sediment layers (0–2 cm) had an average of 39 ± 13 ASVs assigned to heterotrophic protists and the highest mean number of ASVs was found at station 15-5 with 44 ± 18 ASVs, while the lowest number was found at station 2-4 with 33 ± 3 ASVs (Figure 3). Differences in ASV numbers between the stations regarding these sediment layers were not significant (one-way ANOVA, $p > 0.5$). The Shannon index as an alpha-diversity measure ranged between 2.9 and 3.4, showing no significant difference between the stations (Kruskal–Wallis test, $p > 0.5$) in the upper sediment layers. In the uppermost 2 cm sediment, the highest mean number of reads was detected at station 18-6 with 363,238 ± 254,045 and the lowest at station 2-4 with 111,097 ± 56,751 reads at 0–2 cm sediment depth (Figure 3). Comparing the depth layers (0–1 cm, 1–2 cm, 2–4 cm, 4–6 cm, 6–10 cm, 10–15 cm, and 15–20 cm) of cores for stations 17-6 and 18-6, the highest mean number of ASVs was found at 6–10 cm sediment depth for both stations with 61 ± 23 ASVs at station 17-6 and 82 ± 9 ASVs at station 18-6 (Figure 3). The Shannon index for the different layers ranged between 2.5 and 4.1 (Figure 3), the differences were found to be significant (Kruskal–Wallis test, $p < 0.05$).

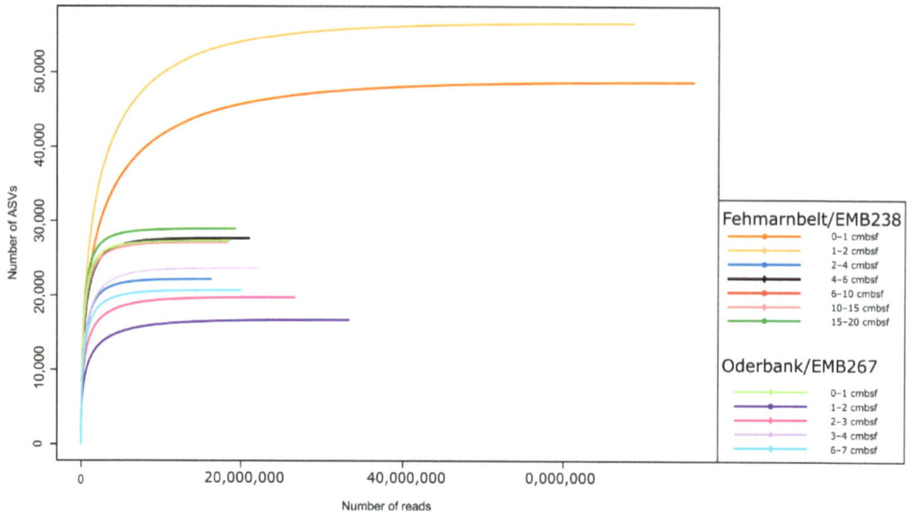


**Figure 2.** Rarefaction curves of samples from Fehmarnbelt and Oderbank region summed for sediment depth.

The highest mean number of reads was found for station 17-6 with 417,669 ± 464,385 (high standard deviation results from one sample with only 89,299 reads) at 15–20 cm sediment depth and the lowest number of reads was found for station 17-6 in the 2–4 cm sediment layer with 137,696 ± 8444 ASVs (Figure 3). For station 18-6, the highest mean number of reads was detected at 15–20 cm with 947,200 ± 540,776 ASVs and the lowest at 2–4 cm sediment depth with 333,787 ± 104,046 ASVs.

For station 17-6 from Fehmarnbelt, we found that for RNA, the mean ASV number was 54 ± 11, and for station 18-6, 46 ± 18. For RNA at station 18-6, the mean read numbers were 364,111 ± 349,887, and for station 17-6, 340,954 ± 184,354.

In the Oderbank region, the stations had an average number of 26 ± 10 ASVs, the highest number of ASVs was detected at station 19-2 with 42 ± 9 ASVs, and the lowest value at station 10-3 with 20 ± 4 ASVs. The differences in these numbers were not significant (Kruskal–Wallis test, $p > 0.5$). The Shannon index between the stations ranged between 2.0 and 3.6 but showed no significant differences (one-way ANOVA, $p > 0.5$). The lowest mean number of reads was detected at station 25-5 with 413,163 ± 191,111 reads, and the highest at station 19-2 with 639,953 ± 304,727 reads (Figure 3).

Looking at the different depth layers in the Oderbank region (0–1 cm, 1–2 cm, 2–3 cm, 3–4 cm, 6–7 cm) the lowest mean number of ASVs was found in layer 2–3 cm with 25 ± 6 ASVs, and the highest in layer 6–7 cm with 27 ± 10, as well as in 1–2 cm with 27 ± 13 ASVs. The differences in the numbers were significant (Kruskal–Wallis test, $p < 0.05$). The Shannon index ranged between 2.0 and 3.6 and was found to not be significantly different (one-way ANOVA, $p > 0.05$). The lowest mean number of reads could be detected at 2–3 cm sediment depth with 428,497 ± 292,736 and the highest at 0–1 cm with 699,095 ± 342,518 reads (Figure 3).

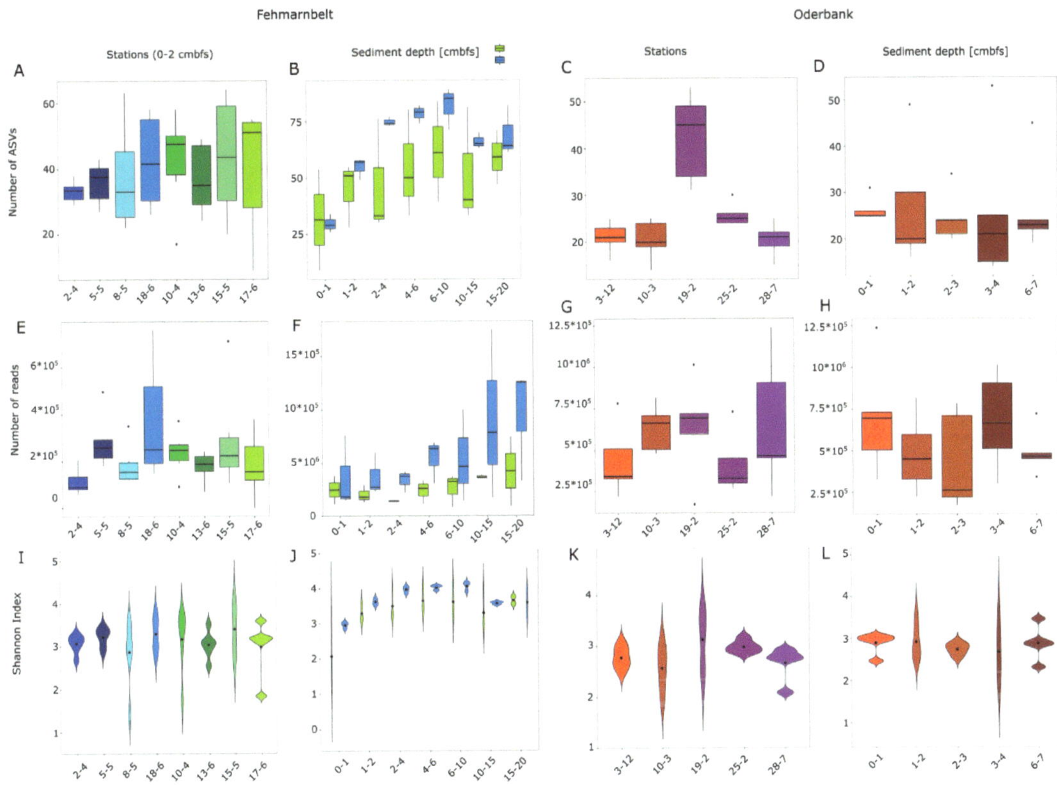


**Figure 3.** Results of metabarcoding studies of two regions in the western Baltic, Fehmarnbelt, and Oderbank. (**A**) Number of ASVs per station for Fehmarnbelt; (**B**) number of ASVs per sediment depth at Fehmarnbelt; (**C**) number of ASVs per station for Oderbank; (**D**) number of ASVs per sediment depth at Oderbank; (**E**) number of reads per station for Fehmarnbelt and (**G**) for Oderbank; (**F**) number of reads per sediment depth at Fehmarnbelt and (**H**) at Oderbank; (**I**) Shannon index per station at Fehmarnbelt and (**K**) at Oderbank; and (**J**) Shannon index per sediment depth at Fehmarnbelt and (**L**) at Oderbank.

### *3.2. Protist Community Composition at Different Regions and Sediment Depths*

The data for the uppermost 2 cm of sediment at all stations of the Fehmarnbelt were dominated by Ciliophora (Figure 4A) with relative proportions of ASVs between 24% (station 17-6) and 38% (station 2-4), followed by Dinoflagellata with 13% (station 10-4) up to 21% (station 13-6), followed by Cercozoa with relative proportions between 11% (station 13-6) and 17% (station 17-6). The largest proportions of ciliate ASVs belong to the Litostomatea (18%), Spirotrichea (18%), and Oligohymenophorea (20%). Among the Dinoflagellata, almost 80% of taxa belong to the Dinophyceae, and among the Cercozoa, most belong to the Filosa-Thecofilosea (55%).

The depth profiles (seven layers) for all three cores of stations 17-6 and 18-6 in Fehmarnbelt were compared. At station 17-6, almost all layers were again dominated by ASVs belonging to Ciliophora with 23% to 44% of relative proportions of ASVs, followed by either Stramenopiles (non-Ochrophyta) with 20% to 23%, Cercozoa with 18–20%, or Dinoflagellata with 17–24% of ASVs. Much lower proportions were reached by Katablepharidophyta, mainly in the lower sediment layers (3–15%, Figure 4B). The largest proportion of ciliate taxa belonged to the Litostomatea (20%), Labyrinthulea were most abundant among Stramenopiles (38%), and Dinophyceae dominated the Dinoflagellata (80%).

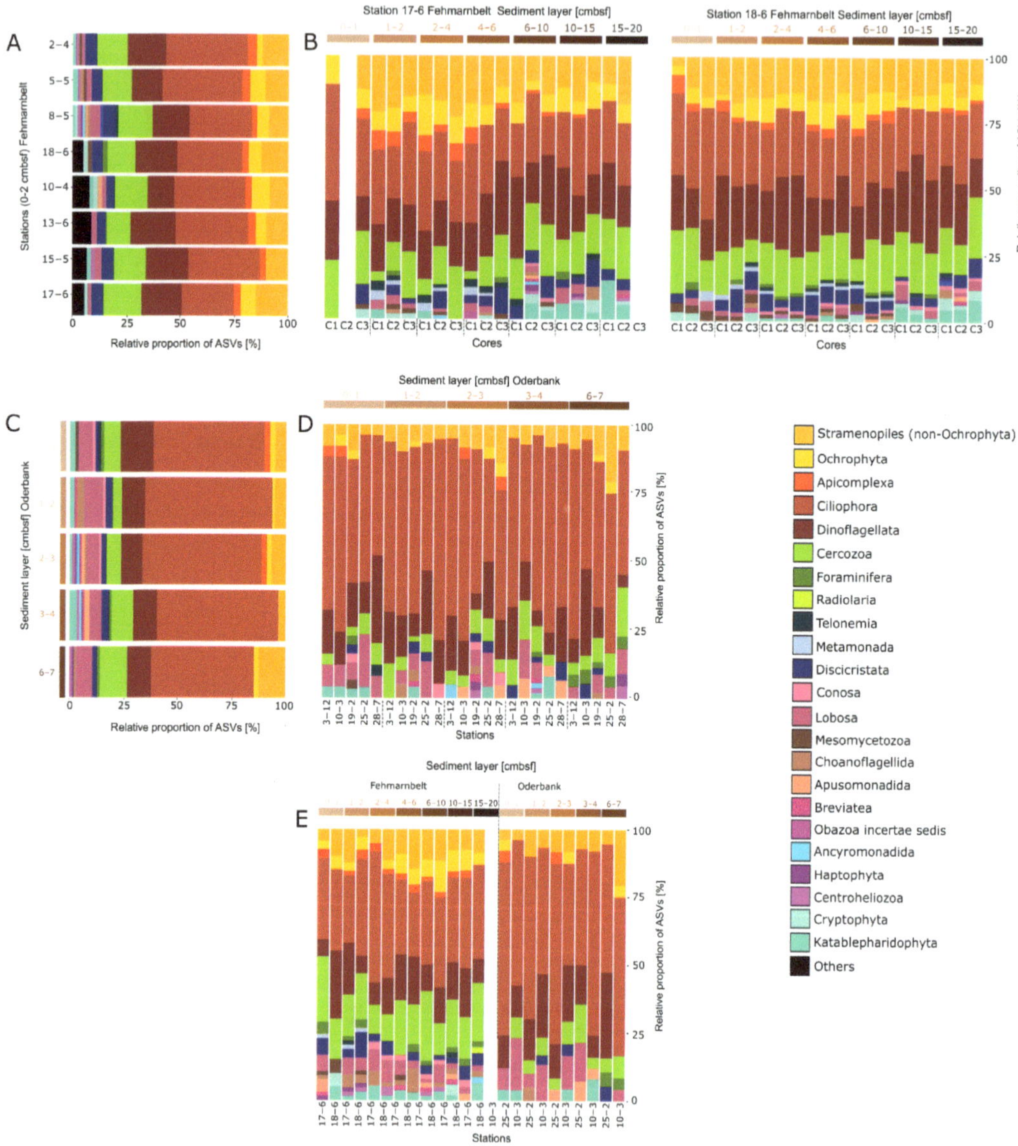


**Figure 4.** Community composition of benthic protists at Fehmarnbelt and Oderbank showing the relative proportion of ASVs assigned to taxonomic groups. (**A**) Comparison of all samples from 0–2 cmbsf at Fehmarnbelt; (**B**) protist community structure obtained from depth profiles of cores (7 different depths, each with 3 replicates) at stations 17-6 and 18-6 at Fehmarnbelt; (**C**) vertical distribution of the protist community structure at Oderbank (summed for all stations from 5 different depth layers); (**D**) vertical changes in community structure for the different stations at Oderbank; and (**E**) direct comparison of the protist community structure of two stations from both regions, Fehmarnbelt and Oderbank, regarding different sediment layers based on RNA ASVs.

A similar pattern was obtained for station 18-6. The most dominant groups of ASVs belonged to the Ciliophora (17–42%, with Oligohymenophorea and Spirotrichea both 19%),

Dinoflagellata (13–33%, with 80% Dinophyceae), Cercozoa (8–24%, 70% of which belong to Filosa-Thecofilosea) and Stramenopiles (non-Ochrophyta) with 3–18% (49% of which belong to Labyrinthulea). Again, a rise of Katablepharidopyhta taxa was observed towards deeper sediment layers. They contributed 5–9% of ASVs in the deepest layer 15–20 cm (vs. ~1% in the upper layers; Figure 4B).

In the Oderbank region, Ciliophora taxa were even more dominant in all sediment layers compared to the Fehmarnbelt. With relative proportions between 38 and 74%, they made up a large proportion of the whole community (Figure 4C,D). The largest proportion (20%) of ASVs belonged to Spirotrichea, followed by Karyorelicta (18%). The second largest relative proportion of ASVs was contributed by Dinoflagellata with 6–32% (with 70% belonging to the Dinophyceae), followed by Stramenopiles (5–21%, of which 30% belong to the MAST groups and 30% to bicosoecids) and Cercozoa (2–18%, with 62% belonging to the group of Filosa-Thecofilosea). Apart from that, no clear pattern of taxa distribution in relation to sediment depth was visible. Summing up ASVs of all stations sorted for sediment depth, one group (Telonemia) was only present in the uppermost layer, whereas the taxa belonging to the Katablepharidophyta were absent in the deepest layer (6–7 cm), while Cercozoa were present to a larger proportion in the deepest layer.

Apart from these minor differences in community composition, both regions did not show remarkable differences in the vertical distribution of protists when only large taxonomic groups are considered. As we could only obtain RNA data for two stations of the Fehmarnbelt region (stations 17-6 and 18-6, Figure 4E), we randomly chose two stations of the Oderbank region (stations 10-3 and 25-2) for direct comparison. While Fehmarnbelt samples were dominated by several groups in more similar proportions (Ciliophora 24–49%, Dinoflagellata 6–25%, Cercozoa 10–26%), Oderbank was highly dominated by Ciliophora (up to 74%). Samples from Fehmarnbelt seem to show a larger variety of taxonomic groups (22, compared to 12 at Oderbank). In this sense, a taxonomic group represents the taxonomic rank of division, above the class rank.

### 3.3. *Protist Beta-Diversity in Relation to Sediment Depth*

NMDS analyses revealed a higher resolution of taxonomic composition regarding sampling stations and sediment depth. In the 0–2 cm layer of the Fehmarnbelt samples, the protist communities formed significantly separate clusters regarding the different stations (permANOVA, $p = 0.001$, Figure 5A).

Stations 17-6 and 18-6, in particular, cluster quite separately, with almost no overlap with the other stations (Figure 5A). This phenomenon is still visible when the depth profiles of stations 17-6 and 18-6 are compared, where no significant differences between sediment depth but between the two stations were recorded (permANOVA, $p < 0.01$, Figure 5D,E). In the uppermost sediment layers, we found the highest number of unique ASVs at station 17-6 with a relative proportion of 16%, followed by station 18-6, with a relative proportion of 12%. Overall, the stations only shared 1% of ASVs, divided between the most dominant groups, with ASVs of Stramenopiles (non-Ochrophyta) at 33.3% and Ciliophora as well as Dinoflagellata both at 25%. With regard to the depth layers, the highest numbers of unique ASVs were detected in the deeper sediment layers, with 11% of unique ASVs found in 15–20 cm depth at station 18-6 followed by station 17-6 in 6–10 cm sediment depth, also with 11%. Overall, the two stations shared only 0.5% ASVs, with cercozoans being the most dominant group (33.3%).

For the Oderbank region, the NMDS analysis showed significant differences in community composition between the sediment layers (permANOVA, $p = 0.001$, Figure 5G), but not between the stations within the region. Therefore, the layers were summed up for all stations for comparison. With 16%, the highest number of unique ASVs was found in 6–7 cm depth, directly followed by 14% of unique ASVs in 0–1 cm (Figure 5H), explaining the significant differences between the layers. While the layers overall only shared 0.4% of ASVs, most taxa were shared out of the clade of Dinoflagellata at 60% (Figure 5I).

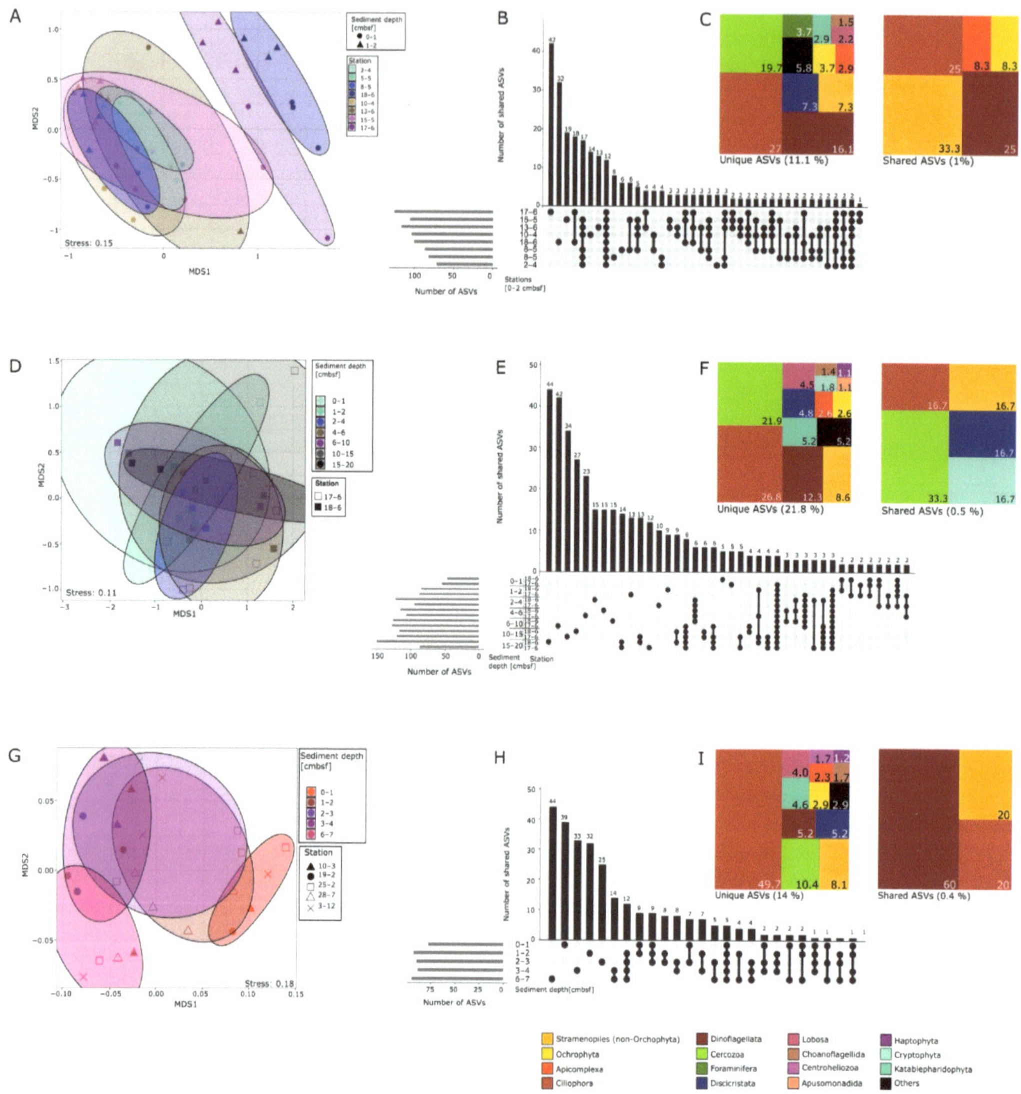


**Figure 5.** NMDS plot based on the Jaccard Index comparing benthic protist communities of the different stations in the western Baltic. (**A**) NMDS plot comparing all stations at Fehmarnbelt for the surface sediment layer 0–2 cmbsf. (**B**) Upset plot showing the number of shared ASVs between the different Fehmarnbelt stations (top bar chart) or unique to one station, as well as the overall number of ASVs (horizontal bars). Connected dots below the bar chart mean ASVs are shared between two or more stations. (**C**) Tree map showing the relative proportion of shared and unique ASVs per taxa group at Fehmarnbelt stations for the 0-2 cmbsf sediment layer. (**D–F**) Comparison of the vertical distribution of ASVs of the two Fehmarnbelt stations 17-6 and 18-6 in a similar manner as for (**A–C**). (**G–I**) Comparison of the community structure regarding all different sediment layers from stations at Oderbank analyzed in a similar manner as for (**A–C**).

The NMDS analysis revealed a clear separation of the compared protist communities from the two stations of Oderbank (stations 10-3 and 25-2) and Fehmarnbelt (stations 17-6 and 18-6) based on RNA-derived data (permANOVA, $p = 0.001$, Figure 6). On the basis of the rigid filtering of the data set using the mock community, the two chosen stations from Oderbank shared no ASVs with the two stations of Fehmarnbelt (Figure 6B,C). They instead displayed a high percentage of unique ASVs that were not shared between all 12 sediment layers (Figure 6B). The highest number of unique ASVs for Oderbank and Fehmarnbelt was found among the Ciliophora group.

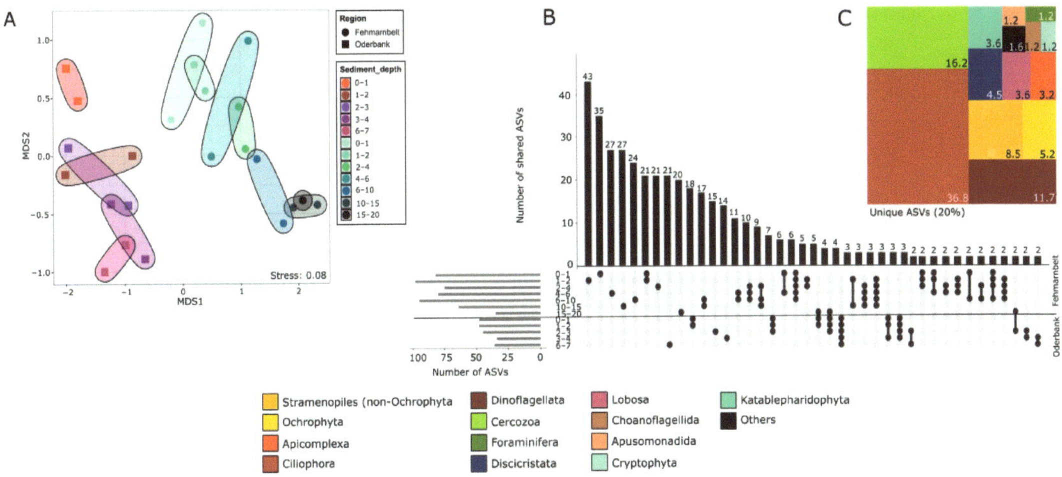


**Figure 6.** NMDS plot based on the Jaccard distance comparing protist communities in different sediment depth layers of two sampling stations in the Oderbank region and two in the Fehmarnbelt region (**A**). (**B**) Upset plot showing the number of shared or unique ASVs for the different sampling depths for both regions (bar chart at the top). Connected dots below each bar show shared ASVs between different depths and stations. Horizontal bars indicate the total number of ASVs for the two stations in each region. (**C**) Tree map showing relative proportions of ASVs for taxa groups unique to one depth and region.

## 4. Discussion

Even though metabarcoding studies of protist communities have become much more frequent in the past 15 years, the majority of studies still concentrate on pelagic protist communities [3]. While studies on benthic communities are scarce, benthic brackish water communities are even more poorly studied and metabarcoding studies of benthic protists of the Baltic Sea are basically non-existent. According to our knowledge, our study represents the first metabarcoding approach to estimate the benthic protist community of sediments in the Baltic Sea and aims to better understand their biodiversity and ecological roles. By targeting the V9 region of the 18S rDNA, we chose a suitable region to estimate the overall richness of the protist community in the Baltic Sea, also including rare taxa [31]. As mentioned above, benthic protist communities for the Baltic Sea have so far received relatively little attention. It is therefore likely that Baltic Sea-specific members of the protist community are underrepresented in the reference databases. To verify and improve the outcome of the analysis, we chose to add an additional filtering step using a mock community. The addition of a mock community as a supplementary sample in a next-generation sequencing run has been recommended by several studies [37–39], especially as a measure to eliminate "noisy" sequences. We adapted those ideas to create individual read thresholds for each library preparation. The rather strict limit values derived in this way served as an additional form of quality control. It has been shown that the overall impression of the community composition does not dramatically change when these thresholds are applied [39]. On the other hand, applying a strict filter increases the likelihood that differences in species composition between stations will be overemphasized. This could be the reason for the relatively high level of uniqueness we found for many species.

As no previous metabarcoding data on benthic protist communities in the Baltic Sea seem to exist, it is hard to compare our results on the basis of molecular data sets. Additionally, the specific nature of the brackish water environment allows only limited comparisons to studies from either marine or freshwater environments, and different bioinformatic pipelines may additionally influence the results. There are only a few quantitative and qualitative studies from the regions based on direct counts using light microscopy. Benthic ciliates were intensively studied at a station in the Kiel Bight [17], in the vicinity of the Fehmarnbelt stations; however, the water depth of the region studied by Sich [17] was much shallower and sandier than the region investigated in the present study. In another study of benthic ciliates in the vicinity of the Oderbank region [22], again the sampling site was shallow, though similar in the sediment quality. In both cited studies, karyorelictid, spirotrich, litostome, and oligohymenophoreans were dominant, comparable to the present investigation using molecular techniques. Regarding benthic flagellated protists, only the shallow-water study near Ruegen Island [22] was available for comparison. The comparison with our metabarcoding studies shows that flagellate groups recorded from live counting were also recovered by the metabarcoding studies.

Regarding the community composition obtained by our metabarcoding study, Ciliophora were the main dominant group regarding the number of ASVs in Baltic Sea sediments (with differing proportions regarding the region), but on the sides of read abundances, the MALV-I clade, a rather poorly studied group of marine Syndiniales with only a few cultured species [24], reached by far the highest read abundances. Sequences of the MALV group are known to dominate in DNA studies, which is, most plausibly, because they have higher rDNA copy numbers [40] and may not reflect actual activity. Still, also in the dataset from Oderbank derived only from RNA, an ASV representing a sequence from the MALV-I 4 group has the highest read abundance. In line with previous studies [24], the largest proportion of ASVs of the MALV clade in the dataset belonged to the MALV-1 group known to be predominant in anoxic environments and hydrothermal vents, and seems to be common in sediments.

Benthic ciliates are known to have a high species richness in brackish water environments [41], especially in the Baltic Sea [42], proposing that salinity can have a negative effect on species richness. It is therefore not surprising that our analyses showed a high proportion of ASVs belonging to ciliates. Similar patterns have been found for planktonic organisms in the Baltic Sea [43], refuting the theory that the taxonomic diversity of organisms is lowest in the horohalinicum [44]. A CCA of the complete dataset has shown that out of several abiotic factors, salinity had a significant effect on the benthic protist community (Monte Carlo permutation, $p$ = 0.001; Figure 7) as well as water depth ($p$ = 0.003) and sediment depth ($p$ = 0.002). For a Pacific littoral region, Gong et al. [45] showed that water depth had the strongest influence on α- and β-diversity of benthic protist communities.

Apart from salinity, grain size not only has an effect on functional ciliate diversity—implying that coarser sediment promotes free-swimming species with an elongated cell form, whereas fine sediment houses species with crawling behavior and flattened cell bodies [41]—but also on the abundance of ciliates, which was shown to be positively correlated to median sediment grain size [46,47]. In contrast to these studies, our results obtained from CCA analysis could not verify that grain size has a significant influence on the protist community as a whole. At Fehmarnbelt, a region with approx. 19 PSU salinity and median grain size of approx. 55 µm, the relative proportion of ASVs belonging to ciliate taxa make up about 44% of the overall number of ASVs, comprising 100 different ASVs of ciliate taxa in total. At Oderbank, we measured a salinity of about 8 PSU and a median grain size of 178 µm. Here, we found a much higher relative proportion of ASVs belonging to ciliates of up to 78%, with 143 different ASVs. Still, one has to keep in mind that the majority of samples from Fehmarnbelt were derived from DNA studies while that from Oderbank originate from RNA. Sediments can act as storages of DNA sunken down from the water column and therefore might contain also DNA of pelagic species, which might lead to an overestimation of diversity [48]. However, most of the dominant ciliate ASVs, for instance, belonged to well-known benthic taxa. Several protist species are known to have remarkably high abilities to adapt to different salinities. We showed that among other Stramenopiles of the genus *Cafeteria*, *Cafeteria baltica*, isolated from sediment of the Fehmarnbelt, can tolerate salinities between 0 and 125 PSU [49]. We, therefore, assume that at least some protist species are ubiquitously dispersed in the Baltic Sea, independent of the salinity. This assumption is supported by the fact that we were able to retrieve several protists from an accompanying cultivation approach that were also recovered from the dataset of Oderbank and Fehmarnbelt (Figure 8).

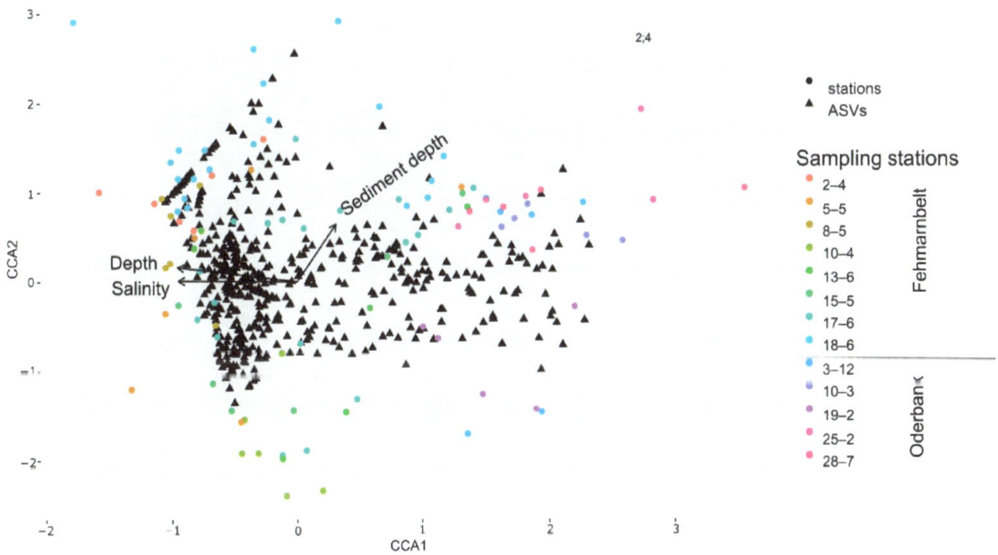


**Figure 7.** CCA analysis including a comparison of the benthic protist community composition based on ASVs including all stations of Fehmarnbelt and Oderbank regions analyzing the influence of water depth, sediment depth, and salinity.

Of course, the results of cultivation approaches may be biased by the fact that mostly generalists or especially robust organisms are easier to cultivate and therefore do not reflect the actual diversity. Nevertheless, it shows that several protist strains isolated from the two study regions are able to live under various abiotic conditions in the laboratory, and, importantly, the recovery of sequences of cultivated protists from the respective region verifies our metabarcoding study.

A large proportion of ASVs, both from Fehmarnbelt and Oderbank, was assigned to the Stramenopiles and Cercozoa. At Fehmarnbelt, the highest proportion of ASVs belonging to the Stramenopiles was assigned to Labyrinthulea, a class of Stramenopiles known mainly from marine and estuarine environments [50]. Labyrinthulids are able to decompose marine detritus by extracellular hydrolytic enzymes [5]. In their role as decomposers, they are typical inhabitants of sediments that are rich in organics [51] and could be typical for the eutrophic environment of the Baltic Sea. At Oderbank, stramenopile sequences mainly belonged to bicosoecids and the MAST group. While the bicosoecids detected in the samples from Oderbank could not be assigned to a level lower than the class level, it is hard to make any specific comments. Sequences of the MAST group (Marine Straminopiles) belonged to different ribogroups regarding their phylogenetic position, but also according to their ecological preferences [25]. At Oderbank, 50% of the

ASVs were assigned to the MAST-2 group, which comprises mainly marine—but also some freshwater—species originating from different geographic regions. This group is known to be exclusively planktonic and mainly occurs in oxic (sometimes also micro-oxic) environments [25]. At Oderbank, ASVs belonging to this group were found in the surface sediment layer, but also in deeper layers (maybe originating from encysted cells). Surprisingly, not one of the ASVs showed 100% identity to the sequences deposited in the reference database (highest identity of 99.2%), this underlines that the Baltic Sea is under-sampled and therefore underrepresented in public databases.

| HFCC no. | Species | isolated at | present [%] in samples from Fehmarnbelt | Oderbank |
|---|---|---|---|---|
| HFCC 1425 | *Cyclidium* sp. | Oderbank | / | 100 % |
| HFCC 1432 | *Neobodo* sp. | Oderbank | 100 % | 80 % |
| HFCC 1429 | *Paraphysomonas* sp. | Oderbank | 1.3 % | 100 % |
| HFCC 1427 | *Holosticha* sp. | Oderbank | / | 56 % |
| HFCC 1447 | *Ochromonas* sp. | Oderbank | / | 100 % |
| HFCC 1414 | *Planomonas* sp. | Fehmarnbelt | 9.2% | 80 % |
| HFCC 1655 | *Neobodo* sp. | Rönnebank | 52.6% | 84 % |
| HFCC 1435 | *Rhynchomonas* sp. | Oderbank | 35.5 % | 12 % |
| HFCC 1438 | *Planomonas* sp. | Oderbank | | 64 % |
| HFCC 1440 | *Neobodo* sp. | Oderbank | 26.3 % | 16 % |
| HFCC 1453 | Kinetoplastida sp. | Bay of Mecklenburg | / | 12 % |
| HFCC 1437 | *Bodo* sp. | Oderbank | / | 16 % |
| HFCC 1415 | *Pedospumella* sp. | Fehmarnbelt | 19.7 % | 4 % |
| HFCC 1434 | Kinetoplastida sp. | Fehmarnbelt | / | 8 % |
| HFCC 1420 | *Rhynchomonas* sp. | Fehmarnbelt | 1.3 % | / |

**Figure 8.** List of genotypes obtained from species that were cultivated from samples collected from the two investigated regions. HFCC stands for the number of the Heterotrophic Flagellate Culture Collection Cologne. The red labeled data indicate that the genotype could not be recovered in metabarcoding studies of the respective region. The number shows the percentage of samples in which they were detected in a regional dataset.

Compared to a metabarcoding study, which investigated the diversity of pico- up to mesoplankton in the Baltic Sea along a salinity gradient [16], there are some similarities to the Arkona Sea, which was their planktonic sampling site closest to our sampling region in the Oderbank. The major taxa groups were composed of similar classes of organisms to the ones in our studies of the sediment, which underlines the idea that the sediment might act as a sink for the planktonic diversity. An example is that, e.g., *Strombidium*, an oligotrich ciliate, was found with high dominance in the planktonic samples, and was also present with high read abundances in Fehmarnbelt sediments. High read numbers of the MAST-2 group occurred in plankton samples and were also found, especially in Oderbank samples.

Another important factor shaping protist communities is the availability of oxygen. While some protist taxa are able to survive both under aerobic and anaerobic conditions, others are sensitive to either one or the other condition [8,52]. Anaerobic ciliates are known to possess certain organelles, called hydrogenosomes, to ferment pyruvate into acetate and $H_2$ [53]. The protist community in the oxidized surface layers of the sediment was found to be different from the deeper sediment layers. While the exact $O_2$ content of the sediment layers was not measured during our study, we observed dark spots in the sediment layers, indicating anaerobic conditions already at 2 cmbsf. Even though the community of the different depth layers at Fehmarnbelt did not show a significantly different community, a high number of unique and unshared ASVs were found, especially in the deeper sediment layers. For Oderbank, we could detect significant differences, even though the sediment layers did not go as deep as those of Fehmarnbelt. At Oderbank, we could detect ciliate species known to be able to survive anaerobic conditions ([53];

e.g., *Trimyema, Lacrymaria, Caenomorpha*) in almost all sediment layers, indicating, at least, anaerobic patches in the sediment. For Fehmarnbelt, we could detect *Trimyema, Metopus,* and also *Lacrymaria* in many layers of the sediment. Besides salinity and water depth, the CCA of the complete dataset showed a significant influence on the sediment depth on the community composition ($p = 0.002$, Figure 7).

As part of the microbial food web, the abundance of protists is closely linked to the predominant bacterial community and abundance [22,54], which are also heavily influenced by abiotic factors and sediment properties [55]. Therefore, it is very likely that the bacterial community in both regions differ, thereby substantially affecting the protist community as well, and vice versa. Studies on prokaryotes are carried out at the moment and might reveal interesting data for comparative analyses in the future.

To analyze if there are differences in community composition between Fehmarnbelt and Oderbank, we compared the RNA-derived dataset of two stations from Fehmarnbelt with two randomly chosen stations at Oderbank. At least for those four stations, we could show that the protist communities form two distinct clusters for the two regions with no shared ASVs, at least not when we use our strict filtering step. Still, we know from our cultivation approach that there are at least a few taxa that appear in both datasets. In terms of the distribution and diversity of protists, several partly contradicting hypotheses have been established during the last years [56] that also addressed the main problems of estimating protist diversity, which includes under-sampling. Other studies have shown that the seafloor can be very heterogeneous regarding protist diversity even at a small spatial scale [57]. More data are needed to draw robust conclusions regarding the differences and similarities of benthic protist communities in the Baltic Sea. Nevertheless, our study might give novel insights into protist diversity for the vastly understudied benthic protist community of the Baltic Sea.

## 5. Conclusions

Our study on the community of benthic protists in the Baltic Sea obtained via metabarcoding of the V9 region of 18S rDNA showed significant differences in community composition not only between the different sampling regions but also between different sediment layers. For both regions, ASVs belonging to Ciliophora dominated the overall community, especially at Oderbank. Dinoflagellata, Stramenopiles, and Cercozoa showed also high diversity, but differed with regard to the lower taxonomic groups between the two regions. We assume that certain abiotic factors such as salinity, sediment grain size, and availability of oxygen are responsible for the differences in the communities, even though there are some taxa being ubiquitously distributed in both regions.

**Author Contributions:** Conceptualization, H.A. and M.S.; methodology, M.S. and M.D.; software, M.S. and M.D.; validation, H.A. and M.S.; formal analysis, M.S.; investigation, M.S., M.D. and H.A.; resources, M.S.; data curation, M.S.; writing—original draft preparation, M.S; writing—review and editing, H.A., M.S. and M.D..; visualization, M.S.; supervision, H.A.; project administration, H.A.; funding acquisition, H.A. All authors have read and agreed to the published version of the manuscript.

**Funding:** This research was funded by the German Federal Ministry for Education and Research (Bundesministerium für Bildung und Forschung, BMBF) within its DAM pilot mission project MGF Baltic Sea (MGF-Ostsee), grant number 03F0848A.

**Institutional Review Board Statement:** Not applicable.

**Informed Consent Statement:** Not applicable.

**Data Availability Statement:** The data presented in this study are available on request from the corresponding author. At the time of publication of this study, the data were submitted to PANGAEA and will be available soon.

**Acknowledgments:** We would like to acknowledge all project partners involved in the project, especially Klaus Jürgens and Christina Schmidt for managing the project as well as Judith Piontek for methodical advice and help through sampling on board. We are very thankful to Julian Wagenhofer, Karla Wiechmann, Ruben Rotmann, Svenja Huth, and Merle Heilmann for their help in isolating and cultivating samples. We express our gratitude towards the captains and crews of R/V Elisabeth Mann Borgese, as well as the BMBF for funding. We would also like to thank the editors of the journal as well as the reviewers for their helpful comments and improvements to this manuscript.

**Conflicts of Interest:** The authors declare no conflict of interest.

## References

1. Marcus, N.H.; Boero, F. Minireview The Importance of Benthic-Pelagic Coupling and the Forgotten Role of Life Cycles in Coastal Aquatic Systems. *Limnol. Oceanogr.* **1998**, *43*, 763–768. [CrossRef]
2. Rodríguez-Martínez, R.; Leonard, G.; Milner, D.S.; Sudek, S.; Conway, M.; Moore, K.; Hudson, T.; Mahé, F.; Keeling, P.J.; Santoro, A.E.; et al. Controlled Sampling of Ribosomally Active Protistan Diversity in Sediment-Surface Layers Identifies Putative Players in the Marine Carbon Sink. *ISME J.* **2020**, *14*, 984–998. [CrossRef] [PubMed]
3. De Vargas, C.; Audic, S.; Henry, N.; Decelle, J.; Mahé, F.; Logares, R.; Lara, E.; Berney, C.; Le Bescot, N.; Probert, I.; et al. Eukaryotic Plankton Diversity in the Sunlit Ocean. *Science* **2015**, *348*, 1261605. [CrossRef]
4. Arndt, H.; Dietrich, D.; Auer, B.; Cleven, E.-J.; Gräfenhan, T.; Weitere, M.; Mynikov, A. Functional Diversity of Heterotrophic Flagellates in Aquatic Ecosystems. In *The Flagellates*; Leadbeater, B.S.C., Green, J.C., Eds.; Taylor & Francis Ltd.: London, UK, 2000; pp. 240–268.
5. Nagano, N.; Matsui, S.; Kuramura, T.; Taoka, Y.; Honda, D.; Hayashi, M. The Distribution of Extracellular Cellulase Activity in Marine Eukaryotes, Thraustochytrids. *Mar. Biotechnol.* **2011**, *13*, 133–136. [CrossRef]
6. Anderson, S.R.; Harvey, E.L. Temporal Variability and Ecological Interactions of Parasitic Marine Syndiniales in Coastal Protist Communities. *mSphere* **2020**, *5*, e00209-20. [CrossRef] [PubMed]
7. Worden, A.Z.; Follows, M.J.; Giovannoni, S.J.; Wilken, S.; Zimmerman, A.E.; Keeling, P.J. Rethinking the Marine Carbon Cycle: Factoring in the Multifarious Lifestyles of Microbes. *Science* **2015**, *347*, 1257594. [CrossRef]
8. Priya, M.; Shibuvardhan, Y.; Manilal, V.B. Response of Anaerobic Protozoa to Oxygen Tension in Anaerobic System. *Int. Microbiol.* **2019**, *22*, 355–361. [CrossRef]
9. Lass, H.-U.; Matthäus, W. General Oceanography of the Baltic Sea. In *State and Evolution of the Baltic Sea, 1952–2005: A Detailed 50-Year Survey of Meteorology and Climate, Physics, Chemistry, Biology, and Marine Environment*; Feistel, R., Naush, G., Wasmund, N., Eds.; Wiley-Interscience: Hoboken, NJ, USA, 2008; pp. 5–44. ISBN 978-0-471-97968-5.
10. Ojaveer, H.; Jaanus, A.; MacKenzie, B.R.; Martin, G.; Olenin, S.; Radziejewska, T.; Telesh, I.; Zettler, M.L.; Zaiko, A. Status of Biodiversity in the Baltic Sea. *PLoS ONE* **2010**, *5*, e12467. [CrossRef]
11. Weber, F.; Mylnikov, A.P.; Jürgens, K.; Wylezich, C. Culturing Heterotrophic Protists from the Baltic Sea: Mostly the "Usual Suspects" but a Few Novelties as Well. *J. Eukaryot. Microbiol.* **2017**, *64*, 153–163. [CrossRef] [PubMed]
12. Arndt, H. On the Importance of Planktonic Protozoans in the Eutrophication Process of the Baltic Sea. *Int. Rev. Der Gesamten Hydrobiol. Hydrogr.* **1991**, *76*, 387–396. [CrossRef]
13. Weber, F.; Del Campo, J.; Wylezich, C.; Massana, R.; Jürgens, K. Unveiling Trophic Functions of Uncultured Protist Taxa by Incubation Experiments in the Brackish Baltic Sea. *PLoS ONE* **2012**, *7*, e41970. [CrossRef]
14. Stock, A.; Jürgens, K.; Bunge, J.; Stoeck, T. Protistan Diversity in Suboxic and Anoxic Waters of the Gotland Deep (Baltic Sea) as Revealed by 18S RRNA Clone Libraries. *Aquat. Microb. Ecol.* **2009**, *55*, 267–284. [CrossRef]
15. Piwosz, K.; Całkiewicz, J.; Gołębiewski, M.; Creer, S. Diversity and Community Composition of Pico- and Nanoplanktonic Protists in the Vistula River Estuary (Gulf of Gdańsk, Baltic Sea). *Estuar. Coast. Shelf Sci.* **2018**, *207*, 242–249. [CrossRef]
16. Hu, Y.O.O.; Karlson, B.; Charvet, S.; Andersson, A.F. Diversity of Pico- to Mesoplankton along the 2000 Km Salinity Gradient of the Baltic Sea. *Front. Microbiol.* **2016**, *7*, 679. [CrossRef]
17. Sich, H. Die Benthische Ciliatenfauna bei Gabelsflach (Kieler Bucht) und deren Beeinflussung durch Bakterien—Eine Studie über Menge, Biomasse, Produktion, Bakterieningestion und Ultrastruktur von Mikroorganismen. Ph.D. Thesis, University of Kiel, Kiel, Germany, 1989. Berichte aus dem Institut für Meereskunde Kiel, Nr. 191. [CrossRef]
18. Garstecki, T.; Verhoeven, R.; Wickham, S.A.; Arndt, H. Benthic-Pelagic Coupling: A Comparison of the Community Structure of Benthic and Planktonic Heterotrophic Protists in Shallow Inlets of the Southern Baltic: Benthic and Pelagic Protistan Community Structure. *Freshw. Biol.* **2000**, *45*, 147–167. [CrossRef]
19. Garstecki, T.; Arndt, H. Seasonal Abundances and Community Structure of Benthic Rhizopods in Shallow Lagoons of the Southern Baltic Sea. *Eur. J. Protistol.* **2000**, *36*, 103–115. [CrossRef]
20. Zhang, J.; Chen, X.; Soetaert, K.; Xu, Y. The Relative Roles of Multiple Drivers on Benthic Ciliate Communities in an Intertidal Zone. *Mar. Pollut. Bull.* **2023**, *187*, 114510. [CrossRef]
21. Lei, Y.; Stumm, K.; Wickham, S.A.; Berninger, U. Distributions and Biomass of Benthic Ciliates, Foraminifera and Amoeboid Protists in Marine, Brackish, and Freshwater Sediments. *J. Eukaryot. Microbiol.* **2014**, *61*, 493–508. [CrossRef] [PubMed]

22. Dietrich, D.; Arndt, H. Biomass Partitioning of Benthic Microbes in a Baltic Inlet: Relationships between Bacteria, Algae, Heterotrophic Flagellates and Ciliates. *Mar. Biol.* **2000**, *136*, 309–322. [CrossRef]
23. Amato, A.; Kooistra, W.; Levialdighiron, J.; Mann, D.; Pröschold, T.; Montresor, M. Reproductive Isolation among Sympatric Cryptic Species in Marine Diatoms. *Protist* **2007**, *158*, 193–207. [CrossRef]
24. Guillou, L.; Viprey, M.; Chambouvet, A.; Welsh, R.M.; Kirkham, A.R.; Massana, R.; Scanlan, D.J.; Worden, A.Z. Widespread Occurrence and Genetic Diversity of Marine Parasitoids Belonging to Syndiniales( Alveolata ). *Environ. Microbiol.* **2008**, *10*, 3349–3365. [CrossRef] [PubMed]
25. Massana, R.; Del Campo, J.; Sieracki, M.E.; Audic, S.; Logares, R. Exploring the Uncultured Microeukaryote Majority in the Oceans: Reevaluation of Ribogroups within Stramenopiles. *ISME J.* **2014**, *8*, 854–866. [CrossRef] [PubMed]
26. Schlitzer, R. Ocean Data View. Available online: odv.awi.de (accessed on 20 May 2021).
27. Danovaro, R. *Methods for the Study of Deep-Sea Sediments, Their Functioning and Biodiversity*; CRC Press, Taylor and Francis Group: Boca Raton, FL, USA, 2009.
28. Fortin, N.; Beaumier, D.; Lee, K.; Greer, C.W. Soil Washing Improves the Recovery of Total Community DNA from Polluted and High Organic Content Sediments. *J. Microbiol. Methods* **2004**, *56*, 181–191. [CrossRef] [PubMed]
29. Amaral-Zettler, L.A.; McCliment, E.A.; Ducklow, H.W.; Huse, S.M. A Method for Studying Protistan Diversity Using Massively Parallel Sequencing of V9 Hypervariable Regions of Small-Subunit Ribosomal RNA Genes. *PLoS ONE* **2009**, *4*, e6372. [CrossRef]
30. Michu, E.; Mráčková, M.; Vyskot, B.; Žlůvová, J. Reduction of Heteroduplex Formation in PCR Amplification. *Biol. Plant.* **2010**, *54*, 173–176. [CrossRef]
31. Choi, J.; Park, J.S. Comparative Analyses of the V4 and V9 Regions of 18S RDNA for the Extant Eukaryotic Community Using the Illumina Platform. *Sci. Rep.* **2020**, *10*, 6519. [CrossRef]
32. Martin, M. Cutadapt Removes Adapter Sequences from High-Throughput Sequencing Reads. *EMBnet J.* **2011**, *17*, 10–12. [CrossRef]
33. Callahan, B.J.; McMurdie, P.J.; Rosen, M.J.; Han, A.W.; Johnson, A.J.A.; Holmes, S.P. DADA2: High-Resolution Sample Inference from Illumina Amplicon Data. *Nat. Methods.* **2016**, *13*, 581–583. [CrossRef]
34. Rognes, T.; Flouri, T.; Nichols, B.; Quince, C.; Mahé, F. VSEARCH: A Versatile Open Source Tool for Metagenomics. *PeerJ* **2016**, *4*, e2584. [CrossRef]
35. Oksanen, J.; Blanchet, F.G.; Friendly, M.; Kindt, R.; Legendre, P.; McGlinn, D.; Minchin, P.R.; O'Hara, R.B.; Simpson, G.L.; Solymos, P.; et al. Vegan: Community Ecology Package. Available online: http://CRAN.Rproject.org/package=vegan (accessed on 21 May 2022).
36. Conway, J.R.; Lex, A.; Gehlenborg, N. UpSetR: An R Package for the Visualization of Intersecting Sets and Their Properties. *Bioinformatics* **2017**, *33*, 2938–2940. [CrossRef]
37. Kozich, J.J.; Westcott, S.L.; Baxter, N.T.; Highlander, S.K.; Schloss, P.D. Development of a Dual-Index Sequencing Strategy and Curation Pipeline for Analyzing Amplicon Sequence Data on the MiSeq Illumina Sequencing Platform. *Appl. Environ. Microbiol.* **2013**, *79*, 5112–5120. [CrossRef] [PubMed]
38. Fiore-Donno, A.M.; Rixen, C.; Rippin, M.; Glaser, K.; Samolov, E.; Karsten, U.; Becker, B.; Bonkowski, M. New Barcoded Primers for Efficient Retrieval of Cercozoan Sequences in High-Throughput Environmental Diversity Surveys, with Emphasis on Worldwide Biological Soil Crusts. *Mol. Ecol. Resour.* **2018**, *18*, 229–239. [CrossRef]
39. Lennartz, A.E.; Nitsche, F.; Schoenle, A.; Voigt, C.; Staubwasser, M.; Arndt, H.; Arndt, H. High Diversity and Isolated Distribution of Aquatic Heterotrophic Protists in Salars of the Atacama Desert at Different Salinities. *Eur. J. Protistol.* **2023**, *89*, 125987. [CrossRef]
40. Weber, A.A.-T.; Pawlowski, J. Can Abundance of Protists Be Inferred from Sequence Data: A Case Study of Foraminifera. *PLoS ONE* **2013**, *8*, e56739. [CrossRef] [PubMed]
41. Xu, Y.; Fan, X.; Warren, A.; Zhang, L.; Xu, H. Functional Diversity of Benthic Ciliate Communities in Response to Environmental Gradients in a Wetland of Yangtze Estuary, China. *Mar. Pollut. Bull.* **2018**, *127*, 726–732. [CrossRef] [PubMed]
42. Azovsky, A.I.; Tikhonenkov, D.V.; Mazei, Y.A. An Estimation of the Global Diversity and Distribution of the Smallest Eukaryotes: Biogeography of Marine Benthic Heterotrophic Flagellates. *Protist* **2016**, *167*, 411–424. [CrossRef]
43. Telesh, I.; Schubert, H.; Skarlato, S. Revisiting Remane's Concept: Evidence for High Plankton Diversity and a Protistan Species Maximum in the Horohalinicum of the Baltic Sea. *Mar. Ecol. Prog. Ser.* **2011**, *421*, 1–11. [CrossRef]
44. Remane, A. Die Brackwasserfauna. In *Verhandlungen der Deutschen Zoologischen Gesellschaft. Deutschen Zoologischen Gesellschaft im Frankfurt am Main*; Harvard University: Cambridge, MA, USA, 1934; pp. 34–74.
45. Gong, J.; Shi, F.; Ma, B.; Dong, J.; Pachiadaki, M.; Zhang, X.; Edgcomb, V.P. Depth Shapes α- and β-Diversities of Microbial Eukayryotes in Surficial Sediments of Coastal Ecosystems. *Environ. Biol.* **2015**, *17*, 3722–3737. [CrossRef]
46. Hamels, I.; Sabbe, K.; Muylaert, K.; Vyverman, W. Quantitative Importance, Composition, and Seasonal Dynamics of Protozoan Communities in Polyhaline versus Freshwater Intertidal Sediments. *Microb. Ecol.* **2004**, *47*, 18–29. [CrossRef]
47. Hamels, I.; Muylaert, K.; Sabbe, K.; Vyverman, W. Contrasting Dynamics of Ciliate Communities in Sandy and Silty Sediments of an Estuarine Intertidal Flat. *Eur. J. Protistol.* **2005**, *41*, 241–250. [CrossRef]
48. Corinaldesi, C.; Barucca, M.; Luna, G.M.; Dell'Anno, A. Preservation, Origin and Genetic Imprint of Extracellular DNA in Permanently Anoxic Deep-Sea Sediments. *Mol. Ecol.* **2011**, *20*, 642–654. [CrossRef]

49. Schoenle, A.; Hohlfeld, M.; Rybarski, A.; Sachs, M.; Freches, E.; Wiechmann, K.; Nitsche, F.; Arndt, H. Cafeteria in Extreme Environments: Investigations on C. Burkhardae and Three New Species from the Atacama Desert and the Deep Ocean. *Eur. J. Protistol.* **2022**, *85*, 125905. [CrossRef]
50. Naganuma, T.; Takasugi, H.; Kimura, H. Abundance of Thraustochytrids in Coastal Plankton. *Mar. Ecol. Prog. Ser.* **1998**, *162*, 105–110. [CrossRef]
51. Bongiorni, L.; Pusceddu, A.; Danovaro, R. Enzymatic Activities of Epiphytic and Benthic Thraustochytrids Involved in Organic Matter Degradation. *Aquat. Microb. Ecol.* **2005**, *41*, 299–305. [CrossRef]
52. Fenchel, T.; Finlay, B. Oxygen and the Spatial Structure of Microbial Communities. *Biol. Rev.* **2008**, *83*, 553–569. [CrossRef]
53. Fenchel, T.; Finlay, B.J. The Biology of Free-Living Anaerobic Ciliates. *Eur. J. Protistol.* **1991**, *26*, 201–215. [CrossRef]
54. Massana, R.; Unrein, F.; Rodríguez-Martínez, R.; Forn, I.; Lefort, T.; Pinhassi, J.; Not, F. Grazing Rates and Functional Diversity of Uncultured Heterotrophic Flagellates. *ISME J.* **2009**, *3*, 588–596. [CrossRef]
55. Zhang, B.; Li, Y.; Xiang, S.-Z.; Yan, Y.; Yang, R.; Lin, M.-P.; Wang, X.-M.; Xue, Y.-L.; Guan, X.-Y. Sediment Microbial Communities and Their Potential Role as Environmental Pollution Indicators in Xuande Atoll, South China Sea. *Front. Microbiol.* **2020**, *11*, 1011. [CrossRef]
56. Foissner, W. Protist Diversity and Distribution: Some Basic Considerations. *Biodivers. Conserv.* **2007**, *17*, 235–242. [CrossRef]
57. Hohlfeld, M.; Schoenle, A.; Arndt, H. Horizontal and Vertical Small-Scale Patterns of Protist Communities at the Atlantic Deep-Sea Floor. *Deep Sea Res. Part I Oceanogr. Res. Pap.* **2021**, *171*, 103515. [CrossRef]

biology

MDPI

Article

# Comparing Seamounts and Coral Reefs with eDNA and BRUVS Reveals Oases and Refuges on Shallow Seamounts

Florian Baletaud [1,2,3,4], Gaël Lecellier [1,5], Antoine Gilbert [2], Laëtitia Mathon [1,6], Jean-Marie Côme [3], Tony Dejean [7], Mahé Dumas [1], Sylvie Fiat [1] and Laurent Vigliola [1,*]

1 ENTROPIE, Institut de Recherche pour le Développement (IRD), UR, UNC, IFREMER, CNRS, Centre IRD de Nouméa, 98848 Noumea, New Caledonia, France; florian.baletaud@soproner.nc (F.B.); gael.lecellier@unc.nc (G.L.); laetitia.mathon@gmail.com (L.M.); mahe.dumas@ird.fr (M.D.); sylvie.fiat@ird.fr (S.F.)

2 GINGER SOPRONER, 98000 Noumea, New Caledonia, France; antoine.gilbert@soproner.nc

3 GINGER BURGEAP, 69000 Lyon, France; jm.come@groupeginger.com

4 MARBEC, University of Montpellier, CNRS, IFREMER, 34000 Montpellier, France

5 ISEA, University of New Caledonia, 98800 Noumea, New Caledonia, France

6 CEFE, University of Montpellier, CNRS, EPHE-PSL, IRD, 34000 Montpellier, France

7 SPYGEN, 73370 Le Bourget-du-lac, France; tony.dejean@spygen.com

* Correspondence: laurent.vigliola@ird.fr

**Simple Summary:** Underwater mountains, or seamounts, are deep-sea habitats collectively forming an area as large as Europe. Yet, they are one of the least studied ecosystems on earth. Known for supporting rich marine life compared to surrounding deep-sea environments, we have no information on how seamounts truly compare to other iconic biodiversity hotspots like shallow coral reefs. To assess the effective ecological value of seamounts, we compared fish communities in coral reefs and seamounts up to 500 m deep using two techniques: environmental DNA to detect the presence of species by filtering fragments of DNA lost by organisms in seawater, and underwater cameras to directly measure fish abundance and size. We found that the deepest seamounts had almost 10 times fewer fish species than coral reefs. However, the shallowest seamounts had larger fish species, including sharks, than coral reefs. We conclude that while seamounts are important and unique ecosystems, they may not be as diverse for fish species as previously thought (diversity hotspots) but rather biomass oases and refuges for endangered species. This study therefore calls for protecting the shallowest seamounts, as they are critical areas for marine life.

**Abstract:** Seamounts are the least known ocean biome. Considered biodiversity hotspots, biomass oases, and refuges for megafauna, large gaps exist in their real diversity relative to other ecosystems like coral reefs. Using environmental DNA metabarcoding (eDNA) and baited video (BRUVS), we compared fish assemblages across five environments of different depths: coral reefs (15 m), shallow seamounts (50 m), continental slopes (150 m), intermediate seamounts (250 m), and deep seamounts (500 m). We modeled assemblages using 12 environmental variables and found depth to be the main driver of fish diversity and biomass, although other variables like human accessibility were important. Boosted Regression Trees (BRT) revealed a strong negative effect of depth on species richness, segregating coral reefs from deep-sea environments. Surprisingly, BRT showed a hump-shaped effect of depth on fish biomass, with significantly lower biomass on coral reefs than in shallowest deep-sea environments. Biomass of large predators like sharks was three times higher on shallow seamounts (50 m) than on coral reefs. The five studied environments showed quite distinct assemblages. However, species shared between coral reefs and deeper-sea environments were dominated by highly mobile large predators. Our results suggest that seamounts are no diversity hotspots for fish. However, we show that shallower seamounts form biomass oases and refuges for threatened megafauna, suggesting that priority should be given to their protection.

**Keywords:** conservation; biomass; biodiversity; hotspot; coral reefs; mesophotic slope

## 1. Introduction

Seamounts are ubiquitous deep-sea habitats that collectively form an area as large as Europe [1]. Although seamounts are often considered biodiversity hotspots [2], oases [3,4], and refuges for marine megafauna [5], they are the least studied landforms of the major biomes of the global ocean. Out of more than 170,000 seamounts around the world, only less than 0.002% have been sampled for scientific purposes [6]. Yet, evidence is accumulating that seamounts are increasingly threatened by anthropogenic impacts such as overfishing, destructive fishing, and marine mining [6,7]. Thus, it is essential to fill the large gaps in our knowledge regarding the diversity and abundance patterns of species on seamounts. This is especially urgent at a time when human-related disturbances are affecting all areas of the world and oceans may experience a mass extinction of sufficient intensity to rank among the major extinctions of the Phanerozoic (541 Ma to present), with vertebrates, including fish, being at the forefront [8].

Seamounts are generally defined as underwater mountains greater than 100 m in relief above the seafloor, with often further subdivision into hills (elevation < 500 m from seafloor), knolls (>500 m from seafloor), and seamounts (>1000 m from seafloor) [9]. In geology, oceanic islands are also considered emerging seamounts [10]. Some ecologists also regard remote atolls as seamounts [11]. Above all, seamounts are geological features that can modify the properties of their surrounding oceanic environment. In particular, some studies indicate higher biodiversity and biomass on seamounts than in surrounding abyssal and pelagic environments [12,13]. Although many factors, including physical and biogenic habitat structure, may explain such patterns, the complex effects of seamounts on ocean circulation, including Taylor column formation, tidal amplification, internal waves, and upwelling formation, are thought to enhance primary production, zooplankton abundance, attract pelagic predators above seamounts, and permit these deep-sea habitats to host large populations of demersal and benthopelagic fish [6,14,15].

Due to their higher biodiversity than their surrounding environment, seamounts have long been considered hotspots, although this claim is now widely debated [16]. The term "biodiversity hotspot" was first introduced in 2000 and defined by delineating subjective regions of high (>1500 species, or 0.5% of the world's richness) endemic vascular plant richness under threat [17]. This analysis was supported by vertebrates' data but clearly warned against different congruence levels depending on the region. The marine realm also adopted the hotspot terminology and used the number of species, proxies of endemism, and global threats [18,19]. Today, "hotspot" has become a broader term that encompasses more biodiversity traits such as functional, beta, and phylogenetic diversity, leading to different definitions and uses. For this study, we consider the original definition of a marine hotspot based on biodiversity, which includes species richness and its previously stated derivatives. However, the identification of hotspots still relies on a comparison with other ecosystems, implying that biodiversity on seamounts should be compared to other recognized hotspots such as coral reefs (i.e., the shallow tropical coral reefs formed by reef-building corals associated with mostly photosynthetic coral assemblages), rather than merely surrounding abyssal and pelagic environments.

Biomass is another index that can refer either to abundance or mass, which directly connects to ecosystem services, for instance, through resources [20]. Temperature, nutrient availability, human influence, and biodiversity (where evolutionary adaptation results in species-rich communities using a larger fraction of available resources) are among the main factors controlling biomass production, at least in marine fish [21]. The concept of an ecosystem with higher biomass than its surroundings typically refers to an "oasis". Originally defining a fertile spot in a desert due to abundant water [22], attempts were made to translate the term to the marine realm with hydrothermal vents in the abyss, as they basically represent an isolated ecosystem lying within a desert, but where previously limiting factors allow a drastic increase in biomass [23]. This definition has further been used to praise coral reefs [24], but also seamounts [3,4] and other deep-sea landforms, e.g., canyons [25]. However, definitions of marine oases and hotspots often mix and shift towards diversity oasis [26], a synonym of the hotspot, or biomass hotspot, a synonym of oasis. Here, we consider the term oasis as the original analogy, which implies that biomass levels are higher in the studied environment (e.g., seamounts) than in their surroundings [3,23].

The definition of an oasis does not have particular regard for anthropogenic threats, but the degree of isolation of an oasis may straddle the line with the concept of "refuge". The original notion of refuge implies the existence of safe havens for species impacted by large-scale disturbances like the effects of climate change [27–29]. This definition has recently been enhanced to accept a smaller scale of direct threat avoidance for mobile species like fish [30,31]. Particularly, the "depth refuge" hypothesis implies that species may use deeper, ecologically less favorable environments at the extreme of their ecological niche as an avoidance pathway to short-time-scaled disturbances like fishing [32] or climatic events like storms [29,33]. Refuges include habitats such as seamounts [5] and remote coral reefs [34,35], where low accessibility to fishing fleets can create an economic barrier to harvesting [36] and induce a high abundance of shallow-water predators such as reef sharks that are otherwise largely extirpated [37].

Considering their wide definition, seamounts include a vast range of environmental conditions, from the warm photic zone at the surface of oceans to the cold aphotic zone on abyssal seafloors, and therefore show high variation in their biological communities, e.g., [38]. The question is thus no longer to know if seamounts are hotspots, oases, or refuges, but to assess which environmental conditions enhance biodiversity, biomass, and threatened species abundance across a wide environmental and human impact gradient and compare various seamount environments with well-studied and recognized hotspots such as coral reefs.

New Caledonia is a vast (1.4 million km$^2$ economic exclusive zone—EEZ) South Pacific archipelago composed of a main island, large surrounding islands (Loyalty Islands), and many islets, approximately 1200 km east of Australia in the Coral Sea [39]. The archipelago has one of the largest barrier reefs in the world, covering 24,000 km$^2$, and hosts one-third of the world's most remote and wilderness reefs [36]. Its extraordinarily rich shallow coral reefs [40] were added to the UNESCO World Heritage List in 2008. The archipelago also includes 80 hills, knolls, and seamounts, with at least 19 larger seamounts >1000 m in height from the seabed (hereafter seamounts), whose summit depths range from near surface at 4 m to 2400 m [41]. Thus, New Caledonia is probably the ideal site to compare seamount biodiversity and biomass with those of another iconic biodiversity hotspot, biomass oasis, and megafauna refuge such as coral reefs.

Comparing biodiversity between shallow and deep environments is challenging due to the use of specialized sampling methods. For example, coral reefs are typically surveyed by divers [31], while seamount fauna is generally surveyed by experimental fishing, acoustic echosounders, and ROVs. However, new technologies such as environmental DNA (eDNA) metabarcoding and video surveys allow the collection of quantitative data in a standardized way in almost all marine ecosystems. The metabarcoding of eDNA is based on the retrieval and analysis of genetic material naturally released by organisms in their environments. It was recently shown to outperform dive and video surveys for estimating marine biodiversity [37,42], with a higher capacity to detect small, cryptic, low-density, and elusive species [43–46]. Yet, the drawback of eDNA metabarcoding is the lack of knowledge about organism size and biomass. Stereo Baited Remote Underwater Video Stations (BRUVS) can efficiently estimate species abundance and biomass in virtually any marine habitat [47,48], so the two methods seem complementary to compare biodiversity across a long depth gradient.

In this study, we collected eDNA and BRUVS data from seven shallow coral reefs, four shallow seamounts, four deep continental slopes, three deep seamounts, and four seamounts of intermediate depth in the New Caledonian EEZ. Taking advantage of this unique dataset, we modeled the effect of key environmental and human variables on fish biodiversity, abundance, and biomass using boosted regression trees (BRT) [49]. Then we compared fish fauna on seamounts of variable summit depth, deep continental slopes, and shallow coral reefs and evaluated if and when seamounts qualify as biodiversity hotspots, biomass oases, and refuges for megafauna.

## 2. Materials and Methods

### 2.1. Data Collection

Data was collected during four oceanographic campaigns aboard the R/V Alis in April and June 2019 and August and September 2020, and during six coastal trips from September to December 2019. We sampled 22 sites, including seven barrier coral reefs, four deep continental slopes along the west coast of Grande Terre, and 11 seamounts (>1000 m in height from the seabed) summits of variable depth (45–511 m) across the New-Caledonian archipelago (Figure 1, Supplementary Section S1, Table S1). All samples were collected at the bottom. The seamounts were chosen to have different summit depths corresponding to euphotic, intermediate, and aphotic zones: four seamounts had summits higher than 200 m depth, four seamounts had summits between 200 and 320 m depth, and three seamounts had summits between 320 and 500 m depth. Coral reefs were sampled between 2 and 28 m depth, virtually covering their full depth range, continental slopes between 80 and 235 m, and seamounts between 45 and 570 m. Altogether, 224 BRUVS were deployed (5–16 per site) and 192 eDNA samples (6–10 per site) were collected in five environmental strata: coral reefs (average sample depth 13 m, SD ± 7 m), shallow seamounts (average 60 m, SD ± 8 m) labeled "Seamount (50 m)", continental slopes (average 142 m, SD ± 45 m) labeled "Continental slope (150 m)", intermediate seamounts (average 265 m, SD ± 36 m) labeled "Seamount (250 m)" and deep seamounts (average 498 m, SD ± 33 m) labeled "Seamount (500 m)".

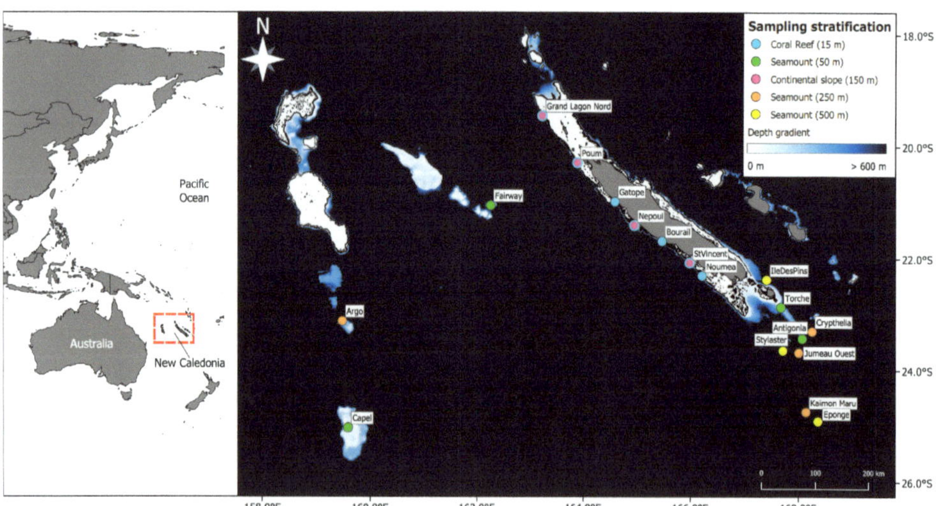


**Figure 1.** Sampling design in the five environmental strata. BRUVS and eDNA samples were collected on seven barrier coral reefs (15 m deep), four shallow seamounts (50 m summit depth), four continental slopes (150 m deep), four intermediate-depth seamounts (250 m summit depth), and three deep seamounts (500 m summit depth). Bathymetry data were derived from [50]. See Table S1 for more details on sampling design.

### 2.2. Stereo Baited Remote Underwater Video Stations (BRUVS)

Sampling started in the daylight morning. Each BRUVS was deployed 300–500 m apart on the coral reef and up to 1 km on seamount summits to prevent fish individuals from appearing on multiple videos and to ensure the independence of samples [48]. A BRUVS rig consisted of a protective metal structure holding two horizontally aligned cameras facing a bait canister attached at the end of a 1.5 m bar [51]. For each deployment, one kilogram of crushed sardines (*Sardinops* spp.) filled the canister. BRUVS were weighted and attached to a rope leading to a surface buoy. The stereo pair of cameras were separated by 800 mm, with a convergent angle of 8°. GoPro Hero 4 cameras were used and set to a medium field of view (FOV) in 1920 × 1080-pixel format, running at 30 frames per second.

Soaking times were calculated from the time BRUVS reached the seabed ($t_0$) to $t_0$ + 60 min. Fish were visually identified and counted on video using the EventMeasure software (version 5.42 (64 bit), released April 2020, www.seagis.com.au). We used the MaxN method (corresponding to the maximum number of fish for each species counted in an image across the video), which is until now the standard method [47,48,52].

Stereo measurement was made available with the recording of three claps before deployment to synchronize frames. Calibration was done using CAL software (version 3.25 (64 bit), released March 2019, www.seagis.com.au). Fork length (FL) of individual fish was measured, when possible, up to a limit of 10 individuals per BRUVS per species to optimize video processing time.

### 2.3. Biomass Estimation

Biomass was calculated for each species on each BRUVS using the length-weight relationship $Weight\ (\mathrm{g}) = a \times Length\ (\mathrm{cm})^b$, with $a$ and $b$ [53] retrieved from FishBase (https://www.fishbase.se (accessed on 12 February 2023)), and fish length calculated as the average length of all measured individuals (up to 10) of a species in a BRUVS [53]. When particular species could not be measured on a single BRUVS, the missing species length was estimated by data imputation using the MissForest algorithm with 999 trees [54]. We imputed the missing length using measured length records of other samples, but also family, genus, maximum size, and size type from Fishbase. The latitude and longitude of the localities where length records were taken were also used to account for the geographic proximity of measured lengths. The MissForest accuracy was tested with a k-fold cross-validation procedure by predicting 5% of the lengths each time by training the missForest on the 95% left of the data and looking at the linear fit between the original and predicted value (see Supplementary Section S2 for details, Figure S1). We also ensured that the imputed length did not exceed Fishbase's maximum reported length.

### 2.4. eDNA Metabarcoding

For each sample, environmental DNA was filtered out of 32 L of seawater in a sterile VigiDNA® 0.2 μm cross-flow filtration capsule with a polyethersulfone membrane (SPYGEN, Le Bourget du Lac, France). Samples were collected as close as possible to the substrate, mostly 5 m above the seafloor. Water was pumped into the filter capsule with a disposable sterile tube connected to an Alexis® peristaltic pump (Proactive Environmental Products LLC, Bradenton, FL, USA; nominal flow of 1.0 L $\mathrm{min}^{-1}$) and a Masterflex™ segment connected to it. On coral reefs, samples could be filtered along transects with a slow forward-going boat (~2 knots) using a reusable, extended, and weighted tube down and close to the substrate. Strict protocols were followed to avoid contamination, which included using the most disposable sterile equipment (surgical gloves, tubes, and tube joints) along with longer reusable, bleached tubes [44,55]. On continental slopes and seamount summits, four 8-L Niskin bottles (Ocean Test Equipment, Ft. Lauderdale, FL, USA) were used to collect 32 L of water at a single point for every sample. Filtration then occurred on the bridge of the ship. When the filtration process ended and all water was expelled from the filter capsules, around 80 mL of CL1 conservation buffer (SPYGEN, Le Bourget du Lac, France) was poured and enclosed in the capsule for storage and transport at room temperature. On coral reefs, eDNA transect itineraries were set to be either parallel (spread out by a few hundred meters) or following each other inside the BRUVS sample's grid at each site. Generally, sampling of BRUVS and eDNA occurred at the same predefined coordinates.

DNA extraction was performed following an existing protocol [46]. Briefly, the DNA extraction was performed using NucleoSpin® Soil (MACHEREY-NAGEL GmbH & Co., Düren, Germany), starting from step 6 and following the manufacturer's instructions. The elution was performed by adding 100 μL of SE buffer twice. The two 50 mL tubes per filtration capsule were extracted separately, and then, the two DNA samples were pooled before the amplification step. A teleost-specific 12S mitochondrial rRNA primer pair (teleo, forward primer—ACACCGCCCGTCACTCT, reverse primer—CTTCCGGTACACTTACCATG) [44] was used for the amplification of metabarcode sequences. Because we analyzed our data using MOTUs as a proxy for species, we chose to amplify only one marker. Twelve DNA amplifications PCR per sample were performed in a final volume of 25 μL, using 3 μL of DNA extract as the template, following the protocol in [56]. The teleo primers were 5′-labeled with an eight-nucleotide tag unique to each PCR replicate with at least three differences between any pair of tags, allowing the assignment of each sequence to the corresponding sample during sequence analysis. The tags for the forward and reverse primers were identical for each PCR replicate. Negative extraction controls and negative PCR controls (ultrapure water) were amplified (with 12 replicates as well) and sequenced in parallel to the samples to monitor possible contaminations. The purified PCR products were pooled in equal volumes, to achieve a theoretical sequencing depth of 1,000,000 reads per sample. Library preparation and sequencing were performed at Fasteris (Geneva, Switzerland). A total of 18 libraries were prepared using the MetaFast protocol. A paired-end sequencing (2 × 125 bp) was carried out using an Illumina MiSeq (2 × 125 bp, Illumina, San Diego, CA, USA) using the MiSeq Flow Cell Kit v3 (Illumina, San Diego, CA, USA) or a NextSeq sequencer (2 × 125 bp, Illumina, San Diego, CA, USA) with the NextSeq Mid kit following the manufacturer's instructions.

### 2.5. eDNA Bioinformatic

Following sequencing, reads were processed using clustering and post-clustering cleaning to remove errors and estimate the number of species using Molecular Operational Taxonomic Units (MOTUs) [57]. The methodology is described elsewhere [46]. Briefly, vsearch [58] and cutadapt [59] were used to assemble and demultiplex reads [58,59]. Swarm v.2 [60] was used to cluster sequences into MOTUs with a minimum distance of 2 mismatch between clusters. The Lower Common Ancestor (LCA) algorithm ecotag implemented in the Obitools toolkit [60] was used for taxonomic assignment of MOTUs [61] using the European Nucleotide Archive (ENA, [62]) as a reference database (release 143, March 2020). We then applied quality filters to be conservative in our estimates. To avoid spurious MOTUs originating from a PCR error, we discarded all sequences with less than 10 reads and presented only one PCR per site. Then, errors generated by tag-jumping and index-hopping [63,64] were corrected using a threshold of 0.001 of occurrence for a given MOTU within a library. Taxonomic assignments at the species level were accepted if the percentage of similarity with the reference sequence was 100%, at the genus level if the similarity was between 90 and 99%, and at the family level if the similarity was ≥85%. If these criteria were not met, the MOTU was left unassigned. The post-LCA algorithm correction threshold of 85% similarity for the family assignment was chosen to include a maximum of correct family assignments while minimizing the risk of adding wrong family identifications. Potential eDNA contamination from BRUVS bait (*Sardinops* spp.) was also removed from eDNA reads.

### 2.6. Environmental Variables

Fourteen environmental variables were identified for each sample (Table S2). They were chosen for their potential influence on fish diversity, biomass, and assemblage structure. They included the mean, minimal, and maximal Sea Surface Temperature (SST) from NASA's

Multiscale Ultrahigh Resolution (MUR) analysis, averaged over the last 10 available years (2009–2019). The potential seafloor temperature from the Mercator global reanalysis of models constructed on satellite and in situ observations (Copernicus—CMEMS) was also used. Temperature is well known to segregate diversity at a large scale across taxa [65]. We used chlorophyll-a, suspended particulate matter, salinity, and current from the Global Ocean Satellite Observations (Copernicus—CMEMS). Chlorophyll-a concentration may indicate regions of higher energy availability and pinpoint the presence of shallow seamounts [14,66]. Suspended particulate matter concentration may differentiate oligotrophic from eutrophic nutrient zones (e.g., lagoon versus open ocean) but also seamounts, which may re-enhance nutrient internal cycling [67]. Salinity may differentiate waters closer to freshwater flux [68] and currents may influence migratory flows for species recruited on seamounts [6]. Depth was recorded for each sample as it highly structures communities, notably through light loss and associated processes [5,69]. Travel time to the nearest fish market, an index of human accessibility to natural resources, was retrieved as human pressure also impacts diversity and biomass [36,70]. The micro-habitat was also included in the BRUVS data. We evaluated, through a semi-quantitative scale [71], the distinct visually observable features (e.g., percent cover of coral, sand, vegetation, and more, see [72] for details on the method). Micro-habitat variables were used to calculate the Shannon habitat diversity index and assess whether micro-habitat diversity would influence fish diversity or biomass [73,74]. Environmental strata ("Stratum") were also considered, as we assumed that while depth may be the main structuring variable, our environmental strata may incorporate a larger spectrum of influence that was not taken into account with the rest of the environmental variables.

### 2.7. Data Analysis

Fish species richness (BRUVS), MOTU richness (eDNA), fish abundance and biomass (BRUVS), and 14 environmental variables (Table S2) were determined for each sample. We also computed the biomass of large predators (>50 cm carnivore or piscivore species) and sharks using our functional traits database [75]. All analyses were performed with R [76].

### 2.8. Diversity and Biomass Modelling

Boosted Regression Trees (BRTs) [49] were used to model species richness, MOTU richness, fish biomass, biomass of large predators, and shark biomass along the matrix of 14 environmental variables. The two advantages of BRTs rely on their ability to assess non-linear relationships between the response and the explanatory variables along with their ability to manage complex interactions between variables. A grid search method [49] was used to determine the best BRT hyper-parameter values (number of trees, tree complexity, learning rate, and bag fraction). BRTs with the best cross-validation (10-fold) correlation were kept and then fitted again keeping only variables with more than 5% importance in the model [49]. Cross-validation correlation was used to assess the accuracy of the models. Variable importance and marginal effects were also computed. Marginal effects allow evaluation of the "pure" effect of an explanatory variable while accounting for the effects of all other variables included in the model. Correlated explanatory variables were removed, and biomass values were transformed prior to modeling. Due to the correlated nature of micro-habitat percent cover and their poor interpretability on such a large ecological gradient (i.e., there is no coral reef at great depth due to lack of light), these were not included in the BRTs. However, habitat was included in the models with the stratum and the Shannon habitat diversity variables.

### 2.9. Comparisons across Strata

Permutational multivariate analyses (PERMANOVAs [77]) were used to compare species richness, MOTU richness, total biomass, biomass of large predators, and biomass of sharks between the five environmental strata. Significant PERMANOVAs were followed by pairwise permutation *t*-tests to identify significant factor levels. Both analyses were done with 9999 permutations. To better assess species and MOTU richness, rarefaction curves were constructed using the Hill-number method with richness as incidence data [78,79]. The Hill number's framework considers sample size to get asymptotic richness estimates that are robust to unbalanced sampling, providing better estimation than other rarefaction methods [72,78].

*2.10. Assemblages Structure*

Principal coordinate analyses (PCoA) were used to determine the assemblages' structure in the five environmental strata [80]. Due to heterogeneous data and the presence of double zeros in our community matrix, we used the Hellinger distance for both abundance data (BRUVS) and presence-absence data (eDNA). The Hellinger distance can accommodate heterogeneous data and allows either presence-absence or abundance data to identify communities [80]. Then, we looked at species that were common to the different environments, especially coral reef species that were also observed in at least one deep-sea environment. We looked at the functional traits of these species to determine the proportion of shared species belonging to large predators. We chose to illustrate species sharing between environments through Euler diagrams. In addition to what Venn diagrams do, Euler diagrams draw ellipses that are proportional to the defined groups of species. Shared species will prompt an intersection between ellipses that is proportional to their list size.

## 3. Results

*3.1. Biodiversity*

A total of 423 species and 791 MOTUs were recorded through 224 BRUVS and 192 eDNA samples. Boosted regression tree modeling fitted well with richness data with values of 0.89 (species richness) and 0.86 (MOTU richness) of cross-validation (CV) correlation. BRTs revealed that depth was the main driver for both species and MOTU richness, with 76.5% variable importance on BRUVS data and 74.8% on eDNA data (Table 1). The analysis of marginal effects further showed a sharp drop in richness with depth, from high richness values at shallow depth to low values at great depth (see Supplementary Section S3, Figures S2 and S3). The pattern was particularly marked for MOTU richness (Figure S3). Habitat diversity was the second most influential variable on BRUVS species richness, with 15.7% importance in the model. Accounting for the effects of other explanatory variables, the marginal effects of habitat diversity indicated higher species richness in more diverse habitats (presence of balanced, multiple feature covers). A slight interaction between depth and habitat diversity was revealed by the BRT, reinforcing the suitability of environments that combine shallow depth and high habitat diversity compared with models in which no interaction effects are allowed (Figure S6). Several other environmental variables were included in the BRT models, however, with weaker importance. Mean sea surface temperature showed 7.8% importance on species richness, with slightly more species at higher temperatures. Travel time (6.7% importance) and northward current (6.6% importance) were also retained in the MOTU richness model. Some interactions were found with these variables but remained anecdotic (Figures S6 and S7).

The major negative effect of depth on richness was reflected in the five studied strata comparison, with a steady decrease in species richness from shallow coral reefs to deep seamounts and a sharp drop of MOTU richness between coral reefs on the one hand and all deep-sea environments on the other hand (Figure 2A). PERMANOVA results on species richness revealed four significantly distinct groups: coral reefs with on average 21.8 species (±18.0 SD), followed by seamounts (50 m) with 11.7 species (±4.7 SD), then the continental slope (150 m) with 8.3 species (±4.1 SD), and finally deeper seamounts (250 and 500 m) with respectively 3.4 (±1.7 SD) and 3.0 (±1.4 SD) species per BRUVS (Figure 2A). MOTU richness was significantly higher on coral reefs (average 71.8 MOTU ± 49.7 SD) than on any deep-sea environment. Little differences were observed between seamounts (50 m) (12.9 MOTUs ± 11.0 SD), continental slopes (150 m) (10.8 MOTUs ± 9.8 SD), seamounts (250 m) (12.2 ± 14.1 SD), and seamounts (500 m) (8.3 MOTUs ± 5.2 SD) (Figure 2B).

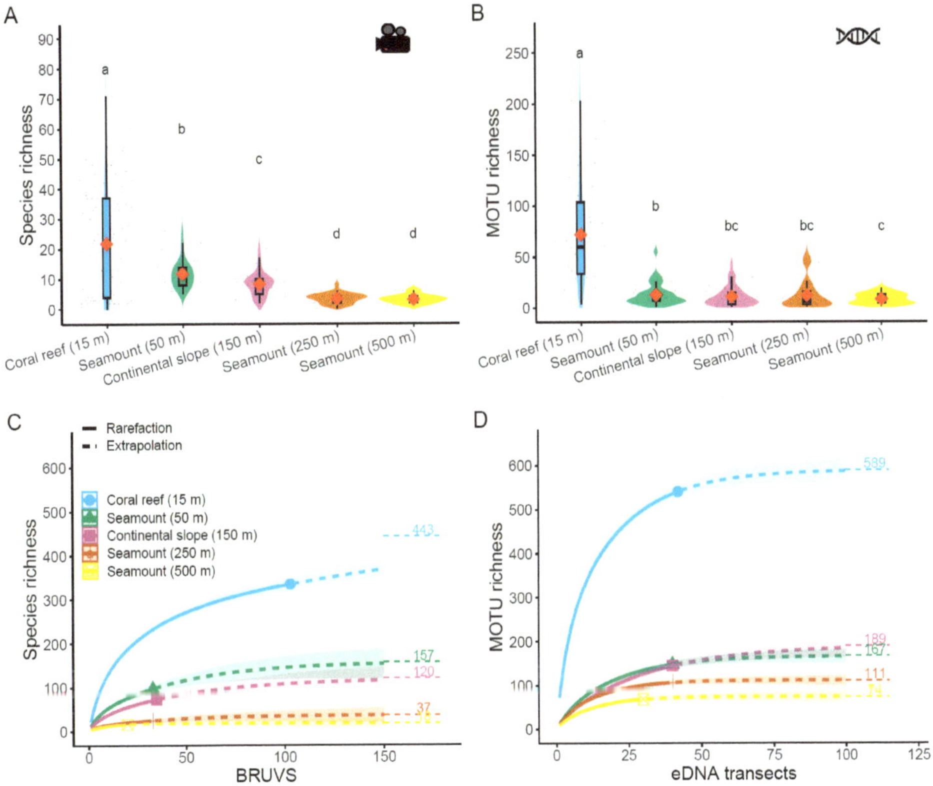


**Figure 2.** Violin plots and superimposed boxplots showing species richness observed on BRUVS (**A**) and MOTU richness in eDNA samples (**B**) for coral reef, seamounts (50, 250, 500 m), and continental slopes (150 m). The mean is represented by the red lozenge. Grey dots represent individual sample values scattered around each distribution. Significant differences at $p < 0.05$ are highlighted by grouping letters (PERMANOVAs and permutational $t$-tests with 9999 permutations). (**C**) Rarefaction curves of species richness from BRUVS and (**D**) MOTU richness from environmental DNA across coral reefs, seamounts, and continental slopes environments. The samples were rarefied (solid line) and extrapolated (dashed line) using the Hill number method [78,79]. 95% confidence intervals (CI) are shown in each respective ribbon. Horizontal lines are asymptote estimates ($\gamma$-diversity).

Rarefaction curves of species and MOTU richness showed dramatically higher biodiversity on coral reefs than on any deep-sea environments, with particularly low richness on the deepest seamounts (Figure 2C,D). The pattern was especially marked for MOTU richness. Asymptotic estimates of richness were 443 species (confidence interval—CI: 403–482) and 589 MOTUs (CI: 570–620) for coral reefs, then half less species (157, CI: 119–195) and even less MOTUs (167, CI: 156–189) on seamounts (50 m), 120 species (CI: 73–168) and 189 MOTUs (CI: 168–229) on continental slopes (150 m), 37 species (CI: 25–62) and 111 MOTUs (CI: 107–122) on seamounts (250 m), and 18 species (CI: 16–27) and 74 MOTUs (CI: 70–88) on the deepest seamounts (500 m).

**Table 1.** Environmental explanatory variables of several fish richness and biomass indices retained in the boosted regression trees modeling. Four different BRTs were run with species (BRUVS) and MOTU (eDNA) richness, followed by biomass (BRUVS) of full assemblages and biomass of large predators (>50 cm carnivorous and piscivorous species). Models were fit with best hyper-parameters (number of trees (NT), tree complexity (TC), learning rate (LR), bag fraction (BF) and evaluated using cross-validation correlation (CV). Explanatory variables retained in each BRT models are ordered by importance.

| BRT Model | NT | TC | LR | BF | CV (SD) | Variables | Variable Importance |
|---|---|---|---|---|---|---|---|
| Species richness (BRUVS) | 1050 | 4 | 0.005 | 0.5 | 0.89 (0.01) | Depth | 76.5% |
| | | | | | | Habitat diversity | 15.7% |
| | | | | | | Mean SST | 7.8% |
| MOTU richness (eDNA) | 700 | 4 | 0.01 | 0.75 | 0.86 (0.05) | Depth | 74.8% |
| | | | | | | Chla | 11.9% |
| | | | | | | Travel time | 6.7% |
| | | | | | | Northward velocity | 6.6% |
| Total biomass (BRUVS) | 2875 | 5 | 0.001 | 0.75 | 0.73 (0.03) | Depth | 36.7% |
| | | | | | | Habitat diversity | 23.6% |
| | | | | | | Travel time | 11.7% |
| | | | | | | Eastward velocity | 8.5% |
| | | | | | | Chla | 8.1% |
| | | | | | | Northward velocity | 5.9% |
| | | | | | | Mean SST | 5.5% |
| Large predators' biomass (BRT) | 2825 | 5 | 0.001 | 0.75 | 0.71 (0.03) | Depth | 26.1% |
| | | | | | | Habitat diversity | 19.6% |
| | | | | | | Travel time | 15.0% |
| | | | | | | Eastward velocity | 8.7% |
| | | | | | | Chla | 8.6% |
| | | | | | | Mean SST | 8.4% |
| | | | | | | Environmental stratum | 6.9% |
| | | | | | | Northward velocity | 6.8% |

### 3.2. Biomass

BRT modeling of fish biomass fitted well with the data, with a cross-validation correlation of 0.73 (Table 1). Depth was again the most important explanatory variable (36.7% importance), followed by habitat diversity (23.6%), travel time (11.7%), Chla (8.1%), currents (eastward: 8.5%, northward: 5.9%), and mean SST (5.5%). The analysis of marginal effects further showed hump-shaped patterns for depth, with the highest biomass observed between approximately 50 and 300 m and the lowest biomass for shallow depth with coral reefs and the deepest strata of seamounts (500 m) (Figure S4). Habitat diversity and SST had an overall positive effect on biomass. Travel time also had a positive effect on fish biomass, with the lowest biomass values recorded near humans, and the highest in remote environments at more than 10 h travel time (Figure S4). BRT modeling of large predator biomass showed similar patterns (Figure S5). Depth was again the most important variable (26.1%), followed by habitat diversity (19.6%), travel time (15.0%), equally eastward current, Chla, and mean SST (8.7, 8.6, and 8.5%, respectively), and finally the environmental stratum (6.9%) and northward current (6.8%). Interactions found by the BRTs involved habitat diversity and northward velocity, and depth and eastward velocity with anecdotic effects (Figures S8 and S9).

Comparison of fish biomass across the five strata showed a dome-shaped pattern corresponding well to the combined effects of environmental variables retained in the BRT model, especially depth and environmental stratum (Figure 3A). PERMANOVAs further revealed that biomass levels on shallow seamounts had significantly the highest biomass (132.9 kg ± 103.4 SD), while coral reefs (54.3 kg ± 71.6 SD) showed similar biomass levels as continental slopes (150 m) (76.9 kg ± 65.7 SD) and seamounts of intermediate depth (250 m) (66.6 kg ± 63.0 SD). The deepest seamounts (500 m) showed the lowest biomass level (7.2 kg ± 5.2 SD). The pattern was identical for large predators' biomass (Figure 3B) and sharks' biomass (Figure 3C).

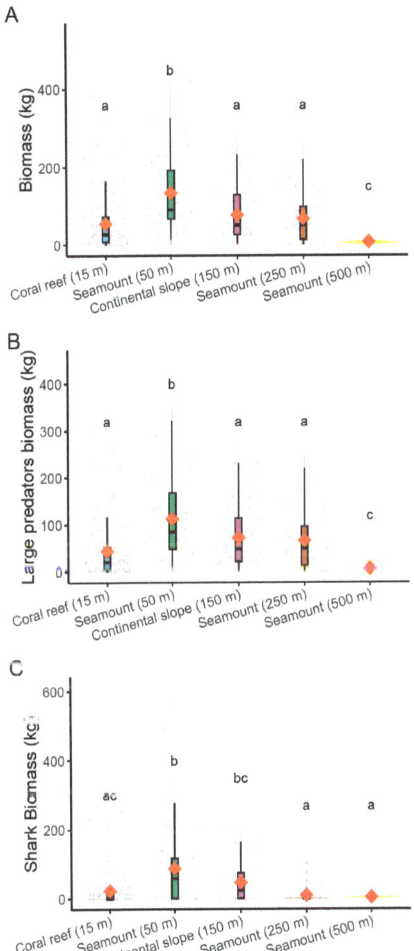


**Figure 3.** Violin plots and superimposed boxplots of biomass (BRUVS) on (**A**) all observed species, (**B**) large predators, and (**C**) sharks across 5 environmental strata: coral reefs, seamounts of variable depths (50, 250, and 500 m), and the continental slope (150 m). The mean is represented by the red lozenge. Grey dots represent individual sample values scattered around each distribution. Significant differences at $p < 0.05$ are highlighted by grouping letters (PERMANOVAs and permutational $t$-tests with 9999 permutations).

### 3.3. Assemblage Structure

Principal Coordinate Analysis showed that the five studied environments were home to relatively distinct assemblages, coral reefs, and deepest seamounts (500 m) showing the highest distinctiveness (see Supplementary Section S4, Figure S10). Coral reefs showed the second highest proportion (77%) of unique species (257 species out of 334) (Figure 4) and the highest proportion (84%) of unique MOTUs (454 MOTUs out of 540) (see Supplementary Section S5, Figure S11). Likewise, the deepest seamounts (500 m) showed the highest proportion (81%) of unique species (13 out of 16 species), although uniqueness was less with eDNA (20 out of 68 MOTUs, 29%). Other deep-sea environments showed relatively mixed assemblages, with 27% unique species (26 out of 97) and 42% unique MOTUS (62 out of 147) on seamounts (50 m), 32% unique species (23 out of 73) and 54% unique MOTUs (78 out of 144) on continental slopes (150 m), 36% unique species (9 out of 25) and 39% unique MOTUs (41 out of 106) on seamounts (250 m).

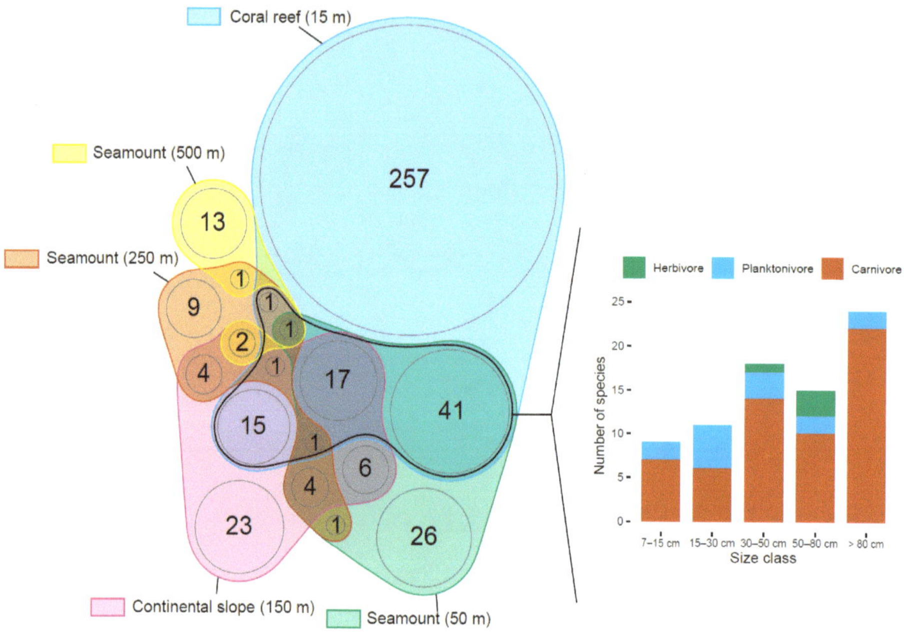


**Figure 4.** Euler diagram of species identified on BRUVS between coral reefs, seamounts of variable summit depths, and the continental slopes. Coral reef species that were also observed in at least one deep-sea environment (black grouping line) are compared in the bar plot through two functional traits: species size class and trophic group.

Interestingly, species that were shared between coral reefs and at least one deep-sea environment, including the deepest seamounts (black contouring in Figure 4), represented 23% of the 334 species observed on coral reefs, corresponding to 77 species. When looking at the functional traits of these coral reef species also observed in deep-sea environments, 77% (59 species) were carnivores, and 42% (32 species) were large-sized (>50 cm) carnivores, i.e., large predators (Barplot, Figure 4).

## 4. Discussion

Our study is one of the few addressing the fish biodiversity of multiple marine ecosystems using two standardized and replicated quantitative methods to provide comparative information. We showed that coral reefs may qualify as biodiversity hotspots with considerably higher species richness than any other deep-sea environments in this study. In turn, seamounts and continental slopes showed comparatively lower biodiversity both at the local scale ($\alpha$-diversity) and the regional scale ($\gamma$-diversity). The deepest seamounts had on average seven times fewer species and nine times fewer MOTUs than coral reefs. Combined with the general negative effect of depth on biodiversity, our results suggest that seamounts are not hotspots for fish diversity. However, shallow seamounts surprisingly showed almost three times higher fish biomass than coral reefs, and biomass levels up to 300 m were at least equivalent. These higher biomass levels in environments between

50 and 300 m depth may represent what would be called oases for fish. Moreover, while species assemblages were distinct among the studied environments, dominant species of shallow seamounts were highly mobile large predators also observed on coral reefs, suggesting that they may use shallow seamounts as refuge from shallow coral reef anthropic pressure. Overall, our results suggest that strong conservation efforts should be prioritized on shallow seamounts and continental slopes where very high fish biomass is observed, especially for threatened large predators such as sharks. Although deeper seamounts are less rich, they are still home to unique fauna that is certainly worth protecting as well.

Our BRUVS records of coral reef fish species richness are impressively in line with previous studies around the world, with around 20–25 species average per BRUVS [81–85]. BRUVS have been deployed near seamounts to assess pelagic diversity [5,12], and on seamounts of abyssal depth, far outside our study's depth range [86]. To our knowledge, this is the first study to explore a seamount's fish biodiversity using BRUVS at summits reaching between 50 and 500 m. However, more work has been done on deep continental slopes, showing usually lower species richness on mesophotic reefs than on shallow coral reefs [74,87–89].

The eDNA method across studies may greatly vary depending on the volume of filtered water, the primers, the laboratory protocols, and the bioinformatic pipelines used to generate the MOTU sequences, making comparisons still binding [57,90,91]. However, the tendency to further use this method is driven by its capacity to establish better levels of species richness by integrating a larger area per sample [37,92]. On coral reefs, a study with a similar protocol highlighted more species found by eDNA than by diver-operated Underwater Visual Censuses (UVC), but with less average MOTU richness per sample than in our study (26.2 ± 12.6 SD against 71.8 ± 49.7 SD in our study) [42]. Sampling deeper strata using eDNA can also become limited when trying to assign MOTUs to referenced species, e.g., [93]. While we partly used assignment to further clean our MOTU list, we did not analyze our assigned species dataset since only 25.6% of the 791 recorded MOTU sequences were assigned to the species level.

Asymptotic estimates of total MOTU richness as well as mean MOTU richness showed unparalleled levels for coral reefs compared to all sampled deep-sea environments. This observation may be explained by the cryptobenthic fish species diversity hosted on coral reefs [94,95]. The cryptobenthic fish diversity is harder for BRUVS to capture. However, this method also showed much higher fish diversity on shallow coral reefs. BRUVS are known to sample a smaller but more representative part of the studied community compared to other methods, meaning the observed assemblages are not necessarily biased [47,48,96]. However, studies of the deep sea using video-assisted methods are bound to strong technical constraints, and our soaking times (60 min) are rarer in the literature, with longer soaking times and time-delayed videos being favored [47]. However, the involved depths in our study (around a 500-m maximum) remained relatively shallow compared to the rest of the deep-sea research that usually works at several thousands of meters.

Defining a biodiversity hotspot comes with defining threats and indices of vulnerability [17,97]. Coral reefs are acknowledged to be globally declining due to anthropic pressure [98,99]. Seamounts have been heavily fished, trawled, and exploited on a global scale [6,100]. The impacts of trawling and dredging are largely documented for seamounts and involve major erosion of biodiversity and habitat complexity [6,101–104]. Regarding our results, coral reefs may then further be praised as hotspots for biodiversity. Seamounts, on the other hand, had much lower fish species richness and therefore were not comparable to a hotspot "reference". However, biodiversity is multifaceted, and other organisms' richness was not studied, which could have a major impact on the overall biodiversity of seamounts. Regarding the hotspot definition applied to the marine realm, the ease of propagules spreading out in the ocean induces much more widespread species compared to land, which limits endemism in marine environments and therefore the delimitations of hotspots [17–19]. Further studies are certainly needed to further compare the whole biodiversity of shallow and deep-sea ecosystems. The metabarcoding of eDNA across all realms has shown promising results in that regard, e.g., [95,105].

Fish biomass followed a dome-shaped relationship with depth. Seamounts at 50 m depth had the largest biomass, followed by 150 m continental slopes and 250 m deep seamounts. Surprisingly, coral reefs showed lower biomass despite the positive effects of their habitat diversity. It is commonly accepted that biomass would decrease with respect to light, primary production, and food availability [106], and indeed, we report decreasing fish biomass between 50 and 500 m with the lowest values on the deepest seamounts. High biomass has recently been reported on continental slopes and the shelf break, e.g., [30,107]. We report even higher levels on shallow seamounts. While coral reefs are characterized by high habitat complexity, associated with important biomass [108], shallow seamounts (50 m) also harbor a strong habitat diversity, notably with extended rhodolite beds composing the substrate along with vegetation and few corals. Rhodolite beds are recognized to host high biomass as they offer substrate complexity and resources, inducing an abundance of predators [109]. Thus, our comparison of seamounts and coral reefs from the same region may promote the vision of shallow seamounts and, to a lesser extent, continental slopes as oases of biomass. Nonetheless, fishing has long exploited the resources of seamounts, and some fisheries even collapsed in the 1980s as stocks could not replenish fast enough due to the slower life cycle of deep-sea species [6,100]. Catches still have increased with time with further demand and technology to go deeper, with targeted species usually being large-bodied predators that are globally declining, e.g., [110–112].

This race to exploit ever deeper resources may have yet left out the shallowest seamounts, as they are not the first target of deep demersal fisheries. We showed that the assemblages of the shallow seamounts at 50 m shared many common species with coral reefs and the other deeper environments (Figure 4). A considerable proportion of these species were large predators. Deep-sea species assemblages often have a high proportion of carnivores [74,87,107], potentially caused by the shift to more heterotrophic environments [106,113]. Yet, the larger biomass of reef-associated large predators, especially sharks, on shallow seamounts should place these environments as refuges from anthropic pressure [35,114]. Seamounts are highly isolated and difficult to access features without large vessels capable of withstanding the open ocean. The travel time for the closest seamount was seven hours and up to two and a half days for the furthest, still inside the new Caledonian EEZ. We suggest that these shallow features may be of crucial ecological importance for endangered and high-value target species also found on coral reefs.

Our results on species richness are consistent across two independent sampling methods. Biomass estimated by BRUVS also seems coherent across functional traits (here size and trophic group), which are known to better reflect assemblages [115]. However, some caveats can be discussed. While we tried to use standardized methods for both shallow and deep environments, small adaptations had to be realized, such as switching between eDNA transects on coral reefs to Niskin bottles in deep environments due to technical limits. The change in protocol may have influenced the observed densities of MOTUs between coral reefs and deep environments. Currents can also increase by an order of magnitude on seamount summits. Coupled with variable degradation times and potentially lower quantities of eDNA produced by deeper species with slower metabolic rates, these factors may have also influenced the detection of MOTUs in eDNA samples [6,90,116,117]. While recent studies show remarkable site fidelity of emitted eDNA [118,119], MOTUs of species associated with the pelagic and reef environment may also not be differentiated as both environments are intimately interacting on the external barrier reef. Soaking time for BRUVS between 60 and 90 min has been estimated to provide the best samples of the shallow communities in the cost/effort ratio [47]. Due to our samples still being relatively shallow (<500 m), the demersal species assemblages remained mostly active and mobile species, with few exceptions in the deeper environments, which supports our standard soaking time of 60 min.

Furthermore, we compared the biodiversity of several environments but used only fish as an indicator. Fish diversity and biomass provide valuable indicators of ecosystem services such as regulation and linkage for ecosystem functioning (e.g., predation, consumption, sediment redistribution, nutrient recycling and redistribution, and more), food security with fisheries, and cultural services like aesthetics [120–124].

## 5. Conclusions

A better understanding of how biodiversity is spatially distributed is fundamental to better addressing ecosystem trajectories and issues caused by large-scale disturbances like climate change [125] or anthropic pressure [126]. The human perspective of nature is mainly utilitarian and economical, implying conservation needs to be better informed through baselines on the priorities for conservation, notably in lesser-known ecosystems like the deep sea [127]. This work aimed at helping to refine fundamental questions underlying these environments, such as seamounts, and better conceive our perception of biodiversity and its distribution across coral reefs and deeper environments. The potential of new technologies with video and eDNA metabarcoding may allow better comparative values to address biodiversity on the same baseline and compare ecosystems, regions, or habitats in order to reprioritize locations of interest for conservation and science. Our study calls for prioritizing the conservation of shallow seamounts and continental slopes since these environments support considerable fish biomass and are a refuge for large predators such as sharks, but are virtually ignored by current management plans, with only 2% of the world's seamounts inside MPAs [5,9]. However, our study was restricted to fish, a crucial yet only small part of biodiversity. Further work looking at the whole biodiversity is warranted and may become possible with the development of key technologies such as eDNA metabarcoding.

**Supplementary Materials:** The following supporting information can be downloaded at: https://www.mdpi.com/article/10.3390/biology12111446/s1, Section S1. General data description: Table S1. BRUVS and eDNA samples collected on 5 different environments: coral reefs, seamounts (50 m), continental slopes (150 m), seamounts (250 m), and seamounts (500 m); Table S2. Environmental variables used in the BRT modeling of species richness, fish biomass and large predator biomass for BRUVS, and eDNA MOTU richness. BRUVS micro-habitat variables were extracted through a semi-quantitative scale. Micro-habitat covers were evaluated on a 0–100% scale at 5% precision. Substrate complexity and topography were evaluated on a scale of 0 to 5 and transformed in percentages. These variables were then used to calculate the Shannon diversity indices used in the boosted regression tree modeling; Section S2. Imputing missing length data on BRUVS: Figure S1. Predicted against true measures of fish for each of the twenty-fold cross-validation results from the 5% artificial imputation using missForest out of the 792-measurement data; Section S3. Boosted regression trees on richness, biomass and biomass of large predators: Figure S2. Variable importance and partial dependence plots from boosted regression trees ran on the reduced number of variables on BRUVS data with richness as the predicted value (response). The cross-validation correlation value was 0.89 for this model; Figure S3. Variable importance and partial dependence plots from boosted regression trees ran on the reduced number of variables through previous boosted trees run on eDNA data with MOTU richness as the predicted value (response). The cross-validation correlation value was 0.86 for this model; Figure S4. Variable importance and partial dependence plots from boosted regression trees ran on the reduced number of variables through previous boosted trees run on BRUVS data with biomass as the predicted value (response). The cross-validation correlation value was 0.71 for this model; Figure S5. Variable importance and partial dependence plots from boosted regression trees ran on the reduced number of variables through previous boosted trees run on BRUVS data with biomass of large predators as the predicted value (response). The cross-validation correlation value was 0.75 for this model; Figure S6. Interaction plots found from boosted regression trees ran on the reduced number of variables on BRUVS data with richness as the predicted value (fitted value). The cross-validation correlation value was 0.89 for this model; Figure S7. Interaction plots found from boosted regression trees ran on the reduced number of variables through previous boosted trees run on eDNA data with MOTU richness as the predicted value (response). The cross-validation correlation value was 0.86 for this model; Figure S8. Interaction plots found from boosted regression trees ran on the reduced number of variables through previous boosted trees run on BRUVS data with biomass as the predicted value (response). The cross-validation correlation value was 0.71 for this model; Figure S9. Interaction plots found from boosted regression trees ran on the reduced number of variables through previous boosted trees run on BRUVS data with biomass of large predators as the predicted value (response). The cross-validation correlation value was 0.75 for this model; Section S4. Assemblage ordination: Figure S10. Ordination using Principal Coordinate Analysis (PCoA) on the Hellinger transformed distance matrix of baited remote underwater video stations (BRUVS) abundance data (A) and presence-absence environmental DNA (eDNA) data (B). Ellipses were calculated from the covariance matrix of each stratum weighted by the number of points; Section S5. sharing of eDNA sequences among the environmental strata: Figure S11. Euler diagram of identified MOTUs from eDNA metabarcoding along 5 environmental strata: coral reefs, seamounts of variable summit depths and continental slopes. References [5,6,14,51,54,65,67–70,72–74,128,129] are cited in the supplementary materials.

**Author Contributions:** Conceptualization, F.B., A.G. and L.V.; methodology, F.B., L.V., L.M. and G.L.; software, F.B., L.V. and G.L.; validation, F.B., L.V. and G.L.; formal analysis, F.B.; investigation, F.B., L.V., M.D., L.M. and S.F.; resources, A.G., T.D., J.-M.C., S.F. and L.V.; data curation, F.B., M.D. and S.F.; writing—original draft preparation, F.B.; writing—review and editing, F.B., G.L., A.G., L.M., J.-M.C., T.D., M.D., S.F. and L.V.; visualization, F.B. and L.V.; supervision, A.G., J.-M.C. and L.V.; project administration, A.G., J.-M.C. and L.V.; funding acquisition, A.G., J.-M.C. and L.V. All authors have read and agreed to the published version of the manuscript.

**Funding:** The study was funded by grant ANR "SEAMOUNTS" #ANR-18-CE02-0016, the French Oceanographic Fleet, and IRD core funding. Florian Baletaud was supported by grant ANRT CIFRE #2019/0105.

**Institutional Review Board Statement:** Not applicable.

**Informed Consent Statement:** Not applicable.

**Data Availability Statement:** The data used for this article are archived in the public repository Zenodo (https://doi.org/10.5281/zenodo.10096023 for BRUVS data, and https://doi.org/10.5281/zenodo.10095584 for eDNA data).

**Acknowledgments:** We would like to thank the many undergrad students and technical staffs as well as the crew of R/V Alis for their help at sea and the lab. We thank SPYGEN staff for technical support during eDNA analysis. Data was collected under permits 2019-733/GNC, 2020-503/GNC and 2020-1077/GNC delivered by the Government of New-Caledonia, 898-2019/ARR/DENV, 3066-2019/ARR/DENV, 844-2020/ARR/DDDT and 1955-2020/ARR/DDDT delivered by the Southern Province of New-Caledonia, and 609011/2019/DEPART/JJC, 609011-18/2019/DEPART/JJC and 609011-39/2020/DEPART/JJC delivered by the Northern Province of New-Caledonia.

**Conflicts of Interest:** T.D. is a research scientist in a private company specializing in the use of eDNA for biodiversity monitoring, with some patent technologies (SPYGEN). The remaining authors declare no conflict of interest.

## References

1. Kvile, K.Ø.; Taranto, G.H.; Pitcher, T.J.; Morato, T. A Global Assessment of Seamount Ecosystems Knowledge Using an Ecosystem Evaluation Framework. *Biol. Conserv.* **2014**, *173*, 108–120. [CrossRef]
2. Morato, T.; Hoyle, S.D.; Allain, V.; Nicol, S.J. Seamounts Are Hotspots of Pelagic Biodiversity in the Open Ocean. *Proc. Natl. Acad. Sci. USA* **2010**, *107*, 9707–9711. [CrossRef] [PubMed]
3. Samadi, S.; Bottan, L.; Macpherson, E.; de Forges, B.R.; Boisselier, M.C. Seamount Endemism Questioned by the Geographic Distribution and Population Genetic Structure of Marine Invertebrates. *Mar. Biol.* **2006**, *149*, 1463–1475. [CrossRef]
4. Rowden, A.A.; Schlacher, T.A.; Williams, A.; Clark, M.R.; Stewart, R.; Althaus, F.; Bowden, D.A.; Consalvey, M.; Robinson, W.; Dowdney, J. A Test of the Seamount Oasis Hypothesis: Seamounts Support Higher Epibenthic Megafaunal Biomass than Adjacent Slopes. *Mar. Ecol.* **2010**, *31*, 95–106. [CrossRef]
5. Letessier, T.B.; Mouillot, D.; Bouchet, P.J.; Vigliola, L.; Fernandes, M.C.; Thompson, C.; Boussarie, G.; Turner, J.; Juhel, J.-B.; Maire, E.; et al. Remote Reefs and Seamounts Are the Last Refuges for Marine Predators across the Indo-Pacific. *PLoS Biol.* **2019**, *17*, e3000366. [CrossRef]
6. Rogers, A.D. The Biology of Seamounts: 25 Years On. In *Advances in Marine Biology*; Elsevier: Amsterdam, The Netherlands, 2018; Volume 79, pp. 137–224, ISBN 9780128151013,

7. Salmerón, F.; Báez, J.; Macías, D.; Fernandez-Peralta, L.; Ramos, A. Rapid Fish Stock Depletion in Previously Unexploited Seamounts: The Case of Beryx Splendens from the Sierra Leone Rise (Gulf of Guinea). *Afr. J. Mar. Sci.* **2015**, *37*, 405–409. [CrossRef]
8. Payne, J.L.; Bush, A.M.; Heim, N.A.; Knope, M.L.; McCauley, D.J. Ecological Selectivity of the Emerging Mass Extinction in the Oceans. *Science* **2016**, *353*, 1284–1286. [CrossRef]
9. Yesson, C.; Clark, M.R.; Taylor, M.L.; Rogers, A.D. The Global Distribution of Seamounts Based on 30 Arc Seconds Bathymetry Data. *Deep. Sea Res. Part I Oceanogr. Res. Pap.* **2011**, *58*, 442–453. [CrossRef]
10. Staudigel, H.; Koppers, A.; Lavelle, J.W.; Pitcher, T.; Shank, T. Defining the Word "Seamount". *Oceanography* **2010**, *23*, 20–21. [CrossRef]
11. Barnett, A.; AbrantesKá, K.G.; Seymour, J.; Fitzpatrick, R. Residency and Spatial Use by Reef Sharks of an Isolated Seamount and Its Implications for Conservation. *PLoS ONE* **2012**, *7*, e0036574. [CrossRef]
12. Cambra, M.; Lara-Lizardi, F.; Peñaherrera-Palma, C.; Hearn, A.; Ketchum, J.T.; Zarate, P.; Chacón, C.; Suárez-Moncada, J.; Herrera, E.; Espinoza, M. A First Assessment of the Distribution and Abundance of Large Pelagic Species at Cocos Ridge Seamounts (Eastern Tropical Pacific) Using Drifting Pelagic Baited Remote Cameras. *PLoS ONE* **2021**, *16*, e0244343. [CrossRef]
13. Zeppilli, D.; Bongiorni, L.; Santos, R.S.; Vanreusel, A. Changes in Nematode Communities in Different Physiographic Sites of the Condor Seamount (North-East Atlantic Ocean) and Adjacent Sediments. *PLoS ONE* **2014**, *9*, e0115601. [CrossRef] [PubMed]
14. Leitner, A.B.; Neuheimer, A.B.; Drazen, J.C. Evidence for Long-Term Seamount-Induced Chlorophyll Enhancements. *Sci. Rep.* **2020**, *10*, 12729. [CrossRef] [PubMed]
15. Lavelle, J.W.; Mohn, C. Motion, Commotion, and Biophysical Connections at Deep Ocean Seamounts. *Oceanography* **2010**, *23*, 90–103. [CrossRef]
16. Rowden, A.A.; Dower, J.F.; Schlacher, T.A.; Consalvey, M.; Clark, M.R. Paradigms in Seamount Ecology: Fact, Fiction and Future. *Mar. Ecol.* **2010**, *31*, 226–241. [CrossRef]
17. Myers, N.; Mittermeier, R.A.; Mittermeier, C.G.; da Fonseca, G.A.B.; Kent, J. Biodiversity Hotspots for Conservation Priorities. *Nature* **2000**, *403*, 853–858. [CrossRef] [PubMed]
18. Roberts, C.M.; McClean, C.J.; Veron, J.E.N.; Hawkins, J.P.; Allen, G.R.; McAllister, D.E.; Mittermeier, C.G.; Schueler, F.W.; Spalding, M.; Wells, F.; et al. Marine Biodiversity Hotspots and Conservation Priorities for Tropical Reefs. *Science* **2002**, *295*, 1280–1284. [CrossRef] [PubMed]
19. Mittermeier, R.A.; Turner, W.R.; Larsen, F.W.; Brooks, T.M.; Gascon, C. *Biodiversity Hotspots*; Zachos, F.E., Habel, J.C., Eds.; Springer: Berlin/Heidelberg, Germany, 2011; ISBN 978-3-642-20991-8.
20. McClanahan, T.R.; Graham, N.A.J.; MacNeil, M.A.; Muthiga, N.A.; Cinner, J.E.; Bruggemann, J.H.; Wilson, S.K. Critical Thresholds and Tangible Targets for Ecosystem-Based Management of Coral Reef Fisheries. *Proc. Natl. Acad. Sci. USA* **2011**, *108*, 17230–17233. [CrossRef]
21. Duffy, J.E.; Lefcheck, J.S.; Stuart-Smith, R.D.; Navarrete, S.A.; Edgar, G.J. Biodiversity Enhances Reef Fish Biomass and Resistance to Climate Change. *Proc. Natl. Acad. Sci. USA* **2016**, *113*, 6230–6235. [CrossRef]
22. Ramdani, M.; Elkhiati, N.; Flower, R.J. Lakes of Africa: North of Sahara. In *Encyclopedia of Inland Waters*; Elsevier: Amsterdam, The Netherlands, 2009; pp. 544–554. ISBN 9780123706263.
23. Carney, R.S. Consideration of the Oasis Analogy for Chemosynthetic Communities at Gulf of Mexico Hydrocarbon Vents. *Geo-Mar. Lett.* **1994**, *14*, 149–159. [CrossRef]
24. Nyström, M.; Folke, C.; Moberg, F. Coral Reef Disturbance and Resilence in a Human-Dominated Environment. *Trends Ecol. Evol.* **2000**, *15*, 413–417. [CrossRef]
25. de Leo, F.C.; Smith, C.R.; Rowden, A.A.; Bowden, D.A.; Clark, M.R. Submarine Canyons: Hotspots of Benthic Biomass and Productivity in the Deep Sea. In Proceedings of the Royal Society B: Biological Sciences, London, UK, 22 September 2010; Volume 277, pp. 2783–2792.
26. Samimi-Namin, K.; Hoeksema, B.W. Hidden Depths: A Unique Biodiversity Oasis in the Persian Gulf in Need of Further Exploration and Conservation. *Diversity* **2023**, *15*, 779. [CrossRef]
27. Sedell, J.R.; Reeves, G.H.; Hauer, F.R.; Stanford, J.A.; Hawkins, C.P. Role of Refugia in Recovery from Disturbances: Modern Fragmented and Disconnected River Systems. *Environ. Manag.* **1990**, *14*, 711–724. [CrossRef]
28. Keppel, G.; van Niel, K.P.; Wardell-Johnson, G.W.; Yates, C.J.; Byrne, M.; Mucina, L.; Schut, A.G.T.; Hopper, S.D.; Franklin, S.E. Refugia: Identifying and Understanding Safe Havens for Biodiversity under Climate Change. *Glob. Ecol. Biogeogr.* **2012**, *21*, 393–404. [CrossRef]
29. Bongaerts, P.; Ridgway, T.; Sampayo, E.M.; Hoegh-Guldberg, O. Assessing the "deep Reef Refugia" Hypothesis: Focus on Caribbean Reefs. *Coral Reefs* **2010**, *29*, 309–327. [CrossRef]
30. Lindfield, S.J.; Harvey, E.S.; Halford, A.R.; McIlwain, J.L. Mesophotic Depths as Refuge Areas for Fishery-Targeted Species on Coral Reefs. *Coral Reefs* **2016**, *35*, 125–137. [CrossRef]
31. Cinner, J.E.; Huchery, C.; MacNeil, M.A.; Graham, N.A.J.; McClanahan, T.R.; Maina, J.; Maire, E.; Kittinger, J.N.; Hicks, C.C.; Mora, C.; et al. Bright Spots among the World's Coral Reefs. *Nature* **2016**, *535*, 416–419. [CrossRef]
32. Lindfield, S.J.; Mcilwain, J.L.; Harvey, E.S. Depth Refuge and the Impacts of SCUBA Spearfishing on Coral Reef Fishes. *PLoS ONE* **2014**, *9*, e92628. [CrossRef]

33. Bongaerts, P.; Smith, T.B. Beyond the "Deep Reef Refuge" Hypothesis: A Conceptual Framework to Characterize Persistence at Depth. In *Mesophotic Coral Ecosystems*; Springer: Berlin/Heidelberg, Germany, 2019; pp. 881–895, ISBN 9783319927350.
34. D'agata, S.; Mouillot, D.; Wantiez, L.; Friedlander, A.M.; Kulbicki, M.; Vigliola, L. Marine Reserves Lag behind Wilderness in the Conservation of Key Functional Roles. *Nat. Commun.* **2016**, *7*, 12000. [CrossRef]
35. Juhel, J.-B.; Vigliola, L.; Mouillot, D.; Kulbicki, M.; Letessier, T.B.; Meeuwig, J.J.; Wantiez, L. Reef Accessibility Impairs the Protection of Sharks. *J. Appl. Ecol.* **2018**, *55*, 673–683. [CrossRef]
36. Januchowski-Hartley, F.A.; Vigliola, L.; Maire, E.; Kulbicki, M.; Mouillot, D. Low Fuel Cost and Rising Fish Price Threaten Coral Reef Wilderness. *Conserv. Lett.* **2020**, *13*, 12706. [CrossRef]
37. Boussarie, G.; Bakker, J.; Wangensteen, O.S.; Mariani, S.; Bonnin, L.; Juhel, J.-B.; Kiszka, J.J.; Kulbicki, M.; Manel, S.; Robbins, W.D.; et al. Environmental DNA Illuminates the Dark Diversity of Sharks. *Sci. Adv.* **2018**, *4*, eaap9661. [CrossRef]
38. Porteiro, F.M.; Gomes-Pereira, J.N.; Pham, C.K.; Tempera, F.; Santos, R.S. Distribution and Habitat Association of Benthic Fish on the Condor Seamount (NE Atlantic, Azores) from in Situ Observations. *Deep. Sea Res. Part II Top. Stud. Oceanogr.* **2013**, *98*, 114–128. [CrossRef]
39. Auzende, J.-M.; Grandperrin, R.; Bouniot, E.; Henin, C.; Lafoy, Y.; Richer de Forges, B.; van de Beuque, S.; Virly, S. Marine Resources of the Economic Zone of New Caledonia. *Oceanol. Acta* **1999**, *22*, 557–566. [CrossRef]
40. Payri, C.E.; Allain, V.; Aucan, J.; David, C.; David, V.; Dutheil, C.; Loubersac, L.; Menkes, C.; Pelletier, B.; Pestana, G. New Caledonia. In *World Seas: An Environmental Evaluation*; Elsevier: Amsterdam, The Netherlands, 2019; pp. 593–618.
41. Allain, V.; Kerandel, J.A.; Andréfouët, S.; Magron, F.; Clark, M.; Kirby, D.S.; Muller-Karger, F.E. Enhanced Seamount Location Database for the Western and Central Pacific Ocean: Screening and Cross-Checking of 20 Existing Datasets. *Deep Sea Res. Part I Oceanogr. Res. Pap.* **2008**, *55*, 1035–1047. [CrossRef]
42. Polanco Fernández, A.; Marques, V.; Fopp, F.; Juhel, J.; Borrero-Pérez, G.H.; Cheutin, M.; Dejean, T.; González Corredor, J.D.; Acosta-Chaparro, A.; Hocdé, R.; et al. Comparing Environmental DNA Metabarcoding and Underwater Visual Census to Monitor Tropical Reef Fishes. *Environ. DNA* **2021**, *3*, 142–156. [CrossRef]
43. Deiner, K.; Yamanaka, H.; Bernatchez, L. The Future of Biodiversity Monitoring and Conservation Utilizing Environmental DNA. *Environ. DNA* **2021**, *3*, 3–7. [CrossRef]
44. Valentini, A.; Taberlet, P.; Miaud, C.; Civade, R.; Herder, J.; Thomsen, P.F.; Bellemain, E.; Besnard, A.; Coissac, E.; Boyer, F.; et al. Next-Generation Monitoring of Aquatic Biodiversity Using Environmental DNA Metabarcoding. *Mol. Ecol.* **2016**, *25*, 929–942. [CrossRef]
45. Stat, M.; John, J.; DiBattista, J.D.; Newman, S.J.; Bunce, M.; Harvey, E.S. Combined Use of EDNA Metabarcoding and Video Surveillance for the Assessment of Fish Biodiversity. *Conserv. Biol.* **2019**, *33*, 196–205. [CrossRef]
46. Mathon, L.; Marques, V.; Mouillot, D.; Albouy, C.; Andrello, M.; Baletaud, F.; Borrero-Pérez, G.H.; Dejean, T.; Edgar, G.J.; Grondin, J.; et al. Cross-Ocean Patterns and Processes in Fish Biodiversity on Coral Reefs through the Lens of EDNA Metabarcoding. *Proc. R. Soc. B Biol. Sci.* **2022**, *289*, 20220162. [CrossRef]
47. Whitmarsh, S.K.; Fairweather, P.G.; Huveneers, C. What Is Big BRUVver up to? Methods and Uses of Baited Underwater Video. *Rev. Fish Biol. Fish* **2017**, *27*, 53–73. [CrossRef]
48. Langlois, T.; Goetze, J.; Bond, T.; Monk, J.; Abesamis, R.A.; Asher, J.; Barrett, N.; Bernard, A.T.F.; Bouchet, P.J.; Birt, M.J.; et al. A Field and Video Annotation Guide for Baited Remote Underwater Stereo-Video Surveys of Demersal Fish Assemblages. *Methods Ecol. Evol.* **2020**, *11*, 1401–1409. [CrossRef]
49. Elith, J.; Leathwick, J.R.; Hastie, T. A Working Guide to Boosted Regression Trees. *J. Anim. Ecol.* **2008**, *77*, 802–813. [CrossRef] [PubMed]
50. Shom-IRD MNT Bathymétrique de Façade de La Nouvelle-Calédonie (Projet TSUCAL). 2021. Available online: http://dx.doi.org/10.17183/MNT_NC100m_TSUCAL_WGS84 (accessed on 1 November 2020).
51. Letessier, T.B.; Juhel, J.B.; Vigliola, L.; Meeuwig, J.J. Low-Cost Small Action Cameras in Stereo Generates Accurate Underwater Measurements of Fish. *J. Exp. Mar. Biol. Ecol.* **2015**, *466*, 120–126. [CrossRef]
52. Cappo, M.; Harvey, E.; Malcolm, H.; Speare, P. Potential of Video Techniques to Monitor Diversity, Abundance and Size of Fish in Studies of Marine Protected Areas. In *Aquatic Protected Areas-What Works Best and How Do We Know?* University of Queensland: Brisbane, Australia, 2003; pp. 455–464.
53. Taylor, R.B.; Willis, T.J. Relationships amongst Length, Weight and Growth of North-Eastern New Zealand Reef Fishes. *Mar. Freshw. Res.* **1998**, *49*, 255–260. [CrossRef]
54. Stekhoven, D.J.; Buhlmann, P. MissForest: Non-Parametric Missing Value Imputation for Mixed-Type Data. *Bioinformatics* **2012**, *28*, 112–118. [CrossRef]
55. Goldberg, C.S.; Turner, C.R.; Deiner, K.; Klymus, K.E.; Thomsen, P.F.; Murphy, M.A.; Spear, S.F.; McKee, A.; Oyler-McCance, S.J.; Cornman, R.S.; et al. Critical Considerations for the Application of Environmental DNA Methods to Detect Aquatic Species. *Methods Ecol. Evol.* **2016**, *7*, 1299–1307. [CrossRef]
56. Mathon, L.; Marques, V.; Manel, S.; Albouy, C.; Andrello, M.; Boulanger, E.; Deter, J.; Hocdé, R.; Leprieur, F.; Letessier, T.B.; et al. The Distribution of Coastal Fish eDNA Sequences in the Anthropocene. *Glob. Ecol. Biogeogr.* **2023**, *32*, 1336–1352. [CrossRef]
57. Marques, V.; Guérin, P.É.; Rocle, M.; Valentini, A.; Manel, S.; Mouillot, D.; Dejean, T. Blind Assessment of Vertebrate Taxonomic Diversity across Spatial Scales by Clustering Environmental DNA Metabarcoding Sequences. *Ecography* **2020**, *43*, 1779–1790. [CrossRef]

58. Rognes, T.; Flouri, T.; Nichols, B.; Quince, C.; Mahé, F. VSEARCH: A Versatile Open Source Tool for Metagenomics. *PeerJ* **2016**, *4*, e2584. [CrossRef]
59. Martin, M. Cutadapt Removes Adapter Sequences from High-Throughput Sequencing Reads. *EMBnet. J.* **2011**, *17*, 10. [CrossRef]
60. Mahé, F.; Rognes, T.; Quince, C.; de Vargas, C.; Dunthorn, M. Swarm v2: Highly-Scalable and High-Resolution Amplicon Clustering. *PeerJ* **2015**, *3*, e1420. [CrossRef]
61. Boyer, F.; Mercier, C.; Bonin, A.; le Bras, Y.; Taberlet, P.; Coissac, E. Obitools: A Unix-Inspired Software Package for DNA Metabarcoding. *Mol. Ecol. Resour.* **2016**, *16*, 176–182. [CrossRef] [PubMed]
62. Leinonen, R.; Akhtar, R.; Birney, E.; Bower, L.; Cerdeno-Tarraga, A.; Cheng, Y.; Cleland, I.; Faruque, N.; Goodgame, N.; Gibson, R.; et al. The European Nucleotide Archive. *Nucleic. Acids Res.* **2011**, *39*, D28–D31. [CrossRef]
63. Schnell, I.B.; Bohmann, K.; Gilbert, M.T.P. Tag Jumps Illuminated-Reducing Sequence-to-Sample Misidentifications in Metabarcoding Studies. *Mol. Ecol. Resour.* **2015**, *15*, 1289–1303. [CrossRef] [PubMed]
64. MacConaill, L.E.; Burns, R.T.; Nag, A.; Coleman, H.A.; Slevin, M.K.; Giorda, K.; Light, M.; Lai, K.; Jarosz, M.; McNeill, M.S.; et al. Unique, Dual-Indexed Sequencing Adapters with UMIs Effectively Eliminate Index Cross-Talk and Significantly Improve Sensitivity of Massively Parallel Sequencing. *BMC Genom.* **2018**, *19*. [CrossRef]
65. Tittensor, D.P.; Mora, C.; Jetz, W.; Lotze, H.K.; Ricard, D.; vanden Berghe, E.; Worm, B. Global Patterns and Predictors of Marine Biodiversity across Taxa. *Nature* **2010**, *466*, 1098–1101. [CrossRef]
66. Currie, D.J.; Mittelbach, G.G.; Cornell, H.v.; Field, R.; Guegan, J.-F.; Hawkins, B.A.; Kaufman, D.M.; Kerr, J.T.; Oberdorff, T.; O'Brien, E.; et al. Predictions and Tests of Climate-Based Hypotheses of Broad-Scale Variation in Taxonomic Richness. *Ecol. Lett.* **2004**, *7*, 1121–1134. [CrossRef]
67. Capblancq, J. Nutrient Dynamics and Pelagic Food Web Interactions in Oligotrophic and Eutrophic Environments: An Overview. *Hydrobiologia* **1990**, *207*, 1–14. [CrossRef]
68. Barletta, M.; Barletta-Bergan, A.; Saint-Paul, U.; Hubold, G. The Role of Salinity in Structuring the Fish Assemblages in a Tropical Estuary. *J. Fish Biol.* **2005**, *66*, 45–72. [CrossRef]
69. Priede, I.G.; Bergstad, O.A.; Miller, P.I.; Vecchione, M.; Gebruk, A.; Falkenhaug, T.; Billett, D.S.M.; Craig, J.; Dale, A.C.; Shields, M.A.; et al. Does Presence of a Mid-Ocean Ridge Enhance Biomass and Biodiversity? *PLoS ONE* **2013**, *8*, e0061550. [CrossRef]
70. Maire, E.; Cinner, J.; Velez, L.; Huchery, C.; Mora, C.; Dagata, S.; Vigliola, L.; Wantiez, L.; Kulbicki, M.; Mouillot, D. How Accessible Are Coral Reefs to People? A Global Assessment Based on Travel Time. *Ecol. Lett.* **2016**, *19*, 351–360. [CrossRef] [PubMed]
71. Clua, E.; Legendre, P.; Vigliola, L.; Magron, F.; Kulbicki, M.; Sarramegna, S.; Labrosse, P.; Galzin, R. Medium Scale Approach (MSA) for Improved Assessment of Coral Reef Fish Habitat. *J. Exp. Mar. Biol. Ecol.* **2006**, *333*, 219–230. [CrossRef]
72. Baletaud, F.; Gilbert, A.; Mouillot, D.; Come, J.-M.; Vigliola, L. Baited Video Reveal Fish Diversity in the Vast Inter-Reef Habitats of a Marine Tropical Lagoon. *Mar. Biodivers.* **2022**, *52*, 16. [CrossRef]
73. Eduardo, L.N.; Frédou, T.; Lira, A.S.; Ferreira, B.P.; Bertrand, A.; Ménard, F.; Frédou, F.L. Identifying Key Habitat and Spatial Patterns of Fish Biodiversity in the Tropical Brazilian Continental Shelf. *Cont. Shelf. Res.* **2018**, *166*, 108–118. [CrossRef]
74. Sih, T.; Daniell, J.; Bridge, T.; Beaman, R.; Cappo, M.; Kingsford, M. Deep-Reef Fish Communities of the Great Barrier Reef Shelf-Break: Trophic Structure and Habitat Associations. *Diversity* **2019**, *11*, 26. [CrossRef]
75. Mouillot, D.; Villeger, S.; Parravicini, V.; Kulbicki, M.; Arias-Gonzalez, J.E.; Bender, M.; Chabanet, P.; Floeter, S.R.; Friedlander, A.; Vigliola, L.; et al. Functional Over-Redundancy and High Functional Vulnerability in Global Fish Faunas on Tropical Reefs. *Proc. Natl. Acad. Sci. USA* **2014**, *111*, 13757–13762. [CrossRef]
76. R Core Team. *R: A Language and Environment for Statistical Computing*; R Foundation for Statistical Computing: Vienna, Austria, 2021.
77. Anderson, M.J. A New Method for Non-Parametric Multivariate Analysis of Variance. *Austral. Ecol.* **2001**, *26*, 32–46. [CrossRef]
78. Chao, A.; Gotelli, N.J.; Hsieh, T.C.; Sander, E.L.; Ma, K.H.; Colwell, R.K.; Ellison, A.M. Rarefaction and Extrapolation with Hill Numbers: A Framework for Sampling and Estimation in Species Diversity Studies. *Ecol. Monogr.* **2014**, *84*, 45–67. [CrossRef]
79. Hsieh, T.C.; Ma, K.H.; Chao, A. INEXT: An R Package for Rarefaction and Extrapolation of Species Diversity (Hill Numbers). *Methods Ecol. Evol.* **2016**, *7*, 1451–1456. [CrossRef]
80. Legendre, P.; Legendre, L. *Numerical Ecology: Second English Edition*; Elsevier Science: Amsterdam, The Netherlands, 1998; Volume 20.
81. Dorman, S.R.; Harvey, E.S.; Newman, S.J. Bait Effects in Sampling Coral Reef Fish Assemblages with Stereo-BRUVs. *PLoS ONE* **2012**, *7*, e0041538. [CrossRef] [PubMed]
82. Cheal, A.J.; Emslie, M.J.; Currey-Randall, L.M.; Heupel, M.R. Comparability and Complementarity of Reef Fish Measures from Underwater Visual Census (UVC) and Baited Remote Underwater Video Stations (BRUVS). *J. Environ. Manag.* **2021**, *289*, 112375. [CrossRef] [PubMed]
83. Abesamis, R.A.; Utzurrum, J.A.T.; Raterta, L.J.J.; Russ, G.R. Shore-Fish Assemblage Structure in the Central Philippines from Shallow Coral Reefs to the Mesophotic Zone. *Mar Biol* **2020**, *167*, 185. [CrossRef]
84. Andradi-Brown, D.A.; Macaya-Solis, C.; Exton, D.A.; Gress, E.; Wright, G.; Rogers, A.D. Assessing Caribbean Shallow and Mesophotic Reef Fish Communities Using Baited-Remote Underwater Video (BRUV) and Diver-Operated Video (DOV) Survey Techniques. *PLoS ONE* **2016**, *11*, e0168235. [CrossRef] [PubMed]

85. Barley, S.; Meekan, M.; Meeuwig, J. Species Diversity, Abundance, Biomass, Size and Trophic Structure of Fish on Coral Reefs in Relation to Shark Abundance. *Mar. Ecol. Prog. Ser.* **2017**, *565*, 163–179. [CrossRef]
86. Leitner, A.B.; Durden, J.M.; Smith, C.R.; Klingberg, E.D.; Drazen, J.C. Synaphobranchid Eel Swarms on Abyssal Seamounts: Largest Aggregation of Fishes Ever Observed at Abyssal Depths. *Deep Sea Res. Part I Oceanogr. Res. Pap.* **2021**, *167*, 103423. [CrossRef]
87. Fitzpatrick, B.M.; Harvey, E.S.; Heyward, A.J.; Twiggs, E.J.; Colquhoun, J. Habitat Specialization in Tropical Continental Shelf Demersal Fish Assemblages. *PLoS ONE* **2012**, *7*, e39634. [CrossRef]
88. Pearson, R.; Stevens, T. Distinct Cross-Shelf Gradient in Mesophotic Reef Fish Assemblages in Subtropical Eastern Australia. *Mar. Ecol. Prog. Ser.* **2015**, *532*, 185–196. [CrossRef]
89. Abesamis, R.A.; Langlois, T.; Birt, M.; Thillainath, E.; Bucol, A.A.; Arceo, H.O.; Russ, G.R. Benthic Habitat and Fish Assemblage Structure from Shallow to Mesophotic Depths in a Storm-Impacted Marine Protected Area. *Coral Reefs* **2018**, *37*, 81–97. [CrossRef]
90. Wang, S.; Yan, Z.; Hänfling, B.; Zheng, X.; Wang, P.; Fan, J.; Li, J. Methodology of Fish EDNA and Its Applications in Ecology and Environment. *Sci. Total Environ.* **2021**, *755*, 142622. [CrossRef]
91. Evans, N.T.; Li, Y.; Renshaw, M.A.; Olds, B.P.; Deiner, K.; Turner, C.R.; Jerde, C.L.; Lodge, D.M.; Lamberti, G.A.; Pfrender, M.E. Fish Community Assessment with EDNA Metabarcoding: Effects of Sampling Design and Bioinformatic Filtering. *Can. J. Fish. Aquat. Sci.* **2017**, *74*, 1362–1374. [CrossRef]
92. Marques, V.; Castagné, P.; Polanco, A.; Borrero-Pérez, G.H.; Hocdé, R.; Guérin, P.É.; Juhel, J.B.; Velez, L.; Loiseau, N.; Letessier, T.B.; et al. Use of Environmental DNA in Assessment of Fish Functional and Phylogenetic Diversity. *Conserv. Biol.* **2021**, *35*, 1944–1956. [CrossRef]
93. McClenaghan, B.; Fahner, N.; Cote, D.; Chawarski, J.; McCarthy, A.; Rajabi, H.; Singer, G.; Hajibabaei, M. Harnessing the Power of EDNA Metabarcoding for the Detection of Deep-Sea Fishes. *PLoS ONE* **2020**, *15*, e0236540. [CrossRef] [PubMed]
94. Goatley, C.H.R.; Brandl, S.J. Cryptobenthic Reef Fishes. *Curr. Biol.* **2017**, *27*, R452–R454. [CrossRef]
95. Stat, M.; Huggett, M.J.; Bernasconi, R.; DiBattista, J.D.; Berry, T.E.; Newman, S.J.; Harvey, E.S.; Bunce, M. Ecosystem Biomonitoring with EDNA: Metabarcoding across the Tree of Life in a Tropical Marine Environment. *Sci. Rep.* **2017**, *7*, 12240. [CrossRef]
96. Cappo, M.; Speare, P.; De'ath, G. Comparison of Baited Remote Underwater Video Stations (BRUVS) and Prawn (Shrimp) Trawls for Assessments of Fish Biodiversity in Inter-Reefal Areas of the Great Barrier Reef Marine Park. *J. Exp. Mar. Biol. Ecol.* **2004**, *302*, 123–152. [CrossRef]
97. Marchese, C. Biodiversity Hotspots: A Shortcut for a More Complicated Concept. *Glob. Ecol. Conserv.* **2015**, *3*, 297–309. [CrossRef]
98. Hughes, T.P.; Barnes, M.L.; Bellwood, D.R.; Cinner, J.E.; Cumming, G.S.; Jackson, J.B.C.; Kleypas, J.; van de Leemput, I.A.; Lough, J.M.; Morrison, T.H.; et al. Coral Reefs in the Anthropocene. *Nature* **2017**, *546*, 82–90. [CrossRef]
99. Halpern, B.S.; Frazier, M.; Afflerbach, J.; Lowndes, J.S.; Micheli, F.; O'Hara, C.; Scarborough, C.; Selkoe, K.A. Recent Pace of Change in Human Impact on the World's Ocean. *Sci. Rep.* **2019**, *9*, 11609. [CrossRef]
100. Morato, T.; Watson, R.; Pitcher, T.J.; Pauly, D. Fishing down the Deep. *Fish Fish.* **2006**, *7*, 24–34. [CrossRef]
101. Tillin, H.; Hiddink, J.; Jennings, S.; Kaiser, M. Chronic Bottom Trawling Alters the Functional Composition of Benthic Invertebrate Communities on a Sea-Basin Scale. *Mar. Ecol. Prog. Ser.* **2006**, *318*, 31–45. [CrossRef]
102. Thrush, S.F.; Dayton, P.K. Disturbance to Marine Benthic Habitats by Trawling and Dredging: Implications for Marine Biodiversity. *Annu. Rev. Ecol. Syst.* **2002**, *33*, 449–473. [CrossRef]
103. Williams, A.; Schlacher, T.A.; Rowden, A.A.; Althaus, F.; Clark, M.R.; Bowden, D.A.; Stewart, R.; Bax, N.J.; Consalvey, M.; Kloser, R.J. Seamount Megabenthic Assemblages Fail to Recover from Trawling Impacts. *Mar. Ecol.* **2010**, *31*, 183–199. [CrossRef]
104. Althaus, F.; Williams, A.; Schlacher, T.A.; Kloser, R.J.; Green, M.A.; Barker, B.A.; Bax, N.J.; Brodie, P.; Schlacher-Hoenlinger, M.A. Impacts of Bottom Trawling on Deep-Coral Ecosystems of Seamounts Are Long-Lasting. *Mar. Ecol. Prog. Ser.* **2009**, *397*, 279–294. [CrossRef]
105. West, K.M.; Stat, M.; Harvey, E.S.; Skepper, C.L.; DiBattista, J.D.; Richards, Z.T.; Travers, M.J.; Newman, S.J.; Bunce, M. EDNA Metabarcoding Survey Reveals Fine-Scale Coral Reef Community Variation across a Remote, Tropical Island Ecosystem. *Mol. Ecol.* **2020**, *29*, 1069–1086. [CrossRef]
106. Danovaro, R.; Snelgrove, P.V.R.; Tyler, P. Challenging the Paradigms of Deep-Sea Ecology. *Trends Ecol. Evol.* **2014**, *29*, 465–475. [CrossRef]
107. Wellington, C.M.; Harvey, E.S.; Wakefield, C.B.; Langlois, T.J.; Williams, A.; White, W.T.; Newman, S.J. Peak in Biomass Driven by Larger-Bodied Meso-Predators in Demersal Fish Communities between Shelf and Slope Habitats at the Head of a Submarine Canyon in the South-Eastern Indian Ocean. *Cont. Shelf. Res.* **2018**, *167*, 55–64. [CrossRef]
108. Rogers, A.; Blanchard, J.L.; Mumby, P.J. Vulnerability of Coral Reef Fisheries to a Loss of Structural Complexity. *Curr. Biol.* **2014**, *24*, 1000–1005. [CrossRef]
109. Moura, R.L.; Abieri, M.L.; Castro, G.M.; Carlos-Júnior, L.A.; Chiroque-Solano, P.M.; Fernandes, N.C.; Teixeira, C.D.; Ribeiro, F.V.; Salomon, P.S.; Freitas, M.O.; et al. Tropical Rhodolith Beds Are a Major and Belittled Reef Fish Habitat. *Sci. Rep.* **2021**, *11*, 794. [CrossRef]
110. Pauly, D.; Christensen, V.; Dalsgaard, J.; Froese, R.; Torres, F., Jr. Fishing down the Food Webs. *Science* **1998**, *279*, 860–863. [CrossRef]
111. Pauly, D.; Zeller, D. Catch Reconstructions Reveal That Global Marine Fisheries Catches Are Higher than Reported and Declining. *Nat. Commun.* **2016**, *7*, 10244. [CrossRef] [PubMed]

112. Ripple, W.J.; Wolf, C.; Newsome, T.M.; Betts, M.G.; Ceballos, G.; Courchamp, F.; Hayward, M.W.; van Valkenburgh, B.; Wallach, A.D.; Worm, B. Are We Eating the World's Megafauna to Extinction? *Conserv. Lett.* **2019**, *12*, 1–10. [CrossRef]
113. Drazen, J.C.; Sutton, T.T. Dining in the Deep: The Feeding Ecology of Deep-Sea Fishes. *Ann. Rev. Mar. Sci.* **2017**, *9*, 337–366. [CrossRef] [PubMed]
114. Bakker, J.; Wangensteen, O.S.; Chapman, D.D.; Boussarie, G.; Buddo, D.; Guttridge, T.L.; Hertler, H.; Mouillot, D.; Vigliola, L.; Mariani, S. Environmental DNA Reveals Tropical Shark Diversity in Contrasting Levels of Anthropogenic Impact. *Sci. Rep.* **2017**, *7*, 16886. [CrossRef]
115. Hemingson, C.R.; Bellwood, D.R. Biogeographic Patterns in Major Marine Realms: Function Not Taxonomy Unites Fish Assemblages in Reef, Seagrass and Mangrove Systems. *Ecography* **2018**, *41*, 174–182. [CrossRef]
116. Collins, R.A.; Wangensteen, O.S.; O'Gorman, E.J.; Mariani, S.; Sims, D.W.; Genner, M.J. Persistence of Environmental DNA in Marine Systems. *Commun. Biol.* **2018**, *1*, 185. [CrossRef]
117. Hansen, B.K.; Bekkevold, D.; Clausen, L.W.; Nielsen, E.E. The Sceptical Optimist: Challenges and Perspectives for the Application of Environmental DNA in Marine Fisheries. *Fish Fish.* **2018**, *19*, 751–768. [CrossRef]
118. Jeunen, G.J.; Knapp, M.; Spencer, H.G.; Lamare, M.D.; Taylor, H.R.; Stat, M.; Bunce, M.; Gemmell, N.J. Environmental DNA (EDNA) Metabarcoding Reveals Strong Discrimination among Diverse Marine Habitats Connected by Water Movement. *Mol. Ecol. Resour.* **2019**, *19*, 426–438. [CrossRef]
119. Nguyen, B.N.; Shen, E.W.; Seemann, J.; Correa, A.M.S.; O'Donnell, J.L.; Altieri, A.H.; Knowlton, N.; Crandall, K.A.; Egan, S.P.; McMillan, W.O.; et al. Environmental DNA Survey Captures Patterns of Fish and Invertebrate Diversity across a Tropical Seascape. *Sci. Rep.* **2020**, *10*, 185. [CrossRef]
120. Tribot, A.-S.; Carabeux, Q.; Deter, J.; Claverie, T.; Villéger, S.; Mouquet, N. Confronting Species Aesthetics with Ecological Functions in Coral Reef Fish. *Sci. Rep.* **2018**, *8*, 11733. [CrossRef]
121. Tribot, A.-S.; Deter, J.; Claverie, T.; Guillhaumon, F.; Villéger, S.; Mouquet, N. Species Diversity and Composition Drive the Aesthetic Value of Coral Reef Fish Assemblages. *Biol. Lett.* **2019**, *15*, 20190703. [CrossRef] [PubMed]
122. Holmlund, C.M.; Hammer, M. Ecosystem Services Generated by Fish Populations. *Ecol. Econ.* **1999**, *29*, 253–268. [CrossRef]
123. Worm, B.; Barbier, E.B.; Beaumont, N.; Duffy, J.E.; Folke, C.; Halpern, B.S.; Jackson, J.B.C.; Lotze, H.K.; Micheli, F.; Palumbi, S.R.; et al. Impacts of Biodiversity Loss on Ocean Ecosystem Services. *Science* **2006**, *314*, 787–790. [CrossRef] [PubMed]
124. Brandl, S.J.; Rasher, D.B.; Côté, I.M.; Casey, J.M.; Darling, E.S.; Lefcheck, J.S.; Duffy, J.E. Coral Reef Ecosystem Functioning: Eight Core Processes and the Role of Biodiversity. *Front. Ecol. Environ.* **2019**, *17*, 445–454. [CrossRef]
125. Pecl, G.T.; Araújo, M.B.; Bell, J.D.; Blanchard, J.; Bonebrake, T.C.; Chen, I.C.; Clark, T.D.; Colwell, R.K.; Danielsen, F.; Evengård, B.; et al. Biodiversity Redistribution under Climate Change: Impacts on Ecosystems and Human Well-Being. *Science* **2017**, *355*, 6332. [CrossRef]
126. Ceballos, G.; Ehrlich, P.R.; Dirzo, R. Biological Annihilation via the Ongoing Sixth Mass Extinction Signaled by Vertebrate Population Losses and Declines. *Proc. Natl. Acad. Sci. USA* **2017**, *114*, E6089–E6096. [CrossRef] [PubMed]
127. Soulé, M.E. What Is Conservation Biology? *Bioscience* **1985**, *35*, 727–734.
128. Boettiger, C.; Lang, D.T.; Wainwright, P.C. rfishbase: Exploring, manipulating and visualizing FishBase data from R. *J. Fish Biol.* **2012**, *81*, 2030–2039. [CrossRef]
129. Woolley, S.N.C.; Tittensor, D.P.; Dunstan, P.K.; Guillera-Arroita, G.; Lahoz-Monfort, J.J.; Wintle, B.A.; Worm, B.; O'Hara, T.D. Deep-sea diversity patterns are shaped by energy availability. *Nature* **2016**, *533*, 393–396. [CrossRef]

*biology* MDPI

*Article*

# Habitats and Biotopes in the German Baltic Sea

**Denise Marx [1,*], Agata Feldens [1], Svenja Papenmeier [1], Peter Feldens [1], Alexander Darr [1], Michael L. Zettler [1] and Kathrin Heinicke [2]**

1 Leibniz Institute for Baltic Sea Research Warnemünde, Seestrasse 15, D-18119 Rostock, Germany; a.feldens@subsea-europe.com (A.F.); svenja.papenmeier@io-warnemuende.de (S.P.); peter.feldens@io-warnemuende.de (P.F.); alexanderdarr@web.de (A.D.); michael.zettler@io-warnemuende.de (M.L.Z.)

2 German Federal Agency for Nature Conservation, Isle of Vilm/Rügen Office, D-18581 Putbus, Germany; kathrin.heinicke@bfn.de

* Correspondence: denise.marx@io-warnemuende.de

**Simple Summary:** This study provides full-coverage maps of the habitats and biotopes in the German Baltic Sea at an unprecedented level of resolution. We combined geological and biological surveys to map the seabed and collected extensive data to classify different habitats and their inhabitants. Using newly established national guidelines and modelling, we produced highly accurate maps. These maps are of practical use in meeting national and regional reporting requirements, facilitating management decisions, supporting marine spatial planning, and answering research questions.

**Abstract:** To maintain or enhance biodiversity and sea floor integrity, mapping benthic habitats is a mandatory requirement in compliance with the Marine Strategy Framework Directive (MSFD). The EU Commission Decision distinguishes between Broad Habitat Types (BHTs) and Other Habitat Types (OHTs). At the regional level, biotopes in the Baltic Sea region are classified according to the HELCOM underwater biotope and habitat classification (HUB). In this study, the habitats and their benthic communities were mapped for the entire German Baltic Sea at a high spatial resolution of 1 km. In two nature conservation areas of the Exclusive Economic Zone (EEZ) as well as selected focus areas in the coastal waters, the resolution we provide is even more detailed at 50 × 50 m. Hydroacoustic data recording and benthological surveys (using bottom grabs, underwater towing camera technology, and diver sampling) helped identify biotopes in high resolution. Based on these data, together with additional data acquired since 2010 (a total of over 7000 stations and transect sections), we were able to spatially delineate benthic biotopes and their communities via predictive habitat modelling. The results are provided as full-coverage maps each for BHT, OHT, and HUB (9 classes of BHTs, 5 classes of OHTs, and 84 classes of HUB) with a level of spatial detail that does not yet exist for the Baltic Sea, and they form an essential basis for future monitoring, status assessments, and protection and management measures.

**Keywords:** habitats; biotopes; mapping; MSFD; broad habitat types—BHTs; other habitat types—OHTs; HELCOM Underwater biotope and habitat classification system—HUB; predictive biotope modelling; Baltic Sea

## 1. Introduction

A good environmental status in our oceans is more important than ever. The Baltic Sea is particularly vulnerable to anthropogenic pressures due to its unique and fragile ecosystem (involving marine and limnic influences, a shallow depth, and limited water exchange through the shallow Strait system) [1,2]. Over the past century, the Baltic Sea has experienced human induced regime shifts towards a eutrophic state with altered species composition [3], affecting entire food webs [4]. To overcome transnational challenges, regional regulatory frameworks were created to protect marine ecosystems. The Marine

Strategy Framework Directive (MSFD, 2008/56/EC [5]) was initiated at the EU level to protect, conserve, and, where feasible, restore the marine environment. Among other things, the Directive requires EU member states to assess and monitor the current ecological status of their marine waters, with the aim to implement measures to reduce pressures as well as monitor their effectiveness. Benthic habitats and their specific benthic communities (together considered as biotopes) are of ecological importance as integral parts of the food web, providing breeding, nursery, and feeding grounds for benthic and pelagic species, as well as migratory birds, and may even represent a biodiversity hotspot [6]. Consequently, the Directive considers the seafloor and its inhabitants under Descriptor 1 ("Benthic Habitats") and Descriptor 6 ("Seafloor Integrity") with the view that benthic ecosystems and physical disturbance as well as loss of the seabed should be avoided, and that they should not be adversely affected (MSFD Annex I). To implement the requirements of the MSFD, but also other EU legislation and regional programmes, e.g., the Habitats Directive (HD, Article 11, 92/43/EEC [7]) and the Baltic Sea Action Plan set by HELCOM (Baltic Marine Environment Protection Commission, or shortly, Helsinki Commission), reliable full coverage maps of the distribution and extent of benthic habitats as well as their changes over time are essential. The respective regulations are based on different habitat and biotope classification systems which have been adapted to each other over time. For the Habitats Directive, delimitation rules for habitat indication had to be created [8], while the MSFD is based on the European Nature Information System (EUNIS), and HELCOM developed its own system (HUB—HELCOM Underwater Biotope and Habitat Classification [9]).

The assessment within the framework of the MSFD must be carried out separately for benthic broad habitat types (BHTs) and other habitat types (OHTs) [10]. Examples of BHTs, which are specified according to the EU Commission Decision (2017/848/EU [11]), are "Infralittoral/circalittoral rock and biogenic reef", "Infralittoral/circalittoral mixed sediment", "Infralittoral/circalittoral coarse sediment", "Infralittoral/circalittoral sand", and "Infralittoral/circalittoral mud". The delimitation of these habitats is based on the definition of the European Nature Information System, EUNIS, and corresponds to EUNIS level 2 (ending at level 3 with regard to the Baltic Sea). EUNIS covers, in a hierarchical way, marine benthic habitats with their sedimentological and biological components. Infralittoral refers to the light-flooded zone that allows for the growth of vascular plants and green algae [12]. In contrast, the circalittoral is not sufficiently flooded with light. EUNIS also distinguishes between the offshore circalittoral and the (nearshore) circalittoral. However, the offshore circalittoral is not relevant in Germany.

Additional benthic habitat types (so-called other habitat types, OHTs) can be selected by the respective EU member state to be assessed separately from BHTs, e.g., due to their ecological relevance as protected biotopes [10]. Such OHTs for German marine waters include biotope types according to §30 of the German Federal Nature Conservation Act (BNatSchG), according to the European Habitats Directive (92/43/EEC), as well as the Baltic Sea-wide HELCOM Red List types [13]. Relevant OHTs of the German Baltic Sea are "species-rich areas of gravel, coarse-sand and shell-gravel areas", "seagrass meadows and other marine macrophyte populations", "reefs", "sandbanks which are slightly covered by sea water all the time", and "Baltic aphotic muddy sediment dominated by ocean quahog (*Arctica islandica*)".

The latter biotope originates from the HELCOM HUB system. It is a hierarchical system, in which the Baltic Sea marine region, the light availability, the structuring habitat, and the associated dominant benthic community are entered and indicated in a uniform code. The definition of the HUB system is largely compatible with EUNIS. In contrast to the BHTs, which are equivalent to EUNIS level 2 (the substrate level), the HUB biotope types in this study are specified down to level 6 (dominating taxa), i.e., the lowest level possible. EUNIS level 2 (or BHT) is consistent with HUB level 3.

Benthic habitat types according to the Habitats Directive [14] and EUNIS [15] as well as benthic biotope types [16,17] or both [18] were previously mapped for selected parts of the Baltic Sea. Habitat types as full-coverage maps were provided by the EMODnet Seabed Habitats project for the entire Baltic Sea on a large scale [19]. A comparable map with HUB biotope types for the German Baltic Sea has previously only been presented by Schiele et al. [20] and modified by Zettler and Darr [21]. However, these maps were based on a limited dataset and took little account of the epibenthic colonisation of the hard substrate. Since the compilation of the map by Schiele et al. in 2015 [20], guidelines were developed on how to geologically map the seafloor in German marine areas in a standardised way, how to compile sediment and boulder distribution maps, and how to delineate geogenic reefs on a large scale, respectively [8,22]. Selected areas in this present study were comprehensively hydroacoustically mapped at a high resolution, and sediment types were delineated using these mapping instructions from the BSH (Federal Maritime and Hydrographic Agency) and the BLANO technical expert group, HyMo (technical expert group "Hydrography, Hydrology and Morphology" of the Federal Government/Federal States Committee on the North Sea and Baltic Sea). Newly developed AI-supported methods for the semi-automatic detection of boulders supported the updated mapping [23–25]. Various sources of information were used in order to create maps that are as coherent as possible and to be able to indicate hard bottom communities in particular. Thus, the BHT, OHT, and HUB maps reach an unprecedented level of detail, combining various classification systems with updated data, which can now be reproduced in a standardised way.

This study provides the basis for the national assessment under MSFD Descriptor 6, taking into account the requirements of various regulations, and it is thus vital for further management decisions and the adaptation of monitoring programmes.

## 2. Materials and Methods

The study area covers the entire German part of the Baltic Sea. However, the map showing HUB biotopes could not be produced for the inner coastal waters of Mecklenburg-Western Pomerania due to a lack of data. The created maps were grid-based with a spatial resolution of at least 1 × 1 km. In areas where seabed sedimentology was fully mapped using hydroacoustic techniques (following section; Figure 1), namely the "Outer Wismar Bay", the "Darss Sill", and the "Plantagenet Ground" in the coastal waters of Mecklenburg-Western Pomerania, as well as the nature conservation areas "Fehmarn Belt" and "Kadet Trench" in the EEZ, a resolution of 50 × 50 m was applied. These high-resolution areas that were mapped with hydroacoustic recordings are referred to as "detail areas" in the following sections. For more information on each detail area (the sediment and boulder distribution maps as well as BHT, OHT, and HUB maps), see [25–29].

### 2.1. Geological Mapping

The sedimentology of the seafloor was mapped by hydroacoustic methods in three areas in the coastal waters of Mecklenburg-Western Pomerania and two nature conservation areas in the EEZ [25,26]. Side-scan sonars (including Marine Klein 4000, Edgetech 4200, Edgetech 4300 MPX, Starfish 450F, Edgetech 4200 HF, and R2Sonic2024) with different frequencies (100–600 kHz) were used for this purpose. External data were acquired (Vermessungsbüro Weigt, BSH, Christian-Albrechts-University Kiel) in order to apply the time-consuming measurements only in areas where data with the required resolution and quality were not yet available. Sediment samples and videos were used to verify the hydroacoustically recorded data within the detail areas (ground truthing, as shown in Section 2.5 and further described together with the benthos sampling in Section 2.4). Data processing was carried out with SonarWiz (Chesapeake Technology Inc., Los Altos, CA, USA) software, which creates backscatter mosaics of the seafloor surface. After setting the bottom tracks, correcting for slant range distortion, and setting the layback, empirical gain normalisation was applied, and the backscatter mosaics were imported into ArcMap 10.7.

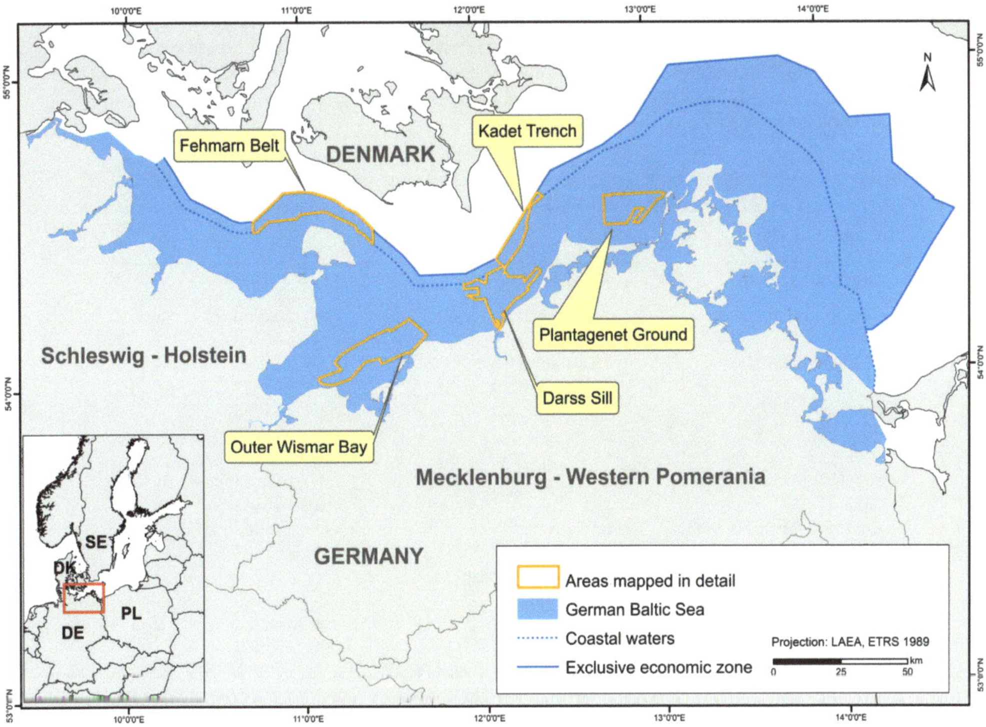


**Figure 1.** The German Baltic Sea including areas which were mapped in detail (resolution: 50 × 50 m).

Sediment analyses from ground truthing samples were performed optically using a Mastersizer 3000 (Malvern Panalytical Ltd., Malvern, UK) as well as by dry and wet sieving due to the heterogeneity of the samples (silt, sand, and coarse sediment). The Mastersizer was used to measure samples up to a maximum grain size of 3.0 mm. Samples with coarser components were sieved. Seemingly fine-grained samples were treated with HCl and $H_2O_2$ before measurement to remove carbonate and organic components, because these compounds impact grain size distributions. For the evaluation of the hydroacoustic data, the results of the sediment analyses were evaluated according to Blott and Pye [30] and fitted to the Folk triangle [31] while considering the BSH hydroacoustic mapping instructions [22].

With the help of the hydroacoustic data, video recordings, sediment samples, and the sediment map according to Tauber [32], sediment distribution maps were created within the detail areas according to a national guideline [22] with the following standards: Sedimentological classification of the areas delineated on the backscatter mosaic was performed for the sediment types at three different levels where possible. Level A includes fine sediments, sands, mixed sediments, coarse sediments (simplified classification according to Folk [31]), and both residual sediments and peat. The term "residual sediment" (lag sediment) is not a clearly defined concept in marine geosciences, but it is nevertheless frequently used for a higher-level description of certain sediment types. Residual sediments cannot be identified by an exact granulometric measurement. Rather, residual sediments describe the remaining part of sediments that have been reworked by natural dynamic processes. Such reworking processes usually result in a granulometric separation/sorting of the sediment components, whereby the less mobile components like gravel, small boulders, or marl remain in the area of the original sediment for longer [33–35]. This distinguishes them from the categories of mixed sediments and coarse sediments defined by Folk [31], which contain only mud, sand, and gravel, but not fractions beyond that. In Level B, the clastic sediment types from Level A are further subdivided according to Folk [31]. Since there is no subdivision of sands in the Folk triangle, Level C was introduced, in which sediments designated as sands in Levels A and B were further subdivided according to Figge [36] (for more information, see [26]; Table 1).

**Table 1.** Level specifications of sediment classifications according to BSH [22]. * Not specified = lack of information and/or knowledge for exact classification. ** Not classified = cannot be classified further in this level.

| Level A | Level B | Level C |
|---|---|---|
| Fine sediment (Fsed) | not specified * | not classified ** |
| | mud (M) | not classified |
| | sandy mud (sM) | |
| | muddy sand (mS) | |
| Sand (S) | sand (S) | not classified |
| | | fine sand (fSa) |
| | | medium sand (mSa) |
| | | mixed sand (mxSa) |
| | | coarse sand (cSa) |
| Coarse sediment (Csed) | not specified | not classified |
| | gravelly sand (gS) | not classified |
| | sandy gravel (sG) | |
| | gravel (G) | |
| Mixed sediments (MxSed) | not specified | not classified |
| | gravelly mud (gM) | not classified |
| | gravelly muddy sand (msG) | |
| | muddy gravel (mG) | |
| Peat | | |
| Lag sediment (LagSed) | not classified | not classified |
| Not specified | not specified | not specified |

The distribution of boulders is displayed in a separate map. The number of boulders in the nature conservation areas in the EEZ ("Fehmarn Belt" and "Kadet Trench") was estimated manually for each 50 × 50 m grid cell by dividing them via subitising (recording the number of boulders at first sight without counting) into three classes, according to the guideline for the large-scale delineation of geogenic reefs in the German Baltic Sea [8]: cells without boulders (Class 1), cells with 1–5 boulders (Class 2), and cells with more than 5 boulders (Class 3). Boulders in the detail areas of the coastal waters in Mecklenburg-Western Pomerania ("Outer Wismar Bay", "Darss Sill", and "Plantagenet Ground") were detected semi-automatically using the methods reported in the study by Feldens et al. [23,24]. A neural network detected individual boulders in the backscatter mosaics [25]. Where possible, mosaics acquired at a frequency between 300 and 500 kHz were chosen as a baseline, as these show better individual object resolution [37]. The results were screened for false detections (mainly caused by water column stratification artefacts in the data) and then classified into the same three classes as described above. A boulder distribution map was used to place a grid over the areas and indicate these three categories per 50 × 50 m cell.

The sediment and boulder distribution maps formed the basis of the subsequent BHT, OHT, and HUB maps, which were created using ArcGIS Desktop 10.7. All maps can be found in the Supplementary Materials and are available for download as an ArcMap package.

### *2.2. Compiling the BHT Map (Benthic Broad Habitat Types according to EU Commission Decision 2017/848/EU)*

The BHT sediment definition according to EUNIS in the area of the German Baltic Sea differentiates types of the infralittoral (light-flooded) and the circalittoral (nonlight-flooded) zones such as "mud", "sand", "coarse sediment", "mixed sediment", and "rocks and biogenic reef". "Mixed sediment" corresponds to a hard substrate fraction of 10–90 % cover. The category of "rocks and biogenic reef" only refers to the occurrence of biogenic reefs because of the lack of information on specific coverages of geogenic hard substrates. A distinction between above 90% ("rocks and biogenic reef") and below 90% hard substrate ("mixed sediment") could thus not be made. However, it cannot be ruled out that the hard substrate cover locally exceeds 90 %. Only peat bottoms entered the category as biogenic reefs because they were covered by mussels. Geogenic hard substrates (categorised as "mixed sediment" in this study) within the detail areas is assigned when >5 boulders are present in a 50 × 50 m cell, or if both 1–5 boulders (as seen from the boulder distribution map) as well as lag sediment (as seen from the sediment distribution map) occur in a cell. Otherwise, the sediment is defined as soft substrate. In addition to the sediment map of Tauber [32], the hard substrate was assigned according to the reef coverage, which, for the first time, was delineated over a large area for reef designation according to [8] within the detail areas. These areas were reported to HOLAS III (HELCOM holistic assessment). Outside of the detail areas, reef coverage in Schleswig-Holstein [38], Mecklenburg-Western Pomerania [39], and the EEZ (BfN) was used for further hard substrate allocation. Mapped reef areas and suspected reef areas were assigned to the BHT "mixed sediment (hard substrate)", and the polygon areas were gridded. Reefs in Mecklenburg-Western Pomerania, in contrast to those in Schleswig-Holstein (at 50 × 50 m), were shown at 1 × 1 km because they were not mapped out. The sediment map of Tauber [32] was used for the soft substrate allocation outside the detail areas. An overview of the data basis entered for the BHT, OHT, and HELCOM HUB maps can be found in Table 2.

**Table 2.** Data basis for the BHT, OHT, and HELCOM HUB maps inside and outside the detail areas. Note the distinction between endobenthos and epibenthos in the predictors used for HUB biotope modelling.

| | BHT | | OHT | | HUB | |
|---|---|---|---|---|---|---|
| | Detail Areas | Outside of Detail Areas | Detail Areas | Outside of Detail Areas | Detail Areas | Outside of Detail Areas |
| Overall resolution | 50 × 50 m | 1 × 1 km | 50 × 50 m and polygons | 1 × 1 km and polygons | 50 × 50 m | 1 × 1 km |
| Map basis for soft bottom | Sediment distribution maps from hydroacoustic surveys (gridded) | Tauber [32] (gridded) | Seagrass meadows and "species-rich areas of gravel, coarse-sand and shell-gravel areas" mapped according to hydroacoustic results; distribution area of "Baltic aphotic muddy sediment dominated by ocean quahog (*Arctica islandica*)" modelled in this study | "Seagrass meadows" modelled by [40,41] (gridded); sandbanks as reported to HOLAS III (polygons); distribution area of "Baltic aphotic muddy sediment dominated by ocean quahog (*Arctica islandica*)" modelled in this study | Sediment distribution maps from hydroacoustic surveys (gridded) | Tauber [32] (gridded) |
| Map basis for hard bottom | Boulder distribution maps from hydroacoustic surveys according to [8] (grids) | Reef areas as reported to HOLAS III (gridded) | Distribution area of "other marine macrophyte populations" modelled in this study; reefs mapped hydroacoustically in this study (gridded) | Reef areas as reported to HOLAS III (polygons) | Boulder distribution maps from hydroacoustic surveys according to [8] (grids) | Reef areas as reported to HOLAS III (gridded) |

**Table 2.** *Cont.*

| | BHT | | OHT | | HUB | |
|---|---|---|---|---|---|---|
| | **Detail Areas** | **Outside of Detail Areas** | **Detail Areas** | **Outside of Detail Areas** | **Detail Areas** | **Outside of Detail Areas** |
| Hard bottom assignment | >5 boulders/50 × 50 m cell or lag sediment and >1 boulder/50 × 50 m cell (from boulder and sediment distribution maps) | Reef areas as reported to HOLAS III (gridded) | Reefs mapped according to [8] | Reef areas as reported to HOLAS III | >5 boulders/50 × 50 m cell or lag sediment and >1 boulder/50 × 50 m cell (from boulder and sediment distribution maps) | Reef areas as reported to HOLAS III (gridded) |
| Biotope classification schemes | EUNIS | EUNIS | "Species-rich areas of gravel, coarse-sand and shell-gravel areas" according to [42]; "Seagrass meadows and other marine macrophyte populations" and "Baltic aphotic muddy sediment dominated by ocean quahog (*Arctica islandica*)" according to HUB; reefs according to [8] | "Seagrass meadows and other marine macrophyte populations" and "Baltic aphotic muddy sediment dominated by ocean quahog (*Arctica islandica*)" according to HUB; reefs according to [8] | HUB | HUB |
| Predictors used for modelling | - | - | Only "Seagrass meadows and other marine macrophyte populations" and "Baltic aphotic muddy sediment dominated by ocean quahog (*Arctica islandica*)" were modelled in this study; the former is equivalent in their spatial extent to HUB class "Baltic photic mixed substrate dominated by perennial non-filamentous corticated red algae" and "Baltic a-/photic mixed substrate/coarse sediment dominated by foliose red algae" (*Zostera* spp. and *Fucus* spp. were not modelled in this study) and only indicated outside the reef areas; for predictors, see HUB entries | See detail areas | **Endobenthos:** sediment distribution map (50 × 50 m), water depth (50 × 50 m), temperature, salinity, current velocity (in directions north/south, east/west, without directional information), bottom shear stress, oxygen concentration, number of hypoxic days, DOC, ammonium, nitrate, phosphate (600 × 600 m) **Epibenthos:** boulder distribution map, water depth, photic zonation, slope gradient (50 × 50 m), temperature, salinity, current velocity (in directions north/south, east/west, without directional information), bottom shear stress, photosynthetically active radiation (PAR), oxygen concentration, number of hypoxic days, DOC, ammonium, nitrate, phosphate (600 × 600 m) | See detail areas; Tauber [32] was used instead of the sediment distribution map for endobenthos modelling, and reef coverage (as reported to HOLAS III) was used instead of boulder distribution map for epibenthos modelling |

**Table 2.** *Cont.*

| | BHT | | OHT | | HUB | |
|---|---|---|---|---|---|---|
| | **Detail Areas** | **Outside of Detail Areas** | **Detail Areas** | **Outside of Detail Areas** | **Detail Areas** | **Outside of Detail Areas** |
| "Seagrass meadows and other marine macrophyte populations" (paragraph §30 Federal Nature Conservation Act) | - | - | Seagrass mapped in the "Plantagenet Ground"; other macrophytes modelled in this study | *Zostera* spp. modelled in Schleswig-Holstein [40] and Mecklenburg-Western Pomerania [41]; *Fucus* spp. modelled in Schleswig-Holstein [40]; other macrophytes modelled in this study | See OHT | See OHT |

The sediment classification schemes used for hydroacoustic interpretation and by Tauber [32] are incompatible with the biotope classification systems; therefore, a translation was necessary. The reclassification for Tauber is documented in Table 3. The silt and gravel/coarse sand content of each sediment class from the map according to Tauber was used from the underlying sediment analyses to categorise the sediment classes according to the definitions by EUNIS. Habitat categorisation according to EUNIS [43] is defined as "muddy sediment" if the mud, silt, or clay (<63 µm) content is at least 20%; "coarse sediment" if the mud/silt/clay fraction is less than 20% and the gravel and pebbles (2–63 µm) exceed 30% of the combined gravel and sand fraction; and "sand" if the mud/silt/clay fraction is less than 20% and the sand (0.063–2 mm) exceeds 70% of the combined gravel and sand fraction. The categories "muddy sediment", "coarse sediment", "sand", or "rock and boulders" (>63 mm) are used when a coverage of at least 90 % is reached. "Mixed sediment" is used if the coverage of hard (rock/boulders/stone) and soft substrata (muddy sediment/sand/coarse sediment) is at least 10–90%. EUNIS is therefore based on the HELCOM HUB classification. The silt fraction (grain size < 63 µm) of the sediment class "very fine sand" from the map according to Tauber, for example, was >20% and was therefore assigned to the EUNIS/BHT type "mud". As the sediment characteristics varied considerably in the detail areas, reclassification from the hydroacoustic surveys to EUNIS sediment types was carried out in an area-specific manner.

**Table 3.** Translation of sediment types classified according to Tauber [32], following Folk [31] and Figge [36], into sediment classification according to EUNIS, on which the BHTs are based.

| Sediment Type Classified according to Tauber (2012) | Sediment Type Reclassified according to EUNIS |
|---|---|
| gravel, very coarse sand | coarse sediment |
| fine sand—coarse sand | sand |
| very fine mud—very fine sand | mud |
| clay, peat, lag sediment/till | mixed sediment (hard substrate) |

For photic zonation, a modelled polygon shape from [44] was used. Photic zonation was assigned in this study to the infralittoral zone, and aphotic zonation was assigned to the circalittoral zone.

For this study, BHTs occupying an area of <1 ha were eliminated and aligned with the surrounding BHTs. Areas within the inner coastal waters were retained as shown in the latest BHT map prepared in 2018 [45].

The nationally protected habitat type involving "species-rich areas of gravel, coarse-sand and shell-gravel areas" (§30 of the German Federal Nature Conservation Act) was specified as BHT "sand" rather than "coarse sediment" because the condition of >30% gravel or coarse sand content (EUNIS) was not met, based on the available sediment distribution maps.

### 2.3. Compiling the OHT Map (Benthic Other Habitat Types according to EU Commission Decision 2017/848/EU)

In contrast to BHTs, OHTs were included in the map as they were. For example, if reef areas were reported as polygons nationally, this area was integrated into the map as it was, and not gridded. Reefs mapped in this study were derived according to the national guideline for large-scale delineation of geogenic reefs in the German Baltic Sea [8]. The guideline specifies certain rules for gap closure within reef occurrences, so that the delineated reefs do not have to completely match with the indication of the BHT "mixed sediment". Thus, in a cell where habitat type 1170 is indicated, the BHT "sand" may occur.

Sandbanks were shown in the same manner as the reefs from Schleswig-Holstein (Schleswig-Holstein State Office for the Environment, LFU [38]), Mecklenburg-Western Pomerania (Leibniz-Institute for Baltic Sea Research Warnemünde, IOW; State Office for the Environment, Nature Conservation and Geology Mecklenburg-Western Pomerania, LUNG [39]) and the EEZ (IOW; Christian-Albrechts-University Kiel, CAU Kiel; Federal Agency for Nature Conservation, BfN; Federal Maritime and Hydrographic Agency, BSH). Sandbanks in "Fehmarn Belt" and "Adler Ground", as identified and described in the study by Boedeker et al. [46], were remapped by CAU Kiel and IOW and intersected with the reef cover from 2022.

"Seagrass meadows and other marine macrophyte populations" (§30 Federal Nature Conservation Act) and "Baltic aphotic muddy sediment dominated by ocean quahog (*Arctica islandica*)" (HELCOM Red List) were modelled in this study in contrast to the other OHTs as HELCOM HUB biotope type and integrated into the OHT map (see Sections 2.6 and 2.7). *Zostera* spp. and *Fucus* spp. distribution areas for Schleswig-Holstein ([40]; data from the State Office for the Environment Schleswig-Holstein, 21 February 2022 and 18 March 2022) that were already modelled and mapped eelgrass beds for Mecklenburg-Western Pomerania [41] were integrated into the modelled HUB map at the end and indicated as OHT. *Fucus* spp. or *Zostera* spp. entered a cell as soon as they were modelled with an occurrence of at least 50 % (this also corresponds to the prediction probability) or mapped with at least 10 individuals/m$^2$. The biotope type "seagrass meadows and other marine macrophyte populations" also includes foliose and corticated red algae, which were not indicated here in favour of the reef indication as habitat type, except when individual occurrences were observed outside the reef cover.

### 2.4. Biological Mapping

A total of 1637 grab samples, 403 station videos, 59 station photos, and 47 photo transects were taken and processed. Dominant benthic communities were classified in preparation for the HELCOM HUB map. Data from grab samples were used to determine dominant endobenthic organisms, and video and photographic records as well as diver samples were used to determine dominant epibenthic organisms.

Grab sampling was conducted using a Van Veen grab (0.1 m$^2$) (Alu-Bau Ltd, Büdelsdorf, Germany) with an additional sediment sample obtained for granulometric analysis. The benthic samples were flushed through a sieve with a mesh size of 1 mm or, in the case of coarser sediment content, suspended in several subsamples, and the supernatant was decanted and poured through a 1 mm sieve again. The sample was fixed using a 4% formalin buffer solution, and marble grit was added to preserve mussel and snail shells. In the laboratory, the specimens were determined to species level, if possible, using a Carl Zeiss Discovery.V8 binocular (Carl Zeiss AG, Oberkochen, Germany). The wet weight was determined. Determination of dry weight and ash-free dry weight was carried out using

Leibniz Institute for Baltic Sea Research (IOW) internal conversion factors [47]. Sampling was carried out according to standard instructions [48,49].

In addition to grab sampling, optical methods (underwater video and photography) were used to record epibenthic colonisation. Simultaneously with grab sampling, video recordings were taken at the grab station sites using a SeaViewer Sea Drop 6000 HD for a minimum of 5 min. For transects (0.3–2 nm), the recordings were obtained using a towed camera system developed at IOW (BaSIS—Baltic Sea Imaging System [50]), towed at ~0.5 kn. This camera system took one image every 15 s, of which one photo per minute that was suitable for analysis (not blurred, no shadows, and no sediment turbulence) was selected. In addition, in one campaign, an external drop camera system was used in the EEZ, which was designed by the German Federal Agency for Nature Conservation (BfN). This drop camera frame, equipped with a GoPro HERO4 Black, took a picture every 5 s at a station.

Video analysis (SeaViewer) was semi-quantitative based on the estimated coverage of epibenthic taxa and substrate according to the ACFOR scale (abundant, common, frequent, occasional, rare), which was visible in 5 min of video recording at the station. Image analysis (BaSIS, BfN drop camera) was performed quantitatively using the open-source software CoralPhotoCount 4.1 with an Excel extension (CPCe [51]), as described in the study by Beisiegel [50].

Furthermore, diver sampling was performed, during which scratch samples were obtained through collection frames (0.1 $m^2$) with attached net bags, where the surface of a stone/boulder was scraped off within the frame, and biomass (dry weight) was determined. In addition, diver photos were used to estimate the degree of coverage of the epibenthic organisms.

Dominant benthic communities were indicated according to the HUB (HELCOM Underwater Biotope and Habitat Classification) system published by HELCOM [9]. Individual HUB classes were assigned manually at each station or georeferenced transect section (still images). As a result, several HUB classes in one cell could be included (as response variables) in the modelling. Both soft- and hard-bottom classifications were carried out separately. Endobenthos classification was assigned first from grab samples, and then epibenthos classification was assigned from video and photographs. For an endobenthic taxon, the critical value was based on a biomass fraction of >50%, and for an epibenthic taxon, the criteria were based on coverage of 10% on the total area or 90% on hard substrate to be considered dominant [9]. Assigned HUB classes were then represented areally by predictive habitat modelling (see Section 2.6).

### 2.5. Data Basis for Modelling

In addition to the data collected in the current study's projects and data from the IOW database, acquired data from the 2010–2021 period were used (~45% internal and ~55% external data). Table 4 shows the amount of data and where it was derived from. These data originate from grab samples provided by authorities and private sector service companies (LUNG, LFU, StALU MM, StALU WM, WSA Stralsund, IfAÖ, Palaemon aquatic service company). A total of 3,628 stations were included in the model for endobenthic communities (Figure 2).

External data used for sessile epibenthos modelling came from photo-recorded diver sampling from management plans [52–56] and diver scratch sampling [57,58]. A total of 3623 stations and transect sections from the 2010–2021 period (~92% internal and ~8% external data) were included in the model (Figure 3). All cells with a larger areal proportion of hard sediment (>5 boulders or lag sediment with at least one boulder per 50 × 50 m cell) to soft sediment within a 1 × 1 km cell were included in the epibenthos modelling. The basis for the hard bottom modelling was the current reef boundaries of the coastal waters of Schleswig-Holstein, Mecklenburg-Western Pomerania, and the EEZ. In the area of Schleswig-Holstein, this includes suspected reef areas and geologically as well as biologically verified reefs [38]. In the area of Mecklenburg-Western Pomerania, the data were derived from the management plans and the suspected habitat type areas (according to the

Habitats Directive) from 2011 [39]. In the EEZ, it consisted of reef areas mapped by CAU Kiel, BSH, and IOW through the EEZ project 6 and the project SEDINO phases I, II, and III (both funded by BfN).

**Table 4.** Number and sampling instruments of internal and external (in brackets) data that were mapped and acquired. Further data were obtained from the Federal Maritime and Hydrographic Agency (BSH), the State Office for the Environment, Nature Conservation and Geology of Mecklenburg-Western Pomerania (LUNG), the Schleswig-Holstein State Office for the Environment (LFU), the State Office for Agriculture and the Environment of Central Mecklenburg (StALU MM) and Western Mecklenburg (StALU WM), the Waterways and Shipping Office Stralsund (WSA Stralsund), the Christian-Albrechts-University Kiel (CAU Kiel), the Institute for Applied Ecosystem Research Ltd. (IfAÖ) (Neu Broderstorf, Germany), and the GEOMAR—Helmholtz Centre for Ocean Research Kiel.

| | Area | Number of Acquired Data Points | Sampling Instruments | References of Used Data |
|---|---|---|---|---|
| Detail areas | Outer Wismar Bay | 85 (18) grab stations, 29 video stations, 6 video transects | Van Veen grab, SeaViewer, BaSIS | IOW, IfAÖ, LUNG, StALU WM, StALU MM |
| | Darss Sill | 73 (106) grab stations, 26 video stations, 4 video transects | Van Veen grab, SeaViewer, BaSIS | IOW, IfAÖ, LUNG, StALU WM, StALU MM |
| | Plantagenet Ground | 49 (67) grab stations, 27 video stations, 4 video transects | Van Veen grab, SeaViewer, BaSIS | IOW, IfAÖ, LUNG, StALU WM, StALU MM |
| | Kadet Trench | 103 (17) grab stations, 37 video stations, 8 video transects, 36 photo stations | Van Veen grab, SeaViewer, BaSIS, BfN drop camera | IOW, CAU Kiel, BSH |
| | Fehmarn Belt | 339 grab stations, 134 video stations, 11 video transects | Van Veen grab, BaSIS | IOW, CAU Kiel, BSH |
| | German Baltic Sea | 1637 (1991) grab stations, 403 video stations, 47 video transects, 59 photo stations, (45) diver stations, 9 (82) diver photo stations | Van Veen grab, SeaViewer, BaSIS, BfN drop camera, diver scratch samples and photos | IOW, BfN, BSH, LFU, LUNG, StALU WM, StALU MM, WSA Stralsund, CAU Kiel, IfAÖ, Geomar |

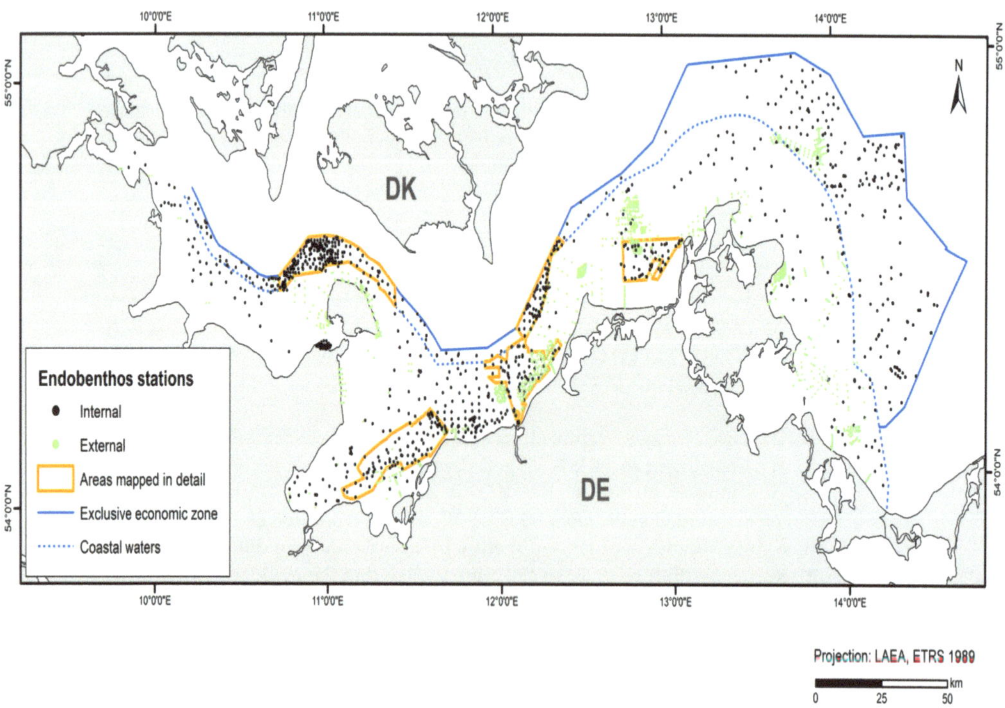


**Figure 2.** Stations and their data sources that entered the endobenthos model.

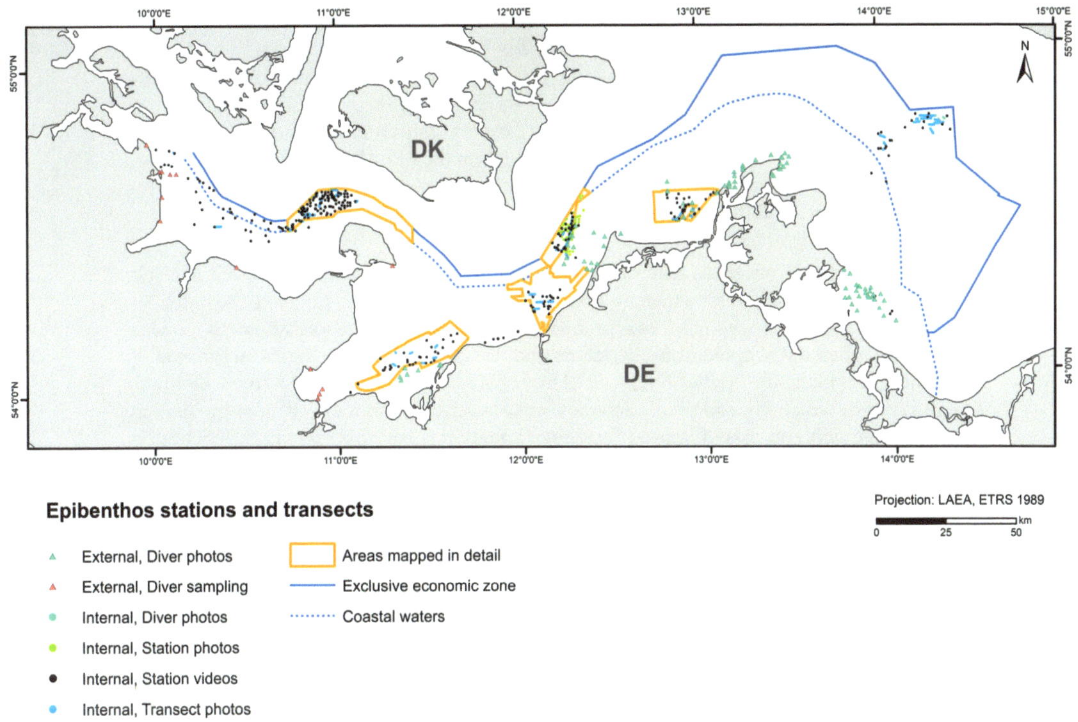


**Figure 3.** Stations and transects that entered the epibenthos model, indicating data source and type.

### 2.6. Predictive Biotope Modelling

The HELCOM HUB map was created using predictive habitat modelling, unlike the BHT and OHT maps (except for HELCOM HUB biotope types included therein). In preparation for the modelling and subsequent HUB biotope map, a grid of 1 × 1 km grid cells (corresponding to the EEA standard grid) was placed over the coastal waters and EEZ, with each cell assigned a unique entry from the abiotic variables. If a grid cell contained multiple sediment types, the sediment with the higher proportion within the cell was assigned to the cell. The same procedure was used for the detail areas with a 50 × 50 m grid. Both the overview area and the detail areas were each modelled separately. The data used for the detail areas also went into the modelling of the overview map.

A random forest classification model (after [59]) was used to predict HUB biotopes. Modelling was carried out separately for endobenthos and epibenthos using the "randomForest" package (version 4.6–14, [60]) in RStudio 2022.12.0 (R environment version 4.2.2, the R Foundation for Statistical Computing Platform).

First, with respect to modelling, the already assigned HUB classes of each station/transect section (as described in Section 2.4) were specified at levels 4–6 (biotope level, without sediment information), and after modelling, the predicted HUB codes were completed with the found sediment and photic zone in the respective cell (levels 1–3) according to the definitions of HELCOM (for the HUB map). The previous manually assigned HUB classes entered the model as response variables and were used to classify HUB classes in every cell of the German marine waters (for endobenthos) and the hard substrate areas (for epibenthos). This study therefore follows a community-based modelling approach, as described by other authors [61–63]. In addition to the soft-bottom data from the sediment distribution maps of the detail areas, the following raster datasets from both the ERGOM model (Ecological Regional Ocean Model, model run from 2010 to 2017, [64]) and the

GETM model (General Estuarine Transport Model, model run from 2010 to 2020, [65]) were available as predictors at a 600 × 600 m resolution:

- Temperature, salinity, current velocity (in directions of north/south, east/west, without directional information), and bottom shear stress from the GETM model [65];
- Photosynthetically active radiation (PAR), oxygen concentration, number of hypoxic days, DOC, ammonium, nitrate, phosphate from the ERGOM model [64];
- Water depth and sediment type [32];
- Photic zonation (based on ERGOM model, [44]);
- Slope gradient (based on [32]).

The values from these raster datasets were assigned from the centre point of a cell. Outside of the detail areas, where no areal geologic mapping was conducted, the sediment map of Tauber [32] was used as a predictor for soft-bottom categorisation. Slope was only included in the epibenthos modelling and was created from the bathymetric map of Tauber [32] using the Spatial Analyst tool in ArcGIS Desktop 10.7.1. PAR (photosynthetically active radiation), and photic zonation was also included in the epibenthos modelling only. The polygon shapefile used for photic zonation is based on the light penetration depth (PAR) values from the 2000–2010 ERGOM model run [44]. To separate the photic and aphotic zones, the 1% light penetration depth (averaged over the growing season from March to October) was coupled with bathymetry [66]. The initial dataset was randomly divided into a training dataset (70%) and testing dataset (30%). To improve model performance, hyperparameters (number of trees and number of predictors at each decision node) were tuned until lowest out-of-bag (OOB) error was found, and model adjustments were made if the dataset was imbalanced (using downsampling, balanced random forest, upsampling, and the SMOTE algorithm).

### *2.7. HUB Map Modelling Limitations and Conventions*

In general, the model performance decreases when modelling classes are very similar to each other, for example, when separating and predicting a biotope class of a dominant specific species from a biotope class of a community containing exactly the same species. Therefore, the following conventions had to be adopted in the modelling process (based on Sections 2.5 and 2.6):

- Elimination of outliers:
- Before modelling the endobenthos in the whole German Baltic Sea, stations dominated by taxa that rarely occurred in the area and that accounted for max. 1% of the total number of stations were eliminated. Such outliers were Actiniaria and oligochaetes (in HELCOM HUB they are classified as meiofauna).
- *Ophelia* spp./*Travisia* spp. could not be separated from other communities by the random forest (RF) model and therefore were not reliably predicted, so stations with dominant *Ophelia* spp./*Travisia* spp. were also deleted.
- Less frequent dominant taxa were assigned to a higher category:
- Dominant *Mya arenaria* and *Astarte* spp. were assigned to the community with multiple infaunal bivalve species, because being a part of the overarching community, they were poorly separable from each other. Because the polychaete communities (partly with dominating *Scoloplos armiger*, *Marenzelleria* spp., *Pygospio elegans*, and *Hediste diversicolor*) were difficult to separate from the other communities; they were grouped together as the community with macroscopic infaunal biotic structures (HUB Level 4), as were stations ending at HUB level 5 (e.g., dominant bivalves/polychaetes/crustaceans). Therefore, the community with macroscopic infaunal biotic structures includes not only communities without dominant taxa, but also those previously mentioned that are too unspecific in their occurrence, leading to improved model performance
- Non-dominant communities were indicated as dominant:
- Epibenthos-dominated stations that ended up at HUB level 5 were indicated as HUB level 6 (e.g., foliose red algae were treated as dominant even though they had < 50%

cover), because the model cannot separate dominant and non-dominant communities, in order for those stations to be included in the model. This means that in areas where epibenthic communities are predicted, they do not need to be dominant, but they are more likely to occur than other communities.

- Mixed communities were indicated as non-mixed communities:
- Mixed communities that are very similar in species composition (e.g., foliose red algae, foliose red algae/sponges, foliose red algae/filamentous red algae, foliose red algae/bryozoans, and foliose red algae/sponges/kelp) cannot be clearly delineated by the model. Therefore, these mixed communities were assigned to those taxa that play a superior role in the biotope function (structuring, long-lived, and geographically dominant). For example, the classes listed above were assigned to dominant foliose red algae. This means that epibenthic mixed communities can always occur, even when indicated otherwise. Transitions cannot be modelled with the procedure chosen here because the model considers each class as distinct.

The predictions of endobenthos and epibenthos from the models were intersected eventually, in the sense that the epibenthic community in a cell was indicated at the sites where hard substrate dominates. Unlike the BHT map, the HUB map was not generalised (i.e., areas < 1 ha were not matched to surrounding sediment).

Benthic broad habitat types and other habitat types according to the Commission Decision [11] are aligned with HELCOM HUB biotope types. This means that habitats (i.e., sediment information) coincide, except for the indication of OHT "reefs" and BHT "mixed sediment", as different delimitation rules underlie here (see Section 2.3).

## 3. Results

### 3.1. Broad Habitat Type (BHT) Map

The map in Figure 4 shows the broad habitat types at a 1 × 1 km resolution with the incorporation of the areas mapped at a 50 × 50 m resolution in this study ("Outer Wismar Bay", "Darss Sill", and "Plantagenet Ground" in the coastal waters of Mecklenburg-Western Pomerania, as well as the nature conservation areas of the EEZ, "Fehmarn Belt" and "Kadet Trench"). Infralittoral sand and circalittoral mud occupy the largest areas in the German Baltic Sea, with each being >20% of the total area (Table 5), followed by circalittoral sand, infralittoral mixed sediment (hard substrate), infralittoral mud, and circalittoral mixed sediment (hard substrate). The remaining BHT categories amount to less than 1% of the total area.

Major differences to the previous version of the map from 2018 [45] are the update of hard-bottom areas and the detailed representations of sediment compositions in selected areas. However, the assignment of sediment types (Table 3) shown according to [32] also differs from the sediment reclassification in the map submitted to HOLAS II; for example, mudflats (e.g., west of Fehmarn, in the Plantagenet Ground, east of the Isle of Rugia) are more widespread or larger than in the 2018 map. Another difference is the photic zonation. While a layer from the EUSeaMap was used for the previous map, a more detailed shapefile from [44] was used here for the classification into infralittoral (photic) and circalittoral (aphotic) zones. The area of the infralittoral zone is larger in the shapefile used in this study, with the Kiel Bight, in particular, differing on a large scale, and the rest differing on a rather small scale. The inner coastal waters are consistent with the 2018 map. The only change was made in the Szczecin Lagoon, where circalittoral mud and sand were changed to infralittoral mud and sand.

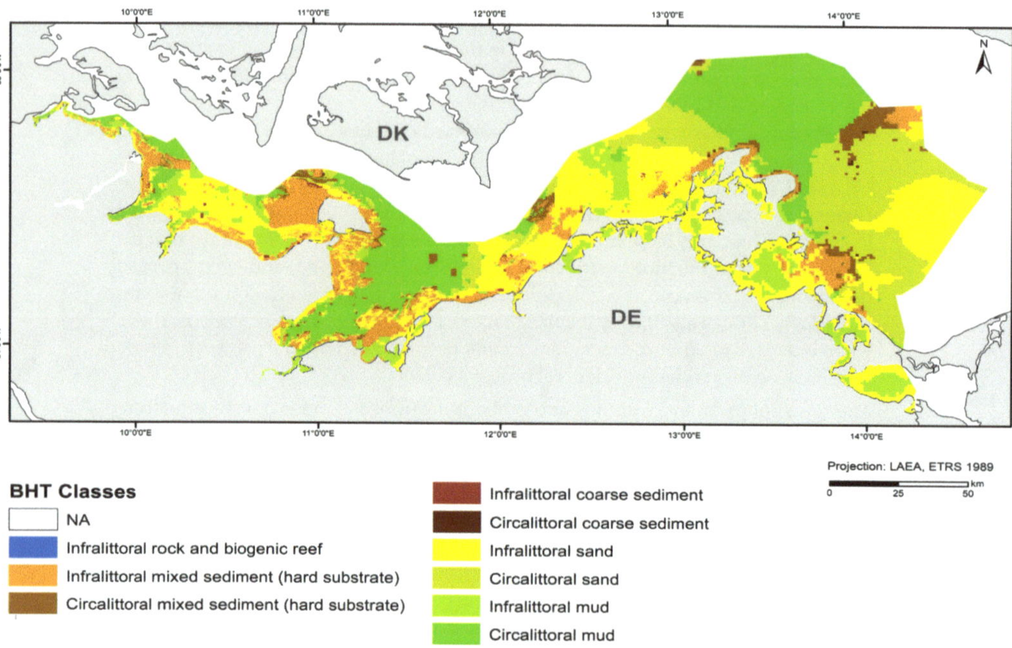


**Figure 4.** Distribution of broad habitat types (BHT) in the German Baltic Sea.

**Table 5.** Areas and their proportions of individual broad habitat types (BHT) in the German Baltic Sea.

| BHT | Area ($km^2$) | Area (%) |
|---|---|---|
| Infralittoral rock and biogenic reef | 1.0 | 0.007 |
| Infralittoral mixed sediment (hard substrate) | 1785.3 | 11.6 |
| Circalittoral mixed sediment (hard substrate) | 488.3 | 3.2 |
| Infralittoral coarse sediment | 35.5 | 0.2 |
| Circalittoral coarse sediment | 16.2 | 0.1 |
| Infralittoral sand | 4600.0 | 29.8 |
| Circalittoral sand | 3010.4 | 19.5 |
| Infralittoral mud | 1393.8 | 9.0 |
| Circalittoral mud | 4115.1 | 26.6 |

### *3.2. Other Habitat Type (OHT) Map*

Reefs occupy the largest area of all OHTs with 2183.5 $km^2$ (Figure 5, Table 6). They consist mainly of boulder fields and extend mostly on abrasion platforms that continuously expose boulders during the ongoing erosion of glacial till [34]. So far, only small areas of pure lag sediment reefs have been mapped. Biogenic reefs (pure mussel beds) have not yet been observed.

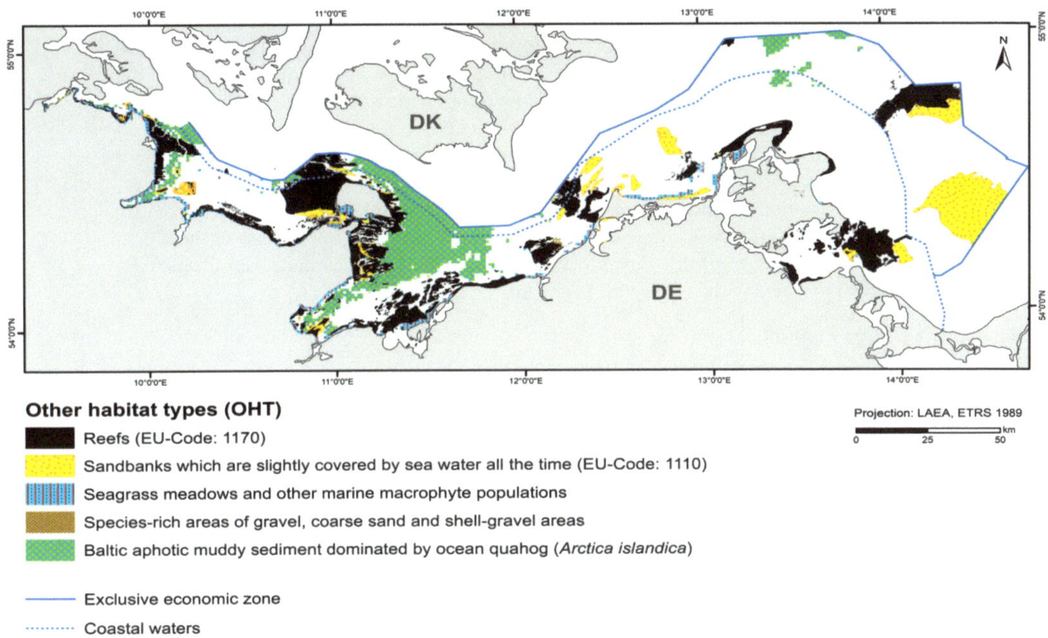


**Figure 5.** Distribution of other habitat types (OHT) in the German Baltic Sea that are protected under EU-/national law or included in the HELCOM Red List.

**Table 6.** Areas and their proportions of other individual habitat types (OHT) in the German Baltic Sea.

| OHT | Area (km$^2$) | Area within the German Baltic Sea (%) |
|---|---|---|
| Reefs (habitat type 1170) | 2183.5 | 14.1 |
| Sandbanks (habitat type 1110) | 875.6 | 5.7 |
| Seagrass meadows and other marine macrophyte populations | 321.4 | 2.1 |
| Species-rich areas of gravel, coarse-sand, and shell-gravel areas | 5.9 | 0.04 |
| Baltic aphotic muddy sediment dominated by ocean quahog (*Arctica islandica*) | 1417.6 | 9.2 |
| Non-OHT | 10,641.6 | 68.9 |

The nationally protected habitat type involving "species-rich areas of gravel, coarse-sand and shell-gravel areas" consists of a suspected area found in the detail area of the "Darss Sill". The total area covered is 9 km$^2$. Reef areas and "species-rich areas of gravel, coarse-sand and shell-gravel areas" partly overlap. However, since both habitat types cannot be designated as protected biotopes at the same time, the area is reduced to almost 6 km$^2$.

"Seagrass meadows and other marine macrophyte populations" (§30 Federal Nature Conservation Act) containing modelled eelgrass is found near the coast in the light-flooded areas. However, they have not yet been mapped for the inner coastal waters of Mecklenburg-Western Pomerania. The only mapped seagrass meadow is located in the "Plantagenet Ground". Seagrass occurrences were recorded using hydroacoustic data (side scan sonar) and could be verified using video footage. They occur in the east of the detail area on fine sand. The stock thins out to the north. Delineation to the 10 % cover is not possible using side-scan sonar mosaics due to shadow formation. Since very shallow areas could not be approached by the research vessel, it is uncertain whether this nationally protected biotope type extends over a larger area towards the west. Seaweeds (*Fucus* spp.) occur more frequently in denser populations in the coastal waters of Schleswig-Holstein. Foliose red algae, such as *Delesseria sanguinea*, occur in marine areas approximately as far as the Darss Sill and where more saline water can flow through the Kadet Trench to the east. They largely dominate the reefs, but are not specified in favour of habitat type 1170. Corticated red algae, such as *Furcellaria lumbricalis*, could only be mapped sporadically as the dominant stock (see also Section 3.3).

"Baltic aphotic muddy sediment dominated by ocean quahog (*Arctica islandica*)" are found in deep basin areas where currents are low enough to allow for fine sediments to deposit, such as Eckernförde Bay, Mecklenburg Bay, and Arkona Basin. It occupies the second largest area of the OHT with 1417.6 km$^2$.

### 3.3. HELCOM HUB Map

A total of 84 HUB biotope types could be modelled in the detail areas and the entire German Baltic Sea (Table 7). Figure 6 shows the HUB biotope map for the German Baltic Sea, and Figures 7–9 show the HUB biotope maps for the detail areas in Mecklenburg-Western Pomerania (for more information on the detail areas in the EEZ, see [26,27]). The colours represent sediment types and the shadings represent benthic communities. NAs result from non-evaluable data in the boulder distribution maps and from sediment types that were not included in the model (due to missing benthological ground truthing) and therefore could not be predicted. However, this accounts for only about 8 km$^2$ in the detail areas of "Kadet Trench", "Outer Wismar Bay", "Darss Sill", and "Plantagenet Ground".

**Table 7.** Mapped HELCOM HUB biotopes and their respective areas with colour indication, as shown in the HUB map (Figure 6). The question marks (?) represent unclassifiable sediment areas. Unclassifiable sediment areas, presumably representing lag sediment areas and/or mussel beds with or without glacial till, were labelled as "AA.I1E1?" (without boulders) or as "AA.M1E1?" (with boulders). The codes marked with an asterisk (*) were introduced in this study and do not yet exist in the HUB classification.

| Colour Coding HUB Map | HUB Code | HUB Biotope | Area (km$^2$) |
|---|---|---|---|
| | AA.? | Baltic photic benthos | 0.4 |
| | AB.? | Baltic aphotic benthos | 0.5 |
| | AA.?1E1 | Baltic photic unknown substrate dominated by *Mytilidae* | 0.5 |
| | AA.?3 | Baltic photic unknown substrate characterised by macroscopic infaunal biotic structures | 2.1 |
| | AA.?3L3 | Baltic photic unknown substrate dominated by ocean quahog (*Arctica islandica*) | 0.003 |
| | AB.?3L3 | Baltic aphotic unknown substrate dominated by ocean quahog (*Arctica islandica*) | 0.1 |
| | AA.?3L4 | Baltic photic unknown substrate dominated by sand gaper (*Mya arenaria*) | 0.4 |
| | AA.?3L9 | Baltic photic unknown substrate dominated by multiple infaunal bivalve species: *Cerastoderma* spp., *Mya arenaria*, *Astarte borealis*, *Arctica islandica*, *Macoma balthica* | 2.4 |
| | AA.M | Baltic photic mixed substrate | 0.02 |
| | AB.M | Baltic aphotic mixed substrate | 0.04 |
| | AA.M1 | Baltic photic mixed substrate characterised by macroscopic epibenthic biotic structures | 14.4 |
| | AB.M1 | Baltic aphotic mixed substrate characterised by macroscopic epibenthic biotic structures | 15.3 |

**Table 7.** *Cont.*

| Colour Coding HUB Map | HUB Code | HUB Biotope | Area (km$^2$) |
|---|---|---|---|
| | AA.M1C1 | Baltic photic mixed substrate dominated by *Fucus* spp. | 102.8 |
| | AA.M1C2 | Baltic photic mixed substrate dominated by perennial non-filamentous corticated red algae | 16.3 |
| | AA.M1C3 | Baltic photic mixed substrate dominated by foliose red algae | 840.8 |
| | AB.M1C3 * | Baltic aphotic mixed substrate dominated by foliose red algae | 0.9 |
| | AA.M1C5 | Baltic photic mixed substrate dominated by perennial filamentous algae | 24.2 |
| | AA.M1E1 | Baltic photic mixed substrate dominated by *Mytilidae* | 540.6 |
| | AA.M1E1? | Baltic photic mixed substrate dominated by *Mytilidae*? | 0.2 |
| | AB.M1E1 | Baltic aphotic mixed substrate dominated by *Mytilidae* | 302.9 |
| | AA.M1G1 | Baltic photic mixed substrate dominated by hydroids (*Hydrozoa*) | 80.4 |
| | AB.M1G1 | Baltic aphotic mixed substrate dominated by hydroids (*Hydrozoa*) | 109.0 |
| | AA.M1H2 | Baltic photic mixed substrate dominated by erect moss animals (*Flustra foliacea*) | 0.02 |
| | AB.M1I1 | Baltic aphotic mixed substrate dominated by barnacles (*Balanidae*) | 0.01 |
| | AA.M1S1 | Baltic photic mixed substrate dominated by filamentous annual algae | 75.0 |
| | AA.M1V | Baltic photic mixed substrate characterised by mixed epibenthic macrocommunity | 0.1 |
| | AB.M1V | Baltic aphotic mixed substrate characterised by mixed epibenthic macrocommunity | 10.9 |
| | AA.M2T | Baltic photic mixed substrate characterised by sparse epibenthic macrocommunity | 36.0 |
| | AB.M2T | Baltic aphotic mixed substrate characterised by sparse epibenthic macrocommunity | 28.9 |
| | AB.M4U | Baltic aphotic mixed substrate characterised by no macrocommunity | 3.0 |
| | AA.G+AA.J1E1 | Baltic photic peat bottoms + Baltic photic sand dominated by *Mytilidae* | 1.0 |
| | AA.I | Baltic photic coarse sediment | 0.006 |
| | AA.I1E1 | Baltic photic coarse sediment dominated by *Mytilidae* | 11.4 |
| | AA.I1E1? | Baltic photic coarse sediment dominated by *Mytilidae*? | 0.7 |
| | AB.I1E1 | Baltic aphotic coarse sediment dominated by *Mytilidae* | 9.0 |
| | AA.I1C3 | Baltic photic coarse sediment dominated by foliose red algae | 0.1 |
| | AA.I3 | Baltic photic coarse sediment characterised by macroscopic infaunal biotic structures | 10.5 |
| | AB.I3 | Baltic aphotic coarse sediment characterised by macroscopic infaunal biotic structures | 3.9 |
| | AA.I3L3 * | Baltic photic coarse sediment dominated by ocean quahog (*Arctica islandica*) | 4.3 |
| | AB.I3L3 * | Baltic aphotic coarse sediment dominated by ocean quahog (*Arctica islandica*) | 2.0 |
| | AA.I3L4 * | Baltic photic coarse sediment dominated by sand gaper (*Mya arenaria*) | 0.2 |
| | AA.I3L9 * | Baltic photic coarse sediment dominated by multiple infaunal bivalve species: *Cerastoderma* spp., *Mya arenaria*, *Astarte borealis*, *Arctica islandica*, *Macoma balthica* | 4.1 |
| | AB.I3L9 * | Baltic aphotic coarse sediment dominated by multiple infaunal bivalve species: *Cerastoderma* spp., *Mya arenaria*, *Astarte borealis*, *Arctica islandica*, *Macoma balthica* | 1.7 |

**Table 7.** *Cont.*

| Colour Coding HUB Map | HUB Code | HUB Biotope | Area ($km^2$) |
|---|---|---|---|
| | AA.I3L10 | Baltic photic coarse sediment dominated by multiple infaunal bivalve species: *Macoma calcarea, Mya truncata, Astarte* spp., *Spisula* spp. | 4.8 |
| | AB.I3L10 | Baltic aphotic coarse sediment dominated by multiple infaunal bivalve species: *Macoma calcarea, Mya truncata, Astarte* spp., *Spisula* spp. | 1.0 |
| | AA.I3L11 | Baltic photic coarse sediment dominated by multiple infaunal polychaete species including *Ophelia* spp. | 0.7 |
| | AB.I3M6 * | Baltic aphotic coarse sediment dominated by multiple infaunal polychaete species | 0.01 |
| | AA.J | Baltic photic sand | 0.1 |
| | AB.J | Baltic aphotic sand | 0.005 |
| | AA.J1B7 | Baltic photic sand dominated by common eelgrass (*Zostera marina*) | 223.1 |
| | AA.J1E1 | Baltic photic sand dominated by *Mytilidae* | 141.3 |
| | AB.J1E1 | Baltic aphotic sand dominated by *Mytilidae* | 196.2 |
| | AA.J1S | Baltic photic sand characterised by annual algae | 4.0 |
| | AA.J3 | Baltic photic sand characterised by macroscopic infaunal biotic structures | 425.9 |
| | AB.J3 | Baltic aphotic sand characterised by macroscopic infaunal biotic structures | 121.5 |
| | AA.J3L1 | Baltic photic sand dominated by Baltic tellin (*Macoma balthica*) | 8.2 |
| | AB.J3L1 | Baltic aphotic sand dominated by Baltic tellin (*Macoma balthica*) | 60.7 |
| | AA.J3L3 | Baltic photic sand dominated by ocean quahog (*Arctica islandica*) | 367.4 |
| | AB.J3L3 | Baltic aphotic sand dominated by ocean quahog (*Arctica islandica*) | 252.4 |
| | AA.J3L4 | Baltic photic sand dominated by sand gaper (*Mya arenaria*) | 15.7 |
| | AB.J3L4 | Baltic aphotic sand dominated by sand gaper (*Mya arenaria*) | 0.1 |
| | AA.J3L9 | Baltic photic sand dominated by multiple infaunal bivalve species: *Cerastoderma* spp., *Mya arenaria, Astarte borealis, Arctica islandica, Macoma balthica* | 2338.7 |
| | AB.J3L9 | Baltic aphotic sand dominated by multiple infaunal bivalve species: *Cerastoderma* spp., *Mya arenaria, Astarte borealis, Arctica islandica, Macoma balthica* | 2381.1 |
| | AA.J3L10 | Baltic photic sand dominated by multiple infaunal bivalve species: *Macoma calcarea, Mya truncata, Astarte* spp., *Spisula* spp. | 1.1 |
| | AB.J3L10 | Baltic aphotic sand dominated by multiple infaunal bivalve species: *Macoma calcarea, Mya truncata, Astarte* spp., *Spisula* spp. | 1.2 |
| | AA.J3L11 | Baltic photic sand dominated by multiple infaunal polychaete species including *Ophelia* spp. | 5.7 |
| | AA.J3M6* | Baltic photic sand dominated by multiple infaunal polychaete species | 0.005 |
| | AB.J3M6* | Baltic aphotic sand dominated by multiple infaunal polychaete species | 0.3 |
| | AA.H1B7 | Baltic photic muddy sediment dominated by common eelgrass (*Zostera marina*) | 69.0 |
| | AA.H1E1 | Baltic photic muddy sediment dominated by *Mytilidae* | 46.6 |
| | AB.H1E1 | Baltic aphotic muddy sediment dominated by *Mytilidae* | 15.0 |
| | AA.H1S | Baltic photic muddy sediment characterised by annual algae | 1.0 |
| | AA.H3 | Baltic photic muddy sediment characterised by macroscopic infaunal biotic structures | 65.2 |
| | AB.H3 | Baltic aphotic muddy sediment characterised by macroscopic infaunal biotic structures | 546.4 |

**Table 7.** *Cont.*

| Colour Coding HUB Map | HUB Code | HUB Biotope | Area ($km^2$) |
|---|---|---|---|
| | AB.H3L1 | Baltic aphotic muddy sediment dominated by Baltic tellin (*Macoma balthica*) | 1131.6 |
| | AA.H3L3 | Baltic photic muddy sediment dominated by ocean quahog (*Arctica islandica*) | 249.9 |
| | AB.H3L3 | Baltic aphotic muddy sediment dominated by ocean quahog (*Arctica islandica*) | 1435.5 |
| | AA.H3L4 * | Baltic photic muddy sediment dominated by sand gaper (*Mya arenaria*) | 0.003 |
| | AB.H3L4 * | Baltic aphotic muddy sediment dominated by sand gaper (*Mya arenaria*) | 0.005 |
| | AA.H3L9 * | Baltic photic muddy sediment dominated by multiple infaunal bivalve species: *Cerastoderma* spp., *Mya arenaria, Astarte borealis, Arctica islandica, Macoma balthica* | 307.9 |
| | AB.H3L9 * | Baltic aphotic muddy sediment dominated by multiple infaunal bivalve species: *Cerastoderma* spp., *Mya arenaria, Astarte borealis, Arctica islandica, Macoma balthica* | 991.0 |
| | AA.H3L10 * | Baltic photic muddy sediment dominated by multiple infaunal bivalve species: *Macoma calcarea, Mya truncata, Astarte* spp., *Spisula* spp. | 0.003 |
| | AB.H3L10 * | Baltic aphotic muddy sediment dominated by multiple infaunal bivalve species: *Macoma calcarea, Mya truncata, Astarte* spp., *Spisula* spp. | 0.1 |
| | AB.H3M6 | Baltic aphotic muddy sediment dominated by multiple infaunal polychaete species | 3.2 |
| | | NA | 8.0 |

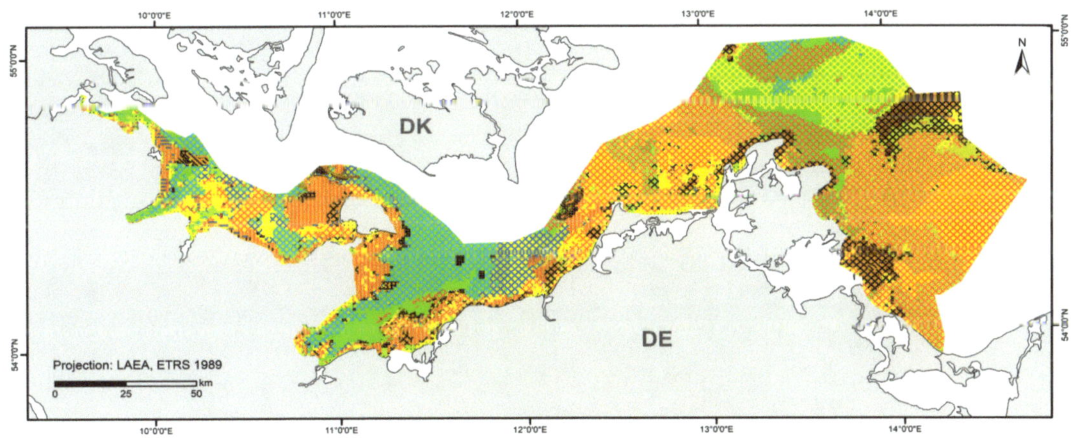


**Figure 6.** Distribution of HELCOM HUB Biotopes in the German Baltic Sea. For legend of colours, see Table 7.

The selected models for the detail areas and the overall area are shown in Table 8. The modelling of the endobenthic communities in the areas of "Plantagenet Ground", "Kadet Trench", and "Fehmarn Belt" achieved a higher model goodness of fit (AUC = 0.79–0.8) than those for the coastal areas, "Outer Wismar Bay" and "Darss Sill" (AUC = 0.65–0.76), where the biotope classes were more difficult to distinguish from each other. The values for the overall German Baltic Sea model were in the middle range (AUC = 0.70). The results of the modelling of the epibenthic communities showed a very high model goodness of fit (AUC > 0.9) for the "Outer Wismar Bay", the "Darss Sill", and the "Fehmarn Belt" areas. In contrast, the values of the "Kadet Trench" (AUC = 0.71) and the overall area (AUC = 0.81) were lower. The epibenthos in the "Plantagenet Ground" was not modelled, as only mussels were observed on the hard substrate in the entire area.

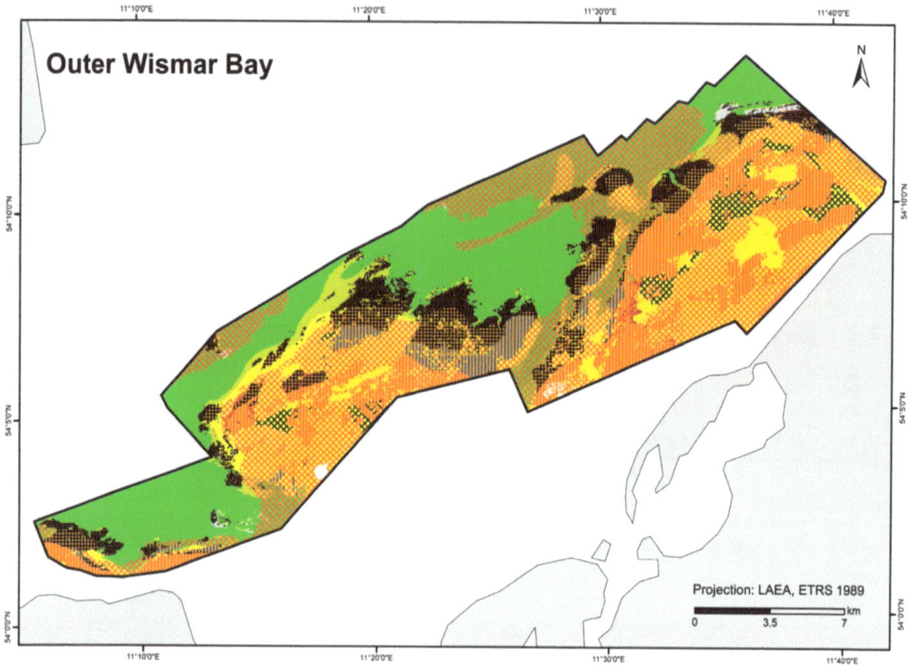


Figure 7. Distribution of HELCOM HUB biotopes in the "Outer Wismar Bay" area. For legend of colours, see Table 7.

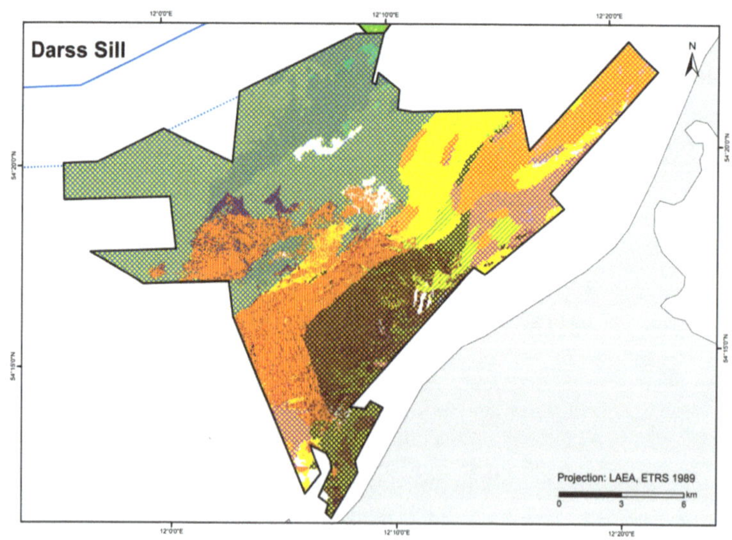


Figure 8. Distribution of HELCOM HUB biotopes in the "Darss Sill" area. For legend of colours, see Table 7.

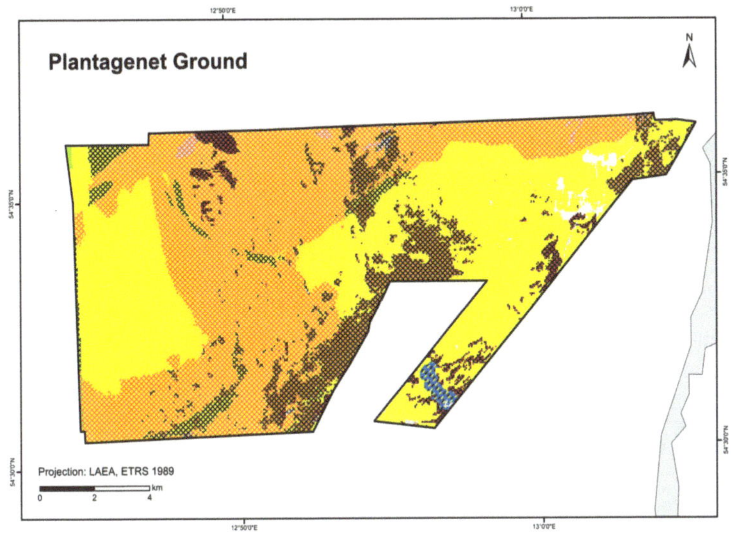


**Figure 9.** Distribution of HELCOM HUB biotopes in the "Plantagenet Ground" area. For legend of colours, see Table 7.

**Table 8.** Overall performance of the selected model in the respective detail areas and the German Baltic Sea; 95% CI: 95 % confidence interval of overall accuracy, AUC: area under (receiver operating characteristic) curve. Epibenthic communities did not need to be modelled in the "Plantagenet Ground", as only mussels dominated the hard bottoms in this eastern area.

| | | Endobenthos | | | | | Epibenthos | | | | |
|---|---|---|---|---|---|---|---|---|---|---|---|
| | **Area** | **Overall Accuracy** | **95% CI** | **AUC** | **Kappa** | **Most Important Variables** | **Overall Accuracy** | **95% CI** | **AUC** | **Kappa** | **Most Important Variables** |
| Detail areas | Outer Wismar Bay | 0.393 | 0.215–0.594 | 0.758 | 0.035 | current velocity (10th percentile) | 0.98 | 0.893–1 | 0.975 | 0.96 | DOC (10th percentile), $O_2$ (10th percentile) |
| | Darss Sill | 0.564 | 0.423–0.7 | 0.648 | 0.453 | temperature (10th percentile) | 1 | 0.936–1 | 1 | 1 | DOC (mean) |
| | Plantagenet Ground | 0.719 | 0.533–0.863 | 0.797 | 0.559 | sediment | NA | NA | NA | NA | NA |
| | Kadet Trench | 0.759 | 0.565–0.9 | 0.786 | 0.576 | shear stress (mean), current velocity N/S (90 percentile) | 0.657 | 0.556–0.748 | 0.716 | 0.485 | depth |
| | Fehmarn Belt | 0.763 | NA | 0.788 | 0.563 | sediment, depth | 0.926 | NA | 0.915 | 0.83 | DOC (mean), depth |
| | German Baltic Sea | 0.666 | 0.636–0.695 | 0.704 | 0.535 | sediment | 0.797 | 0.770–0.821 | 0.805 | 0.712 | depth, salinity (mean) |

The wide sandy areas in the Pomeranian Bay and the Rugia-Falster Plateau are colonised by multiple infaunal bivalve species (consisting of *Cerastoderma glaucum*, *Macoma balthica*, *Mya arenaria*, *Astarte borealis*, and *Arctica islandica*). Silty areas in the Arkona Basin are particularly dominated by *Macoma balthica*, which, although also a component of the aforementioned community, is the main dominant species, especially in this area. Other basins where mud is deposited, such as the Mecklenburg Bay, the Fehmarn Belt, the Eckernförde Bay, and parts of the Arkona Basin, are dominated by *Arctica islandica*. Especially in the first two areas, the prediction confidence that ocean quahog is dominant is high (>80%). Mussels (also as part of the endobenthos) are correctly predicted where reef structures or hard bottoms are present. The habitats characterised by macroscopic infaunal biotic structures (ending on HUB level 4) covers not only communities where no taxa dominate, but also polychaete communities, and generally dominant bivalves and crustaceans. Particularly, in the areas where the sediment is heterogeneously distributed on a small scale (nearshore areas off northwest Mecklenburg and the Rostock district, the Darss Sill, and the coastal waters of Schleswig-Holstein) or due to the lack of data in the nearshore areas in Schleswig-Holstein or in the southwestern Arkona Sea, the prediction probability of the model is low (<50%).

The biotope map shows that not only does sediment influence the spatial distribution of benthic communities, but also salinity, which is observable at the Darss Sill, which is a natural barrier. In front of it (in the western Baltic Sea), a wide variety of marine communities occur, whilst behind it, specialists adapted to brackish water have established themselves. The salinity gradient is also visible in the spatial distribution of the epibenthic communities (see also Table 8). While there are still numerous mixed communities of various colonisers off the Darss Sill, the number of species decreases steadily towards the east. In the Bay of Kiel, communities with non-filamentous corticated red algae, such as *Furcellaria lumbricalis*, dominate the coastal waters. Towards Fehmarn and Mecklenburg Bight up to the Kadet Trench, predominant communities are foliose red algae and mussels, while the deeper, poorly lit areas are colonised only by hydrozoans or are sparsely colonised. In the eastern Baltic Sea, mainly *Mytilus edulis* communities dominate the hard substrates. Mixed communities are rarely found here anymore.

In the "Plantagenet Ground", an area of peat with a thin sand layer was identified using a video transect (Figure 9). It is colonised by mussels (with a cover of filamentous algae) and was therefore classified as peat bottom with mussels on sand (AA.G+AA.J1E1). Thus, peat is also considered a reef-building substrate.

Benthic communities modelled in both the study by Schiele et al. [20] and the current study show similar spatial distributions. This is found for *Arctica islandica*, the multiple infaunal bivalve community (HUB code L9), the Mytilidae community, and *Macoma balthica*, whereby the latter's distribution range extends further south according to Schiele et al. [20] than in this current map. However, with *Macoma balthica* being the dominant species and also occurring within the (L9) community adjacent in the south, the boundaries of its distribution range are likely to be fluid. A difference in the degree of detail between the two maps is further evident at the outer edge of the Arkona Basin on the German side, where Schiele et al. [20] indicate Bivalvia (ending at HUB level 5), whilst here, the multiple infaunal bivalve community, the *Arctica islandica* community, and macroscopic infaunal biotic structures (ending at HUB level 4), respectively, were modelled. The reef structure east of the Bay of Greifswald was mapped only after the publication of the 2015 biotope map, so that mussel occurrence increases here.

Another significant difference is that additional epibenthic communities, such as red algae, hydrozoans, barnacles, and moss animals, were modelled here, and *Zostera* spp. and *Fucus* spp. were added. Due to the fact that reef structures were further mapped and sampled after 2015, additional epibenthic biotope types, including sparse (2T) and non-existent colonisation (4U), could be indicated here.

Macrophytes or algae were not differentiated into perennial or annual macrophytes in the study by Schiele et al. [20], whereas in this study, perennial macrophytes and algae were differentiated. This provides a more accurate picture of the occurrence of specific morpho-species. Nevertheless, the distribution areas are similar. Furthermore, it should be mentioned that annual filamentous algae are often associated with other taxa, and such mixed communities were assigned to another taxon at the expense of the algae in this work. This means, e.g., that when annual filamentous algae co-occurred with mussels, the biotope was assigned to mussel-dominated areas. Thus, depending on the season, annual filamentous algae can also be found more widely distributed than shown in this HUB map (Figure 6).

The HUB, BHT, and OHT maps are largely congruent. The photic zonation and substrate allocation are the same for the HUB and BHT map. The OHT map partly includes results of the HUB modelling ("seagrass meadows and other marine macrophyte populations" and "Baltic aphotic muddy sediment dominated by ocean quahog (*Arctica islandica*)"). Sandbank areas are also shown as "sand" in the HUB and BHT maps. Only "species-rich areas of gravel, coarse-sand and shell-gravel areas" are indicated as "sand" in the BHT and HUB maps, as this substrate does not correspond to the EUNIS or HUB type "coarse sediment" according to the sediment analysis, but to "sand", as already described in Section 2.2. Other exceptions are the reef areas, which do not fully correspond to the BHT/HUB substrate "mixed sediment", as described in Section 2.3.

## 4. Discussion

The new maps now integrate the latest mapping results of widespread habitats, their benthic communities, and protected habitats and biotope types in one map package. Due to the, in some part, high-resolution, standardised, up-to-date mapping and improved modelling through a larger data basis, a more precise picture of the state of the seafloor in the German Baltic Sea is now provided (Figure 10). The map according to Tauber [32] is a pure soft-bottom sediment map interpolated from a large dataset of grab samples. A separate hard bottom map created by the same author only gives roughly drawn polygons from point observations [32]. Although the sediment information gives a correct representation of the seabed when viewed over a large scale, it is too inaccurate when viewed over a small scale. The habitat maps from the current study can now replace the previous sediment and boulder maps according to Tauber [32] in selected areas (two nature conservation areas in the German EEZ and three detail areas in Mecklenburg-Western Pomerania) as well as the biotope map of Schiele et al. [20]. This is due to an improved methodology and more recent mapping results, which increase the level of detail compared to earlier maps, and it should not be interpreted as an indicator of temporal habitat change. [20] used a dataset from 2004 to 2013, while in this modelling, data from 2010 to 2021 were used, partially overlapping with Schiele's dataset. The temporal factor (as well as seasonality) was neglected in this work, as the focus was on the spatial distribution of habitats and biotopes. For a more accurate assessment of potential habitat changes, it is recommended to conduct precise mapping (using hydroacoustics and ground truthing) and delineation of an area already known and, ideally, captured according to national standards (such as habitat types according to the HD). The maps serve as the basis for this purpose.

### 4.1. Modelling Biotope Distributions

Benthic communities settle on certain substrate types under specific conditions of salinity, light availability, exposure, etc. [67–69], which was reflected in the importance of variables in our model building, where parameters such as sediment, salinity, and depth played major roles, especially at large scales (Table 8). On the other hand, at small scales, the most important predictors were the dissolved organic carbon content ("Darss Sill", "Outer Wismar Bay", and "Fehmarn Belt"), sediment ("Plantagenet Ground" and "Fehmarn Belt"), current velocity ("Outer Wismar Bay" and "Kadet Trench"), depth ("Fehmarn Belt"), bottom temperature ("Darss Sill"), oxygen content ("Outer Wismar Bay"), and bottom shear stress ("Kadet Trench"). Small-scale processes, which can overlap large-scale ones, are relevant in the detail areas, which are reflected in the formation of different biocenosis. For example, in the channel system of the "Kadet Trench", the bottom shear stress and current velocity play major roles in the distribution of endobenthic communities within the channels or on the flanks and reef flats. However, both parameters are also related to water depth and sediment. In the shallower reef areas, the currents reach a higher velocity than in the deeper channels. Fine sediment is washed away from the lag sediment areas above and deposited within the deep channels, where the current velocity decreases. On these fine sediments, a multiple infaunal bivalve species community settles with *Arctica islandica* or *Macoma balthica* dominating in certain areas, respectively. This community was easily distinguishable for the random forest model from the community characterised by macroscopic infaunal biotic structures predicted in the shallower areas with higher current velocity, following the simulated current velocity by the GETM model [65]. The interaction of several environmental parameters defines the benthic community formation. Large-scale gradients such as salinity, which cause a shift in benthic community composition in the Baltic Sea [2,70] are replaced on a small scale by other environmental factors that influence the diversity of community structures through their local heterogeneity [71,72].

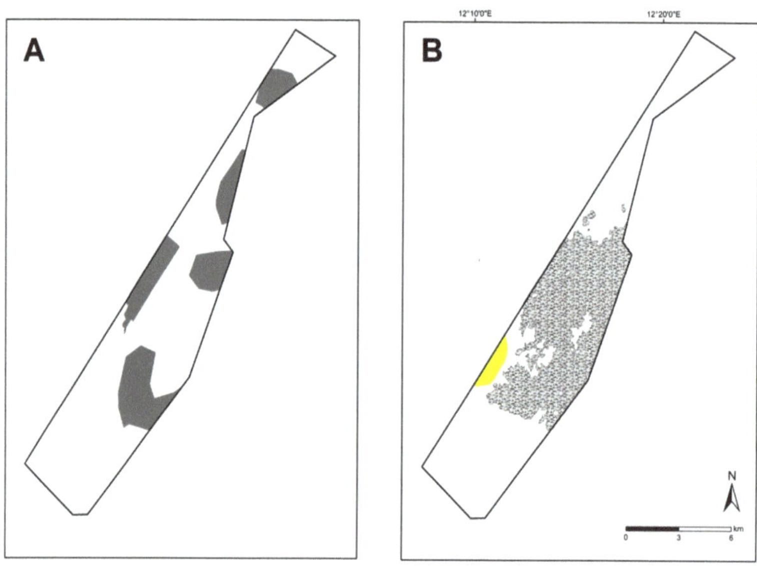


**Figure 10.** Previous suspected reef areas (**A**) and newly mapped OHT (**B**) reefs (grey) and sandbanks (yellow) in the Kadet Trench.

In addition to model statistics (Table 8), prediction probabilities (not shown here) and the resulting biotope maps also determined model selection. A model was chosen if the biotope classes were predicted where they were actually found, and the biotope map generally showed high confidence (at least a 67% prediction probability). The dominant taxa were superimposed on the biotope maps for additional validation and compared. With very few exceptions, the dominant taxa found in ground truthing coincided with those predicted by the model as biotope class. The "Plantagenet Ground", which is located east of the Darss Sill barrier and where the number of marine species is thus strongly reduced due to reduced salinity, has a higher goodness of fit (endobenthos AUC = 0.80) with its very homogeneous sediment composition than the more diverse "Outer Wismar Bay" (endobenthos AUC = 0.76). Where the distribution of a community is limited by a boundary of divergent abiotic conditions, thus favouring biotope delimitation, biotope classes can be clearly distinguished from each other, as was the case at the "Darss Sill" (epibenthos AUC = 1). There, mussels dominate the southern part of the hard substrate, which is shallower with a higher temperature, lower salinity, and lower DOC content than the northerly deeper areas, where foliose red algae prevail. The extent of foliose red algae with a transition to filamentous red algae or hydrozoans in the "Outer Wismar Bay" also follows the simulated distribution of dissolved organic carbon from the ERGOM model and therefore reaches a high goodness of fit (AUC = 0.98). These degraded areas, where turbidity and organic sedimentation are high and oxygen depletion occurs, can only be successfully colonised by hydrozoans, whilst other epibenthic colonisers struggle to survive.

### 4.2. OHT "Species-Rich Areas of Gravel, Coarse-Sand and Shell-Gravel Areas"

The biotope type "species-rich areas of gravel, coarse-sand and shell-gravel areas" is present if, among others, the indicator organisms *Ophelia* spp./*Travisia forbesii* occur at three stations within an occurrence area, according to [42]. This condition was met for the sediment type gS-mxSa (gravelly sand to mixed sand) in the "Darss Sill" area where the biomass fraction of a single taxon was at least 10% or a combined biomass fraction (of both *Ophelia* spp./*Travisia forbesii*) comprising at least 5% of the total biomass. In addition, there were two stations on LagSed+mSa (lag sediment and medium sand), in close proximity to gS-mxSa, where both taxa occur. Furthermore, *Ophelia* spp./*Travisia forbesii* dominate at two stations on mxSa-gS, close to LagSed (lag sediment). Due to the similarity of mxSa-gS and gS-mxSa, both substrate types in the "Darss Sill" are considered as potential "species-rich areas of gravel, coarse-sand and shell-gravel areas". However, both sediment types do not comprise >50 % of gravel, coarse sand, and shell fraction and therefore do not meet the conditions to be designated as "species-rich areas of gravel, coarse-sand and shell-gravel areas" [42]. Due to the heterogeneity of sediment composition during sampling, it is open to question whether thin covers of sand in this area are positionally stable or if they might instead cover coarse sediment within the suspected substrate types. Further, the area is in spatial proximity to lag sediment and is dominated by reef structures, which is considered an indication of the protected habitat type according to [42]. Towards the east/northeast, *Ophelia* spp./*Travisia forbesii* occurrence reaches into fine sand areas (with a transition to mixed and medium sand). Therefore, the high density of dominant indicator organisms generally supports the plausibility of this suspected area in the "Darss Sill" area.

### 4.3. Methodological Review

Methodologies differ per state for the external data obtained. Regarding the coastal waters of Schleswig-Holstein, it is above all the mapping of the epibenthic organisms that differs from the procedure at IOW (and thus also Mecklenburg-Western Pomerania). At IOW, mainly video and photo techniques are used, which are supported by diver sampling, while in Schleswig-Holstein, mainly diving is used. Therefore, there were differences in the data availability during the adaptation of these external data to our approach. In some cases, only biomass was taken, but no coverage was recorded. In order to preserve these data, the epibenthic colonisation shown in the biotope maps is therefore based on the coverage of video, photo, and diver samples, but also on the biomass fraction of the total biomass at a station. The external data, where only macrophytes were mapped as a part of MSFD monitoring, could unfortunately not be included in the epibenthic modelling.

Little benthos data were collected in the nearshore area, and the raster data of the predictors are also less reliable in this zone. For methodological reasons, this area has been included in the maps but should be treated with caution.

The random forest method used here is considered very well suited for biotope classification because it is a nonparametric, robust algorithm showing high performances in supervised machine learning methods [73–75] that can handle outliers and noisy or redundant input features [59,76]. The algorithm selects a variable out of a random subset of predictors that is most important for decision formation at each node of a branch of each decision tree [60]. Aggregating the outcome of many random trees leads to an increase in generalisation power [76]. The overall AUC (0.7–1) and Cohen's Kappa values (0.5–1) indicated a good to excellent prediction [77,78]. Nevertheless, the endobenthos model of the "Outer Wismar Bay" only showed very poor performance regarding Cohen's Kappa (0.035). However, a visual examination showed that even with the perceived poor model performance, the random forest algorithm provides biotope maps that can approximate reality.

The model also reaches its limits as the number of response variables (biotope classes) that are closely interrelated increases. Naturally, there are no strictly delineated sediment types, each with different biocenosis. Transitions of different grain size fractions are fluent, or sediment mixtures can cause benthic communities to overlap [67]. Particularly in the case of differently composed sedimentological substrates, high biodiversity can occur, where it becomes difficult for the Random Forest model to find patterns and boundaries, as was the case, e.g., in the "Outer Wismar Bay" area. However, this is a general difficulty in modelling and not a question of methodology.

Furthermore, an assemblage always consists of different community-associated species. The dominant species given here, which give a biotope its name, therefore always occur with other associated species. Thus, the maps do not lay claim to the exclusive occurrence of individual habitat-determining species, nor do these species occur with absolute confidence in certain areas, nor are these distribution areas fixed. Rigid boundaries, as the maps suggest, do not exist, and are instead fluid transitions. Sediment types, like their inhabitants, are subject to natural dynamics. Rather, the maps are intended to provide indications of the likely spatial distribution patterns of benthic communities at large and small scales, and they do not give any indication of the status of biotopes.

### 4.4. Outlook

The spatial extent of habitats and protected biotopes in high resolution is still unknown in vast areas of the Baltic Sea. The demand for biotope maps is increasing, so monitoring and mapping is ongoing and will continue, not only to detect spatial changes but also to detect temporal changes due to natural and anthropogenic causes. By observing habitat changes over time, possible habitat loss can be detected (with regard to MSFD Descriptor 6), impacts can be assessed, and measures can be taken. The interplay between applied and basic research can contribute to the direct implementation of nature conservation measures.

Comprehensive hydroacoustic mapping provides new insights, particularly in the area of suspect reef areas, which can now be identified more accurately and in much greater detail. This does not only fulfil national mapping requirements at high resolution, but also national reporting requirements (monitoring of spatio-temporal changes and improvement and restoration measures) for the implementation of relevant directives, as well as for marine spatial planning and specific projects. Particularly, in light of the forthcoming Nature Restoration Law, which includes the introduction of restoration measures for at least 20% of the EU's marine and terrestrial ecosystems by 2030 and for all ecosystems in need of restoration by 2050, these maps are an important tool, e.g., for identifying potential restoration areas or assessing the success of measures. The comprehensive HUB map is also a valuable tool for monitoring and assessment under HELCOM and the MSFD, and for deriving targeted management measures. However, as the modelling of community distribution is highly dependent on the dataset used, the maps should be used with caution as a basis for detecting changes in biotope distribution and for projects in small-scaled areas.

## 5. Conclusions

For the first time, habitats and biotopes in the German Baltic Sea have been mapped at a level of detail that has not been available before. In this study, we mapped specific sediment types in their actual extent using side scan sonar and on a larger scale for the first time using neural networks for stone detection [23,24]. Furthermore, the latest mapping results from federal and state governments have been incorporated into the maps. Based on national guidelines that have been developed over the past seven years to standardise sediment and boulder distribution maps and the large-scale mapping of reefs [8,22], these maps have been improved and updated with the latest available data.

The spatial distributions of the protected biotope types here show that specifically, the geogenic reefs (HD, §30 Federal Nature Conservation Act) can now be exactly reported. The biological verification of these geogenic reefs was essential and paves the way for a subsequent designation of this protected habitat type at an official level. The same applies to the protected seagrass meadows (§30 Federal Nature Conservation Act) in the "Plantagenet Ground" and the sandbanks (HD, §30 Federal Nature Conservation Act) in the "Kadet Trench" and "Fehmarn Belt". For the first time, a suspected area of the nationally protected habitat type "species-rich areas of gravel, coarse-sand and shell-gravel areas" was found and mapped in the "Darss Sill".

**Supplementary Materials:** The following supporting information can be downloaded at https://www.mdpi.com/article/10.3390/biology13010006/s1, Map Package S1: Maps showing broad habitat types (BHTs), other habitat types (OHTs), and HELCOM HUB biotope types in the German Baltic Sea and the location of the underlying observation data for endobenthos and epibenthos sampling.

**Author Contributions:** Conceptualisation, S.P., A.D., P.F., M.L.Z. and K.H.; methodology, S.P., A.D., P.F. and D.M.; validation, A.F., S.P., P.F., A.D., M.L.Z., K.H. and D.M.; formal analysis, A.F., S.P., P.F., A.D. and D.M.; investigation, A.F., S.P., P.F., A.D. and D.M.; data curation, A.F., S.P., P.F., A.D. and D.M.; software, S.P., P.F., A.D. and D.M.; writing—original draft preparation, D.M.; writing—review and editing, A.F., S.P., P.F., A.D., M.L.Z. and K.H.; visualisation, A.F., S.P., P.F., A.D., K.H. and D.M.; project administration, M.L.Z. and K.H.; supervision, K.H.; funding acquisition, P.F., A.D. and M.L.Z. All authors have read and agreed to the published version of the manuscript.

**Funding:** This work was carried out within the framework of the ATLAS project ("Mapping of habitats (biotopes) and their communities on the seabed in the coastal waters of Mecklenburg-Western Pomerania", grant number MV-VIII.1-LM-001; EU funding programme 7306), which was funded by the State Agency for the Environment, Nature Conservation and Geology of Mecklenburg-Western Pomerania (LUNG) and the European Maritime and Fisheries Fund (EMFF), as well as the framework of the LEGRA project ("Live along gradients—Analyses of the impact of environmental parameters on the distribution, diversity and function of benthic communities and their habitats in the southern Baltic Sea and its implication within the European marine conservation directives", grant number 3 519 532 202), funded by the German Federal Agency for Nature Conservation (BfN).

**Institutional Review Board Statement:** Not applicable.

**Informed Consent Statement:** Not applicable.

**Data Availability Statement:** The BHT, OHT, and HUB maps are available as a map package (.mpk file) in the Supplementary Materials. Soon, they will also be uploaded on EMODnet.

**Acknowledgments:** We would like to thank Kiel University (CAU) for the realisation of the sediment and boulder distribution maps in the "Fehmarn Belt" and the "Kadet Trench" areas and Jennifer Valerius of the Federal Maritime and Hydrographic Agency (BSH) for the subsequent quality check. Furthermore, we thank Anne Herbst for processing the ATLAS project in the initial period, Katharina Romoth, Iris Schaub, Louisa Schulz, Sarah Pirrung, and Philipp Hoy for processing the epibenthic and endobenthic data, the IOW Benthos laboratory for the identification of the benthic organisms, the technicians and the captain and crew of the research vessel *Elisabeth Mann Borgese* for supporting the sampling. We are very grateful for the provision of external data by BSH, BfN, LFU, LUNG, StALU WM, StALU MM, WSA Stralsund, IfAÖ, Geomar, Ingenieurbüro Weigt, CAU Kiel, and the University of Rostock. In particular, we would like to thank Mario von Weber (LUNG) and Simone Eisenbarth (BfN) for their excellent support during the project phases. Further thanks go to Claudia Morys and Simone Eisenbarth (both BfN) for their helpful comments on the manuscript. Last but not least, we would like to thank the anonymous reviewers as well as Sophia Bertullies and Darren Hughes, whose numerous constructive comments and English language corrections significantly improved the manuscript.

**Conflicts of Interest:** The authors declare no conflict of interest.

## References

1. Bogalecka, M.; Kołowrocki, K. The Baltic Sea circumstances significant for its critical infrastructure networks. *J. Polish Saf. Reliab. Assoc. Summer Saf. Reliab. Semin.* **2016**, *7*, 37–42.
2. Snoeijs-Leijonmalm, P.; Andrén, E. Why is the Baltic Sea so special to live in? In *Biological Oceanography of the Baltic Sea*; Snoeijs-Leijonmalm, P., Schubert, H., Radziejewska, T., Eds.; Springer: Dordrecht, The Netherlands, 2017; pp. 23–84.

3. Österblom, H.; Hansson, S.; Larsson, U.; Hjerne, O.; Wulff, F.; Elmgren, R.; Folke, C. Human-induced trophic cascades and ecological regime shifts in the Baltic Sea. *Ecosystems* **2007**, *10*, 877–889. [CrossRef]
4. Tomczak, M.T.; Müller-Karulis, B.; Blenckner, T.; Ehrnsten, E.; Eero, M.; Gustafsson, B.; Norkko, A.; Otto, S.A.; Timmermann, K.; Humborg, C. Reference state, structure, regime shifts, and regulatory drivers in a coastal sea over the last century: The Central Baltic Sea case. *Limnol. Oceanogr.* **2022**, *67*, S266–S284. [CrossRef]
5. European Commission. *Directive 2008/56/EC of the European Parliament and of the Council of 17 June 2008*; Establishing a Framework for Community Action in the Field of Marine Environmental Policy (Marine Strategy Framework Directive). *Off. J. Eur. Union* **2008**, *L 164*, 19–40.
6. Salomidi, M.; Katsanevakis, S.; Borja, A.; Braeckman, U.; Damalas, D.; Galparsoro, I.; Mifsud, R.; Mirto, S.; Pascual, M.; Pipitone, C.; et al. Assessment of goods and services, vulnerability, and conservation status of European seabed biotopes: A stepping stone towards ecosystem-based marine spatial management. *Mediterr. Mar. Sci.* **2012**, *13*, 49–88. [CrossRef]
7. European Commission. Council Directive 92/43/EEC of 21 May 1992 on the Conservation of Natural Habitats and of Wild Fauna and Flora. *Off. J. Eur. Union* **1992**, *206*, 50.
8. Heinicke, K.; Bildstein, T.; Reimers, H.-C.; Boedeker, D. Leitfaden zur großflächigen Abgrenzung und Kartierung des Lebensraumtyps "Riffe" in der deutschen Ostsee (EU-Code 1170; Untertyp: Geogene Riffe). *BfN-Skripten* **2021**, *612*, 46.
9. HELCOM. HELCOM HUB—Technical Report on the HELCOM Underwater Biotope and Habitat Classification; Baltic Sea Environment Proceedings No. 139. 2013. Available online: https://www.academia.edu/32554320/HELCOM_HUB_Technical_Report_on_the_HELCOM_Underwater_Biotope_and_habitat_classifi_cation (accessed on 12 December 2023).
10. European Commission. *MSFD CIS Guidance Document No. 19*; Article 8 MSFD; European Commission: Brussels, Belgium, 2022.
11. European Commission. Commission Decision (EU) 2017/848 of 17 May 2017 Laying down Criteria and Methodological Standards on Good Environmental Status of Marine Waters and Specifications and Standardised Methods for Monitoring and Assessment, and Repealing Decision 2010/477/EU. *Off. J. Eur. Union* **2017**, *L 125*, 43–74.
12. Evans, D.; Aish, A.; Boon, A.; Condé, S.; Connor, D.; Gelabert, E.; Michez, N.; Parry, M.; Richard, D.; Salvati, E.; et al. Revising the Marine Section of the EUNIS Habitat Classification—Report of a Workshop Held at the European Topic Centre on Biological Diversity, 12 & 13 May 2016; ETC/BD Report to the EEA. 2016. Available online: https://www.eionet.europa.eu/etcs/etc-bd/products/etc-bd-reports/revising_marine_section_eunis_hab_classification (accessed on 12 December 2023).
13. HELCOM. *Red List of Baltic Sea Underwater Biotopes, Habitats and Biotope Complexes*; Baltic Sea Environment Proceedings No. 138; HELCOM: Helsinki, Finland, 2013.
14. Torn, K.; Herkül, K.; Martin, G.; Oganjan, K. Assessment of quality of three marine benthic habitat types in northern Baltic Sea. *Ecol. Indic.* **2017**, *73*, 772–783. [CrossRef]
15. Sokołowski, A.; Jankowska, E.; Balazy, P.; Jędruch, A. Distribution and extent of benthic habitats in Puck Bay (Gulf of Gdańsk, southern Baltic Sea). *Oceanologia* **2021**, *63*, 301–320. [CrossRef]
16. Olenin, S.; Daunys, D. Coastal typology based on benthic biotope and community data: The Lithuanian case study. *Coastline Rep.* **2004**, *4*, 65–83.
17. Weslawski, J.M.; Kryla-Straszewska, L.; Piwowarczyk, J.; Urbański, J.; Warzocha, J.; Kotwicki, L.; Wlodarska-Kowalczuk, M.; Wiktor, J. Habitat modelling limitations—Puck Bay, Baltic Sea—A case study. *Oceanologia* **2013**, *55*, 167–183. [CrossRef]
18. Wikström, S.A.; Enhus, C.; Fyhr, F.; Näslund, J.; Sundblad, G. Distribution of Biotopes, Habitats and Biological Values at Holmöarna and in the Kvarken Archipelago. 2013. Available online: https://aquabiota.se/wp-content/uploads/ABWR_Report2013-06_SUPERB.pdf (accessed on 12 December 2023).
19. Vasquez, M.; Allen, H.; Manca, E.; Castle, L.; Lillis, H.; Agnesi, S.; Al Hamdani, Z.; Annunziatellis, A.; Askew, N.; Bekkby, T.; et al. EUSeaMap 2021. A European Broad-Scale Seabed Habitat Map; D1.13 EASME/EMFF/2018/1.3.1.8/Lot2/SI2.810241—EMODnet Thematic Lot n° 2—Seabed Habitats EUSeaMap 2021—Technical Report. 2021. Available online: https://archimer.ifremer.fr/doc/00723/83528/ (accessed on 12 December 2023).
20. Schiele, K.S.; Darr, A.; Zettler, M.L.; Friedland, R.; Tauber, F.; von Weber, M.; Voss, J. Biotope map of the German Baltic Sea. *Mar. Pollut. Bull.* **2015**, *96*, 127–135. [CrossRef] [PubMed]
21. Zettler, M.L.; Darr, A. Benthic Habitats and Their Inhabitants. In *Southern Baltic Coastal Systems Analysis*; Ecological Studies; Schubert, H., Müller, F., Eds.; Springer: Cham, Switzerland, 2023; Volume 246, pp. 97–102. ISBN 9783031136825.
22. BSH. Anleitung zur Kartierung des Meeresbodens Mittels Hochauflösender Sonare in den Deutschen Meeresgebieten. BSH Nr. 7201. 2016. Available online: https://www.bsh.de/download/Kartierung-des-Meeresboden.pdf (accessed on 12 December 2023).
23. Feldens, P.; Darr, A.; Feldens, A.; Tauber, F. Detection of boulders in side scan sonar mosaics by a neural network. *Geoscience* **2019**, *9*, 159. [CrossRef]
24. Feldens, P.; Westfeld, P.; Valerius, J.; Feldens, A.; Papenmeier, S. Automatic detection of boulders by neural networks: A comparison of multibeam echo sounder and side-scan sonar performance. *Hydrogr. Nachrichten* **2021**, *119*, 6–17. [CrossRef]
25. Feldens, A.; Marx, D.; Herbst, A.; Darr, A.; Papenmeier, S.; Hinz, M.; Zettler, M.L.; Feldens, P. Distribution of boulders in coastal waters of Western Pomerania, German Baltic Sea. *Front. Earth Sci.* **2023**, *11*, 1155765. [CrossRef]
26. Darr, A.; Heinicke, K.; Meier, F.; Papenmeier, S.; Richter, P.; Schwarzer, K.; Valerius, J.; Boedeker, D. Die Biotope des Meeresbodens im Naturschutzgebiet Fehmarnbelt. *BfN-Schriften* **2022**, *636*, 1–99.

27. Marx, D.; Romoth, K.; Papenmeier, S.; Valerius, J.; Darr, A.; Eisenbarth, S.; Heinicke, K. Die Biotope des Meeresbodens im Naturschutzgebiet "Kadetrinne". 2024; *in preparation*.
28. Richter, P.; Schwarzer, K.; Feldens, A.; Valerius, J.; Thiesen, M.; Mulckau, A. Map of Sediment Distribution in the German EEZ (1:10.000). 2020. Available online: www.geoseaportal.de (accessed on 12 December 2023).
29. Richter, P.; Höft, D.; Feldens, A.; Schwarzer, K.; Diesing, M.; Valerius, J.; Mulckau, A. Map of Sediment Distribution in the German EEZ (1:10.000). 2021. Available online: https://www.geoseaportal.de/mapapps/resources/apps/sedimentverteilung_auf_dem_meeresboden/index.html?lang=de (accessed on 12 December 2023).
30. Blott, S.J.; Pye, K. GRADISTAT: A grain size distribution and statistics package for the analysis of unconsolidated sediments. *Earth Surf. Process. Landforms* **2001**, *26*, 1237–1248. [CrossRef]
31. Folk, R.L. The distinction between grain size and mineral composition in sedimentary-rock nomenclature. *J. Geol.* **1954**, *62*, 344–359. [CrossRef]
32. Tauber, F. *Seabed Sediments and Seabed Relief in the German Baltic Sea: Data Positions and Sheet Index, Map no. 2930 [Map] 1: 550 000, 54° N*; Federal Maritime and Hydrographic Agency: Hamburg, Germany, 2012.
33. Diesing, M.; Schwarzer, K. Identification of submarine hard-bottom substrates in the German North Sea and Baltic Sea EEZ with high-resolution acoustic seafloor imaging. In *Progress in Marine Conservation in Europe: NATURA 2000 Sites in German Offshore Waters*; Von Nordheim, H., Boedeker, D., Krause, J.C., Eds.; Springer: Berlin, Germany; New York, NY, USA, 2006; pp. 111–125. ISBN 9783540332909.
34. Bohling, B.; May, H.; Mosch, T.; Schwarzer, K. Regeneration of submarine hard-bottom substrate by natural abrasion in the western Baltic Sea. *Marbg. Geogr. Schr.* **2009**, *145*, 66–79.
35. Schwarzer, K.; Bohling, B.; Heinrich, C. Submarine hard-bottom substrates in the western Baltic Sea—Human impact versus natural development. *J. Coast. Res.* **2014**, *70*, 145–150. [CrossRef]
36. Figge, K. *Sedimentverteilung in der Deutschen Bucht (Blatt: 2900, Maßstab: 1:250.000)*; Deutsches Hydrographisches Institut: Hamburg, Germany, 1981.
37. Papenmeier, S.; Darr, A.; Feldens, P.; Michaelis, R. Hydroacoustic mapping of geogenic hard substrates: Challenges and review of German approaches. *Geosci.* **2020**, *10*, 100. [CrossRef]
38. LANIS-SH, Schleswig-Holstein State Office for the Environment. Maritime Daten Ostsee LRT 1110 und 1170, State: December 2021. Available online: https://opendata.schleswig-holstein.de/dataset/maritime-daten-ostsee-lrt-1110-und-1170 (accessed on 12 December 2023).
39. LUNG MV. *State Office for the Environment Nature Conservation and Geology Mecklenburg-Western Pomerania: HELCOM HOLAS III Data Natura Habitats MV, State: 10.06.2022*; LUNG MV: Güstrow, Germany, 2022.
40. Bobsien, I.C.; Hukriede, W.; Schlamkow, C.; Friedland, R.; Dreier, N.; Schubert, P.R.; Karez, R.; Reusch, T.B.H. Modeling eelgrass spatial response to nutrient abatement measures in a changing climate. *Ambio* **2021**, *50*, 400–412. [CrossRef] [PubMed]
41. Schubert, H.; Schygulla, C. *Abschlussbericht des Fachprojektes: "Die Erfassung rezenter Zostera-Bestände und weiterer Makrophyten in den Küstengewässern MV" (ZOSINF). Im Auftrag des Landesamtes für Umwelt, Naturschutz und Geologie Mecklenburg-Vorpommern (LUNG 100G-30.15/16)*; Landesamt für Umwelt, Naturschutz u. Geologie: Güstrow, Germany, 2017; p. 76.
42. BfN. *Artenreiche Kies-, Grobsand- und Schillgründe im Meeres- und Küstenbereich*; Definition und Kartieranleitung; BfN: Bonn, Germany, 2011.
43. EUNIS Marine Habitat Classification 2022 including Crosswalks. Available online: https://www.eea.europa.eu/data-and-maps/data/eunis-habitat-classification-1/eunis-marine-habitat-classification-review-2022/eunis-marine-habitat-classification-2022 (accessed on 4 October 2023).
44. Friedland, R.; Neumann, T.; Schernewski, G. Climate change and the Baltic Sea action plan: Model simulations on the future of the western Baltic Sea. *J. Mar. Syst.* **2012**, *105–108*, 175–186. [CrossRef]
45. Zustand der Deutschen Ostseegewässer 2018. Aktualisierung der Anfangsbewertung nach § 45c, der Beschreibung des Guten Zustands der Meeresgewässer nach § 45d und der Festlegung von Zielen nach § 45e des Wasserhaushaltsgesetzes zur Umsetzung der Meeresstrategie-Rahmenrichtlinie; BMU Hintergrunddokument: Methodik der Benthosbewertung in den Deutschen Ostseegewässern. 2018. Available online: https://dmkn.de/wp-content/uploads/2018/03/2018_Zustand_Ostsee_Entwurf.pdf (accessed on 12 December 2023).
46. Boedeker, D.; Krause, J.C.; von Nordheim, H. Interpretation, identification and ecological assessment of the NATURA 2000 habitats "sandbank" and "reef.". In *Progress in Marine Conservation in Europe*; Von Nordheim, H., Boedeker, D., Krause, J.C., Eds.; Springer: Berlin/Heidelberg, Germany, 2006; ISBN 978-3-540-33290-9.
47. Gogina, M.; Zettler, A.; Zettler, M.L. Weight-to-weight conversion factors for benthic macrofauna: Recent measurements from the Baltic and the North seas. *Earth Syst. Sci. Data* **2022**, *14*, 1–4. [CrossRef]
48. Bundesamt für Seeschifffahrt und Hydrographie (BSH). *Standard Untersuchung der Auswirkungen von Offshore-Windenergieanlagen auf die Meeresumwelt (StUK 4)*; BSH: Munich, Germany, 2013.
49. HELCOM Balsam Project 2013–2015. In *Recommendations and Guidelines for Benthic Habitat Monitoring with Method Descriptions for Two Methods for Monitoring of Biotope and Habitat Extent*; HELCOM: Helsinki, Finland, 2015; Available online: https://helcom.fi/wp-content/uploads/2019/08/Recommendations-and-guidelines-for-benthic-habitat-monitoring-in-the-Baltic-Sea.pdf (accessed on 12 December 2023).

50. Beisiegel, K. Picturing the Seafloor: Shedding Light on Offshore Rock Habitats in the Baltic Sea using Imaging Systems. Ph.D. Thesis, The University of Rostock, Rostock, Germany, 2019.
51. Kohler, K.E.; Gill, S.M. Coral Point Count with Excel extensions (CPCe): A Visual Basic program for the determination of coral and substrate coverage using random point count methodology. *Comput. Geosci.* **2006**, *32*, 1259–1269. [CrossRef]
52. StALU WM. *Kartierbericht für den Managementplan für das FFH-Gebiet DE 1934-303 "Erweiterung Wismarbucht"*; StALU WM: Schwerin, Germany, 2017.
53. StALU WM. *Managementplan für das Gebiet von gemeinschaftlicher Bedeutung DE 1540-302 Darßer Schwelle*; StALU WM: Schwerin, Germany, 2019.
54. StALU WM. *Managementplan für das Gebiet von gemeinschaftlicher Bedeutung DE 1345-301 Erweiterung Libben, Steilküste und Blockgründe Wittow und Arkona*; StALU WM: Schwerin, Germany, 2019.
55. StALU WM. *Managementplan für das Gebiet von gemeinschaftlicher Bedeutung DE 1343-301 Plantagenetgrund*; StALU WM: Schwerin, Germany, 2019.
56. StALU WM. *Managementplan für das Gebiet von gemeinschaftlicher Bedeutung (GGB) DE 1749-302 "Greifswalder Boddenrandschwelle und Teile der Pommerschen Bucht"*; StALU WM: Schwerin, Germany, 2020.
57. Hiebenthal, C.; Bock, G. *Maßnahme: Meeresstrategie Rahmenrichtlinie (MSRL) Biodiversität Makrophyten Ostseeküste. [Aktenzeichen 0608-451219]—Abschlussbericht-.Im Auftrag des Landesamtes für Landwirtschaft, Umwelt und ländliche Räume (Schleswig-Holstein)*; GEOMAR: Kiel, Germany, 2013; p. 134.
58. Hiebenthal, C. *Forschungskooperation: "Konzept zum Monitoring der Entwicklung von Flachwasser-Hartbodengemeinschaften in der s.-h. Ostsee" [Aktenzeichen 0608.451722]–Endbericht [Projektteile 1 und 2]—Im Auftrag des Landesamtes für Landwirtschaft, Umwelt und ländliche Räume (Schleswig-Holstein)*; GEOMAR: Kiel, Germany, 2021; p. 76.
59. Breiman, L. Random forests. *Mach. Learn.* **2001**, *45*, 5–32. [CrossRef]
60. Liaw, A.; Wiener, M. Classification and Regression by randomForest. *R News* **2002**, *2*, 18–22.
61. Gonzalez-Mirelis, G.; Bergström, P.; Lindegarth, M. Interaction between classification detail and prediction of community types: Implications for predictive modelling of benthic biotopes. *Mar. Ecol. Prog. Ser.* **2011**, *432*, 31–44. [CrossRef]
62. Miller, K.; Huettmann, F.; Norcross, B.; Lorenz, M. Multivariate random forest models of estuarine-associated fish and invertebrate communities. *Mar. Ecol. Prog. Ser.* **2014**, *500*, 159–174. [CrossRef]
63. Galaiduk, R.; Radford, B.; Harries, S.; Case, M.; Williams, D.; Low Choy, D.; Smit, N. *Technical Report: Darwin–Bynoe Harbours Predictive Mapping of Benthic Communities*; Australian Institute of Marine Science: Perth, Australia, 2019; p. 42. Available online: https://www.researchgate.net/publication/340374082_Technical_Report_Darwin_-_Bynoe_Harbours_Predictive_Mapping_of_Benthic_Communities_httppidgeosciencegovaudatasetga127389 (accessed on 12 December 2023).
64. Leipe, T.; Naumann, M.; Tauber, F.; Radtke, H.; Friedland, R.; Hiller, A.; Arz, H.W. Regional distribution patterns of chemical parameters in surface sediments of the south-western Baltic Sea and their possible causes. *Geo-Mar. Lett.* **2017**, *37*, 593–606. [CrossRef]
65. Gräwe, U.; Naumann, M.; Mohrholz, V.; Burchard, H. Anatomizing one of the largest saltwater inflows into the Baltic Sea in December 2014. *J. Geophys. Res. Oceans* **2015**, *120*, 7676–7697. [CrossRef]
66. Schiele, K. Benthic Community and Habitat Analysis towards an Application in Marine Management. Ph.D. Thesis, Universität Rostock, Rostock, Germany, 2014.
67. Van Hoey, G.; Degraer, S.; Vincx, M. Macrobenthic community structure of soft-bottom sediments at the Belgian Continental Shelf. *Estuar. Coast. Shelf Sci.* **2004**, *59*, 599–613. [CrossRef]
68. Coleman, N.; Cuff, W.; Moverley, J.; Gason, A.S.H.; Heislers, S. Depth, sediment type, biogeography and high species richness in shallow-water benthos. *Mar. Freshw. Res.* **2007**, *58*, 293–305. [CrossRef]
69. Kröncke, I.; Reiss, H.; Eggleton, J.D.; Aldridge, J.; Bergman, M.J.N.; Cochrane, S.; Craeymeersch, J.A.; Degraer, S.; Desroy, N.; Dewarumez, J.M.; et al. Changes in North Sea macrofauna communities and species distribution between 1986 and 2000. *Estuar. Coast. Shelf Sci.* **2011**, *94*, 1–15. [CrossRef]
70. Bonsdorff, E. Zoobenthic diversity-gradients in the Baltic Sea: Continuous post-glacial succession in a stressed ecosystem. *J. Exp. Mar. Bio. Ecol.* **2006**, *330*, 383–391. [CrossRef]
71. Bonsdorff, E.; Laine, A.O.; Hänninen, J.; Vuorinen, I.; Norkko, A. Zoobenthos of the outer archipelago waters (N. Baltic Sea)—The importance of local conditions for spatial distribution patterns. *Boreal Environ. Res.* **2003**, *8*, 135–145.
72. Nyström Sandman, A.; Wikström, S.A.; Blomqvist, M.; Kautsky, H.; Isaeus, M. Scale-dependent influence of environmental variables on species distribution: A case study on five coastal benthic species in the Baltic Sea. *Ecography (Cop.)* **2013**, *36*, 354–363. [CrossRef]
73. Peterson, A.; Herkül, K. Mapping benthic biodiversity using georeferenced environmental data and predictive modeling. *Mar. Biodivers.* **2019**, *49*, 131–146. [CrossRef]
74. Wicaksono, P.; Aryaguna, P.A.; Lazuardi, W. Benthic habitat mapping model and cross validation using machine-learning classification algorithms. *Remote Sens.* **2019**, *11*, 1279. [CrossRef]
75. Janowski, L.; Wroblewski, R.; Dworniczak, J.; Kolakowski, M.; Rogowska, K.; Wojcik, M.; Gajewski, J. Offshore benthic habitat mapping based on object-based image analysis and geomorphometric approach. A case study from the Slupsk Bank, Southern Baltic Sea. *Sci. Total Environ.* **2021**, *801*, 149712. [CrossRef]

76. Stephens, D.; Diesing, M. A comparison of supervised classification methods for the prediction of substrate type using multibeam acoustic and legacy grain-size data. *PLoS ONE* **2014**, *9*, e93950. [CrossRef]
77. Hosmer, D.W.; Lemeshow, S. *Applied Logistic Regression*; Wiley Interscience: New York, NY, USA, 2000.
78. Landis, J.R.; Koch, G.G. Measurement of observed agreement for categorical data. *Biometrics* **1977**, *33*, 159–174. [CrossRef]

*biology* MDPI

*Article*

# Baseline Inventory of Benthic Macrofauna in German Marine Protected Areas (2020–2022) before Closure for Bottom-Contact Fishing

**Mayya Gogina [1,*,†], Sarah Joy Hahn [2,†], Ramona Ohde [3,†], Angelika Brandt [2,4], Stefan Forster [5], Ingrid Kröncke [3], Martin Powilleit [5], Katharina Romoth [1], Moritz Sonnewald [2] and Michael L. Zettler [1]**

1 Leibniz Institute for Baltic Sea Research Warnemünde, Seestrasse 15, D-18119 Rostock, Germany; katharina.romoth@io-warnemuende.de (K.R.); michael.zettler@io-warnemuende.de (M.L.Z.)

2 Senckenberg Research Institute and Museum of Nature, D-60325 Frankfurt am Main, Germany; sarah.joy.hahn@gmx.de (S.J.H.); angelika.brandt@senckenberg.de (A.B.); moritz.sonnewald@senckenberg.de (M.S.)

3 Senckenberg am Meer, Department for Marine Research, D-26382 Wilhelmshaven, Germany; ramona.ohde@senckenberg.de (R.O.); ingrid.kroencke@senckenberg.de (I.K.)

4 Department of Biological Sciences, Institute of Ecology, Evolution and Diversity, Johann Wolfgang Goethe University, D-60439 Frankfurt am Main, Germany

5 Faculty of Mathematics and Natural Sciences, Institute for Biosciences (IfBi), University of Rostock, D-18059 Rostock, Germany; stefan.forster@uni-rostock.de (S.F.); martin.powilleit@uni-rostock.de (M.P.)

* Correspondence: mayya.gogina@io-warnemuende.de

† These authors contributed equally to this work.

**Simple Summary:** This study illustrates a baseline biodiversity snapshot of macrofauna inhabiting the seafloor in German marine protected areas (MPAs) if the North and Baltic Seas in 2020–2022, before the full closure for bottom-contact fishing. While the closure is now in place in some MPAs, it is still planned for the near future in others. The analyses included different habitats in nine Natura 2000 MPAs. We provide essential data and comprehensive macrofauna species lists per area, relevant for the joint future conservation efforts and effective management. We explore environmental drivers of community structure and touch upon suggested effects of bottom-contact fishing in both geographic regions. Despite the expectation of more limited connectivity between MPAs in the Baltic Sea compared to the North Sea, the degree of community differentiation between MPAs was higher in the North Sea. Alpha diversity generally increased towards the open North Sea, and gamma diversity seemed comparable for these two regions. The Baltic Sea dataset unexpectedly contained a higher number of taxa, including Red List species. Achieving homogeneity of monitoring data and joint assessment even within one national program and biological compartment between different geographic regions, research institutions and fields remain challenging. This joint work appeals for flexible data sharing and prioritizing informal intersessional communication. Such a baseline is important for assessing future faunal changes.

**Abstract:** The response of benthic habitats and organisms to bottom-contact fishing intensity is investigated in marine protected areas (MPAs) of the German EEZ in the North and Baltic Seas. We examined the current state of macrofauna biodiversity in 2020–2022. Comparative analysis for macrofauna (in- and epifauna) inhabiting nine Natura 2000 MPAs constitutes a baseline to assess the effects of bottom-contact fishing exclusion in the future. Aspects of spatial and temporal variability are briefly summarized and discussed. We provide a species list for each region, including 481 taxa, of which 79 were found in both regions, 183 only in the North Sea, and 219 only in the Baltic Sea. The Baltic Sea dataset surprisingly included higher numbers of taxa and revealed more Red List species. The share of major taxonomic groups (polychaetes, bivalves and amphipods) in species richness showed peculiar commonalities between the two regions. In the North Sea, multivariate analysis of community structure revealed significantly higher within-similarity and stronger separation between the considered MPAs compared to the Baltic MPAs. Salinity, temperature and sediment fractions of sand were responsible for over 60% of the variation in the North Sea macrofauna occurrence data. Salinity, mud fraction and bottom-contact fishing were the most important drivers in the Baltic Sea and, together with other considered environmental drivers, were responsible for 53% of the variation. This study identifies aspects of macrofauna occurrence that may be used to assess (causes of) future changes.

**Keywords:** Natura 2000; Baltic Sea; North Sea; benthic habitats; invertebrates; fishing intensity; variability; environmental drivers; diversity; macrozoobenthos

## 1. Introduction

Germany borders on two semi-enclosed seas, the North Sea [1] and the more continental brackish Baltic Sea [2]. These two seas, separated only by a few hundred kilometers of land mass (at the narrowest point only 33 km wide), possess distinct characteristics. Linked by the narrow passage of the Skagerrak and Kattegat [3], both seas offer a variety of ecosystem services necessary for humans and provide habitats or breeding grounds for hundreds of species including benthos [4,5]. Although they are both part of the Atlantic Ocean and are geographically close, they show remarkable differences in macrobenthic ecology [6].

The North Sea (NS), located between the coasts of Norway, Sweden, Denmark, Germany, the Netherlands, Belgium and the UK, is characterized by higher salinity and greater water movement. Powerful tides and strong currents ensure a high rate of water exchange with the open Atlantic. These dynamic conditions shape the diversity of habitats from the sandy coastal areas to the deep underwater trenches. In contrast, the Baltic Sea is semi-enclosed, surrounded by the coasts of Denmark, Sweden, Germany, Finland, Estonia, Latvia, Lithuania, Russia and Poland. The Baltic Sea (BS) is characterized by lower salinity and less water movement, barely affected by any tides [7]. This results in more stratified water and a limited exchange rate with the open ocean. These features have led to a unique evolution and adaptation of inhabiting organisms [8].

The large-scale distribution of the North Sea macrofauna communities was intensively studied since the last century (e.g., [9–14]). These studies confirmed a generally depth based structure of three benthic zones: less than 50 m, between 50 and 100 m, and beyond 100 m [15]. Other environmental drivers of this zonation were sediment composition, depth, salinity, tidal patterns, sea surface temperature (SST), and primary productivity (PP). Comparable spatial studies of the ICES (International Council for the Exploration of the Seas) NS Benthos Survey (NSBS) in 1986 and the NS Benthos Project (NSBP) in 2000 (e.g., [10,12,16]) and recent studies of Fiorentino et al. [17] and Meyer et al. [14] identified four southeastern North Sea macrofauna communities at a depth < 50 m. These are the *Amphiura filiformis* community, the *Tellina (Fabulina) fabula* (or *Bathyporeia-Tellina*) community, the *Goniadella-Spisula* community, and the *Nucula nitidosa* community. Between 1986 and 2000, the spatial distribution of the four communities was stable [18]. However, structural changes within each of the southeastern North Sea macrofauna communities were since found in small-scale studies (e.g., [19–21]). These changes were often thought to be—directly or indirectly—driven by an increase in SST by 1.1 °C for the whole North Sea since 1950 [22] and about 2 °C for the southern North Sea [20].

Benthic macrofauna in the German waters of the Baltic Sea was systematically investigated since the 18th century (e.g., [23–29]). The distribution and dynamics of macrozoobenthos east of Fehmarn Belt were summarized in [30] based on data from 1839 to the 2000s. High temporal fluctuations in the occurrence, abundance, and biomass of macrozoobenthos were linked to (albeit natural irregular) saltwater intrusions and oxygen deficiency. The latter likely caused declines of some relict species, including the amphipods *Pontoporeia femorata* and *Monoporeia affinis*, the mussels *Macoma calcarea* and *Astarte* spp., or the isopod *Saduria entomon*. For other species, like the lugworm *Arenicola marina*, data suggested an eastward expansion. Relying on the findings by Zenkevitch [31], Schiewer [32] listed the most important species assemblages for the Baltic Sea, including the *Abra alba*-coenosis (with *Varicorbula gibba*, *Arctica islandica*, *Lagis koreni*, *Nephtys* spp., *Diastylis rathkei*, and *Ophiura albida*) dominating the western part, *Arctica-Astarte* assemblages found eastwards and *Macoma balthica*-coenosis dominating the shallow part of the Baltic Proper. Recent studies of spatial distribution on the large and medium scales suggest stability of community structure over time for some areas, higher fluctuations or even regime shifts due to species invasions for others, and increasing variability towards the entrance to the North Sea [33,34].

### *1.1. Habitat Protection*

The German seas and MPAs therein are protected by various conservation measures to ensure their ecological integrity, biodiversity and the sustainable use of resources [35]. The German Exclusive Economic Zones (EEZ) of the North and Baltic Seas include ten nominated Natura 2000 sites within the EU Natura 2000 protected areas network [36]. The main international EU legislative drivers that regulate protection of endangered wild plants and animals in those special natural habitats are the Birds Directive and the Habitats Directive (Council Directive 92/43/EEC), as well as OSPAR and HELCOM regulations; nationally, they have the status of protected nature reserves [37]. The EU Marine Strategy Framework Directive (MSFD) divides benthic habitats into broad habitat types (BHTs) and, in accordance with the EU Commission, into biotope classes and other habitat types (OHT). In addition to the protected areas designated on the basis of the Habitats Directive, the OHTs thereby also include the particularly endangered biotope types such as species-rich gravel, coarse sand and gravel beds, or mudflats with drilling megafauna, based on OSPAR or national law (Section 30 BNatSchG).

In the North Sea, the "Dogger Bank" (DGB), "Borkum Reef Ground" (BRG), and the "Sylt Outer Reef–Eastern German Bight" (that comprise two sites included in this study: the "Sylt Outer Reef" (SAR) and the "Amrum Bank" (AMB)) cover an area of 7920 km$^2$ (28% of the EEZ). In the Baltic Sea, the "Fehmarnbelt" (FB), "Kadetrinne" (KT) and "Pomeranian Bay—Rønne Bank" (including Adler Ground (AG), Odra Bank (OB) and Western Rønne Bank (RB)) have a total area of 2472 km$^2$, which constitutes 55% of the EEZ [38,39].

### *1.2. MGF and BfN Monitoring Projects and Aims of This Study*

Here, we aim to summarize baseline macrofauna biodiversity data gained within two research projects (MGF North Sea and MGF Baltic Sea, financed by the Federal Ministry for Education and Research (BMBF)) that investigate the effects of the exclusion of mobile bottom-contact fishing in MPAs of the German EEZ.

We complement it with external data to enhance spatial consistency across regions. In the North Sea, MGF-project data were collected from larger areas, while grid-based sampling was employed in designated MPAs (Figure 1A). In the Baltic Sea, sampling focused on specific areas within and outside future exclusion zones of MPAs (Figure 1B).

Macrofauna communities of the Western Baltic Sea and the North Sea have high exposure to natural and anthropogenic stressors [18,40] and especially to bottom-contact fishing [41–46]. Historically, in both regions, macrofauna monitoring and assessment are well-covered by established programs, such as projects in MPAs funded by the Federal Agency for Nature Conservation (BfN) like LABEL [47,48], CLUSTER and LEGRA [49], the project ATLAS (funded by the State Agency for Environment, Nature Conservation and Geology Mecklenburg-Vorpommern (LUNG MV) [50]), as well as long-term research studies (e.g., [18,51]).

Mobile bottom-contact fishing impacts on macrofauna have been studied in the North Sea in EU projects like IMPACT I-II [52,53], MAFCONS [54], and recently BENTHIS [55]; in ICES actions [37,56,57]; as well as in national projects (such as those named above or the recently launched CRANIMPACT, which investigates the effects of shrimp fishing on the seabed).

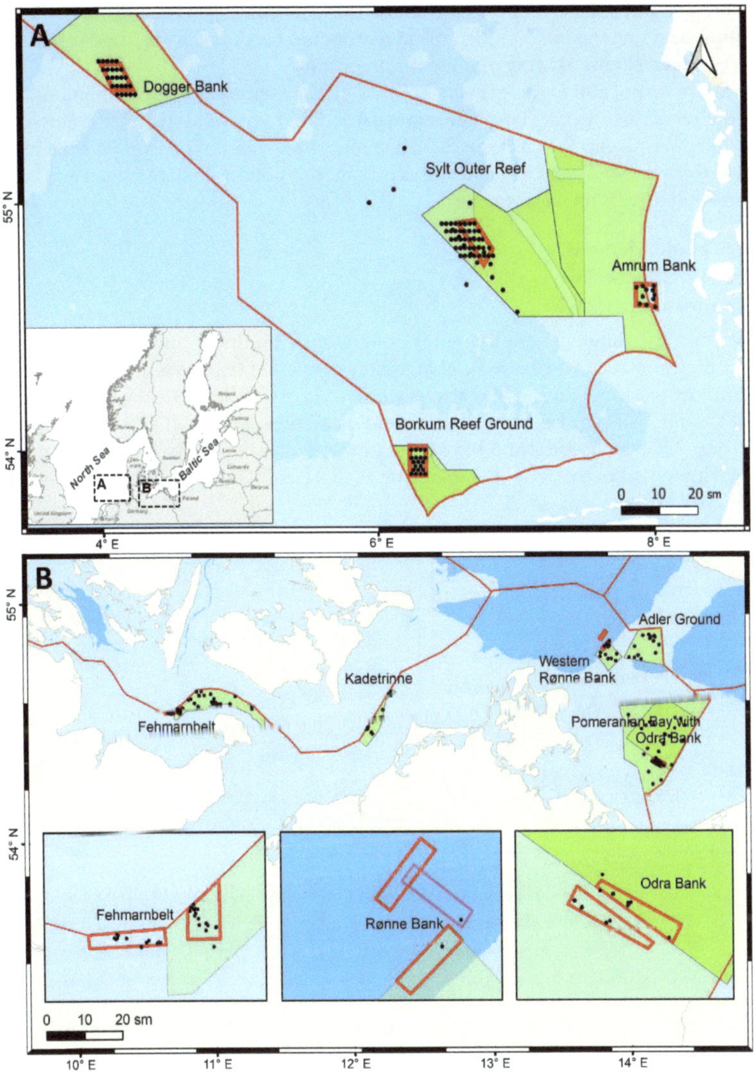


**Figure 1.** Maps of Natura 2000 sites (green polygons) and the MGF focus areas (thick red line boxes) in (**A**) the North Sea and (**B**) the Baltic Sea. The small grayscale inlet (inserted in (**A**)) shows a general view of the North and Baltic Seas. The thin red line marks the boundaries of the German Exclusive Economic Zone (EEZ). Black dots show the sampled stations. Focus areas in the Baltic Sea are zoomed in on the three small inlet maps. Dots inside the focus areas are stations sampled within the MGF Baltic Sea project, whereas other stations were mostly visited within the LEGRA and ATLAS projects. The half-transparent red line outlines the initial focus area in Rønne Bank, later shifted due to proximity to wind farms that inhibited later sampling. Intense green background outlines the future OB closure area if it will only take place in part of the MPA.

In the Baltic Sea, there have been no targeted studies since the 1990s [58–60]. The planned closure of MPAs for mobile bottom-contact fishing requires scientific evaluation of its efficiency. Such evaluation implies the development of optimal methods and monitoring concepts, particularly targeting those aspects that relate to potential changes, and is impossible without sufficient knowledge of the present standing stock and variability in macrofauna and understanding of its role in maintaining ecosystem services. Both MGF projects complement the existing monitoring programs.

Here, we do not aim for an explicit report of all the investigated macrofauna-related aspects but rather give a joint status quo summary to build upon and discuss the emerging peculiarities. We do aim to synthesize baseline macrofauna biodiversity data from MGF projects in German MPAs and evaluate impacts of mobile bottom-contact fishing and other environmental drivers on macrofauna.

## 2. Materials and Methods

### 2.1. The North Sea Case Study

#### 2.1.1. North Sea Study Areas

The sampled stations were located within or near the focus areas defined by the MGF North Sea project in order to be able to investigate the regions with strong anthropogenic influence before, during, and after the exclusion of mobile bottom-contact fisheries (Figures 1A and 2A). We also included all sampled stations at the MPA Sylt Outer Reef (SAR). These MGF focus areas and the SAR stations allow a comparison of different subsamples with regard to the in- and epifauna, as well as temporal comparison with earlier collected data for certain areas [20,51,61].

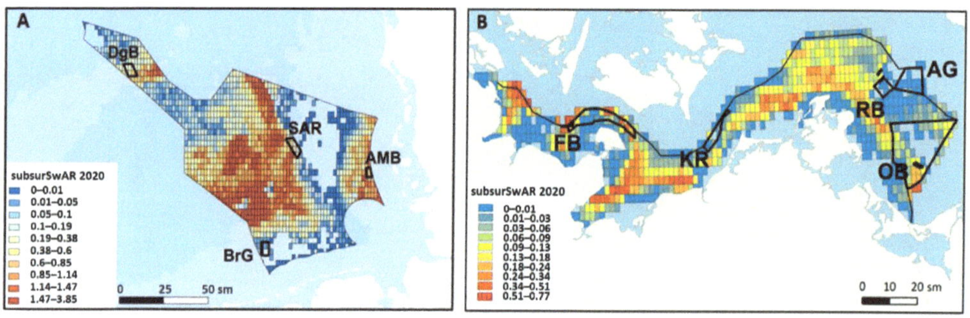


**Figure 2.** Overview of the mobile bottom-contact fishing intensity, (**A**) subsurface swept area ratio (>2 cm, subsurSwAR) in 2020 in the North Sea EEZ, based on ICES [62,63], and (**B**) subsurSwAR in 2020 in each 0.05° × 0.05°-degree c-square from ICES [64] data in the Baltic Sea EEZ; red = high mobile bottom-contact fishing intensity; blue = non or low mobile bottom-contact fishing intensity.

The Dogger Bank (DGB) is a shallow, 300 km-wide sandbank in the central North Sea [20], interesting due to its faunal composition: in the north, it is characterized by species typical for the northern North Sea, while in the south, species typical for the southern North Sea are common [20,65]. However, the German MPA at DGB only covers a comparatively small part of the whole DGB. Borkum Reef Ground (BRG) is characterized by reef structures surrounded by sandbanks. It is located in the southern North Sea, relatively close to the coast of the East Frisian island of Borkum [66]. The Sylt Outer Reef (SAR) area shows a variety of sediment structures with reefs, gravel areas and sandbanks [67]. The Amrum Bank (AMB) is mainly characterized by sandy substrate. In addition, due to strong wind conditions, the sandbank is used for wind turbine installation. Exclusion of mobile bottom-contact fishing in 2023 took place in a large part of SAR and the entire area of BRG.

#### 2.1.2. North Sea Data Collection

A total of 150 stations were sampled within the four study areas in the North Sea (Figure 1A) with RV "Senckenberg" in 2020–2022 in order to study the in- and epifauna biodiversity along the gradients of decreasing bottom-contact fishing intensity (Table 1).

**Table 1.** Sampling in 2020–2022 in the North Sea Natura 2000 areas, sampling method (infauna = 0.1 $m^2$ van Veen grab, epifauna = ring dredge and beam trawl), number of sampled stations and sampling month. Note: # only four stations had the same locations for the in- and epifauna sampling, but were sampled at different research cruises; * in- and epifauna sampling took place at the same research cruise: at each station, infauna was sampled first, then epifauna.

| Study Site | Sampling Methods | No. of Stations | Month |
|---|---|---|---|
| | 2020 | | |
| Sylt Outer Reef (SAR) # | In- and Epifauna | 20 | May |
| Borkum Reef Ground (BRG) * | In- and Epifauna | 14 | July |
| | 2021 | | |
| Dogger Bank (DGB) | Infauna | 20 | May |
| Sylt Outer Reef (SAR) | Infauna | 20 | May |
| Dogger Bank (DGB) | Epifauna | 25 | July |
| Amrum Bank (AMB) | Epifauna | 11 | August |
| | 2022 | | |
| Sylt Outer Reef (SAR) | Epifauna | 15 | May |
| Borkum Reef Ground (BRG) | Epifauna | 14 | July |
| Amrum Bank (AMB) | Epifauna | 11 | July |

The sampling for in- and epifauna took place at the same research cruise only in 2020 at the MPA BRG (14 stations). During this research cruise at these stations the infauna sampling was performed, and next, the epifauna sampling was performed. The other stations were sampled in separate cruises. At the MPA DGB 20 stations sampled in 2021 had the same location but were visited on different cruises in different months (Table 1). Only four stations in 2020 at the MPA SAR had the same locations for the in- and epifauna sampling (Table 1). For the infauna, two replicate samples were collected with a 0.1 $m^2$ van Veen grab at each station and were sieved through a 1 mm mesh size. The samples were preserved on board in a 4% buffered formaldehyde seawater solution. Retained material was identified to the lowest possible taxonomic level. The taxonomy (also for the Baltic Sea) was harmonized following the World Register of Marine Species [68].

During epifauna sampling, at each station first, the water temperature and salinity were determined using a CTD probe (Sea and Sun technologies). After measurement of water parameters, a ring dredge (diameter 1 m, mesh size 1 $cm^2$) was lowered to the seafloor for sampling the main taxa of the in- and epifauna. The dredge was slowly pulled by the ship in a constant direction for 3–5 min (depending on the prevailing sediment). The ring dredge penetrated about 5 cm into the sediment of the seafloor (also depending on the sediment type) so that after retrieval, the main in- and epifauna could first be documented photographically, sorted and identified to the finest taxonomic level possible. The identified species were recorded in a presence/absence matrix and then released directly back into their natural habitat to ensure their survival. Additionally, the epifauna was subsequently sampled using a 2 m beam trawl (rump mesh size 1 $cm^2$). The beam trawl was lowered to the seafloor and then towed in one direction at 2 knots over a distance of 1 nautical mile (=1.85 km). The sample was then documented photographically on board, and the fine fraction (>1 mm) was separated from the larger sieve fraction (<1 cm) using a sieve barrel. After sorting and identification, caught species were quantitatively recorded in the case of non-colonial forms. The sieve fractions (>1 mm) of the beam trawl sampling and species that could not be determined directly on board were fixed in 96% ethanol or in a ~5% formaldehyde-seawater solution to ensure later final determination in the Senckenberg Research Institute's laboratory.

### 2.2. The Baltic Sea Case Study

#### 2.2.1. The Baltic Sea Study Areas

The three MGF Baltic Sea focus areas (Figure 1B) selected in the Fehmarnbelt (FB), Rønne Bank (RB) and Odra Bank (OB) are characterized by different sediments. While the FB focus area is located on muddy sediment with a fine sand component, the area selected at RB is covered by fine, organically rich mud, and the OB focus area is a typical sand bank. Moreover, they are also home to different communities due to the gradient of salinity, which, in the Baltic Sea, decreases sharply from west to east. Thus, different responses to bottom-contact fishing intensity and termination thereof are expected in the three areas. Not all MPAs in the Baltic Sea are equally affected by bottom-contact fishing. Two MPAs, Kadetrinne (where highly intensive ship traffic takes place) and the Adlergrund, both characterized by reef structures (not favored by trawling fishers due to the risk of fishing gear damage), were excluded from the MGF investigation as less relevant in order to keep the efforts feasible (see Figure 2B). However, on the larger scale, the condition of these reef MPAs and inhabiting benthic fauna was annually monitored within the LEGRA project.

#### 2.2.2. Data Collection: Baltic Sea

A total of 222 stations were sampled in the Baltic Sea MPAs in 2020–2022, and 35 more stations in close vicinity (Table 2, Figure 1B). At each station and visit, for quantitative macrofauna data, three replicate samples were commonly collected with a 0.1 $m^2$ van Veen grab (weight about 75 kg, sediment penetration depth of up to 15–20 cm) and washed through a 1 mm sieve. Remaining animals were preserved in a 4% formaldehyde-seawater solution buffered with marble chippings; material was sorted in the laboratory at the Leibniz Institute for Baltic Sea Research, Warnemünde, with a stereomicroscope and identified to the lowest possible taxonomic level. Organisms were counted and weighed to obtain estimates of species abundance and biomass per square meter. At three specific areas—FB, RB, and OB—we identified the "key species" defined here as those having a substantial contribution to biomass, an extended lifespan, a high potential for bioturbation, and a pivotal role in the local food web [69,70].

**Table 2.** Number of sampled stations as well as the year and month of sampling of the Baltic Sea MPAs in 2020–2022—always in- and epifauna. * Single stations were visited in other months.

| Study Site | No. of Stations in MPA (Close Outside) | Month |
|---|---|---|
| | 2020 | |
| Fehmarnbelt (FB) | 29 (12) | June * |
| Western Rønne Bank (RB) | 1 | July |
| Pomeranian Bay with Odra Bank (OB) | 4 | June–July |
| Kadetrinne (KR) | 6 | June |
| Adler Ground (AG) | 8 | July |
| | 2021 | |
| Fehmarnbelt (FB) | 13 (3) | June * |
| Western Rønne Bank (RB) | 6 | June |
| Pomeranian Bay with Odra Bank (OB) | 40 | June * |
| Kadetrinne (KR) | 11 | June |
| Adler Ground (AG) | 14 | July–Aug |
| | 2022 | |
| Fehmarnbelt (FB) | 33 (16) | March, June |
| Western Rønne Bank (RB) | 7 (4) | April, June |

**Table 2.** *Cont.*

| Study Site | No. of Stations in MPA (Close Outside) | Month |
|---|---|---|
| Pomeranian Bay with Odra Bank (OB) | 40 | March, June |
| Kadetrinne (KR) | 4 | June |
| Adler Ground (AG) | 6 | June |

We also took 0.00785 $m^2$ sediment core samples with a multicorer. The number of cores per station varied from one to six. Cores were sliced for macrofauna vertical distribution (using 7 intervals of 0–2, 2–4, 4–6, 6–8, 8–10, 10–15, and >15 cm sediment depth) and each slice was sieved separately with a 0.5 mm sieve.

Additionally, the Kieler Kinderwagen dredge was used to qualitatively assess quick-moving, rare or large species [71]. The dredge has a 92 cm inner opening, and 5 mm mesh; it was towed with 1 knot over the ground for about 1 min (=31 m) in mud and 5 min in sand (=155 m), penetrating the sediment to roughly 5 cm in mud and only scraping the sediment surface in sand.

Epifauna in the studied habitats was additionally investigated using an underwater video system (only a hand-held SeaViewer HD camera could be used on MGF transects due to logistical limitations, whereas in LEGRA campaigns, the BaSIS system that is suitable for gathering quantitative coverage data [72] was also applied).

### 2.3. *Environmental Drivers, Mobile Bottom-Contact Fishing Data and Statistical Analysis*

#### 2.3.1. Temperature, Salinity, and Sediment Data

CTD near-bottom water measurements were conducted at each location before biological sampling in order to obtain relevant abiotic parameters (including near-bottom water temperature and salinity (for North Sea and Baltic Sea) and oxygen concentrations (only for Baltic Sea)).

A surface sediment sample (upper 2 cm) was taken from one additional grab replicate at each location for later sediment granulometry. The North Sea sediment samples were sieved using a 63 µm mesh size to determine the mud content (<63 µm, in %). The shell content (>2000 µm, in %) was determined by wet dry sieving over a 2 mm mesh. In addition, the % of gravel debris was measured. For the Baltic Sea samples, dry sieving was used for sands, and a Mastersizer 3000 was used for finer sediments.

For the Baltic Sea, mean near-bottom temperature values for 2010–2020 available from the GETM model [73] were extracted using ArcGIS for each sampling location to illustrate general longer-term conditions.

#### 2.3.2. Bottom-Contact Fishing Intensity

Data describing mobile bottom-contact fishing intensity originated from ICES for both the North Sea [62,63] and the Baltic Sea regions [64,74]. The intensity of bottom-contact fishing was calculated based on VMS and linked logbook data submitted by EU member states to ICES and aggregated consistently across years and quarters for 2016–2020 for the North Sea and 2016–2021 for the Baltic Sea period. Intensity is expressed in either kilowatt fishing hours (kwfhr) or as surface or subsurface swept area ratio (surSwAR or subsurSwAR) at the spatial resolution of c-square with the extension of 0.05° × 0.05° degrees. A SwAR value of 1 implies that the sediment of the entire area was trawled once or, e.g., that half of the area was swept over twice within a period of time (here recalculated to multiannual values). Surface SwAR (surSwAR) reflects the potential impacts on benthic epifauna, considering the surface penetration depth (<2 cm) of the gear components. The impact on benthic infauna is reflected in the Subsurface SwAR (subsurSwAR), considering the subsurface penetration depths (≥2 cm) of each gear, assuming no differences across sediment types [75].

#### 2.3.3. Statistical Analysis

To avoid the bias related to the differences in sampling methods and efforts applied in the studied regions, we focused here on multivariate statistical techniques, such as ordination, to explore patterns in species composition and occurrence within each area rather than abundance.

A non-metric multidimensional scaling (nMDS) and similarity profile analysis (SIMPROF) were accomplished for each MPA, based on a Bray–Curtis resemblance matrix of the Presence/Absence transformed abundance data, separately for each region, using PRIMER 6 for the North Sea and PRIMER 7 for the Baltic Sea Data. The SIMPROF analysis is a permutation test analyzing the statistical significance of groups. Characteristic taxa for each MPA were identified by similarity percentage analysis (SIMPER), using the Bray–Curtis similarity matrix. The defined clusters (representing the MPAs) were confirmed by Analysis of Similarities (ANOSIM), which is a permutation test, analyzing the statistical significance of a priori divided clusters. ANOSIM reveals a global R for the whole dataset and a pairwise R, testing between the clusters [76].

For the North Sea statistical analyses, we only used the data from the stations for which we had abundance data for both in- and epifauna. For the MPA SAR there were only four stations sampled in 2020 for which both data sets were available. In the MPA BRG, 14 stations were sampled in 2020, and in the MPA DGB, 20 stations in 2021. In the Baltic Sea, all stations were included, since sampling was targeting all macrofauna without distinguishing between in- and epifauna.

To determine the set of environmental drivers that best explain the variation of benthic macrofauna at each MPA, we performed a distance-based linear model permutation test (DistLM) based on a significant RELATE analysis employing the routine from the software PRIMER 6 with PERMANOVA+ add-on [77]. Predictor variables were subjected to a sequential stepwise selection procedure using Akaike's information criterion with a correction for finite sample size (AICc). To calculate resemblance in DistLM, the Bray–Curtis similarity was used. We included several environmental drivers (Table 3) that can affect macrofauna and the available corresponding mobile bottom-contact fishing intensity data at each location as additional independent variables. Predictor pairs were tested for collinearity.

For both the NS and BS, based on a correlation threshold of 0.8 (with higher values suggesting multicollinearity) and the marginal test results, only subsurface SwAR was retained in DistLM as the most influential out of three initially considered fishing intensity parameters. For the BS, some variables (% mud, % gravel and subsurface SwAR) were square root transformed to remove right-skewness in the raw data in case it was observed on Draftsman's plots. For the NS, only the subsurface SwAR was square root transformed.

**Table 3.** Overview of the nine MPAs in the North and Baltic Sea, including the total taxa number, the abiotic factors as depth (m), mud (%), gravel (%), temperature (°C), salinity (psu) and oxygen content (mL/L) over the time period 2020–2022, and the bottom-contact fishing intensity (surSwAR and subsurSwAR). SurSwAR and subsurSwAR are averaged per year based on ICES 2016–2020 data [62,63] for the North Sea and 2016–2021 data [64,74] for the Baltic Sea. * At the focus area AMB, only epifauna sampling took place. Values are average per MPA stations ± standard deviation.

| Sea | Area | Total Taxa No. | Depth, m | Mud, % | Gravel, % | Temp, °C | Temp Mod, °C | Sal, (psu) | $O_2$, ML/L | Trawling | |
|---|---|---|---|---|---|---|---|---|---|---|---|
| | | | | | | | | | | sur SwAR | Subsurswar |
| North Sea | SAR | 187 | 43 ± 1.2 | 16.2 ± 8.3 | 2.6 ± 9.9 | 7.5 ± 0.4 | / | 34.3 ± 0.2 | / | 0.61 ± 0.62 | 0.40 ± 0.32 |
| | BRG | 135 | 30 ± 1.2 | 0.5 ± 0.5 | 1.2 ± 2.2 | 16.4 ± 0.6 | / | 33.4 ± 0.2 | / | 0.2 ± 0.22 | 0.03 ± 0.02 |
| | DGB | 143 | 43.3 ± 0.6 | 0.2 ± 0.1 | 0 | 11 ± 0.6 | / | 34.5 ± 0.03 | / | 1.08 ± 0.2 | 0.68 ± 0.3 |
| | AMB * | 50 | 11.4 ± 2.7 | / | / | 18.2 ± 0.2 | / | 30.9 ± 0.6 | / | 3.04 ± 0.96 | 1.58 ± 0.5 |
| Baltic Sea | FB | 264 | 22 ± 5 | 33.5 ± 23.5 | 2.8 ± 8.2 | 8.[illegible] ± 4 | 7.9 ± 0 | 19.3 ± 2.9 | 6 ± 1.8 | 2.64 ± 2.45 | 0.21 ± 0.19 |
| | RB | 58 | 34 ± 5 | 15 ± 15.8 | 5.8 ± 14.3 | 6.9 ± 2.2 | 5.7 ± 0.1 | 10.7 ± 1.6 | 6 ± 0.7 | 0.85 ± 1.05 | 0.07 ± 0.08 |
| | OB | 56 | 14 ± 2 | 0.3 ± 0.4 | 0 ± 0 | 13.1 ± 4.4 | 8.4 ±0.1 | 8.2 ± 1.6 | 6.9 ± 1.3 | 0.88 ± 0.67 | 0.07 ± 0.05 |
| | KR | 141 | 19 ± 4 | 26.6 ± 33.8 | 2.3 ± 7.6 | 10.5 ± 1 | 7.4 ± 0 | 16.5 ± 2.4 | 5.2 ± 0.9 | 0.21 ± 0.22 | 0.01 ± 0.01 |
| | AG | 62 | 16 ± 7 | 0 ± 0 | 3 ± 9.5 | 12.9 ± 3.4 | 7.6 ± 0.9 | 7.8 ± 0.7 | 6.6 ± 0.5 | 0.03 ± 0.04 | 0.00 ± 0.00 |

## 3. Results

### *3.1. Biodiversity and Community Analysis*

#### 3.1.1. Species Richness and Major Groups

A total of 481 taxa were found in all nine MPAs of the Baltic and North Sea during 2020–2022 (see full taxa list in Supplementary Material S1; there, all the scientific names are provided with authorities, whereas for the sake of readability, authorities are mostly omitted here in the main text). The 481 taxa belonged to the phylum Annelida (162 taxa), Arthropoda (126 taxa), Mollusca (100 taxa), Cnidaria (33 taxa), Echinodermata (19 taxa), Bryozoa (16 taxa), Porifera (7 taxa), Chordata (6 taxa), Nemertea (6 taxa), Priapulida (3 taxa), Phoronida (1 taxon), Platyhelminthes (1 taxon) and Entoprocta (1 taxon).

Only in the North Sea MPAs, 183 of the 481 (38.0%) taxa were found, for example the polychaete *Aonides paucibranchiata*, the crustacean *Urothoe elegans*, or the bivalve *Gari fervensis*.

Exclusively in the Baltic Sea MPAs, 219 of the 481 (45.5%) taxa were found. The polychaete *Dipolydora quadrilobata*, the gastropod *Alvania punctura*, and the echinoderm *Ekmania barthii* were examples of such taxa found in the MPAs of the Baltic Sea but not in the North Sea MPAs.

79 of the 481 (16.4%) taxa occurred in the North Sea and in the Baltic, for example the polychaete *Eteone longa*, the crustacean *Pagurus bernhardus*, or the gastropod *Aporrhais pespelecani*. Only one of the 481 (0.2%) taxa was found in all of the nine MPAs: the crustacean *Crangon crangon*.

180 of the total of 481 (37.4%) taxa were found in only one of the nine MPAs in the North and Baltic Sea. For example, the crustacean *Tryphosites longipes* was present only in the SAR, and the echinoderm *Echinocardium flavescens* only in the AMB in the North Sea, whereas the mollusc *Lamprops fasciatus* was found only in the FB, and the gastropod *Ecrobia ventrosa* only in the OB in the Baltic Sea.

North Sea

A total of 262 taxa were found in the four MPAs in the NS during 2020–2022 (Supplementary Material S1). The 262 taxa belonged to the phylum Annelida (93 taxa), Arthropoda (76 taxa), Mollusca (50 taxa), Echinodermata (17 taxa), Cnidaria (12 taxa), Bryozoa (5 taxa), Chordata (2 taxa), Nemertea and Porifera (2 taxa each), as well as 1 taxon each of Phoronida, Platyhelminthes and Priapulida (Figure 3).

The most taxa were found in the SAR with 187 taxa (Table 3). Only 17 of the 262 (6.5%) taxa were found in all four MPAs of the North Sea, for example, the polychaete *Nephtys hombergii* and the decapod *Corystes cassivelaunus*. 106 of the 262 taxa (40.5%) were present in only one of the four MPAs of the North Sea. The holothurian *Leptosynapta inhaerens*, the gastropod *Epitonium clathrus* and the decapod *Goneplax rhomboides* appeared only in the MPA SAR. In the MPA BRG, the polychaete *Hesionura elongata* and the bivalve *Lutraria lutraria* occurred exclusively. The echinoderm *Amphipholis squamata*, the gastropod *Euspira montagui* and the polychaete *Hydroides norvegica* were only found in the MPA DGB (see full species list in Supplementary Material S1).

Baltic Sea

A total of 298 taxa were found in the five MPAs of the Baltic Sea during 2020–2022 (Supplementary Material S1). These taxa belonged to the phylum Annelida (96 taxa), Mollusca (69 taxa), Arthropoda (66 taxa), Cnidaria (23 taxa), Bryozoa (15 taxa), Echinodermata (7 taxa), Porifera (7 taxa), Chordata (5 taxa), Nemertea (5 taxa) and one taxon each of Phoronida, Platyhelminthes and Entoprocta (Figure 3).

The most taxa were found in the focus area FB with 264 taxa (Table 3). Only 19 (6.4%) of taxa were found in all five MPAs of the Baltic Sea: for example, the polychaete *Hediste diversicolor* and the bryozoan *Einhornia crustulenta*. 145 of the 298 taxa (48.7%) were present in only one of the five MPAs. The bivalve *Tellimya ferruginosa* and the amphipod *Aora gracilis* appeared in the MPA FB only. The gastropod *Theodoxus fluviatilis*, the tanaid *Heterotanais oerstedii* and the fish leech *Piscicola* sp. occurred exclusively in the MPA AG. The polychaete

*Marenzelleria neglecta* and the crustacean *Rhithropanopeus harrisii* were only found on the OB (see full species list in Supplementary Material S1).

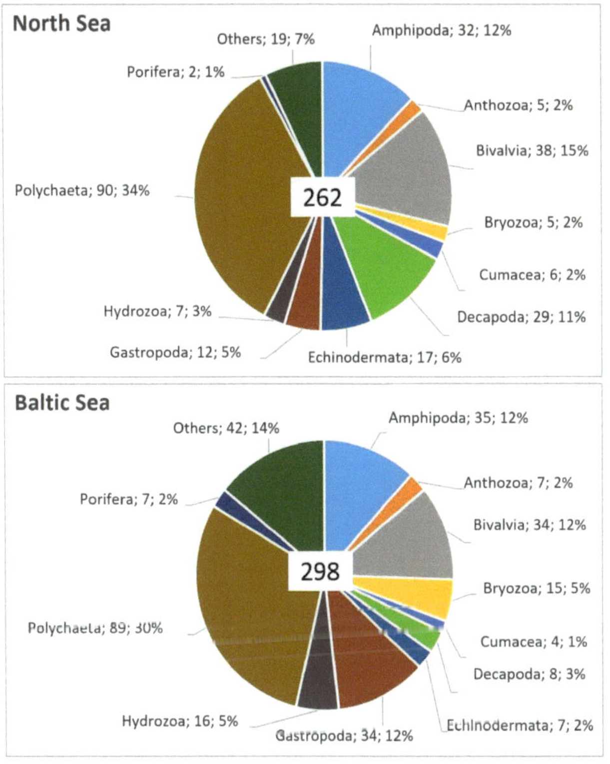


**Figure 3.** Number and (after semicolon) percentage of taxa found per group in the MPAs in the North Sea (**upper pane**) and the Baltic Sea (**lower pane**). The groups used here in order to facilitate the summary should be rather considered as functional, i.e., not strictly taxonomic, as they vary in rank ranging from Phylum to Order level. In the North Sea MPAs (upper pane), the category "other" includes Isopoda (4), Cirripedia (3 taxa), Nemertea (2), Sipuncula (2) and single taxa of Ascidiacea, Leptocardii, Oligochaeta, Phoronida, Platyhelminthes, Priapulida, Pycnogonida and Tanaidacea. In the Baltic Sea MPAs (lower pane), the category "other" includes Oligochaeta (6), Isopoda (5), Mysida (5), Nemertea (5), Ascidiacea (4 taxa), Cirripedia (4), Priapulida (2), Pycnogonida (2), Tanaidacea (2) and single taxa of Arachnida, Entoprocta, Hirudinea, Leptocardii, Phoronida, Platyhelminthes and Polyplacophora.

#### 3.1.2. Community Structure

North Sea

The analysis of community structure based on presence/absence transformed data for the NS showed distinct differences between the MPAs (Figure 4, upper pane). The MPAs in the North Sea can be significantly separated in terms of the macrofauna (ANOSIM: R-value = 0.991; $p$-value = 0.001), at least based on stations where in- and epifauna data were available. Some exemplary most frequent species found in the study region are shown in Figure 5.

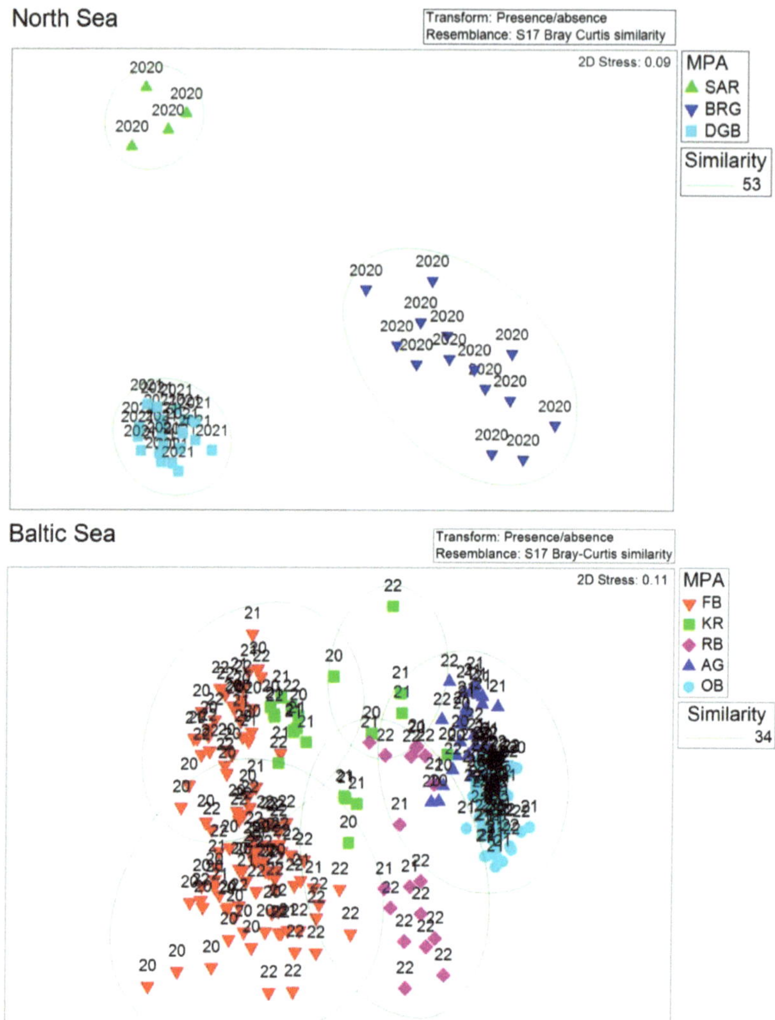


**Figure 4.** Multidimensional scaling (MDS) plots for the North Sea and the Baltic Sea areas based on presence/absence transformed data. The North Sea plot only includes stations where both data sets for in- and epifauna were available. Labeling is according to the MPAs and the sampling years.

**Figure 5.** Most common species in the North Sea MPAs. (**A**) *Echinocardium cordatum* (Pennant, 1777), (**B**) *Spiophanes bombyx* (Claparède, 1870), (**C**) *Liocarcinus holsatus* (Fabricius, 1798), (**D**) *Aonides paucibranchiata* Southern, 1915, (**E**) *Asterias rubens* Linnaeus, 1758, (**F**) *Bathyporeia elegans* Watkin, 1940, (**G**) *Abra alba* (W. Wood, 1802), (**H**) *Ophiura ophiura* (Linnaeus, 1758), (**I**) *Spisula solida* (Linnaeus, 1758). Indicated sizes are approximate total lengths (of longest dimension) for all species, with two exceptions: for *L. holsatus* (**C**), the value corresponds to carapace length, and for *O. ophiura*, the disc diameter is specified. These sizes were measured with calipers and are provided only for visualization and to show scale differences between species; they are not relevant for any other reported results.

The SIMPER analysis revealed the mean similarity of MPAs (Table 4). The MPA SAR had a mean similarity of 62.3%. It was mainly characterized by the bivalve *Abra alba*, the echinoderm *Amphiura filiformis*, the echinoderm *Astropecten irregularis*, the bivalve

*Chamelea striatula*, and the decapod *Corystes cassivelaunus*. The mean similarity in the MPA BRG was 57.5%. Characterizing taxa were the polychaete *Aonides paucibranchiata*, the echinoderm *Astropecten irregularis*, *Ensis* spp. bivalves, the polychaete *Lanice conchilega*, and the swimming crab *Liocarcinus holsatus*. The MPA DGB had a mean similarity of 67.5% and was mainly characterized by the echinoderm *Amphiura filiformis*, Cnidaria (Anthozoa indet.), the amphipod *Aora gracilis*, as well as by the echinoderms *Asterias rubens* and *Astropecten irregularis*.

**Table 4.** Results of the SIMPER analysis: ten characteristic taxa contributing most to the average similarity within the MPAs in the North Sea.

| | SAR | | BRG | | DGB |
|---|---|---|---|---|---|
| Mean similarity: 62.3% | *Abra alba* | Mean similarity: 57.5% | *Aonides paucibranchiata* | Mean similarity: 67.5% | *Amphiura filiformis* |
| | *Amphiura filiformis* | | *Astropecten irregularis* | | Anthozoa |
| | *Astropecten irregularis* | | *Ensis* spp. | | *Aora gracilis* |
| | *Chamelea striatula* | | *Lanice conchilega* | | *Asterias rubens* |
| | *Corystes cassivelaunus* | | *Liocarcinus holsatus* | | *Astropecten irregularis* |
| | *Cylichna cylindracea* | | *Spio symphyta* | | *Bathyporeia elegans* |
| | *Echinocardium cordatum* | | *Spiophanes bombyx* | | *Bathyporeia guilliamsoniana* |
| | *Eudorella truncatula* | | *Thia scutellata* | | *Dosinia lupinus* |
| | *Hyala vitrea* | | *Asterias rubens* | | *Echinocyamus pusillus* |
| | *Kurtiella bidentata* | | *Bathyporeia guilliamsoniana* | | *Euspira nitida* |

Baltic Sea

The analysis of community structure for the BS based on presence/absence data (Figure 4, lower pane) showed less difference between MPAs compared to the NS dataset. The MPAs in the BS could still be significantly separated in terms of the macrofauna (ANOSIM: R-value = 0.784; $p$-value = 0.001). Some exemplary most frequent species found in the study region are shown in Figure 6.

The SIMPER analysis revealed the mean similarity of each MPA (Table 5). The FB MPA had the lowest mean similarity of 38.8% among the Baltic Sea MPAs. In terms of presence/absence, FB was mainly characterized by the polychaetes *Aricidea suecica* and *Scoloplos armiger*, the echinoderm *Ophiura albida*, the bivalve *Varicorbula gibba* and the cumacean *Diastylis rathkei*. The mean similarity in the KR MPA was 46.9%. Characterizing taxa were the bivalve *Mytilus edulis*, the gastropod *Peringia ulvae*, the polychaetes *Bylgides sarsi* and *Pygospio elegans*, and the bryozoan *Eucratea loricata*. The MPA RB had a mean similarity of 49.2% and was mainly characterized by the bivalve *Macoma balthica*, the gastropod *P. ulvae*, the polychaete *S. armiger*, the cumacean *Diastylis rathkei* and the amphipod *Pontoporeia femorata*. Within the stations of AG MPA, a mean similarity of 60.2% was observed, mainly driven by the bivalve *M. edulis*, the gastropod *P. ulvae*, the amphipod *Gammarus salinus*, the bryozoan *Einhornia crustulenta*, and the polychaete *P. elegans*. In the OB MPA, the mean similarity was 64%, and characteristic species were the gastropod *P. ulvae*, the polychaete *P. elegans*, the oligochaetes of subfamily Tubificinae, as well as the bivalves *M. edulis* and *Mya arenaria*.

Additionally, within the three MGF focus areas in the Baltic Sea, it is worth noting the "key species", which we defined as biomass-dominant bivalves possessing a long lifespan, playing a crucial role in the local food web, and making significant contributions to bioturbation. For FB, it is the ocean quahog *Arctica islandica*, with biomasses > 90% of the total macrozoobenthos biomass; for RB, it is the Baltic tellin *Macoma balthica*, with biomasses of about 23% of the total biomass. For OB, there are two key species: the sand gaper *Mya arenaria* as well as *M. balthica*, with biomasses of about 54% and 18% of the total biomasses, respectively.

**Figure 6.** Most common species in the Baltic Sea MPAs. Size measures for each species are given in mm. (**A**) *Mytilus edulis* Linnaeus, 1758, (**B**) *Mya arenaria* Linnaeus, 1758, (**C**) *Cerastoderma glaucum* (Bruguière, 1789), (**D**) *Peringia ulvae* (Pennant, 1777), (**E**) *Macoma balthica* (Linnaeus, 1758), (**F**) *Abra alba* (W. Wood, 1802), (**G**) *Diastylis rathkei* (Krøyer, 1841), (**H**) *Scoloplos armiger* (Müller, 1776), (**I**) *Carcinus maenas* (Linnaeus, 1758), (**J**) *Pygospio elegans* Claparède, 1863, (**K**) *Crangon crangon* (Linnaeus, 1758). Indicated sizes are approximate total lengths (of longest dimension) for all species but I (for *C. maenas*, the value corresponds to carapace length). These sizes were measured with calipers and are provided only for visualization and to show scale differences between species; they are not relevant for any other reported results.

**Table 5.** Results of the SIMPER analysis: ten characteristic taxa contributing most to the average similarity within the MPAs in the Baltic Sea.

| FB | KR | RB | AG | OB |
|---|---|---|---|---|
| Mean similarity: 38.8% | Mean similarity: 46.9% | Mean similarity: 49.2% | Mean similarity: 60.2% | Mean similarity: 64.0% |
| *Aricidea suecica* | *Mytilus edulis* | *Macoma balthica* | *Mytilus edulis* | *Peringia ulvae* |
| *Scoloplos armiger* | *Peringia ulvae* | *Peringia ulvae* | *Peringia ulvae* | *Pygospio elegans* |
| *Ophiura albida* | *Bylgides sarsi* | *Scoloplos armiger* | *Gammarus salinus* | Tubificinae |
| *Varicorbula gibba* | *Pygospio elegans* | *Diastylis rathkei* | *Einhornia crustulenta* | *Mytilus edulis* |
| *Diastylis rathkei* | *Eucratea loricata* | *Pontoporeia femorata* | *Pygospio elegans* | *Mya arenaria* |
| *Kurtiella bidentata* | *Kurtiella bidentata* | *Halicryptus spinulosus* | *Hediste diversicolor* | *Hediste diversicolor* |
| Tubificinae | *Asterias rubens* | *Bylgides sarsi* | Tubificinae | *Marenzelleria viridis* |
| *Levinsenia gracilis* | *Mya arenaria* | *Hediste diversicolor* | *Jaera albifrons* | *Macoma balthica* |
| *Abra alba* | *Nephtys caeca* | *Capitella capitata* | *Amphibalanus improvisus* | *Cerastoderma glaucum* |
| *Paradoneis eliasoni* | *Diastylis rathkei* | *Mya arenaria* | *Bylgides sarsi* | *Streblospio shrubsolii* |

## 3.2. Variation Explained by Environmental Drivers and Trawling Intensity

### 3.2.1. North Sea

The results of the dbRDA (Figure 7) show the relationship between the environmental drivers that best explain the variability in the macrofauna communities in the three North Sea MPAs. The RELATE analysis revealed a significant relation of the environmental drivers and bottom-contact fishing to the presence/absence macrofauna data with a Rho of 0.815 (significance level of 0.1%).

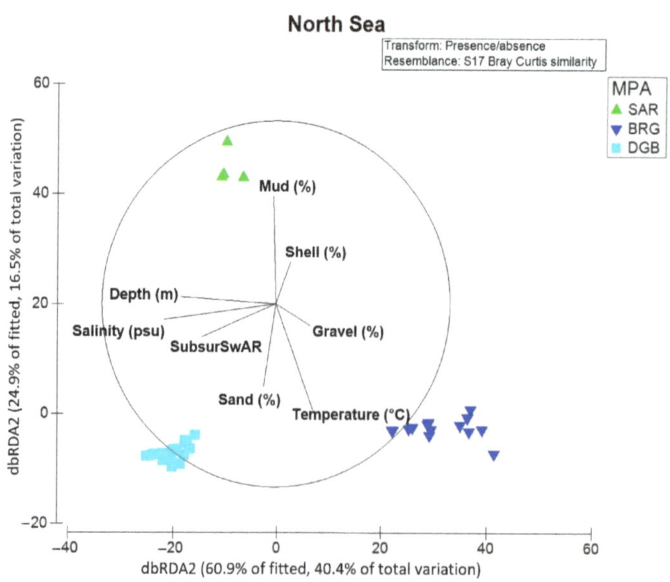


**Figure 7.** dbRDA ordination of stations in the North Sea MPAs along environmental (depth (m), sediment parameters (shell fraction > 2 mm, sand fraction < 2 mm to >0.063 mm, mud fraction < 0.063 mm, and gravel fraction), temperature (°C) and salinity (psu)) and anthropogenic (bottom-contact fishing expressed as subsusSwAR) drivers. Labeling according to the MPAs.

In the North Sea, the set of considered abiotic predictors explained together 66.3% of the total variation in the presence/absence macrofauna data. Based on results of the marginal test, salinity explained 39.7%, the depth (m) of the stations was responsible for 38.7% of changes in community structure, and the bottom-contact fishing explained 33.6% (SubsurSwAR) of variation in presence/absence data (see "DistLM results North Sea" tab in the Supplementary Material S1). In the sequential test (see "DistLM results North Sea" tab in the Supplementary Material S1), salinity (psu) as the most important predictor was followed by temperature and fraction of sand that explained an additional 20.5% of variation in macrofauna.

### 3.2.2. Baltic Sea

The results of the dbRDA (Figure 8) show the relationship between the environmental drivers that best explain the variation in the macrofauna composition in the five sampled Baltic Sea MPAs. Here, the set of considered abiotic predictors explained 53.1% of the total variation in the presence/absence macrofauna data. The dbRDA1 was mainly driven by salinity and % mud in sediment, and the dbRDA2 by bottom-contact fishing (subsurface SwAR), % mud, % gravel, and measured near-bottom temperature. Based on results of the marginal test, salinity of the near-bottom water alone was responsible for 33.3% of changes in community structure, mud content in sediments explained 24.3%, and among fishing parameters (that were highly correlated with each other), subsurface SAR explained 13.7% of variation in presence/absence data. In the sequential test (see "DistLM results Baltic Sea" tab in the Supplementary Material S1), bottom-contact fishing expressed in subsurface SwAR was the second most important predictor, explaining an additional 8.2% of variation in macrofauna. Modeled near-bottom water temperature and % gravel showed the least direct effect on community structure among considered predictors (based on the marginal test), but were still significant and retained in the final model.

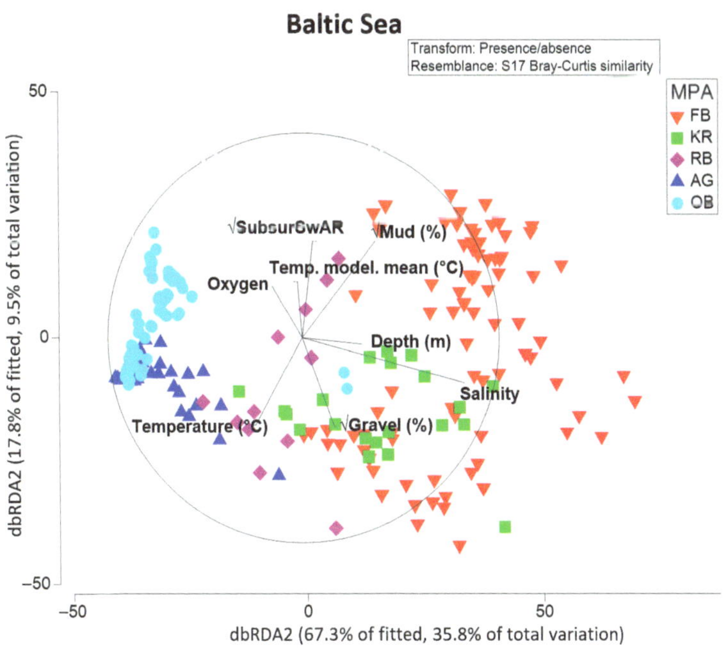


**Figure 8.** dbRDA ordination of stations in the Baltic Sea MPAs along environmental and anthropogenic (bottom-contact fishing expressed as subsurSwAR) drivers. Labeling according to the MPAs.

### 3.3. Endangered Species

In total, 110 of the 481 (22.9%) taxa found in all considered MPAs of the North and the Baltic Seas were at or near risk of various degrees of extinction (Table 6). In addition, 47 of the 481 (9.8%) taxa are endangered with the status "Threat of unknown Extent" according to the Red List based on [78,79]. Some examples were the chordate *Branchiostoma lanceolatum*, the echinoderm *Astropecten irregularis*, the polychaete *Fabriciola baltica* and the bivalve *Musculus discors*.

**Table 6.** Number of Taxa with a critical Red List status in the focus areas of the North Sea and Baltic Sea (source: [78,79]).

| Status | Both | North Sea | Baltic Sea |
|---|---|---|---|
| Near Threatened | 16 | 7 | 9 |
| Extremely Rare | 26 | 12 | 16 |
| Threat of Unknown Extent | 47 | 28 | 30 |
| Threatened | 10 | 6 | 5 |
| Highly Threatened | 8 | 5 | 5 |
| Threatened with Extinction | 3 | 1 | 2 |

#### 3.3.1. North Sea

Fifty-nine of the 262 taxa found in the North Sea were at or near risk of various degrees of extinction, accounting for 22.5% of the taxa (Table 6). Five taxa are endangered with the status "Highly Threatened": the gastropod *Buccinum undatum*, the polychaete *Sabellaria spinulosa*, and the bivalves *Ensis ensis*, *Mya truncate*, and *Spisula elliptica*. The crustaceans *Ebalia tumefacta* and *Lepas anatifera*, the bivalves *Arctica islandica*, *Ensis magnus*, and *Goodallia triangularis*, and the cnidarian *Alcyonium digitatum* are endangered with the status "Threatened". The status "Near Threatened" is allocated to the polychaete *Polygordius appendiculatus*, the decapod *Galathea intermedia*, the amphipod *Megaluropus agilis*, the bryozoan *Membranipora membranacea*, the echinoderms *Amphipholis squamata* and *Ophiothrix fragilis* and the gastropod *Acteon tornatilis*.

#### 3.3.2. Baltic Sea

In total, 67 of the 298 taxa found in the Baltic Sea were at or near risk of various degrees of extinction, accounting for 22.5% of the taxa (Table 6). Two taxa endangered with the status "Threatened with Extinction" are the cnidarian *Halcampa duodecimcirrata* and the bivalve *Macoma calcarea*. The bivalves *Modiolus modiolus* and *Mya truncata*, the gastropods *Boreotrophon truncatus* and *Buccinum undatum* and the polychaete *Euchone papillosa* are endangered with the status "Highly Threatened". The status "Threatened" is allocated to the amphipod *Monoporeia affinis*, the bivalves *Arctica islandica* (though not uncommon in the typical *Arctica* community in aphotic muddy sediment) and *Astarte montagui*, the hydrozoan *Halitholus yoldiaearcticae* and the polyplacophoran *Lepidochitona cinerea*.

## 4. Discussion

### 4.1. Species Richness and Composition in MPAs

We found in a total of 481 taxa in all nine MPAs in the North and the Baltic Seas (see full taxa list in Supplementary Material S1). Contrary to our expectation and the literature [40], 37.9% of the taxa were found exclusively in the North Sea MPAs, while 45.8% of our taxa were found exclusively in the Baltic Sea MPAs, and only 16.3% were shared by the two seas. According to the annotated checklist from Zettler et al. [40], 36.6% of taxa were shared between two seas, 48.7% occurred only in the NS, while 14.7% occurred only in the BS. Among taxa found only in the NS, both in Zettler et al. [40] and in our study are the polychaete *Aphrodita aculeata* and *Nephtys cirrosa*, the echinoderm *Echinocardium cordatum*, the decapod *Necora puber*, the amphipod *Megaluropus agilis* and the bivalve *Dosinia lupinus*. Example taxa specific to the BS in both studies are the bivalve *Astarte borealis*, the polychaetes *Fabriciola baltica* and *Spio arndti*, the gastropod *Lacuna parva* and the amphipod *Pontoporeia femorata*.

Some examples of species found in both areas and confirmed by both studies are the echinoderms *Amphiura filiformis* and *Echinocyamus pusillus*, the crustaceans *Carcinus maenas* and *Pagurus bernhardus*, the polychaetes *Eteone longa* and *Lanice conchilega* and the bivalves *Arctica islandica* and *Kurtiella bidentata*.

Multivariate analysis of the North Sea macrofauna data in our study revealed low similarity between the three considered MPAs, DGB, SAR and BRG. Based on ring dredge and beam trawl data from 33 stations in SAR and BRG sampled in 2020, Hahn et al. [80] published a checklist of benthic fauna that comprised 99 species from the phyla Mollusca, Arthropoda, Echinodermata, Annelida, Cnidaria, and Bryozoa (listed according to descending species number per group). In line with our results, Hahn et al. [80] also reported clear separation of species composition between the two areas. In contrast to our results, the higher species diversity found at BRG (compared to SAR) in that dataset was associated to lower bottom-contact fishing pressure. In our study, community composition within one area was more similar in DGB and SAR compared to the BRG. In the southeastern North Sea, differences in macrofauna biodiversity, intensively studied since the last century (e.g., [9–14,81]), are well represented by the four distinct macrofauna communities already mentioned in the introduction. The macrofauna communities of the DGB, BRG and SAR are assigned to the *Bathyporeia-Tellina*, the *Goniadella-Spisula*, and the *Amphiura-filiformis* communities, respectively [17]. According to Fiorentino et al. [17] taxa identified as characteristic for the *Bathyporeia-Tellina* community are *Bathyporeia elegans*, *Lanice conchilega*, *Tellina (Fabulina) fabula* and *Spiophanes bombyx*. These macrofauna species were also found in DGB MPA in our study. Among characteristic taxa of the *Goniadella-Spisula* community, Fiorentino et al. [17] listed *Aonides paucibrachiata*, *Branchiostoma lanceolatum*, *Pisione remota* and *Echinocyamus pusillus*, reported in our list for BRG MPA. In agreement with our results for SAR MPA, taxa listed as characteristic for the *Amphiura filiformis* community, apart from the name-giving species, were *Kurtiella bidentata*, *Nucula nitidosa* and *Phaxas pellucidus*.

Multivariate analysis for the Baltic Sea macrofauna suggested less distinct and therefore more similar community structure between the five MPAs compared to those in the North Sea. This is likely due to the inclusion of multiple habitats and biotopes with patchy distribution in each of the German Baltic Sea MPAs [50], and due to less distinct boundaries and gradual ecological transitions along environmental gradients between the adjacent communities. In agreement with previously reported increasing variability towards the entrance to the North Sea [33,34], our results showed gradually increasing similarity within the Baltic Sea MPAs with decreasing salinity from west to east. The FB MPA had the lowest mean similarity of 38.8% among the Baltic Sea MPAs as it includes stations from broad habitat types ranging from circa- and infralittoral mixed sediment (hard substrate) to sand and mud [50]. Characteristic species in the FB MPA apart from *A. islandica* included infaunal polychaete species like *Aricidea suecica* and *Levinsenia gracilis*, as well as *Diastylis rathkei*, *Varicorbula gibba* and *Abra alba*, in line with Schiewer [32], Gogina et al. [34] and Marx et al. [50]. Blue mussel *Mytilus edulis* and infaunal bivalve species like sand gaper *Mya arenaria* were among characteristic taxa for KR MPA, confirming the habitat distribution reported in [50]. Also, in accordance with Marx et al. [50], muddy sediment at RB MPA was dominated by the Baltic tellin *M. balthica*, whereas blue mussels were common in mixed and sand substrate. *Mytilus edulis* as well as *Gammarus salinus* associated with benthic vegetation were typical macrofauna species for AG. At sands of OB, characteristic infaunal bivalve species were *M. arenaria*, *M. balthica* and *Cerastoderma glaucum*; Mytilidae were also common.

The inventory presented here is just a snapshot documenting and comparing macrofauna diversity sampled within two years before the expected official closure for bottom-contact fishing took place. It seemed interesting to check how well our taxa list, which consists of two years of sampling, matches the compilation of the study from Zettler et al. [40], who included long-term databases provided by 11 marine research institutes and private consultancies. We have identified 9 taxa not recorded in the earlier published annotated checklist (marked in yellow in the full species list in Supplementary Material S1). Newly recorded 8 taxa were found in samples from the North Sea MPAs (*Epizoanthus papillosus* (SAR and DGB), *Macropodia tenuirostris* (BRG), *Lepas anatifera* (DGB), *Gilvossius tyrrhenus* (SAR and BRG), *Epimeria cornigera* (DGB), *Malmgrenia lunulata* and *Loimia ramzega* (BRG and DGB) and *Clymenura lankesteri* (DGB)), and one taxon occurred in the MPA Fehmarnbelt (*Alvania punctura*) in the Baltic Sea. *Cylista* sp. (formerly *Sagartia* sp.) was recorded only in the North Sea according to the checklist issued in 2018, whereas we recorded this taxon in our FB samples in 2020 and 2021. Such additional records in new studies were expected and discussed in [40], on the one hand through the introduction of new species and on the other through the spread of marine species from neighboring areas. In their review of non-indigenous species, Lackschewitz et al. [82] reported 159 marine and estuarine taxa, including both macrofauna and macroflora introduced by anthropogenic vectors as well as cryptic species. The number of introduced species detections increased from 9 before, to 48 within, and 65 after the 20th century, partly due to ship traffic, but also due to targeted monitoring programs and growth of taxonomic expertise. The highest number of macrofauna neobiota in the North Sea was represented by bryozoans and tunicates, in the Baltic Sea—by Ponto-Caspian amphipods and mysids. In addition to the actual immigration or even introduction of species, taxonomic revisions are also responsible for the fact that nomenclatures change or species were split up or deleted. One example of such a recently described species is the polychaete *L. ramzega* [83]. Spatial expansion of warm-temperate non-native species into German waters due to water temperature rise or changes in (de-)eutrophication [14,84,85], increasing number of newly introduced species [86], and disappearance of some native taxa are among expected drivers of future changes in species compositions. After closure for bottom-contact fishing in MPAs, those drivers might superimpose on effects of vanished bottom-contact fishing, and will act alongside high natural variability and unpredictable recruitment events particularly relevant in the young and temporally less stable Baltic Sea ecosystem.

### 4.2. Environmental Drivers

Salinity, temperature, sediment parameters describing fractions of sand, mud, gravel and shell, depth and bottom-contact fishing were together responsible for over 68% of the variation in the presence/absence structure of macrofauna data from the North Sea MPAs. The MPA SAR showed the highest taxa number (187 taxa) among North Sea MPAs. The MPA SAR covers 28% of the German EEZ in the North Sea and is characterized by different kinds of sediment parameters [67]. The correlation between sediment parameters and the distribution of macrofauna communities was found in many previous studies [18,40,51,81], although it is often linked to food supply. The distribution of sediments in the German EEZ is heterogeneous, consisting mainly of sand, mud or a mixture of both [87,88]. The mud content correlated, for example, with the abundances of the decapods *Nephrops norvegicus* and *Goneplax rhomboides* [87]. These two species together with high numbers of the holothurian *Paraleptopentacta elongata* were found at stations in the MPA SAR, where the mud content was higher (species list in Supplementary Material S1). MPAs DGB, BRG and AMB are sandbanks and are normally not characterized by a high biodiversity [89]. Due to its geographical position, the DGB MPA had a higher number of taxa than the other two, comprising species typical for the northern North Sea together with species typical for the southern North Sea [20,65]. Typical northern arctic-boreal species were the bivalve *Abra prismatica* and the polychaete *Ophelia limacina* [20]. However, climate change had led to a community shift even on the Doggerbank, and the abundance of northern species decreased at the MPA DGB [20].

Measured near-bottom salinity was by far the strongest factor of changes in (presence/absence-based) community structure on the scale of five studied Baltic Sea MPAs, alone responsible for over 33% of variation (in agreement with multiple studies reporting higher diversity with high salinity ([34,72] and references therein)). Without salinity, mud fraction could explain over 24%. Surprisingly, fishing intensity (subsurface SwAR) was the second most important predictor, explaining 8% of the cumulative effect, followed by modeled long-term averaged temperature (4%), whereas sediment parameters describing fractions of gravel and mud as well as depth each added no more than 2% to the cumulative explained variation. On one hand, this indirect confirmation of trawling impact on macrofaunal biodiversity is in line with recent findings of Bradshaw et al. [90]. This study highlighted for the Swedish part of the southern Baltic Sea that environmental variables (including salinity) affected fauna more than trawling (we will discuss this further in the next section). On the other hand, sediment parameters and depth are commonly considered major environmental forcing factors for macrofauna distribution in the Baltic; therefore, such a small partial effect here, outperformed even by modeled temperature, was somewhat unexpected in the context of previous findings (see e.g., [34]). This discrepancy is likely explained by certain redundancy of variation in those drivers and variation in salinity as the main predictor in this particular dataset, meaning they are capturing similar aspects of the variation in the presence/absence-based community structure (presumably, those covariates would also have more additional explanatory power for abundance- or biomass-based structure).

### 4.3. Bottom-Contact Fishing Intensity

Overall bottom-contact fishing intensity in the considered North Sea MPAs was highest at AMB (Table 3). Bottom-contact fishing seemed to change in ranking between the MPAs over time. Based on the spatial distribution of the surSwAR and subsurSwAR data [62,63], the bottom-contact fishing at the DGB was second highest, followed by the SAR and BRG (Table 3). Even though there is a significant difference between the bottom-contact fishing in the North Sea MPAs, it alone explained over 33.6% (subsurSwAR) of variation in the presence/absence data. In the Baltic Sea, bottom-contact fishing alone could explain 13% of macrofauna variation. It was highest in the Fehmarnbelt MPA (Table 3), followed by Western Rønne Bank (where trawling was mainly active on the muddy northwestern side of the major MPA area) and Odra Bank. It was substantially lower in the Kadetrinne (where shipping traffic is particularly intense) and in the Adler Ground MPA (characterized by riffs avoided by fishers due to gear damage risk). Where bottom-contact fishing occurs, it is often found to be among the most significant disturbances of macrofauna taxonomic [39] and functional composition [91], resulting in clear declines in benthic abundance and species richness [18,92]. Remarkably, despite including the most heavily trawled spots for our Baltic Sea study area, the Fehmarnbelt MPA provided home for the highest number of recorded species (264 taxa) and showed the largest variation in assemblage composition between its stations, not least due to its transition position, variety of habitat types, and highest salinity among Baltic Sea focus areas (see [93]). Generally, in the North and the Baltic Seas, the mobile bottom-contact fishing intensity has more or less decreased since the early 2000s (see Supplementary Material S2 refs. [62–64,74]). In the North Sea, the implementation of the Natura 2000 directive seems to be having an effect, but no uniform consensus has yet been implemented to bring fisheries and nature conservation together in a coherent way (see also [94]).

Here we have not studied the influence of other anthropogenic drivers, but among other factors communities are also affected by the construction and exploitation of offshore wind farms (OWF), marine traffic, heavy metal pollution, and changes in oxygen conditions [95]. Large-scale development of OWF has an impact on marine biodiversity due to changes in sediment characteristics and the creation of artificial reefs, the latter causing a doubling of species richness and an increase of abundance by two orders of magnitude. Furthermore, it leads to a decrease or cessation of bottom-contact fishing, prohibited in many OWFs, though fishing avoidance benefits there have yet to be proved [95].

Though temporal variability could not be explicitly analyzed based on datasets considered in our study, it is worth noting that the decline of bottom-contact fishing may have different effects on the macrofauna, depending on how bottom-contact fishing and environmental drivers change and how quickly the communities can adapt. In protected areas where no bottom-contact fishing took place, certain fish species and their macrofauna prey have the opportunity to recover. However, the cumulative effect of decreased direct bottom-contact fishing, changes of the environmental drivers and increased predation pressure may be difficult to disentangle. For macrofauna, especially for epifauna, improved food supply can have a positive impact on population dynamics. Decreased bottom-contact fishing can allow the restoration of mussel banks and seagrass beds.

Some species can adapt quickly to changing environmental conditions (climatic changes), while others are more sensitive [96,97]. The response of the species is driven by the different life histories (growth or age at maturity), differences in morphology (shape and structure) and ecological attributes (like mobility and position on/within the sediment) [42]. The effects of extreme events, for example cold winters in the North Sea can influence habitats for several years, and recovery time of the macrofauna communities from trawling in such disturbed habitats may take just as long [87,98]. When bottom-contact fishing activities change, e.g., in response to sustainable management, changes in community composition and density are expected to follow, but how quick and strong the response will be is dependent on the region. Species that can recover quickly after bottom-contact fishing distribution are decapods like *Crangon crangon*, *Carcinus maenas*, *Corystes cassivelaunus* and *Pagurus bernhardus* [42]. Species with a long lifespan and fragile morphology like bivalves and sessile species (ascidians and bryozoans) showed no short-term recovery after bottom-contact fishing events [42,86,98]. Long-term monitoring is thus crucial to understand the ecosystem changes and to develop appropriate protection measures for benthic macrofauna and their habitats.

To assess the possible impact of bottom-contact fishing on the populations of "key species" [69,70] in the MGF focus areas of the Baltic Sea (*A. islandica* in the FB, *M. balthica* in the RB, and *M. balthica* and *M. arenaria* in the OB), their size-frequency distributions were documented to allow comparison with data planned to be obtained after the fishery exclusion. In FB, the absence of medium size classes (10–30 mm) of *A. islandica* was noticeable, suggesting only occasional mass recruitment success. This lack of cohorts may threaten the continuous development of a stable population, with large, old mussels ensuring its continued existence and dominance. In addition, shell damage from trawling was evaluated [99], suggesting a significant negative impact caused particularly to larger individuals of *A. islandica* by high mechanical forces (traction, pressure) while towing or hauling up the net.

ICES datasets are only of limited use for ecosystem impact studies due to limited spatial resolution. In relation to bottom-contact fishing, these datasets only map the biomass brought ashore, not the biomass (including bycatch) actually removed from the habitat [100–102]. For the scale of this study, it currently remains the best available data source, though for small case studies, acoustically derived bottom-contact fishing indices can be the best alternative [103].

Limitations and possible biases of our results should draw attention to differences in methodologies used (overall standardized, but featuring regional details, particularly as in the Baltic in- and epifauna were always sampled together, whereas separate targeting of one or another was more common for the North Sea, as explained in detail in the Materials and Methods section) and the temporal and spatial distribution of stations. The relatively small number of environmental factors used in our statistical analyses, as well as the absence of some other important factors, such as sediment organic matter content or chlorophyll A (that were not available for both studied regions), could have implications for our results, and would likely explain additional variation. Two factors were included for only one of the two regions, namely mean multiannual near-bottom water temperature from the GETM model for the Baltic [73] and the % of shell content > 2000 µm in the sediment for the North Sea. In contrast to the North Sea, where temperature measured at stations during sampling added over 16% to cumulative explained variation, in the Baltic Sea, modelled temperature explained an additional 4% of the cumulative effects on benthic macrofauna structure, whereas measured temperature, though significant, added only 0.7%. As samples in both regions were collected in different seasons and months of the year, these differences most likely reflect variations in thermal regimes and species composition. As for % of shell, it had no significance in the cumulative effect for community composition in the North Sea after considering the effects of salinity, temperature and % of sand fraction.

Syntheses of long-term monitoring data collected over the last two decades under the terms of BfN and other initiatives (already highlighted in the Introduction) should provide a more thorough description and understanding through future research. There is an intense discussion that future data collection, particularly in MPAs, should be revised and developed towards non-invasive sampling. Possibilities here include, for example, the collection of eDNA from water samples (e.g., [104]) or the observation of certain areas using underwater video [72,105]. Even though these methods were confirmed to have great complementary value (e.g., [105]), the taxonomic and quantitative resolution and reliability of data that they can deliver [106] seem, for now, to be insufficient, even for a snapshot assessment like ours. The development of new emerging methods (including the assessment of eDNA persistence and spatial representability, upbuilding of reference libraries and automation of imagery data processing) should go along with their comparison with traditional morphological approaches to support consistency [107,108]. For solid scientific comparison before and after fishing closure, it is still inevitable and essential to use bottom-contacting and dragged scientific equipment (at least in a limited amount) even after the closure. For now, keeping "invasive" scientific gear out of the MPAs may only hamper efficient monitoring. In particular, the importance of long-term ecological research sites in these areas should also be emphasized.

## 5. Conclusions

The baseline inventory of macrobenthic species presented here is important for assessing future faunal changes. Studying biodiversity across German NS and BS MPAs in a collective approach is particularly important for understanding ecological connectivity, integrating conservation strategies, and robustly evaluating the resilience of these ecosystems. A collaborative viewpoint enables the identification of shared species, fostering a more precise understanding of species migration, interactions, and their contributions to overall ecosystem vitality. By adopting a joint perspective, conservation efforts can be strategically enhanced on a broader scale, taking into consideration common threats and species distribution patterns across both regions, thus facilitating more effective planning and management strategies.

**Supplementary Materials:** The following supporting information can be downloaded at: https://www.mdpi.com/article/10.3390/biology13060389/s1, Supplementary Materials S1: Joint species list for the German MPAs in the North and Baltic Seas in 2020–2022. Temporarily resolved list for the Baltic Sea. Results of DistLM. Supplementary Materials S2: Temporal dynamics of trawling activity. Biodiversity values expressed as mean Shannon–Wiener indices.

**Author Contributions:** Conceptualization, M.G., S.J.H., R.O. and I.K.; methodology, M.G., S.J.H. and R.O.; validation, M.G. and R.O.; formal analysis, M.G., R.O. and S.J.H.; investigation, M.G., R.O. and S.J.H.; data curation, all; writing—original draft preparation, M.G., R.O. and S.J.H.; writing—review and editing, M.G., S.J.H., R.O., A.B., S.F., I.K., M.P., K.R., M.S. and M.L.Z.; visualization, M.G., R.O., S.J.H. and M.L.Z.; project administration, I.K., M.G., K.R., M.L.Z., S.F., A.B. and M.S.; funding acquisition, I.K., M.G., K.R., M.L.Z., S.F., A.B. and M.S. All authors have read and agreed to the published version of the manuscript.

**Funding:** This research was funded by the German Federal Ministry of Education and Research grants 03F0848A and 03F0937A (MGF Baltic Sea projects), as well as grant 03F0847D (MGF North Sea projects), and the Federal Agency for Nature Conservation (BfN), grant number 3 519 532 202 (LEGRA).

**Institutional Review Board Statement:** Not applicable.

**Informed Consent Statement:** Not applicable.

**Data Availability Statement:** The original contributions presented in the study are included in the article/Supplementary Material, further inquiries can be directed to the corresponding author/s. The LEGRA infauna data from the Baltic Sea are also going to be available via an online portal of the Federal Agency for Nature Conservation at the end of 2024 (contact: kathrin.heinicke@bfn.de), and the MGF data are going to be available via an online portal, PANGEA, by the end of 2024 (contact: mayya.gogina@io-warnemuende.de). CTD data from the DGB of the MGF-North Sea project benthos and macrofauna team by Senckenberg are available at https://doi.pangaea.de/10.1594/PANGAEA.964365 (accessed on 22 March 2024).

**Acknowledgments:** We gratefully acknowledge the motivations from BfN, BMBF and DAM (Deutsche Alliance Meeresforschung), the valuable work of all colleagues involved in the sampling campaigns and laboratory analyses, including the staff of research vessels Senckenberg, Elisabeth Mann Borgese and Alkor, and Benthos Laboratories of the Leibniz Institute for Baltic Sea Research Warnemünde and Senckenberg. We express our thanks to the anonymous reviewers for numerous constructive comments that greatly helped to improve the earlier version of the manuscript.

**Conflicts of Interest:** The authors declare no conflicts of interest. The funders had no role in the design of the study; in the collection, analyses, or interpretation of data; in the writing of the manuscript; or in the decision to publish the results.

## References

1. Ducrotoy, J.P.; Elliott, M.; de Jonge, V.N. The North Sea. *Mar. Pollut. Bull.* **2000**, *41*, 5–23. [CrossRef]
2. Feistel, R.; Weinreben, S.; Wolf, H.; Seitz, S.; Spitzer, P.; Adel, B.; Nausch, G.; Schneider, B.; Wright, D. Density and Absolute Salinity of the Baltic Sea 2006–2009. *Ocean. Sci.* **2009**, *6*, 3–24. [CrossRef]
3. Bendtsen, J.; Gustafsson, K.E.; Söderkvist, J.; Hansen, J.L.S. Ventilation of bottom water in the North Sea–Baltic Sea transition zone. *J. Mar. Syst.* **2009**, *75*, 138–149. [CrossRef]
4. Inácio, M.; Karnauskaitė, D.; Baltranaitė, E.; Kalinauskas, M.; Bogdzevič, K.; Gomes, E.; Pereira, P. Ecosystem services of the Baltic Sea: An assessment and mapping perspective. *Geogr. Sustain.* **2020**, *1*, 256–265. [CrossRef]
5. Selim, S.A.; Blanchard, J.L.; Bedford, J.; Webb, T.J. Direct and indirect effects of climate and fishing on changes in coastal ecosystem services: A historical perspective from the North Sea. *Reg. Environ. Chang.* **2014**, *16*, 341–351. [CrossRef]
6. Geburzi, J.C.; Heuer, N.; Homberger, L.; Kabus, J.; Moesges, Z.; Ovnbeck, K.; Brandis, D.; Ewers, C. An environmental gradient dominates ecological and genetic differentiation of marine invertebrates between the North and Baltic Sea. *Ecol. Evol.* **2022**, *12*, e8868. [CrossRef] [PubMed]
7. Medvedev, I.P.; Rabinovich, A.B.; Kulikov, E.A. Tidal oscillations in the Baltic Sea. *Oceanology* **2013**, *53*, 526–538. [CrossRef]
8. Johannesson, K.; Smolarz, K.; Grahn, M.; André, C. The future of Baltic Sea populations: Local extinction or evolutionary rescue? *Ambio* **2011**, *40*, 179–190. [CrossRef] [PubMed]
9. Duineveld, G.C.A.; Künitzer, A.; Niermann, U.; de Wilde, P.A.W.J.; Gray, J.S. The macrobenthos of the north sea. *Neth. J. Sea Res.* **1991**, *28*, 53–65. [CrossRef]
10. Heip, C.; Basford, D.; Craeymeersch, J.A.; Dewarumez, J.M.; Dorjes, J.; Wilde, P.; Duineveld, G.; Eleftheriou, A.; Herman, P.M.J.; Niermann, U.; et al. Trends in biomass, density and diversity of North Sea macrofauna. *ICES J. Mar. Sci.* **1992**, *49*, 13–22. [CrossRef]
11. Heip, C.; Craeymeersch, J.A. Benthic community structures in the North Sea. *Helgol. Meeresunters.* **1995**, *49*, 313–328. [CrossRef]
12. Künitzer, A.; Basford, D.; Craeymeersch, J.A.; Dewarumez, J.M.; Dörjes, J.; Duineveld, G.C.A.; Eleftheriou, A.; Heip, C.; Herman, P.; Kingston, P.; et al. The benthic infauna of the North Sea: Species distribution and assemblages. *ICES J. Mar. Sci.* **1992**, *49*, 127–143. [CrossRef]
13. Reiss, H.; Degraer, S.; Duineveld, G.C.A.; Kröncke, I.; Aldridge, J.; Craeymeersch, J.A.; Eggleton, J.D.; Hillewaert, H.; Lavaleye, M.S.S.; Moll, A.; et al. Spatial patterns of infauna, epifauna, and demersal fish communities in the North Sea. *ICES J. Mar. Sci.* **2009**, *67*, 278–293. [CrossRef]
14. Meyer, J.; Nehmer, P.; Moll, A.; Kröncke, I. Shifting south-eastern North Sea macrofauna community structure since 1986: A response to de-eutrophication and regionally decreasing food supply? *Estaurine Coast. Shelf Sci.* **2018**, *213*, 115–127. [CrossRef]
15. Glémarec, M. The benthic communities of the European North Atlantic continental shelf. *Oceanogr. Mar. Biol. Annu. Rev.* **1973**, *11*, 263–289.
16. Craeymeersch, J.A.; Heip, C.H.R.; Buijs, J. Atlas of North Sea benthic infauna. Based on the 1986 North Sea Benthos Survey. *ICES Coop. Res. Rep.* **1997**, *218*, 90. [CrossRef]
17. Fiorentino, D.; Pesch, R.; Guenther, C.P.; Gutow, L.; Holstein, J.; Dannheim, J.; Ebbe, B.; Bildstein, T.; Schroeder, W.; Schuchardt, B.; et al. A 'fuzzy clustering' approach to conceptual confusion: How to classify natural ecological associations. *Mar. Ecol. Prog. Ser.* **2017**, *574*, 17–30. [CrossRef]
18. Kröncke, I.; Reiss, H.; Eggleton, J.D.; Aldridge, J.; Bergman, M.J.N.; Cocrane, S.; Craeymeersch, J.A.; Degraer, S.; Desroy, N.; Dewaumez, J.M.; et al. Changes in North Sea macrofauna communities and species distribution between 1986 and 2000. *Estaurine Coast. Shelf Sci.* **2011**, *94*, 1–15. [CrossRef]

19. Shojaei, M.G.; Gutow, L.; Dannheim, J.; Pehlk, H.; Brey, T. Functional Diversity and Traits Assembly Patterns of Benthic Macrofaunal Communities in the Southern North Sea. In *Towards an Interdisciplinary Approach in Earth System Science*; Lohmann, G., Meggers, H., Unnithan, V., Wolf-Gladrow, D., Notholt, J., Bracher, A., Eds.; Springer: Berlin/Heidelberg, Germany, 2015; 251p. [CrossRef]
20. Kröncke, I. Changes in Dogger Bank macrofauna communities in the 20th century caused by fishing and climate. *Estaurine Coast. Shelf Sci.* **2011**, *94*, 234–245. [CrossRef]
21. Rombouts, I.; Beaugrand, G.; Dauvin, J.C. Potential changes in benthic macrofaunal distributions from the English Channel simulated under climate change scenarios. *Estaurine Coast. Shelf Sci.* **2012**, *99*, 153–161. [CrossRef]
22. Edwards, M.; Beaugrand, G.; Reid, P.C.; Rowden, A.A.; Jones, M.B. Ocean climate anomalies and the ecology of the North Sea. *Mar. Ecol. Prog. Ser.* **2002**, *239*, 1–10. [CrossRef]
23. Dethardíng, G.G. *Systematisches Verzeichnis der Mecklenburgischen Conchylien*; Siemssen, A.C., Ed.; W. Bärensprung: Schwerin, Germany, 1794; p. 40.
24. Arnold, C.; Lenz, H. Erster allgemeiner Bericht über die im Jahre 1872 angestellten zoologisch-botanischen Untersuchungen der Travemünder Bucht. *Lübeckische Blätter* **1873**, *15*, 213–216.
25. Möbius, K. Die wirbellosen Thiere der Ostsee. *Jahresber. Komm. Wiss. Unters. Dtsch. Meere Kiel* **1873**, *1*, 97–144.
26. Lenz, H. Die wirbellosen Thiere der Travemünder Bucht. Theil II. Resultate der im Auftrage der Freien und Hanse-Stadt Lübeck angestellten Schleppnetzuntersuchungen. In *Vierter Bericht der Commission zur Wissenschaftlichen Untersuchung der Deutschen Meere, in Kiel Für Die Jahre 1877 bis 1881 7.-11. Jahrgang*; I. Abth. Paul Parey: Berlin, Germany, 1882; Volume 4, pp. 169–184.
27. Hagmeier, A.V. Die Bodenfauna der Ostsee im April 1929 nebst einigen Vergleichen mit April 1925 und juli 1926. *Berichte Dtsch. Wiss. Kommision Meeresforsch. NF* **1930**, *5*, 156–173.
28. Schulz, S. Benthos and sediment in Mecklenburg Bay. *Beitr. Meereskunde* **1969**, *24–25*, 15–55.
29. Löwe, F.K. Quantitative Benthosuntersuchungen in der Arkonasee. *Mitteilungen Mus. Naturkunde Berlin. Zool. Mus. Inst. Spez. Zool.* **1963**, *39*, 247–349. [CrossRef]
30. Zettler, M.L.; Röhner, M. Verbreitung und Entwicklung des Makrozoobenthos der Ostsee zwischen Fehmarnbelt und Usedom—Daten von 1839 bis 2001. In *Die Biodiversität in Nord- und Ostsee. Band 3. Bericht BfG, 1421*; Leuchs, H., König, B., Eds.; Bundesanstalt für Gewässerkunde: Koblenz, Germany, 2004; pp. 1–175.
31. Zenkevitch, L. *Biology of the Seas of the U.S.S.R.*; Allen & Unwin: London, UK, 1963; p. 953. [CrossRef]
32. Schiewer, U. *Ecology of Baltic Coastal Waters*, 1st ed.; Ecological Studies; Springer: Berlin/Heidelberg, Germany, 2008; Volume 19, p. 430.
33. Zettler, M.L.; Karlsson, A.; Kontula, T.; Gruszka, P.; Laine, A.O.; Herkül, K.; Schiele, K.S.; Maximov, A.; Haldin, J. Biodiversity gradient in the Baltic Sea: A comprehensive inventory of macrozoobenthos data. *Helgol. Mar. Res.* **2014**, *68*, 49–57.
34. Gogina, M.; Nygård, H.; Blomqvist, M.; Daunys, D.; Josefson, A.B.; Kotta, J.; Maximov, A.; Warzocha, J.; Yermakov, V.; Gräwe, U.; et al. The Baltic Sea scale inventory of benthic faunal communities. *ICES J. Mar. Sci.* **2016**, *73*, 1196–1213. [CrossRef]
35. Salomon, M.; Schumacher, J. Maritime spatial planning: Germany as a forerunner in ecosystem-based management? *Plan. Pract. Res.* **2022**, *37*, 739–750. [CrossRef]
36. Krause, J.C.; Boedeker, D.; Backhausen, I.; Heinicke, K.; Groß, A.; von Nordheim, H. Rationale behind site selection for the NATURA 2000 network in the German EEZ. In *Progress in Marine Conservation in Europe*; von Nordheim, H., Boedeker, D., Krause, J.C., Eds.; Springer: Berlin/Heidelberg, Germany, 2006; pp. 65–96. [CrossRef]
37. Pedersen, S.A.; Fock, H.; Krause, J.; Pusch, C.; Sell, A.L.; Böttcher, U.; Rogers, S.I.; Sköld, M.; Skov, H.; Podolska, M.; et al. Natura 2000 sites and fisheries in German offshore waters. *ICES J. Mar. Sci.* **2009**, *66*, 155–169. [CrossRef]
38. Krause, J.; Wollny-Goerke, K.; Boller, F.; Hauswirth, M.; Heinicke, K.; Herrmann, C.; Körber, P.; Narberhaus, I.; Richter-Kemmermann, A. The German marine protected areas in the North Sea and Baltic Sea. *Nat. Landsch.* **2011**, *9–10*, 397–409.
39. Bildstein, T.; Schuchardt, B.; Bleich, S.; Bennecke, S.; Schückel, S.; Huber, A.; Dierschke, V.; Koschinski, S.; Darr, A. Die Meeresschutzgebiete in der deutschen AWZ der Ostsee—Beschreibung und Zustandsbewertung. *BfN-Skripten* **2020**, *553*, 1–497. [CrossRef]
40. Zettler, M.L.; Beermann, J.; Dannheim, J.; Ebbe, B.; Grotjahn, M.; Günther, C.P.; Gusky, M.; Kind, B.; Kröncke, I.; Kuhlenkamp, R.; et al. An annotated checklist of macrozoobenthic species in German waters of the North and Baltic Seas. *Helgol. Mar. Res.* **2018**, *72*, 5. [CrossRef]
41. Kaiser, M.J.; Ramsay, K.; Richardson, C.A.; Spence, F.E.; Brand, A.R. Chronic fishing disturbance has changed shelf sea benthic community structure. *J. Anim. Ecol.* **2000**, *69*, 494–503. [CrossRef]
42. Sciberras, M.; Hiddink, J.G.; Jennings, S.; Szostek, C.L.; Hughes, K.M.; Kneafsey, B.; Clarke, L.J.; Ellis, N.; Rijnsdorp, A.D.; McConnaughey, R.S.; et al. Response of benthic fauna to experimental bottom fishing: A global meta-analysis. *Fish Fish.* **2018**, *19*, 698–715. [CrossRef]
43. Hiddink, J.G.; Kaiser, M.J.; Sciberras, M.; McConnaughey, R.A.; Mazor, T.; Hilborn, R.; Collie, J.S.; Pitcher, C.R.; Parma, A.M.; Suuronen, P.; et al. Selection of indicators for assessing and managing the impacts of bottom trawling on seabed habitats. *J. Appl. Ecol.* **2020**, *57*, 1199–1209. [CrossRef]
44. Mazor, T.; Pitcher, C.R.; Rochester, W.; Kaiser, M.J.; Hiddink, J.G.; Jennings, S.; Amoroso, R.; McConnaughey, R.A.; Rijnsdorp, A.D.; Parma, A.M.; et al. Trawl fishing impacts on the status of seabed fauna in diverse regions of the globe. *Fish Fish.* **2021**, *22*, 72–86. [CrossRef]

45. Rijnsdorp, A.D.; Bolam, S.G.; Garcia, C.; Hiddink, J.G.; Hintzen, N.T.; van Denderen, P.D.; van Kooten, T. Estimating sensitivity of seabed habitats to disturbance by bottom trawling based on the longevity of benthic fauna. *Ecol. Appl.* **2018**, *28*, 1302–1312. [CrossRef] [PubMed]
46. van Denderen, P.D.; Törnroos, A.; Sciberras, M.; Hinz, H.; Friedland, R.; Lasota, R.; Mangano, M.C.; Robertson, C.; Valanko, S.; Hiddink, J.G. Effects of bottom trawling and hypoxia on benthic invertebrate communities. *Mar. Ecol. Prog. Ser.* **2022**, *694*, 13–27. [CrossRef]
47. Beermann, J.; Gutow, L. Langzeitbeobachtung benthischer Lebensräume in küstenfernen Gebieten der deutschen Nordsee (LABEL). In Proceedings of the Koordinatorentreffen des Bundesamtes für Naturschutz (BfN), Vilm, Germany, 18–20 March 2019; hdl:10013/epic.09150f67-8b69-43f6-925d-33a1fc0eac5a.
48. Bonthond, G.; Beermann, J.; Gutow, L.; Neumann, A.; Barboza, F.R.; Desiderato, A.; Fofonova, V.; Helber, S.; Khodami, S.; Kraan, C.; et al. Benthic microbial biogeographic trends in the North Sea are shaped by an interplay of environmental drivers and bottom trawling effort. *ISME Commun.* **2023**, *3*, 132. [CrossRef]
49. Gutow, L.; Günther, C.P.; Ebbe, B.; Schückel, S.; Schuchardt, B.; Dannheim, J.; Darr, A.; Pesch, R. Structure and distribution of a threatened muddy biotope in the south-eastern North Sea. *J. Environ. Manag.* **2020**, *255*, 109876. [CrossRef] [PubMed]
50. Marx, D.; Feldens, A.; Papenmeier, S.; Feldens, P.; Darr, A.; Zettler, M.L.; Heinicke, K. Habitats and Biotopes in the German Baltic Sea. *Biology* **2024**, *13*, 6. [CrossRef] [PubMed]
51. Meyer, J.; Kröncke, I.; Bartholomä, A.; Dippner, J.W.; Schückel, U. Long-term changes in species composition of demersal fish and epibenthic species in the Jade area (German Wadden Sea/southern North Sea) since 1972. *Estaurine Coast. Shelf Sci.* **2016**, *181*, 284–293. [CrossRef]
52. de Groot, S.J.; Lindeboom, H.J. Environmental Impact of bottom gears on benthic fauna in relation to natural resources management and protection of the North Sea. *NIOZ Rapp.* **1994**, *11*, 254.
53. Lindeboom, H.J.; de Groot, S.J. The effects of different types of fisheries on the North Sea and Irish Sea benthic ecosystems (IMPACT II). *NIOZ-Rapp.* **1998**, *1*, 404.
54. Greenstreet, S.; Robinson, L.; Piet, G.; Craeymeersch, J.; Callaway, R.; Reiss, H.; Ehrich, S.; Kröncke, I.; Fraser, H.; Lancaster, J.; et al. The ecological disturbance caused by fishing in the North Sea. *FRS Collab. Rep.* **2007**, *04/07*, 169.
55. Sköld, M.; Göransson, P.; Jonsson, P.; Bastardie, F.; Blomqvist, M.; Agrenius, S.; Hiddink, J.G.; Nilsson, H.C.; Bartolino, V. Effects of chronic bottom trawling on soft-seafloor macrofauna in the Kattegat. *Mar. Ecol. Prog. Ser.* **2018**, *586*, 41–55. [CrossRef]
56. Rumohr, H.; Kujawski, T. The impact of trawl fishery on the epifauna of the southern North Sea. *ICES J. Mar. Sci.* **2000**, *57*, 1389–1394. [CrossRef]
57. Duineveld, G.C.A.; Bergman, M.J.N.; Lavaleye, M.S.S. Effects of an area closed to fisheries on the composition of the benthic fauna in the southern North Sea. *ICES J. Mar. Sci.* **2007**, *64*, 899–908. [CrossRef]
58. Arntz, W.E.; Weber, W. *Cyprina islandica* (Mollusca: Bivalvia) als Nahrung von Dorsch und Kliesche in der Kieler Bucht. *Berichte Dtsch. Wiss. Komm. Meeresforsch.* **1970**, *21*, 193–209.
59. Krost, P. The Impact of Otter-Trawl Fishery on Nutrient Release from the Sediment and Macrofauna of Kieler Bucht (Western Baltic). Ph.D. Thesis, Christian-Albrechts-Universität Kiel, Kiel, Germany, 1990.
60. Rumohr, H.; Krost, P. Experimental evidence of damage to benthos by bottom trawling with special reference to Arctica islandica. *Meeresforsch. Rep. Mar. Res.* **1991**, *33*, 340–345.
61. Sonnewald, M.; Türkay, M. Composition of the Epibenthic Decapod Crustacean Megafauna of the German Exclusive Economic Zone: Comparison and Analysis of Past and Recent Surveys. *Crustaceana* **2017**, *90*, 1193–1210. [CrossRef]
62. ICES. Data for OSPAR request on the production of spatial data layers of fishing intensity/pressure. *Data Outputs* **2021**. [CrossRef]
63. ICES. ICES Data Call for VMS/Log Book Data 2009–2020. *Data Calls* **2021**. [CrossRef]
64. ICES. HELCOM request 2022 for spatial data layers on effort, fishing intensity and fishing footprint for the years 2016–2021. *Data Outputs* **2022**. [CrossRef]
65. Diesing, M.; Ware, S.; Foster-Smith, R.; Stewart, H.; Long, D.; Vanstaen, K.; Forster, R.; Morando, A. *Understanding the Marine Environment—Seabed Habitat Investigations of the Dogger Bank Offshore Draft SAC*; JNCC Report No. 429; Joint Nature Conservation Committee: Peterborough, UK, 2009; p. 89. Available online: http://www.jncc.gov.uk/ (accessed on 22 March 2024).
66. Coolen, J.W.P.; Bos, O.G.; Glorius, S.; Lengkeek, W.; Cuperus, J.; van der Weide, B.; Agüera, A. Reefs, sand and reef-like sand: A comparison of the benthic biodiversity of habitats in the Dutch Borkum Reef Grounds. *J. Sea Res.* **2015**, *103*, 84–92. [CrossRef]
67. Galvez, D.S.; Papenmeier, S.; Sander, L.; Hass, H.C.; Fofonova, V.; Bartholomä, A.; Wiltshire, K.H. Ensemble Mapping and Change Analysis of the Seafloor Sediment Distribution in the Sylt Outer Reef, German North Sea from 2016 to 2018. *Water* **2021**, *13*, 2254. [CrossRef]
68. Ahyong, S.; Boyko, C.B.; Bailly, N.; Bernot, J.; Bieler, R.; Brandão, S.N.; Daly, M.; De Grave, S.; Gofas, S.; Hernandez, F.; et al. World Register of Marine Species. Available online: https://www.marinespecies.org/imis.php?dasid=1447&doiid=170 (accessed on 4 January 2024).
69. Joseph, L.; Cusson, M. Resistance of benthic intertidal communities to multiple disturbances and stresses. *Mar. Ecol. Prog. Ser.* **2015**, *534*, 49–64. [CrossRef]
70. Karlson, A.M.L.; Pilditch, C.A.; Probert, P.K.; Leduc, D.; Savage, C. Large Infaunal Bivalves Determine Community Uptake of Macroalgal Detritus and Food Web Pathways. *Ecosystems* **2021**, *24*, 384–402. [CrossRef]
71. Rees, H.L. (Ed.) Guidelines for the study of the epibenthos of subtidal environments. *ICES Tech. Mar. Environ. Sci.* **2009**, *42*, 88

72. Beisiegel, K.; Darr, A.; Zettler, M.L.; Friedland, R.; Gräwe, U.; Gogina, M. Spatial variability in subtidal hard substrate assemblages across horizontal and vertical gradients: A multi-scale approach by seafloor imaging. *Mar. Ecol. Prog. Ser.* **2020**, *633*, 23–36. [CrossRef]
73. Gräwe, U.; Naumann, M.; Mohrholz, V.; Burchard, H. Anatomizing one of the largest saltwater inflows into the Baltic Sea in December 2014. *J. Geophys. Res. Ocean.* **2015**, *120*, 7676–7697. [CrossRef]
74. ICES. Baltic Sea Ecoregion—Ecosystem overview. ICES Advice: Ecosystem Overviews. *Report* **2022**. [CrossRef]
75. Clarke, K.; Warwick, R. *An Approach to Statistical Analyses and Interpretation. Change in Marine Communities*, 2nd ed.; RIMER-E Ltd.; Plynouth Marine Laboratory: Plymouth, UK, 2001; pp. 6-1–6-14.
76. Eigaard, O.E.; Bastardie, F.; Breen, M.; Dinesen, G.E.; Hintzen, N.T.; Laffargue, P.; Mortensen, L.O.; Nielsen, J.R.; Nilsson, H.C.; O'Neill, F.G.; et al. Estimating seabed pressure from demersal trawls, seines, and dredges based on gear design and dimensions. *ICES J. Mar. Sci.* **2016**, *73*, i27–i43. [CrossRef]
77. McArdle, B.H.; Anderson, M.J. Fitting multivariate models to community data: A comment on distance-based redundancy analysis. *Ecology* **2001**, *82*, 290–297. [CrossRef]
78. Rachor, E.; Bönsch, R.; Boos, K.; Gosselck, F.; Grotjahn, M.; Günther, C.P.; Gusky, M.; Gutow, L.; Heiber, W.; Jantschik, P.; et al. Rote Liste der bodenlebenden wirbellosen Meerestiere. *Naturschutz Biol. Vielfalt* **2013**, *70*, 81–176.
79. Rote-Liste-Zentrum.de. Available online: https://www.rote-liste-zentrum.de/en/Categories-1711.html (accessed on 27 September 2023).
80. Hahn, S.J.; Brandt, A.; Sonnewald, M. Annotated checklist and biodiversity analysis of benthic fauna at Sylt Outer Reef and Borkum Reef Ground (North Sea). *Check List* **2022**, *18*, 593–628. [CrossRef]
81. Rachor, E.; Nehmer, P. *Erfassung und Bewertung Ökologisch Wertvoller Lebensräume in der Nordsee*; Wegener Institute for Polar and Marine Research: Bremerhaven, Germany, 2003; pp. 1–175.
82. Lackschewitz, D.; Reise, K.; Buschbaum, C.; Karez, R. *Neobiota der Deutschen Nord- und Ostseeküste. Eingeschleppte Arten in Deutschen Küstengewässern*; Landesamt für Landwirtschaft, Umwelt und ländliche Räume Schleswig-Holstein: Flintbek, Germany, 2022; pp. 1–394.
83. Lavesque, N.; Bonifácio, P.; Londoño-Mesa, M.H.; Le Garrec, V.; Grall, J. *Loimia ramzega* sp. nov.; a new giant species of Terebellidae (Polychaeta) from French waters (Brittany, English Channel). *J. Mar. Biol. Assoc. United Kingd.* **2017**, *97*, 935–942. [CrossRef]
84. Viitasalo, M.; Bonsdorff, E. Global climate change and the Baltic Sea ecosystem: Direct and indirect effects on species, communities and ecosystem functioning. *Earth Syst. Dynam.* **2022**, *13*, 711–747. [CrossRef]
85. Singer, A.; Bijleveld, A.I.; Hahner, F.; Holthuijsen, S.J.; Hubert, K.; Kerimoglu, O.; Kleine Schaars, L.; Kröncke, I.; Lettmann, K.A.; Rittweg, T.; et al. Long-term response of coastal macrofauna communities to deeutrophication and sea level rise mediated habitat changes (1980s versus 2018). *Front. Mar. Sci.* **2023**, *9*, 963325. [CrossRef]
86. Zenetos, A.; Tsiamis, K.; Galandi, M.; Carvalho, N.; Bartilotti, C.; Canning-Clode, J.; Castriota, L.; Chainho, P.; Comas-González, R.; Costa, A.C.; et al. Status and Trends in the Rate of Introduction of Marine Non-Indigenous Species in European Seas. *Diversity* **2022**, *14*, 1077. [CrossRef]
87. Neumann, H.; Kröncke, I. The effect of temperature variability on ecological functioning of epifauna in the German Bight. *Mar. Ecol.* **2011**, *32*, 49–57. [CrossRef]
88. Laurer, W.-U.; Naumann, M.; Zeiler, M. *Erstellung der Karte zur Sedimentverteilung auf dem Meeresboden in der Deutschen Nordsee nach der Klassifikation von Figge (1981)*; Modul B—Dokumentation Nr. 1: LBEG—BSH—BGR; Geopotenzial Deutsche Nordsee: Hannover/Hamburg, Germany, 2012; p. 19. Available online: https://www.gpdn.de (accessed on 22 March 2024).
89. Beermann, J.; Gutow, L.; Wührdemann, S.; Heinicke, K.; Bildstein, T.; Jaklin, S.; Gusky, M.; Zettler, M.L.; Dannheim, J.; Pesch, R. Characterization and differentiation of sublittoral sandbanks in the southeastern North Sea. *Biodivers. Conserv.* **2023**, *32*, 2747–2768. [CrossRef]
90. Bradshaw, C.; Iburg, S.; Morys, C.; Sköld, M.; Pusceddu, A.; Ennas, C.; Jonsson, P.; Nascimento, F.J.A. Effects of bottom trawling and environmental factors on benthic bacteria, meiofauna and macrofaunal communities and benthic ecosystem processes. *Sci. Total. Environ.* **2024**, *921*, 171076. [CrossRef] [PubMed]
91. Beauchard, O.; Bradshaw, C.; Bolam, S.; Tiano, J.; Garcia, C.; De Borger, E.; Laffargue, P.; Blomqvist, M.; Tsikopoulou, I.; Papadopoulou, N.K.; et al. Trawling-induced change in benthic effect trait composition—A multiple case study. *Front. Mar. Sci.* **2023**, *10*, 1303909. [CrossRef]
92. McLaverty, C.; Eigaard, O.R.; Olsen, J.; Brooks, M.E.; Petersen, J.K.; Erichsen, A.C.; van der Reijden, K.; Dinesen, G.E. European Coastal Monitoring Programmes May Fail to Identify Impacts on Benthic Macrofauna Caused by Bottom Trawling. *J. Environ. Manag.* **2023**, *334*, 117510. [CrossRef] [PubMed]
93. Romoth, K.; Darr, A.; Papenmeier, S.; Zettler, M.L.; Gogina, M. Substrate heterogeneity as a trigger for species diversity in marine benthic assemblages. *Biology* **2023**, *12*, 825. [CrossRef] [PubMed]
94. Fock, H.O. Natura 2000 and the European Common Fisheries Policy. *Mar. Policy.* **2011**, *35*, 181–188. [CrossRef]
95. Li, C.; Coolen, J.W.P.; Scherer, L.; Mogollón, J.M.; Braeckman, U.; Vanaverbeke, J.; Tukker, A.; Steubing, B. Offshore Wind Energy and Marine Biodiversity in the North Sea: Life Cycle Impact Assessment for Benthic Communities. *Environ. Sci. Technol.* **2023**, *57*, 6455–6464. [CrossRef]

96. González-Irusta, J.M.; de la Torriente, A.; Punzón, A.; Blanco, M.; Serrano, A. Determining and mapping species sensitivity to trawling impacts: The BEnthos Sensitivity Index to Trawling Operations (BESITO). *ICES J. Mar. Sci.* **2018**, *75*, 1710–1721. [CrossRef]
97. Serrano, A.; de la Torriente, A.; Punzón, A.; Blanco, M.; Bellas, J.; Durán-Muñoz, P.; Murillo, F.J.; Sacau, M.; García-Alegre, A.; Antolínez, A.; et al. Sentinels of Seabed (SoS) indicator: Assessing benthic habitats condition using typical and sensitive species. *Ecol. Ind.* **2022**, *140*, 108979. [CrossRef]
98. Schröder, A. Community Dynamics and Development of Soft Bottom Benthos in the German Bight (North Sea) 1969–2000. Ph.D. Thesis, Universität Bremen, Bremen, Germany, 2005.
99. Werner, O. Charakterisierung des Aktuellen Zustandes der Arctica Islandica—Population im Meeresschutzgebiet Fehmarnbelt (Westliche Ostsee), Einem von Grundschleppnetzfischerei Beeinflussten Schlickigen Meeresboden. Master's Thesis, Rostock University, Rostock, Germany, 2021.
100. Zeller, D.; Rossing, P.; Harper, S.; Persson, L.; Pauly, D. The Baltic Sea: Estimates of total fisheries removals 1950–2007. *Fish. Res.* **2011**, *108*, 356–363. [CrossRef]
101. Couce, E.; Schratzberger, M.; Engelhard, G.H. Reconstructing three decades of total international trawling effort in the North Sea. *Earth Syst. Sci. Data* **2020**, *12*, 373–386. [CrossRef]
102. Global Fishing Watch. Available online: https://globalfishingwatch.org (accessed on 19 September 2023).
103. Schönke, M.; Clemens, D.; Feldens, P. Quantifying the Physical Impact of Bottom Trawling Based on High-Resolution Bathymetric Data. *Remote Sens.* **2022**, *14*, 2782. [CrossRef]
104. Ohnesorge, A.; John, U.; Taudien, S.; Neuhaus, S.; Kuczynski, L.; Laakmann, S. Capturing drifting species and molecules—Lessons learned from integrated approaches to assess marine metazoan diversity in highly dynamic waters. *Environ. DNA* **2023**, *5*, 1541–1556. [CrossRef]
105. Jac, C.; Desroy, N.; Duchêne, J.-C.; Foveau, A.; Labrune, C.; Lescure, L.; Vaz, S. Assessing the impact of trawling on benthic megafauna: Comparative study of video surveys vs. scientific trawling. *ICES J. Mar. Sci.* **2021**, *78*, 1636–1649. [CrossRef]
106. Pawlowski, J.; Bonin, A.; Boyer, F.; Cordier, T.; Taberlet, P. Environmental DNA for biomonitoring. *Mol. Ecol.* **2021**, *30*, 2931–2936. [CrossRef] [PubMed]
107. Derycke, S.; Maes, S.; Van den Bulcke, L.; Vanhollebeke, J.; Wittoeck, J.; Hillewaert, H.; Ampe, B.; Haegeman, A.; Hostens, K.; De Backer, A. Detection of Macrobenthos Species with Metabarcoding Is Consistent in Bulk DNA but Dependent on Body Size and Sclerotization in eDNA From the Ethanol Preservative. *Front. Mar. Sci.* **2021**, *8*, 637858. [CrossRef]
108. McEver, R.A.; Zhang, B.; Levenson, C.; Iftekhar, A.S.M.; Manjunath, B.S. Context-Driven Detection of Invertebrate Species in Deep-Sea Video. *Int. J. Comput. Vis.* **2023**, *131*, 1367–1388. [CrossRef]

*biology*

MDPI

*Article*

# Seasonal Dynamics of Benthic Infauna Communities in *Zostera marina* Meadows: Effects of Plant Density Gradients

**Natalia Anna Gintowt *, Halina Kendzierska and Urszula Janas**

Faculty of Oceanography and Geography, University of Gdańsk, al. Marszałka Piłsudskiego 46, 81-378 Gdynia, Poland; halina.kendzierska@ug.edu.pl (H.K.); urszula.janas@ug.edu.pl (U.J.)

* Correspondence: natalia.gintowt@phdstud.ug.edu.pl; Tel.: +48-58-523-6866

**Simple Summary:** Seagrass meadows, especially those formed by eelgrass *Zostera marina*, are very important biotopes, crucial to the healthy functioning of the Baltic coastal ecosystems. These meadows provide feeding and breeding areas for many marine organisms. Plants and animals that form these biotopes play a key role in transforming organic matter and nutrients in the marine ecosystem. Organisms dwelling in the sediments of seagrass meadows alter them through bioturbation (sediment mixing) and bioirrigation (water flow in the sediment). The objective of our study was to determine how the density of plants that form a meadow and the season of the year affect the species composition, density, and activity of these organisms. The results show that the presence of seagrass increases the number of species in the meadows and that their abundance boosts the activity of organisms. The season also plays an important role in shaping the functioning of the meadows, with the majority of organisms found in autumn; benthic communities have also been proven to have the highest potential for activity during this season compared to the rest of the year. Overall, the presence of seagrass helps maintain coastal ecosystems much more effectively than bare sand.

**Abstract:** *Zostera marina* meadows play a key role in the Baltic Sea ecosystem. They are characterized by high primary production and provide feeding and reproduction grounds for organisms. These characteristics vary due to year-round environmental changes and may be due to the characteristics of the meadows themselves. Organisms inhabiting seagrass meadows are involved in the transformation of substances from terrestrial runoff, and, through bioturbation and bioirrigation, affect biogeochemical processes in the sediments. This study aimed to determine the structure of benthic communities inhabiting *Z. marina* meadows and their bioturbation ($BP_C$) and bioirrigation ($IP_C$) potential as affected by seagrass density and seasonal changes. This study shows a positive correlation between the density of *Z. marina* and the structure of macrozoobenthos, as well as the bioturbation and bioirrigation potential of the studied communities. The autumn season stimulated the density of macrofauna and recorded the highest values of their potential activities indices. The presence of *Z. marina* positively affects macrozoobenthic communities and their functioning regardless of seagrass density, indicating that seagrass meadows inhabited by macrofauna are key biotopes that can support biogeochemical processes in the coastal zone more effectively than bare sand.

**Keywords:** seagrass; macrozoobenthos; bioturbation; bioirrigation; marine biodiversity; coastal zone; Baltic Sea

## 1. Introduction

Coastal zones represent one of the most important areas in marine ecosystems. They are characterized by specific physicochemical conditions that uniquely shape the diversity of habitats and the organisms living in them [1]. These zones are under very high water dynamics, and their environmental conditions are strongly affected by rivers and anthropogenic factors. All of these factors, combined with conditions such as the type of sediment in a given habitat, strongly affect the formation of unique and heterogeneous habitats in these zones [2]. Various types of unique habitats can be observed on the coast of the Baltic Sea [1,3–5]. These habitats—including the seagrass meadows of *Zostera marina* Linnaeus, 1753—are characterized by high variability in environmental conditions and perform different functions in the ecosystem [6,7].

Seagrasses play an important role as an engineering biotope; as meadows, they can modify the direction of ocean currents and stabilize sediments, thus preventing bottom erosion, and are involved in the circulation of elements and matter and the flow of energy [8,9]. In addition, they play a crucial role as stores of chemical substances, e.g., carbon. Even though they occupy less than 0.2% of the ocean's surface, they store up to 10% of the carbon entering the oceans annually [10,11]. The high capacity of seagrasses to absorb carbon compounds from the environment can result in a local reduction in the level of water acidification [12]. However, high heterogeneity is observed among submerged meadows. This can include both the species composition of the macrophytes that make up the meadows and the prevailing conditions, namely the physicochemical parameters and the density of shoots. Each meadow is different and is characterized by unique environmental conditions. It has also been shown that even meadows in the same area can be distinguished by different functionality [13–15].

Organisms living in underwater meadows perform diverse activities and contribute to the proper functioning of these ecosystems [16]. Meadows are home to organisms of various sizes and taxonomic groups and representatives of all kinds of trophic guilds, from filter feeders and grazing organisms to predators [17,18]. Benthic organisms are essential to the cycling of chemical elements and nutrients, which they affect directly through activities such as feeding, respiration, and excretion, or indirectly by altering the sediment structure through activities such bioturbation and bioirrigation [6,19]. Both of these activities positively affect sediment conditions: water flowing through the burrows increases sediment oxygenation, and the burrows themselves create an increased surface area for colonization by microorganisms. In addition, intensive bioturbation and bioirrigation cause an exchange of substances between water and sediment and can stimulate the transformation of organic matter in the sediment [20–25]. Due to their important role in ecosystems, bioturbation and bioirrigation have been relatively extensively studied over the years. Various types of methods have been used for this purpose, including the estimation of bioturbation as its actual measurement. Bioturbation intensity is estimated by calculating Bioturbation ($BP_C$) and Bioirrigation ($IP_C$) Potential Indices, and is a relatively simple and widely used method for determining the functionality of given benthic species or communities [26–30].

Bioturbation and bioirrigation are important processes, but neither has been extensively studied for underwater meadows. One of the few studies conducted to date in these ecosystems involved *Zostera noltei* Hornemann, 1832 meadows in France, and showed an inhibitory effect from the meadows on the intensity of bioturbation activity of organisms [31]. The Baltic underwater meadows remain unexplored in terms of the intensity of bioturbation taking place there, but are fairly well studied in such aspects as the taxonomic composition of plants and animals that form and inhabit underwater meadows, the functioning of the meadows as trophic networks observed in them, and the biological traits of the macrofauna species living there. The studies were carried out on a single- or multi-seasonal basis [32–36].

The research presented here focuses on the effects of *Z. marina* density on macrofauna communities and their bioturbation and bioirrigation activities, both of which affect the functioning of coastal habitats, but are so far poorly researched in the Baltic Sea. The research was conducted on a seasonal basis to determine the effect of environmental conditions on the benthic communities of seagrass meadows and their activities.

## 2. Materials and Methods

### *2.1. Study Area and Sampling Design*

Macrozoobenthos samples were collected seasonally in November 2021 (Autumn) and February (Winter), May (Spring), and August (Summer) 2022 by divers at four density treatments in the *Z. marina* meadow (54°40.352′ N; 18°41.754′ E) on Długa Mielizna, a sandy shoal along the Hel Peninsula (Puck Bay) (Figure 1).

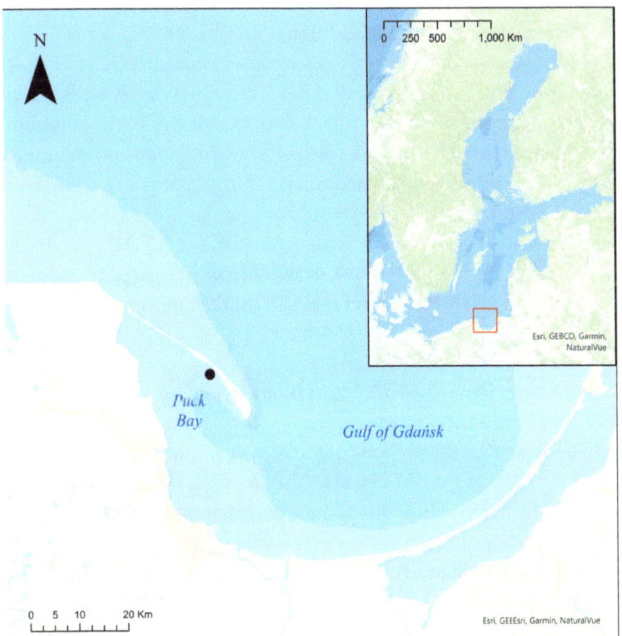


**Figure 1.** Study area with sampling site. The red rectangle indicates the location of the study area on a map of the Baltic Sea.

The unvegetated (UnV) site was located on bare sand about 2 m from a vegetated area (outside of the meadow) and three remaining sites—Low-Density Seagrass (SLD), Medium-Density Seagrass (SMD), and High-Density Seagrass (SHD)—were selected and delineated by a diver in the increased density gradient of *Z. marina* (Figure 2). Each replicate of macrofauna cores was located approx. 50–70 cm away from each other; meanwhile, sediment samples were subtracted approx. 30 cm from the fauna cores. Each *Zostera* treatment was taken from different *Z. marina* patches; the patches were about 4–6 $m^2$ in size. A 50 × 50 cm frame was placed at each treatment site five times, and the number of *Zostera* shoots was counted. Bottom water temperature, salinity, and dissolved oxygen (DO) concentration were measured at each site. Five sediment cores (10 cm inner diameter, 25 cm of sediment) were collected at each site for macrofauna analysis.

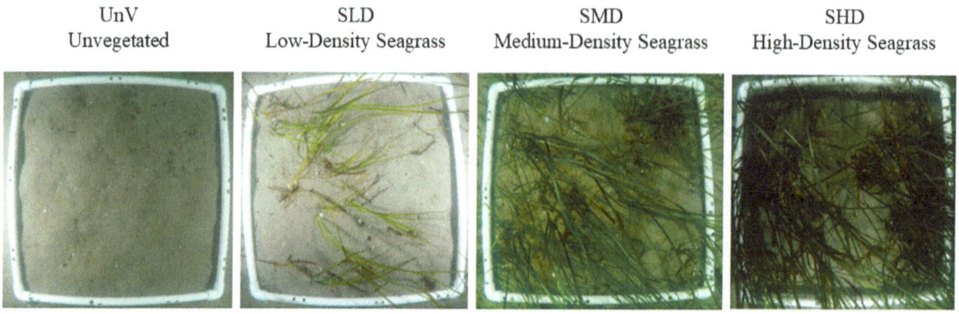


**Figure 2.** Photographs of the sampling sites with an increasing gradient of *Z. marina* shoots. The treatments were as follows: Unvegetated (UnV), Low-Density Seagrass (SLD), Medium-Density Seagrass (SMD), and High-Density Seagrass (SHD).

### 2.2. Sediment Characteristics

Fifteen (three per macrofaunal treatment) additional intact sediment cores (3.5 cm inner diameter, 8 cm length) were collected to determine sediment parameters at each station. From five of these cores, the surface layer of the sediment (0–1 cm) was sliced and collected, and samples were subsequently frozen after being transported to the laboratory. The organic matter (OM) content was determined as the percentage of mass loss on ignition (LOI) (450 °C, 5 h) of the dried, homogenized sediment. The remaining two cores per macrofaunal treatment were used for grain size analysis. Samples were sieved using a shaker and a set of standard test sieves with mesh diameters of 2, 1, 0.5, 0.25, 0.125, and 0.063 mm [37]. Sediments were classified according to the Udden–Wentworth grain-size scale [38] based on the percentage of each class in the total sample mass.

### 2.3. Macrofauna

Animals were selected from the sediment, bypassing the sieving step that usually precedes the analysis of the macrofauna in a sample. The organisms were then sorted, and all taxa, except for Oligochaeta and *Marenzelleria* spp., were identified to the species level using specialistic identification guides [39–41]. The taxa were counted, and their wet mass was measured to determine their abundance and biomass per square meter. Shannon-Wiener's (H') diversity index and Pielou's (J) evenness index were calculated using PRIMER 6 software (PRIMER-E Ltd., Ivybridge, UK).

### 2.4. Bioturbation Potential and Irrigation Potential

The wet mass of the organisms was used to calculate the Bioturbation Potential and Bioirrigation Potential indices. Despite the fact that many researchers use ash-free dry weight to calculate potentials, our previous studies of Baltic organisms showed that values of the indices calculated from wet weight and ash-free dry weight are in perfect agreement with each other r = 0.999 [26]. The Bioturbation Potential Index ($BP_C$) at each site was the sum of the bioturbation potentials of individual taxa ($BP_i$) [28,29] calculated according to the following equation:

$$BP_c = \sum BP_i \text{ where } BP_i = \left(\frac{B_i}{A_i}\right)^{0,5} * A_i * M_i * R_i \quad (1)$$

where $B_i$ is the biomass (wet mass g·m$^{-2}$) and $A_i$ is the abundance (ind.·m$^{-2}$) of taxon $i$ in each sample, while mobility $M_i$ and sediment reworking $R_i$ are categorical scores assigned to species *i* (Appendix A). In the scores assigned we focused on the macrofaunal impact on the sediment including epifauna living and influencing the sediment surface [26,28,29,42,43].

The Irrigation Potential Community Index ($IP_C$) at each site was calculated by summing the irrigation potentials ($IP_i$) calculated for each taxon [44]:

$$IP_c = \sum IP_i \text{ where } IP_i = \left(\frac{B_i}{A_i}\right)^{0,75} * A_i * BT_i * FT_i * ID_i \tag{2}$$

where $B_i$ is the biomass ($g \cdot m^{-2}$) and $A_i$ is the abundance ($ind. \cdot m^{-2}$) for taxon *i* in each sample, while the feeding type $FT_i$, burrow type $BT_i$ and depth $ID_i$ are scores for the trait categories assigned to each species.

Exponent 0.5 used in $BP_C$ emphasizes the importance of organisms with high density and relatively low biomass, while exponent 0.75 used in $IP_C$ emphasizes the activity of organisms with larger sizes but lower densities [30].

### 2.5. Data Analysis

A Principal Component Analysis (PCA) was conducted to determine the relationship between physicochemical conditions in sediments and bottom water and the variability between the sites. Data normality was tested using a Shapiro–Wilk test. A matrix with normalized data on bottom water temperature, surface sediment organic matter content, and the number of *Zostera* shoots was used in the analysis. The amount of <63 µm fraction and the biomass of *Z. marina* were not taken into account in this analysis due to the high correlation with organic matter content (r = 0.699) and the number of *Z. marina* shoots (r = 0.615), respectively. The abundance of macrofauna was square root transformed, and cluster analysis (Bray–Curtis similarity) was used to determine the similarity between macrofauna samples. Differences in total abundance, H', J, $BP_C$, and $IP_C$ between the sampling sites and seasons were tested using PERmutational Multivariate Analysis of VAriances (PERMANOVAs) [45]. Data visualization and statistical analyses were performed in Microsoft Office 365 ProPlus, RStudio v4.30 (Venn diagram), and PRIMER v7 with PERMANOVA+ (PRIMER-E Ltd., Plymouth, UK).

## 3. Results

### 3.1. Environmental Conditions

Due to the close proximity of the sites, no differences in the parameters of bottom water were observed in a given season at the study sites. Variability in these parameters was observed only between seasons. The lowest temperature (3.9 °C) and the highest concentration of oxygen in the benthic water (13.97 mL/L) was observed in winter (Table 1). An inverse relationship was observed in summer when the temperature was highest (20.5 °C) and dissolved oxygen concentration was lowest (9.81 mL/L). Due to the shallow depth of the study sites (3 m), the water was well mixed and well oxygenated. Salinity at the surveyed sites was relatively stable, ranging from 6.8 in winter to 7.5 in autumn and summer. The organic matter content was relatively uniform among all sites during the studied seasons, and so was the sediment type, which was medium sand at all sites in all seasons. In all seasons, we were able to find seagrass patches at our designated density. The SLD treatment covered a density of 24–64 *Z. marina* shoots$\cdot m^{-2}$ and the SHD treatment covered a density of 148–240 shoots$\cdot m^{-2}$.

**Table 1.** Sediment characteristics and environmental parameters measured in bottom waters at the study sites in all seasons at the studied treatments—Unvegetated (UnV), Low-Density Seagrass (SLD), Medium-Density Seagrass (SMD), and High-Density Seagrass (SHD).

| Season | Site | Temperature [°C] | Oxygen [mL/L] | Salinity | LOI [%] 0–1 cm | Number of *Zostera* shoots |
|---|---|---|---|---|---|---|
| Autumn | UnV | 7.8 | 11.86 | 7.5 | 0.26 | 0 |
| | SLD | 7.8 | 11.86 | 7.5 | 0.20 | 48–64 |
| | SMD | 7.8 | 11.86 | 7.5 | 0.26 | 116–144 |
| | SHD | 7.8 | 11.86 | 7.5 | 0.24 | 164–240 |
| Winter | UnV | 3.9 | 13.97 | 7.3 | 0.27 | 0 |
| | SLD | 3.9 | 13.97 | 7.3 | 0.28 | 28–52 |
| | SMD | 3.9 | 13.97 | 7.3 | 0.32 | 112–132 |
| | SHD | 3.9 | 13.97 | 7.3 | 0.28 | 148–184 |
| Spring | UnV | 9.9 | 12.56 | 6.8 | 0.23 | 0 |
| | SLD | 9.9 | 12.56 | 6.8 | 0.27 | 24–44 |
| | SMD | 9.9 | 12.56 | 6.8 | 0.29 | 98–140 |
| | SHD | 9.9 | 12.56 | 6.8 | 0.27 | 148–180 |
| Summer | UnV | 20.5 | 9.81 | 7.5 | 0.25 | 0 |
| | SLD | 20.5 | 9.81 | 7.5 | 0.29 | 28–44 |
| | SMD | 20.5 | 9.81 | 7.5 | 0.28 | 96–124 |
| | SHD | 20.5 | 9.81 | 7.5 | 0.32 | 156–192 |

The PC1 axis resulting from PCA explains 41.1% of the total variance (eigenvalue 1.23), with OM and the number of shoots being the most important explanatory factors (Table 2). PC1 and PC2 (eigenvalue 1.02) together explain 75.1% of the total variance (Figure 3). Temperature, with a coefficient of 0.936, contributed the most to the distribution along the PC2 axis.

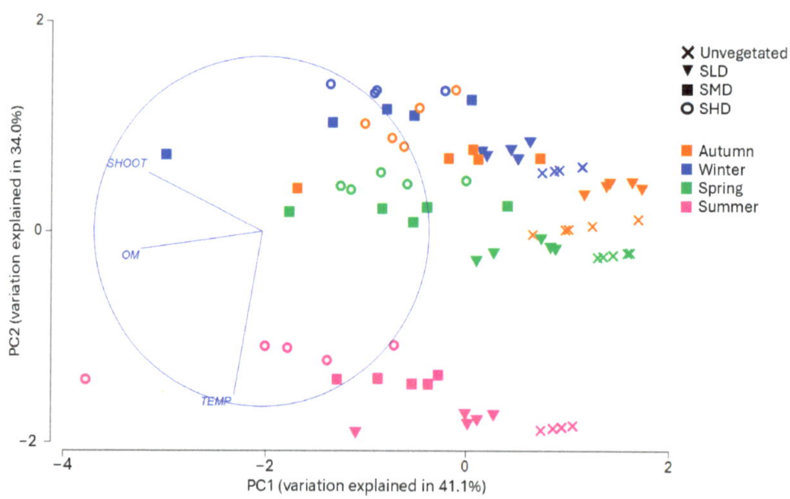


**Figure 3.** PCA results. Variables included in the PCA are bottom water temperature (TEMP), organic matter (OM) content of surface sediments and the number of *Zostera* shoots (SHOOT). Unvegetated (UnV), Low-Density Seagrass (SLD), Medium-Density Seagrass (SMD), and High-Density Seagrass (SHD)

**Table 2.** Percentage of variation and coefficients in linear combinations of variables forming PCs.

| Variable | PC1 | PC2 | PC3 |
|---|---|---|---|
| Variation [%] | 41.1 | 34.0 | 24.9 |
| Bottom water temperature | −0.166 | −0.936 | −0.310 |
| Organic matter content of surface sediments | −0.718 | −0.101 | 0.689 |
| Number of *Zostera* shoots | −0.676 | 0.337 | −0.655 |

### 3.2. *Macrofauna*

In the present study, we observed 29 taxa of benthic macrofauna in *Z. marina* meadows in Długa Mielizna in 2021–2022 (listed in Appendix Table A1). The species with the highest frequency of occurrence were the polychaete *Hediste diversicolor* (O.F. Müller, 1776) (99% of all collected cores) and the mud snails *Peringia ulvae* (Pennant, 1777) (97%) and *Ecrobia ventrosa* (Montagu, 1803) (85%). Other species present in more than 70% of the samples included Oligochaetes, polychaetes of the genus *Marenzelleria*, the bivalves *Cerastoderma glaucum* (Bruguière, 1789) and *Mya arenaria* Linnaeus, 1758, and the crustacean *Cyathura carinata* (Krøyer, 1847). In all of the studied seasons, the lowest species richness was noted at the unvegetated site. The greatest difference between the number of species at the surveyed sites was observed in summer, when only eight taxa were found at the site without vegetation, while 21 species were found at the SMD and SHD sites (Figure 3). The highest number of taxa at a single site was 23, observed in spring at the SMD site. At none of the bare sand treatments did the number of taxa exceed 10.

The highest H′ values were recorded at the sites covered with vegetation in spring (Figure 4). Statistically significant differences were found between the studied seasons (Table 3). There were no significant differences in the *Z. marina* gradient treatments. The Pielou index showed statistically significant differences between the sites and the seasons, as well as in their interaction—season x site. The autumn season was characterized by the lowest evenness values, which were due to the high dominance of *P. ulvae* in the macrofauna communities

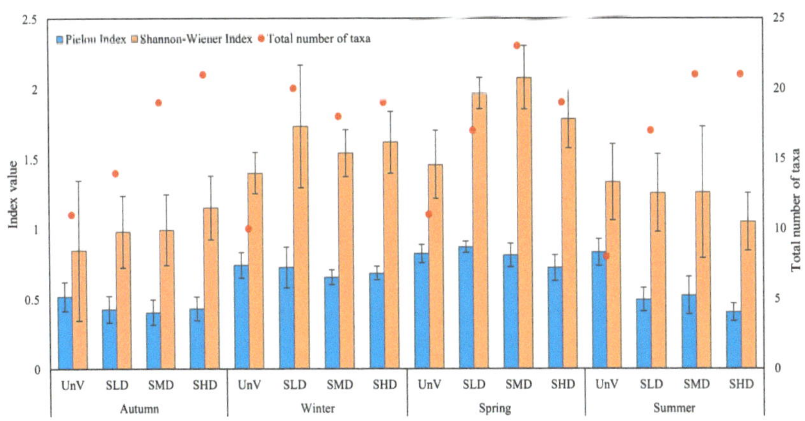


**Figure 4.** Macrofauna community descriptors (mean ± SD), Pielou index, Shannon–Wiener and total number of taxa diversity index by season following a spatial gradient of increasing shoot density: Unvegetated (UnV), Low-Density Seagrass (SLD), Medium-Density Seagrass (SMD), and High-Density Seagrass (SHD).

**Table 3.** Results of PERMANOVA analysis for differences in Abundance, $BP_C$, $IP_C$, Pielou Index, and Shannon–Wiener Index. Bold—statistically significant differences. df stands for deegrees of freedom.

| | | Abundance | Pielou Index | Shannon Index | $BP_C$ | $IP_C$ |
|---|---|---|---|---|---|---|
| Site | df | 3 | 3 | 3 | 3 | 3 |
| | MS | 5838.3 | $1.0872 \times 10^7$ | $1.2916 \times 10^7$ | 12,109 | $1.9954 \times 10^7$ |
| | Pseudo-F | 9.6171 | 3.0852 | 1.5442 | 24.513 | 4.0247 |
| | p(perm) | **0.001** | **0.035** | 0.184 | **0.001** | **0.001** |
| Season | df | 3 | 3 | 3 | 3 | 3 |
| | MS | 10,678 | $5.0784 \times 10^7$ | $7.4832 \times 10^7$ | 10,608 | $3.4079 \times 10^7$ |
| | Pseudo-F | 17.589 | 14.411 | 8.947 | 21.475 | 6.8738 |
| | p(perm) | **0.001** | **0.001** | **0.002** | **0.001** | **0.001** |
| Site x Season | df | 3 | 9 | 9 | 9 | 9 |
| | MS | 1797.1 | $8.3333 \times 10^6$ | $1.6627 \times 10^7$ | 2137.2 | $5.7653 \times 10^6$ |
| | Pseudo-F | 2.9603 | 2.3647 | 1.988 | 4.3267 | 1.1629 |
| | p(perm) | **0.001** | **0.025** | 0.057 | **0.001** | 0.239 |

Analysis of the Venn diagram showed that 11 of the observed taxa were present in both the meadows and the bare sand area (Figure 5). Furthermore, 13 taxa were unique to *Z. marina* meadows. Four taxa were observed only in areas of dense seagrass: *Rhithropanopeus harrisii* (Gould, 1841), *Fabricia stellaris* (Müller, 1774), *Corophium volutator* (Pallas, 1766), and *Gammarus zaddachi* Sexton, 1912. The taxon found only on bare sand and at medium seagrass density was the amphipod *Bathyporeia pilosa* Lindström, 1855.

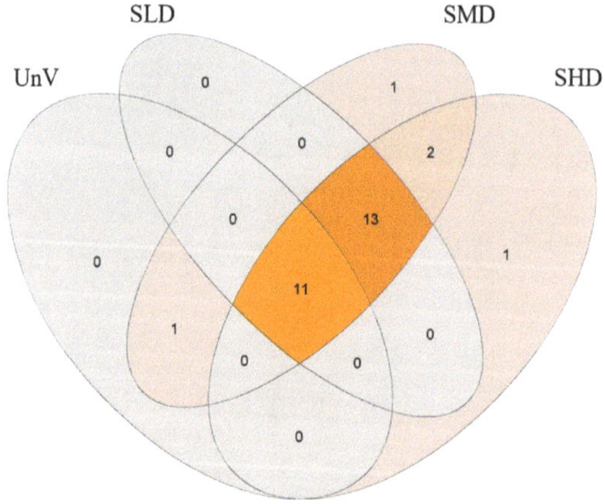


**Figure 5.** Venn diagram showing the number of taxa unique or common to different *Zostera* treatments—Unvegetated (UnV), Low-Density Seagrass (SLD), Medium-Density Seagrass (SMD), and High-Density Seagrass (SHD).

Sites with the lowest biodiversity, i.e., those that were not overgrown with seagrass in any season, were also characterized by the lowest density of macrofauna. Our research indicates a small number of epifaunal species. At a few sites, less than 1% of the total macrofauna belonged to epifauna, but in most treatments, epifauna accounted for approx. 5% of all the fauna abundance. In terms of density, snails of the species *P. ulvae* dominated in all the study sites (12–76%, Figure 6). Mud snails *E. ventrosa* and clams *C. glaucum* contributed up to 14% and 10%, respectively, to the total abundance of organisms. We observed large differences between the abundance of organisms in all seasons (PERMANOVA, $p < 0.01$;

Table 3). The highest abundance of organisms was observed in autumn, and large numbers of organisms were also observed in the meadows in summer. The fewest organisms were observed at all sites in spring. The difference between the abundance of macrofauna at the UnV and *Zostera* sites was statistically significant (PERMANOVA, $p < 0.001$; Table A3). In some months, lower densities of organisms were observed at the SMD sites than at the SLD and SHD (e.g., in autumn and summer).

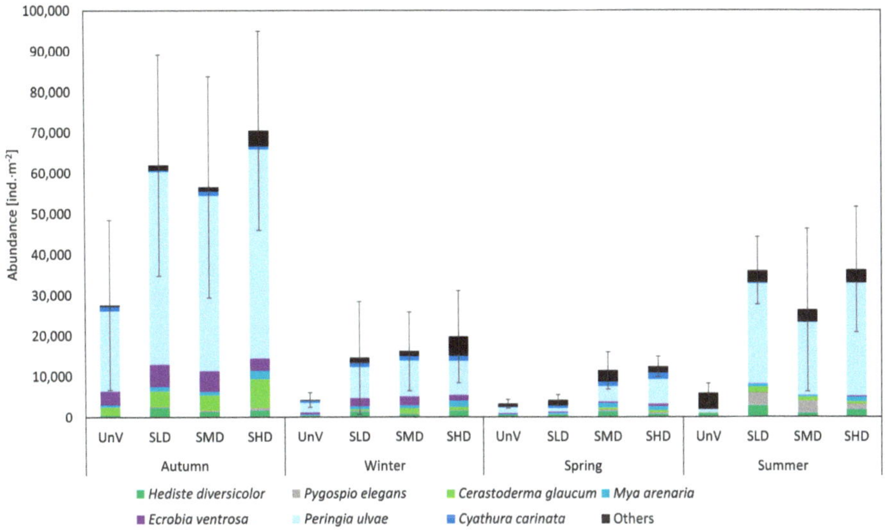


**Figure 6.** Abundance of macrofauna [ind.·m$^{-2}$] at the sampling sites in four seasons. Values are presented as means with standard deviation indicated. Unvegetated (UnV), Low-Density Seagrass (SLD), Medium-Density Seagrass (SMD), and High-Density Seagrass (SHD). 'Others' include 22 species with the lowest abundance (less than 5% of the total abundance).

### 3.3. Bioturbation and Bioirrigation

Both the bioturbation potential index and the bioirrigation potential index were higher at sites with higher seagrass density than at bare sand (Figure 7). In most seasons, the highest values of the bioturbation index were determined at the SHD site with the highest seagrass density. The bivalves *C. glaucum* and *M. arenaria* and *P. ulvae* snail were the main contributors to $BP_C$, followed by the fourth most abundant taxon, *H. diversicolor*. The lowest index was recorded at the sites not covered with seagrass, and was relatively similar (<5000) in all seasons except autumn; the difference between UnV and all *Zostera* treatments was statistically significant. Autumn was also the only season that was different from all of the other studied seasons (PERMANOVA, $p < 0.001$; Table A3). Other taxa than those responsible for $BP_C$ were involved in the formation of bioirrigation potential. Polychaetes of the species *H. diversicolor* contributed the most to the formation of $IP_C$, followed by the snails *P. ulvae*. *C. glaucum*, although also present, was not as important at most sites, except those surveyed in autumn. The highest values of the indices were observed in autumn, with the highest $BP_C$ (35 162) and $IP_C$ (11 071) being at the SHD site (Figure 7). As with $BP_C$, the only season that differed statistically significantly from the other seasons was autumn, and only the UnV site differed from the other sites in terms of *Zostera* treatments (PERMANOVA, $p < 0.001$; Table A3).

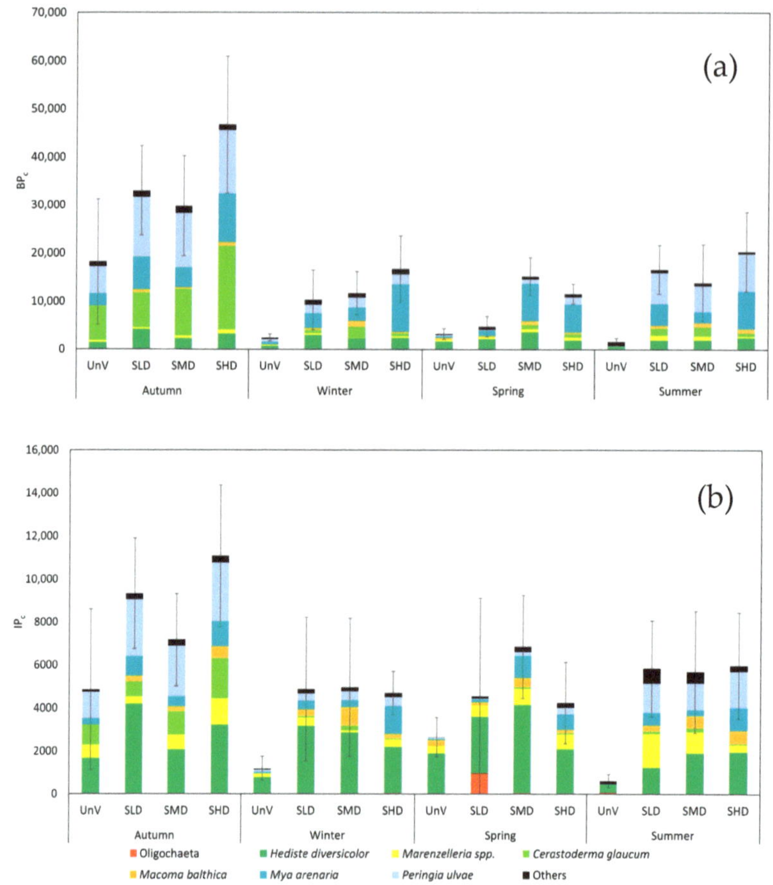


**Figure 7.** Functional indices (**a**) $BP_C$ and (**b**) $IP_C$ of the benthic community in cores collected from the sampling sites (n = 5). Values are presented as means with standard deviation indicated. Unvegetated (UnV), Low-Density Seagrass (SLD), Medium-Density Seagrass (SMD), and High-Density Seagrass (SHD). 'Others' include 22 species with the lowest abundance (less than 5% of the total abundance).

## 4. Discussion

### 4.1. *Zostera Meadows*

The coastal zone is a dynamic environment characterized by very good mixing and oxygenation of water, resulting in a lower probability of water stratification [46]. The bottom water in the study area was well oxygenated and the salinity was uniform in all of the study sites. It is more likely that other parameters, such as sediment characteristics or temperature, had a greater impact on shaping the diversity of the benthic community. Research suggests that seagrass meadows are particularly attractive habitats for benthic organisms due to the accumulation of large amounts of organic matter in the sediments [11,47,48]. This is largely due to the filtering function of meadows in the coastal zone; they have the ability to trap and retain organic matter [49] and prevent sediment erosion [50]. Our results, however, did not confirm this relationship. A possible explanation for this is the fact that our research focused on seagrass patches rather than large seagrass meadows, which may accumulate less organic matter. In the past, Jankowska et al. [47] indicated that the organic matter content in the surface layer (up to 10 cm) of the sediments was not correlated with seagrass density in this area.

Our research has shown that *Zostera* meadows occur in the bay all year round, and in each season of the year, we are able to find fragments of meadows with a relatively high density of seagrass. Although the density of grass at each site varied from season to season in our study, we observed seagrass densities reaching up to at least 180 shoots·m$^{-2}$, even in the winter season, considered unfavorable for seagrass growth. Thus, the density of underwater meadows observed in our study appears to be relatively typical of Puck Bay. To the best of our knowledge, however, meadows of varying density and species composition of the macrophytes forming them are also observed in Puck Bay. Jankowska et al. [51] observed the highest density of seagrass in summer, and the density of *Z. marina* reached 200 shoots·m$^{-2}$, while in winter, the observed meadows were much sparser, and the number of shoots was only about 55 m$^{-2}$. However, compared to other *Z. marina* sites in the Baltic Sea, it can be concluded that our meadows are characterized by relatively low densities. While conducting research in the Baltic Archipelago, Rodil et al. [35] observed that meadows there could reach a density of up to 800 shoots·m$^{-2}$, but those that were sheltered had lower densities of 150–250 shoots·m$^{-2}$.

### 4.2. Macrozoobenthos

The study of the macrofauna of underwater *Z. marina* meadows in Puck Bay presented here revealed the presence of 29 taxa at the surveyed sites, indicating the high taxonomic richness of the area. Many of these species were observed both in the underwater meadows and at the sites without vegetation, but the vegetated sites were always characterized by higher species diversity. This phenomenon underscores the importance of *Z. marina* meadows in shaping the coastal biodiversity, as described previously [3,32,35,52,53]. Thirteen of the observed taxa were specific to submerged meadows, regardless of their density. A similar pattern of the presence of specific species associated solely with vegetation was reported previously [16,33,54].

Interestingly, our study does not show a clear effect of meadow density on species composition; most species were simply present in vegetated areas, and only a few species were associated with specific levels of seagrass density: *R. harrisii* (SHD), *C. volutator* and *G. zaddachii* (SMD + SHD), and *B. pilosa* (UnV + SMD). Some researchers have shown a positive correlation between seagrass complexity and the diversity of macrofaunal communities [35,55]. In Puck Bay, *Z. marina* forms meadows together with other macrophyte species, such as *Potamogeton* spp., *Zannichelia palustris* L., *Stuckenia* spp., and *Ruppia* spp. [13,33,56]. We chose monospecific meadows for our study because we wanted sites to be homogeneous in terms of meadow density and species composition, even though mixed meadows may harbor a more diverse macrofaunal community [35,57].

Regarding the density of macrofauna, we found a typical pattern: higher abundance and biomass of organisms were recorded at vegetated sites compared to areas with bare sand. Overall, as indicated by other researchers [16,33,52,58], seagrass had a positive effect on the abundance of benthic taxa, but we did not observe a correlation between the density of macrofauna and the density of plants. Dense underwater meadows are not necessarily an indicator of diverse and densely populated macrofauna biotopes, and there are even studies showing a positive effect of habitat fragmentation on the density and species richness of macrofauna [59]. Rodil et al. [35], on the other hand, showed that the density of organisms and grasses is not a linear relationship.

Temperature was the second factor that explained most of the differences in macrozoobenthic communities and was closely related to the sampling season. Studies have shown large changes throughout the year in the number and diversity of species [51,58,60]. Some mobile macrofauna species are known to migrate to deeper parts of the coastal zone in search of more favorable environmental conditions during the year [61]. Ten of the species observed in submerged meadows occurred throughout the year; these were mainly the most abundant species with limited mobility, such as *C. glaucum* and *Macoma balthica* (Linnaeus, 1758). The highest number of organisms was found in autumn, which may be related to favorable environmental conditions for life and reproduction. Previous studies [51] have shown that the highest density and biomass of organisms in seagrass meadows were observed in the summer of 2011, so high biodiversity in the autumn season may indicate an extension or shift in favorable living conditions for organisms. These findings underscore the key role of temperature in shaping the diversity of macrozoobenthos communities.

The H′ index showed no differences between the treatments, but there were differences between seasons. Previous studies have shown differences in H′ between bare sand and meadow [33,58] and a relationship between H′ and the number of grass seedlings [62]. The lack of differences in the values of our indices may be due to the relatively homogeneous and similar environmental conditions prevailing at the treatments studied. However, in most cases, we observed larger differences in the J index at sandy sites than in the meadow. This is due to the fact that the meadows were often dominated by *P. ulvae*, a snail species that is typical and common in meadows.

### 4.3. Bioturbation and Bioirrigation

The occurrence of seagrass meadows affected both indices of organism activity, i.e., $BP_C$ and $IP_C$. In both cases, we observed higher values of the indices for the vegetated treatments than for those without vegetation, and there was also a strong seasonal variation between the values of the indices. The $BP_C$ values recorded for the *Zostera* meadows are slightly lower in autumn and significantly lower in the other seasons compared to $BP_C$ (calculated for wet mass) determined for the Vistula Plume [26]. This study was conducted on a similar type of sediment, but, due to the greater depth, a large number of intensely bioturbating taxa (*M. balthica*, *Marenzelleria* spp., *H. diversicolor*, and *M. arenaria*) dominated in the plum areas. Queirós et al. [63] showed that, in communities of muddy bottoms, $BP_C$ was the highest in summer and autumn. High seasonal variation in the $BP_C$ index was also observed in the western part of the Baltic Sea [64].

Our results indicate that autumn was a favorable season for the development of seagrass meadows and the activity of macrofauna due to favorable environmental conditions. As a result, we recorded high values of $BP_C$ and $IP_C$ indices, which reflect the potential for intense activity of organisms during this season.

The research presented here shows significant differences in the contribution of individual macrofaunal species to the formation of bioturbation and bioirrigation potential of benthic communities in the studied biotopes. The bioturbation potential is most strongly shaped by bivalves of the species *C. glaucum* and *M. arenaria*, which were the dominant species in the macrofauna biomass. Previous studies conducted in the Gulf of Gdańsk have shown a strong correlation between the $BP_C$ index and the biomass of the observed organisms [26]. Our previous studies have also shown the dominance of bivalves in creating bioturbation potential. Although both indices, $BP_C$ and $IP_C$, are calculated based on the abundance and biomass of organisms, the coefficients used in the calculations to determine the aforementioned processes emphasize the contribution of specific species to the formation of the indices, and not necessarily those that dominated the abundance or biomass of organisms in the meadows. We observed such a phenomenon in the case of the index of bioirrigation potential, $IP_C$. For this index, we observed a strong dominance of the polychaete *H. diversicolor* in its formation, as well as a relatively high proportion of mud snails *P. ulvae*. Numerous studies have shown that *H. diversicolor* is a very effective bioturbator and bioirrigator [65,66], even though, in our research, it did not significantly contribute to the abundance of organisms in the meadows. On the other hand, the high contribution of snails to the bioirrigation potential of organisms is more surprising. *P. ulvae* occurs relatively shallowly in the sediment and burrows only a few centimeters deep into the sediment without forming burrows. However, the high density of these snails indicates that this species can have a major impact on bioturbation and bioirrigation in sediments. The study by Andersen et al. [67] showed that the high density of *P. ulvae* causes a significant increase in the rate of sediment erosion. This was due to the intense movement of snails on the sediment surface and in the sediment, resulting in bioturbation and increased the permeability of the sediment to water. Thus, although *P. ulvae* does not create typical burrows, its presence in high densities can significantly affect bioturbation and bioirrigation.

$BP_C$ and $IP_C$ indices indicated both seasonal variation and variation related to *Zostera* density. Bioturbation and bioirrigation are very complex processes that are determined by numerous factors affecting the functioning of benthic organisms. In general, it can be said that both indices showed higher values in the meadows than in nearby unvegetated areas. Unfortunately, bioturbation or bioirrigation activity in seagrass meadows has not been studied much to date, neither by direct nor indirect methods. There are few studies focusing on bioturbation in *Z. marina* meadows. Bernard et al. [31] conducted direct experiments using luminescent markers to determine the exact bioturbation in seagrass meadows. However, they showed a correlation opposite to that presented in our study. Bernard et al. [31] showed that sediment particle mixing processes were less intense in meadows than in unvegetated areas, indicating a mitigating effect observed in submerged meadows. In summary, the results of our study indicate that bioturbation and bioirrigation in seagrass meadows are very complex and dynamic processes. They vary depending on local environmental conditions and the selected method of measurement or estimation. The different conclusions of our and Bernard's research underscore the need for further and more comprehensive research in this area. Understanding the processes performed by organisms is crucial to assessing the role of seagrass meadows and seagrass itself in the functioning of benthic coastal ecosystems.

## 5. Conclusions

Our study highlights the key role of seagrasses in shaping the structure and functionality of benthic communities. Even meadows with relatively low densities have a positive impact on the biodiversity and functionality of the biotope. Both the overall density and species richness of organisms and potential for bioturbation $BP_C$ and bioirrigation $IP_C$ were significantly higher in the meadows than in the nearby sandy bottom without vegetation. Moreover, seasonal dynamics play a crucial role in shaping macrobenthic communities and their potential functioning. Autumn proved to be the most stimulating season, having the highest abundance of organisms and favoring increased $BP_C$ and $IP_C$. Our results underscore the ecological importance of seagrass meadows and the importance of these endangered habitats not only as reservoirs of biodiversity, but also as drivers of the ecological functioning of coastal biotopes.

**Author Contributions:** Conceptualization, N.A.G., U.J. and H.K.; methodology, N.A.G., U.J. and H.K.; formal analysis, N.A.G.; investigation, N.A.G. and H.K.; writing—original draft preparation, N.A.G.; writing—review and editing, N.A.G., U.J. and H.K.; visualization, N.A.G.; supervision, U.J.; project administration, N.A.G.; funding acquisition, N.A.G. and U.J. All authors have read and agreed to the published version of the manuscript.

**Funding:** Research work funded by the science budget in 2020, as a research project under the "Diamond Grant" program.

**Institutional Review Board Statement:** Not applicable.

**Informed Consent Statement:** Not applicable.

**Data Availability Statement:** Data are contained within the article.

**Acknowledgments:** We would like to express our sincere thanks to all those who contributed to the present research, especially the students and Ph.D. students involved in the field and laboratory work: Michał Gintowt, Zuzanna Czenczek, Kamila Styrcz-Olesiak, Agnieszka Zimoch, Michał Miernik, Dobromił Skrzypniak, Izabela Orzechowska, and Adam Makatun. We would also like to thank the two reviewers whose comments significantly contributed to improving this article.

**Conflicts of Interest:** The authors declare no conflicts of interest.

## Appendix A

**Table A1.** Categorical scores assigned to each taxon for $BP_C$ and $IP_C$ indices were calculated according to Solan et al. [28], Villnäs et al. [29], Queirós et al. [42], Miernik et al. [26], and Wrede et al. [43] (modified). Modifications to scores were made to focus on the impact of macrofauna on the sediment, including epifauna living and influencing the sediment surface. Mobility ($M_i$) scores: 1—movement on the sediment surface, sessile; 2—limited movement; 3—slow, free movement through sediment; 4—free movement through sediment. Reworking types ($R_i$): 1—epifauna; 2—surficial modifiers; 3—upward or downward conveyors; 4—biodiffusors. Burrow types ($BT_i$): 1—epifauna or internal irrigation (i.e., siphons); 2—open irrigation (i.e., Y- or U-shaped burrow); 3—blind ended burrow. Feeding types ($FT_i$): 1—surface filter feeders; 2—predators; 3—deposit feeders; 4—subsurface filter feeders. Irrigation depth ($ID_i$): 1—0–1 cm; 2—1–3 cm; 3—3–6 cm; 4—6–10 cm; 5—10–15 cm.

| Taxa | $BP_C$ | | $IP_C$ | | |
|---|---|---|---|---|---|
| | $M_i$ | $R_i$ | $Bt_i$ | $Ft_i$ | $ID_i$ |
| Oligochaeta | 3 | 2 | 3 | 3 | 4 |
| *Fabricia stellaris* (Müller, 1774) | 2 | 1 | 3 | 1 | 2 |
| *Hediste diversicolor* (O.F. Müller, 1776) | 4 | 3 | 2 | 3 | 5 |
| *Marenzelleria* spp. | 4 | 4 | 3 | 3 | 5 |
| *Pygospio elegans* Claparède, 1863 | 2 | 2 | 3 | 3 | 3 |
| *Streblospio shrubsolii* (Buchanan, 1890) | 2 | 2 | 3 | 3 | 2 |
| *Cerastoderma glaucum* (Bruguière, 1789) | 3 | 2 | 1 | 1 | 1 |
| *Macoma balthica* (Linnaeus, 1758) | 3 | 4 | 1 | 3 | 4 |
| *Mytilus trossulus* Gould, 1850 | 1 | 1 | 1 | 1 | 1 |
| *Mya arenaria* Linnaeus, 1758 | 3 | 4 | 1 | 1 | 2 |
| *Ecrobia ventrosa* (Montagu, 1803) | 2 | 3 | 1 | 3 | 1 |
| *Peringia ulvae* (Pennant, 1777) | 2 | 3 | 1 | 3 | 2 |
| *Theodoxus fluviatilis* (Linnaeus, 1758) | 1 | 1 | 1 | 3 | 1 |
| *Amphibalanus improvisus* (Darwin, 1854) | 1 | 1 | 1 | 1 | 1 |
| *Bathyporeia pilosa* Lindström, 1855 | 4 | 4 | 3 | 3 | 2 |
| Chironomidae | 2 | 2 | 1 | 1 | 1 |
| *Corophium multisetosum* Stock, 1952 | 2 | 2 | 2 | 3 | 3 |
| *Corophium volutator* (Pallas, 1766) | 2 | 2 | 2 | 3 | 3 |
| *Crangon crangon* (Linnaeus, 1758) | 4 | 2 | 1 | 3 | 1 |
| *Cyathura carinata* (Krøyer, 1847) | 2 | 2 | 1 | 3 | 1 |
| *Gammarus oceanicus* Segerstråle, 1947 | 1 | 1 | 1 | 3 | 1 |
| *Gammarus salinus* Spooner, 1947 | 1 | 1 | 1 | 3 | 1 |
| *Gammarus zaddachi* Sexton, 1912 | 1 | 1 | 1 | 3 | 1 |
| *Idotea balthica* (Pallas, 1772) | 1 | 1 | 1 | 3 | 1 |
| *Idotea chelipes* (Pallas, 1766) | 1 | 1 | 1 | 3 | 1 |
| *Idotea granulosa* Rathke, 1843 | 1 | 1 | 1 | 3 | 1 |
| Insecta larvae | 1 | 1 | 1 | 3 | 1 |
| *Lekanesphaera hookeri* (Leach, 1814) | 1 | 1 | 1 | 3 | 1 |
| *Rhithropanopeus harrisii* (Gould, 1841) | 4 | 2 | 1 | 2 | 1 |

**Table A2.** PERMANOVA pairwise results for differences in $IP_C$ and Shannon–Wiener Index. Unvegetated (UnV), Low-Density Seagrass (SLD), Medium-Density Seagrass (SMD), and High-Density Seagrass (SHD). Bold—statistically significant differences.

| | Pair | $IP_C$ | | Shannon–Wiener | |
|---|---|---|---|---|---|
| | | t | *p* | t | *p* |
| Site | **UnV, SLD** | 2.721 | **0.002** | | |
| | **UnV, SMD** | 3.202 | **0.002** | | |
| | **UnV, SHD** | 3.202 | **0.001** | | |
| | **SLD, SMD** | 0.521 | 0.916 | | |
| | **SLD, SHD** | 1.046 | 0.326 | | |
| | **SMD, SHD** | 1.042 | 0.366 | | |
| Season | **Autumn, Summer** | 2.881 | **0.001** | 0.162 | 0.858 |
| | **Autumn, Winter** | 3.175 | **0.001** | 4.308 | **0.003** |
| | **Autumn, Spring** | 3.105 | **0.001** | 4.308 | **0.002** |
| | **Summer, Winter** | 2.020 | **0.014** | 3.194 | **0.009** |
| | **Summer, Spring** | 2.432 | **0.004** | 3.194 | **0.005** |
| | **Winter, Spring** | 0.960 | 0.338 | 0.584 | 0.781 |

**Table A3.** PERMANOVA pairwise results for differences in the Abundance, Pielou Index, and $BP_C$. Unvegetated (UnV), Low-Density Seagrass (SLD), Medium-Density Seagrass (SMD), and High-Density Seagrass (SHD). Bold—statistically significant differences.

| | | Abundance | | Pielou Index | | $BP_C$ | |
|---|---|---|---|---|---|---|---|
| | Pair | t | *p* | t | *p* | t | *p* |
| UnV | **Autumn, Summer** | 2.891 | **0.012** | 3.624 | **0.011** | 3.224 | **0.005** |
| | **Autumn, Winter** | 1.912 | **0.008** | 2.982 | **0.013** | 3.089 | **0.012** |
| | **Autumn, Spring** | 2.187 | **0.008** | 3.644 | **0.008** | 2.886 | **0.005** |
| | **Summer, Winter** | 4.229 | **0.008** | 1.572 | 0.132 | 1.753 | 0.081 |
| | **Summer, Spring** | 2.973 | **0.013** | 0.190 | 0.858 | 2.699 | 0.051 |
| | **Winter, Spring** | 1.631 | **0.008** | 1.653 | 0.146 | 1.162 | 0.29 |
| SLD | **Autumn, Summer** | 2.774 | **0.008** | 1.384 | 0.216 | 3.625 | **0.013** |
| | **Autumn, Winter** | 2.630 | **0.02** | 3.284 | 0.016 | 3.383 | **0.008** |
| | **Autumn, Spring** | 4.287 | **0.012** | 1.941 | 0.093 | 6.757 | **0.009** |
| | **Summer, Winter** | 2.643 | **0.009** | 3.054 | 0.036 | 1.658 | 0.09 |
| | **Summer, Spring** | 4.038 | **0.008** | 1.338 | 0.256 | 4.735 | **0.01** |
| | **Winter, Spring** | 1.997 | **0.006** | 0.068 | 0.985 | 1.738 | 0.101 |
| SMD | **Autumn, Summer** | 2.350 | **0.008** | 1.653 | 0.145 | 2.076 | **0.033** |
| | **Autumn, Winter** | 1.924 | **0.01** | 0.071 | 0.903 | 3.317 | **0.023** |
| | **Autumn, Spring** | 2.966 | **0.008** | 7.332 | **0.01** | 2.973 | **0.023** |
| | **Summer, Winter** | 1.759 | **0.015** | 0.706 | 0.474 | 0.525 | 0.701 |
| | **Summer, Spring** | 1.923 | **0.016** | 4.000 | **0.009** | 0.877 | 0.457 |
| | **Winter, Spring** | 1.512 | **0.025** | 2.659 | **0.014** | 1.302 | 0.207 |
| SHD | **Autumn, Summer** | 2.864 | **0.005** | 1.136 | 0.325 | 3.517 | **0.012** |
| | **Autumn, Winter** | 2.874 | **0.008** | 1.033 | 0.351 | 4.008 | **0.009** |
| | **Autumn, Spring** | 3.702 | **0.005** | 5.252 | **0.008** | 7.453 | **0.01** |
| | **Summer, Winter** | 2.266 | **0.009** | 1.578 | 0.159 | 0.714 | 0.526 |
| | **Summer, Spring** | 2.713 | **0.006** | 4.420 | **0.011** | 2.493 | **0.021** |
| | **Winter, Spring** | 1.048 | 0.394 | 1.151 | 0.297 | 1.467 | 0.192 |

**Table A3.** *Cont.*

| | Pair | Abundance | | Pielou Index | | $BP_C$ | |
|---|---|---|---|---|---|---|---|
| | | t | *p* | t | *p* | t | *p* |
| **Autumn** | **UnV, SLD** | 1.186 | 0.061 | 0.185 | 0.883 | 1.193 | 0.138 |
| | **UnV, SMD** | 1.085 | 0.168 | 0.537 | 0.565 | 1.004 | 0.337 |
| | **UnV, SHD** | 1.507 | **0.009** | 0.731 | 0.454 | 1.585 | **0.013** |
| | **SLD, SMD** | 0.743 | 0.746 | 0.450 | 0.739 | 0.580 | 0.677 |
| | **SLD, SHD** | 1.719 | **0.039** | 0.719 | 0.634 | 1.833 | 0.113 |
| | **SMD, SHD** | 1.776 | **0.013** | 0.465 | 0.718 | 1.991 | 0.067 |
| **Winter** | **UnV, SLD** | 2.044 | **0.002** | 0.202 | 0.845 | 3.257 | **0.011** |
| | **UnV, SMD** | 2.352 | **0.012** | 2.177 | **0.043** | 5.223 | **0.009** |
| | **UnV, SHD** | 2.496 | **0.013** | 1.292 | 0.18 | 5.893 | **0.008** |
| | **SLD, SMD** | 0.868 | 0.601 | 1.964 | **0.084** | 0.644 | 0.726 |
| | **SLD, SHD** | 1.011 | 0.357 | 1.100 | 0.329 | 1.398 | 0.208 |
| | **SMD, SHD** | 0.891 | 0.583 | 0.794 | 0.302 | 1.191 | 0.274 |
| **Spring** | **UnV, SLD** | 2.024 | **0.006** | 0.674 | 0.883 | 1.425 | 0.164 |
| | **UnV, SMD** | 2.582 | **0.009** | 0.232 | 0.842 | 6.835 | **0.011** |
| | **UnV, SHD** | 3.046 | **0.009** | 1.998 | 0.094 | 6.585 | **0.008** |
| | **SLD, SMD** | 1.930 | **0.019** | 0.599 | 0.893 | 4.774 | **0.005** |
| | **SLD, SHD** | 2.433 | **0.005** | 0.051 | 0.992 | 4.194 | **0.006** |
| | **SMD, SHD** | 0.946 | 0.595 | 1.611 | 0.149 | 1.776 | 0.106 |
| **Summer** | **UnV, SLD** | 5.220 | **0.01** | 5.880 | **0.015** | 6.982 | **0.013** |
| | **UnV, SMD** | 3.559 | **0.007** | 4.136 | **0.024** | 3.971 | **0.008** |
| | **UnV, SHD** | 4.581 | **0.017** | 5.608 | **0.008** | 6.453 | **0.011** |
| | **SLD, SMD** | 1.135 | 0.279 | 0.412 | 0.667 | 0.959 | 0.403 |
| | **SLD, SHD** | 1.408 | 0.068 | 1.884 | 0.064 | 0.756 | 0.478 |
| | **SMD, SHD** | 1.184 | 0.242 | 1.955 | 0.081 | 1.196 | 0.263 |

# References


1. Carstensen, J.; Conley, D.J.; Almroth-Rosell, E.; Asmala, E.; Bonsdorff, E.; Fleming-Lehtinen, V.; Gustafsson, B.G.; Gustafsson, C.; Heiskanen, A.-S.; Janas, U.; et al. Factors Regulating the Coastal Nutrient Filter in the Baltic Sea. *Ambio* **2019**, *49*, 1194–1210. [CrossRef] [PubMed]
2. Romoth, K.; Darr, A.; Papenmeier, S.; Zettler, M.L.; Gogina, M. Substrate Heterogeneity as a Trigger for Species Diversity in Marine Benthic Assemblages. *Biology* **2023**, *12*, 825. [CrossRef] [PubMed]
3. Janas, U.; Kendzierska, H. Benthic Non-Indigenous Species among Indigenous Species and Their Habitat Preferences in Puck Bay (Southern Baltic Sea). *Oceanologia* **2014**, *56*, 603–628. [CrossRef]
4. Sokołowski, A.; Jankowska, E.; Balazy, P.; Jędruch, A. Distribution and Extent of Benthic Habitats in Puck Bay (Gulf of Gdańsk, Southern Baltic Sea). *Oceanologia* **2021**, *63*, 301–320. [CrossRef]
5. Villnäs, A.; Janas, U.; Josefson, A.; Kendzierska, H.; Nygård, H.; Norkko, J.; Norkko, A. Changes in Macrofaunal Biological Traits across Estuarine Gradients: Implications for the Coastal Nutrient Filter. *Mar. Ecol. Prog. Ser.* **2019**, *622*, 31–48. [CrossRef]
6. Janas, U.; Burska, D.; Kendzierska, H.; Pryputniewicz-Flis, D.; Łukawska-Matuszewska, K. Importance of Benthic Macrofauna and Coastal Biotopes for Ecosystem Functioning—Oxygen and Nutrient Fluxes in the Coastal Zone. *Estuar. Coast. Shelf Sci.* **2019**, *225*, 106238. [CrossRef]
7. Moreira-Saporiti, A.; Teichberg, M.; Garnier, E.; Cornelissen, J.H.C.; Alcoverro, T.; Björk, M.; Boström, C.; Dattolo, E.; Eklöf, J.S.; Hasler-Sheetal, H.; et al. A Trait-Based Framework for Seagrass Ecology: Trends and Prospects. *Front. Plant Sci.* **2023**, *14*, 1088643. [CrossRef]
8. Hemminga, M.A.; Duarte, C.M. *Seagrass Ecology*; Cambridge University Press: Cambridge, UK, 2000; ISBN 9780521661843.
9. Temmink, R.J.M.; Christianen, M.J.A.; Fivash, G.S.; Angelini, C.; Boström, C.; Didderen, K.; Engel, S.M.; Esteban, N.; Gaeckle, J.L.; Gagnon, K.; et al. Mimicry of Emergent Traits Amplifies Coastal Restoration Success. *Nat. Commun.* **2020**, *11*, 3668. [CrossRef]
10. Duarte, C.; Middelburg, J.; Caraco, N. Major Role of Marine Vegetation on the Oceanic Carbon Cycle. *Biogeosciences* **2005**, *2*, 1–8. [CrossRef]
11. Fourqurean, J.W.; Duarte, C.M.; Kennedy, H.; Marbà, N.; Holmer, M.; Mateo, M.A.; Apostolaki, E.T.; Kendrick, G.A.; Krause-Jensen, D.; McGlathery, K.J.; et al. Seagrass Ecosystems as a Globally Significant Carbon Stock. *Nat. Geosci.* **2012**, *5*, 505–509. [CrossRef]

12. Purvaja, R.; Robin, R.S.; Ganguly, D.; Hariharan, G.; Singh, G.; Raghuraman, R.; Ramesh, R. Seagrass Meadows as Proxy for Assessment of Ecosystem Health. *Ocean. Coast. Manag.* **2018**, *159*, 34–45. [CrossRef]
13. Boström, C.; Baden, S.; Bockelmann, A.C.; Dromph, K.; Fredriksen, S.; Gustafsson, C.; Krause-Jensen, D.; Möller, T.; Nielsen, S.L.; Olesen, B.; et al. Distribution, Structure and Function of Nordic Eelgrass (*Zostera Marina*) Ecosystems: Implications for Coastal Management and Conservation. *Aquat. Conserv.* **2014**, *24*, 410–434. [CrossRef] [PubMed]
14. Jankowska, E.; Włodarska-Kowalczuk, M. Habitat-Builders Complexity Boosts Associated Fauna Functional Trait Richness (*Zostera marina* Meadows, Baltic Sea). *Ecol. Indic.* **2022**, *144*, 109512. [CrossRef]
15. Ziółkowska, M.; Sokołowski, A.; Richard, P. Spatial and Temporal Variability of Organic Matter Sources and Food Web Structure across Benthic Habitats in a Low Diversity System (Southern Baltic Sea). *J. Sea Res.* **2018**, *141*, 47–60. [CrossRef]
16. Boström, C.; Bonsdorff, E. Community Structure and Spatial Variation of Benthic Invertebrates Associated with *Zostera marina* (L.) Beds in the Northern Baltic Sea. *J. Sea Res.* **1997**, *37*, 153–166. [CrossRef]
17. Jankowska, E.; Michel, L.N.; Lepoint, G.; Włodarska-Kowalczuk, M. Stabilizing Effects of Seagrass Meadows on Coastal Water Benthic Food Webs. *J. Exp. Mar. Biol. Ecol.* **2019**, *510*, 54–63. [CrossRef]
18. Thormar, J.; Hasler-Sheetal, H.; Baden, S.; Boström, C.; Clausen, K.K.; Krause-Jensen, D.; Olesen, B.; Rasmussen, J.R.; Svensson, C.J.; Holmer, M. Eelgrass (*Zostera Marina*) Food Web Structure in Different Environmental Settings. *PLoS ONE* **2016**, *11*, e0146479. [CrossRef]
19. Foshtomi, M.Y.; Braeckman, U.; Derycke, S.; Sapp, M.; Van Gansbeke, D.; Sabbe, K.; Willems, A.; Vincx, M.; Vanaverbeke, J. The Link between Microbial Diversity and Nitrogen Cycling in Marine Sediments Is Modulated by Macrofaunal Bioturbation. *PLoS ONE* **2015**, *10*, e0130116. [CrossRef]
20. Bartoli, M.; Benelli, S.; Magri, M.; Ribaudo, C.; Moraes, P.C.; Castaldelli, G. Contrasting Effects of Bioturbation Studied in Intact and Reconstructed Estuarine Sediments. *Water* **2020**, *12*, 3125. [CrossRef]
21. Benelli, S.; Bartoli, M.; Magri, M.; Brzana, R.; Kendzierska, H.; Styrcz-Olesiak, K.; Janas, U. Spatial and Seasonal Pattern of Microbial Nitrate Reduction in Coastal Sediments in the Vistula River Plume Area, Gulf of Gdańsk. *Front. Mar. Sci.* **2024**, *11*, 1333707. [CrossRef]
22. Biles, C.L.; Paterson, D.M.; Ford, R.B.; Solan, M.; Raffaelli, D.G. Bioturbation, Ecosystem Functioning and Community Structure. *Hydrol. Earth Syst. Sci.* **2002**, *6*, 999–1005. [CrossRef]
23. Borawska, Z.; Szymczycha, B.; Silberberger, M.J.; Koziorowska-Makuch, K.; Szczepanek, M.; Kędra, M. Benthic Fluxes of Dissolved Silica Are an Important Component of the Marine Si Cycle in the Coastal Zone. *Estuar. Coast. Shelf Sci.* **2022**, *273*, 107880. [CrossRef]
24. Braeckman, U.; Provoost, P.; Gribsholt, B.; Van Gansbeke, D.; Middelburg, J.J.; Soetaert, K.; Vincx, M.; Vanaverbeke, J. Role of Macrofauna Functional Traits and Density in Biogeochemical Fluxes and Bioturbation. *Mar. Ecol. Prog. Ser.* **2010**, *399*, 173–186. [CrossRef]
25. Kauppi, L.; Bernard, G.; Bastrop, R.; Norkko, A.; Norkko, J. Increasing Densities of an Invasive Polychaete Enhance Bioturbation with Variable Effects on Solute Fluxes. *Sci. Rep.* **2018**, *8*, 7619. [CrossRef]
26. Miernik, N.A.; Janas, U.; Kendzierska, H. Role of Macrofaunal Communities in the Vistula River Plume, the Baltic Sea—Bioturbation and Bioirrigation Potential. *Biology* **2023**, *12*, 147. [CrossRef]
27. Renz, J.R.; Powilleit, M.; Gogina, M.; Zettler, M.L.; Morys, C.; Forster, S. Community Bioirrigation Potential (BIP c), an Index to Quantify the Potential for Solute Exchange at the Sediment-Water Interface. *Mar. Environ. Res.* **2018**, *141*, 214–224. [CrossRef]
28. Solan, M.; Cardinale, B.J.; Downing, A.L.; Engelhardt, K.A.M.; Ruesink, J.L.; Srivastava, D.S. Extinction and Ecosystem Function in the Marine Benthos. *Science (1979)* **2004**, *306*, 1177–1180. [CrossRef]
29. Villnäs, A.; Norkko, J.; Lukkari, K.; Hewitt, J.; Norkko, A. Consequences of Increasing Hypoxic Disturbance on Benthic Communities and Ecosystem Functioning. *PLoS ONE* **2012**, *7*, e44920. [CrossRef]
30. Wrede, A.; Andresen, H.; Asmus, R.; Wiltshire, K.H.; Brey, T. Macrofaunal Irrigation Traits Enhance Predictability of Nutrient Fluxes across the Sediment-Water Interface. *Mar. Ecol. Prog. Ser.* **2019**, *632*, 27–42. [CrossRef]
31. Bernard, G.; Delgard, M.L.; Maire, O.; Ciutat, A.; Lecroart, P.; Deflandre, B.; Duchêne, J.C.; Grémare, A. Comparative Study of Sediment Particle Mixing in a Zostera Noltei Meadow and a Bare Sediment Mudflat. *Mar. Ecol. Prog. Ser.* **2014**, *514*, 71–86. [CrossRef]
32. Boström, C.; Bonsdorff, E. Zoobenthic Community Establishment and Habitat Complexity—The Importance of Seagrass Shoot-Density, Morphology and Physical Disturbance for Faunal Recruitment. *Mar. Ecol. Prog. Ser.* **2000**, *205*, 123–138. [CrossRef]
33. Dąbrowska, A.H.; Janas, U.; Kendzierska, H. Assessment of Biodiversity and Environmental Quality Using Macrozoobenthos Communities in the Seagrass Meadow (Gulf of Gdańsk, Southern Baltic). *Oceanol. Hydrobiol. Stud.* **2016**, *45*, 286–294. [CrossRef]
34. Gustafsson, C.; Norkko, A. Quantifying the Importance of Functional Traits for Primary Production in Aquatic Plant Communities. *J. Ecol.* **2019**, *107*, 154–166. [CrossRef]

35. Rodil, I.F.; Lohrer, A.M.; Attard, K.M.; Hewitt, J.E.; Thrush, S.F.; Norkko, A. Macrofauna Communities across a Seascape of Seagrass Meadows: Environmental Drivers, Biodiversity Patterns and Conservation Implications. *Biodivers. Conserv.* **2021**, *30*, 3023–3043. [CrossRef]
36. Włodarska-Kowalczuk, M.; Jankowska, E.; Kotwicki, L.; Damrat, M.; Zajączkowski, M. Effects of Fluvial Discharges on Meiobenthic and Macrobenthic Variability in the Vistula River Prodelta (Baltic Sea). *J. Mar. Syst.* **2016**, *157*, 135–146. [CrossRef]
37. Łęczyński, L.; Szymczak, E. *Własności Fizyczne Osadów Dennych. Fizyczne, Biologiczne i Chemiczne Badania Morskich Osadów Dennych*; Wydawnictwo UG: Gdańsk, Poland, 2010.
38. Wentworth, C.K. A Scale of Grade and Class Terms for Clastic Sediments. *J. Geol.* **1922**, *30*, 377–392. [CrossRef]
39. Hayward, P.J.; Ryland, J.S. (Eds.) *Handbook of the Marine Fauna of North-West Europe*; Oxford University Press: Oxford, UK, 2017; ISBN 9780199549443.
40. Zettler, M.; Alf, A. *Bivalvia of German Marine Waters of the North and Baltic Sea*; ConchBooks: Harxheim, Germany, 2021.
41. Zettler, M.; Zettler, A. *Marine and Freshwater Amphipoda from the Baltic Sea and Adjacent Territories*; ConchBooks: Harxheim, Germany, 2017.
42. Queirós, A.M.; Birchenough, S.N.R.; Bremner, J.; Godbold, J.A.; Parker, R.E.; Romero-Ramirez, A.; Reiss, H.; Solan, M.; Somerfield, P.J.; Van Colen, C.; et al. A Bioturbation Classification of European Marine Infaunal Invertebrates. *Ecol. Evol.* **2013**, *3*, 3958–3985. [CrossRef]
43. Wrede, A.; Dannheim, J.; Gutow, L.; Brey, T. Who Really Matters: Influence of German Bight Key Bioturbators on Biogeochemical Cycling and Sediment Turnover. *J. Exp. Mar. Biol. Ecol.* **2017**, *488*, 92–101. [CrossRef]
44. Wrede, A.; Beermann, J.; Dannheim, J.; Gutow, L.; Brey, T. Organism Functional Traits and Ecosystem Supporting Services—A Novel Approach to Predict Bioirrigation. *Ecol. Indic.* **2018**, *91*, 737–743. [CrossRef]
45. Anderson, M.; Gorley, R.; Clarke, K.P. *For PRIMER: Guide to Software and Statistical Methods*; PRIMER-E: Plymouth, UK, 2008.
46. Schernewski, G.; Schiewer, U. *Baltic Coastal Ecosytems. Structure, Function and Coastal Zone Management*; Springer Science & Business Media: Berlin/Heidelberg, Germany, 2001; ISBN 9783642076909.
47. Jankowska, E.; Michel, L.N.; Zaborska, A.; Włodarska-Kowalczuk, M. Sediment Carbon Sink in Low-Density Temperate Eelgrass Meadows (Baltic Sea). *J. Geophys. Res. Biogeosci.* **2016**, *121*, 2918–2934. [CrossRef]
48. McGlathery, K.J.; Sundbäck, K.; Anderson, I.C. Eutrophication in Shallow Coastal Bays and Lagoons: The Role of Plants in the Coastal Filter. *Mar. Ecol. Prog. Ser.* **2007**, *348*, 1–18. [CrossRef]
49. Kuwae, T.; Hori, M. (Eds.) *Blue Carbon in Shallow Coastal Ecosystems*; Springer Nature Singapore Pte Ltd.: Singapore, 2019.
50. Meysick, L.; Infantes, E.; Rugiu, L.; Gagnon, K.; Boström, C. Coastal Ecosystem Engineers and Their Impact on Sediment Dynamics: Eelgrass–Bivalve Interactions under Wave Exposure. *Limnol. Oceanogr.* **2022**, *67*, 621–633. [CrossRef]
51. Jankowska, E.; Włodarska-Kowalczuk, M.; Kotwicki, L.; Balazy, P.; Kuliński, K. Seasonality in Vegetation Biometrics and Its Effects on Sediment Characteristics and Meiofauna in Baltic Seagrass Meadows. *Estuar. Coast. Shelf Sci.* **2014**, *139*, 159–170. [CrossRef]
52. Mattila, J.; Chaplin, G.; Eilers, M.R.; Heck, K.L.; O'neal, J.P.; Valentine, J.F. Spatial and Diurnal Distribution of Invertebrate and Fish Fauna of a Zostera marina Bed and Nearby Unvegetated Sediments in Damariscotta River, Maine (USA). *J. Sea Res.* **1999**, *41*, 321–332. [CrossRef]
53. Zaiko, A.; Olenin, S.; Daunys, D.; Nalepa, T. Vulnerability of Benthic Habitats to the Aquatic Invasive Species. *Biol. Invasions* **2007**, *9*, 703–714. [CrossRef]
54. Jankowska, E.; Jankowska, K.; Włodarska-Kowalczuk, M. Seagrass Vegetation and Meiofauna Enhance the Bacterial Abundance in the Baltic Sea Sediments (Puck Bay). *Environ. Sci. Pollut. Res.* **2015**, *22*, 14372–14378. [CrossRef]
55. Gustafsson, C.; Boström, C. Algal Mats Reduce Eelgrass (*Zostera marina* L.) Growth in Mixed and Monospecific Meadows. *J. Exp. Mar. Biol. Ecol.* **2014**, *461*, 85–92. [CrossRef]
56. Kruk-Dowgiałło, L.; Szaniawska, A. Gulf of Gdańsk and Puck Bay. In *Ecology of Baltic Coastal Waters. Ecological Studies*; Schiewer, U., Ed.; Springer: Berlin/Heidelberg, Germany, 2008; Volume 197.
57. Gustafsson, C.; Boström, C. Biodiversity Influences Ecosystem Functioning in Aquatic Angiosperm Communities. *Oikos* **2011**, *120*, 1037–1046. [CrossRef]
58. Włodarska-Kowalczuk, M.; Jankowska, E.; Kotwicki, L.; Balazy, P. Evidence of Season-Dependency in Vegetation Effects on Macrofauna in Temperate Seagrass Meadows (Baltic Sea). *PLoS ONE* **2014**, *9*, e100788. [CrossRef]
59. Barberá-Cebrián, C.; Sánchez-Jerez, P.; Ramos-Esplá, A.A. Fragmented Seagrass Habitats on the Mediterranean Coast, and Distribution and Abundance of Mysid Assemblages. *Mar. Biol.* **2002**, *141*, 405–413. [CrossRef]
60. Li, Y.; Rárová, L.; Scarpato, S.; Çiçek, S.S.; Jordheim, M.; Štenclová, T.; Strnad, M.; Mangoni, A.; Zidorn, C. Seasonal Variation of Phenolic Compounds in *Zostera marina* (Zosteraceae) from the Baltic Sea. *Phytochemistry* **2022**, *196*, 113099. [CrossRef] [PubMed]
61. Beukema, J.J. *Expected Effects of Changes in Winter Temperatures on Benthic Animals Living in Soft Sediments in Coastal North Sea Areas*; Springer: Berlin/Heidelberg, Germany, 1990.

62. Rodil, I.F.; Cividanes, S.; Lastra, M.; López, J. Seasonal Variability in the Vertical Distribution of Benthic Macrofauna and Sedimentary Organic Matter in an Estuarine Beach (NW Spain). *Estuaries Coasts* **2008**, *31*, 382–395. [CrossRef]
63. Queirós, A.M.; Stephens, N.; Cook, R.; Ravaglioli, C.; Nunes, J.; Dashfield, S.; Harris, C.; Tilstone, G.H.; Fishwick, J.; Braeckman, U.; et al. Can Benthic Community Structure Be Used to Predict the Process of Bioturbation in Real Ecosystems ? *Prog. Oceanogr.* **2015**, *137*, 559–569. [CrossRef]
64. Gogina, M.; Morys, C.; Forster, S.; Gräwe, U.; Friedland, R.; Zettler, M.L. Towards Benthic Ecosystem Functioning Maps: Quantifying Bioturbation Potential in the German Part of the Baltic Sea. *Ecol. Indic.* **2017**, *73*, 574–588. [CrossRef]
65. Hedman, J.E.; Gunnarsson, J.S.; Samuelsson, G.; Gilbert, F. Particle Reworking and Solute Transport by the Sediment-Living Polychaetes Marenzelleria Neglecta and Hediste Diversicolor. *J. Exp. Mar. Biol. Ecol.* **2011**, *407*, 294–301. [CrossRef]
66. Vasquez-Cardenas, D.; Quintana, C.O.; Meysman, F.J.R.; Kristensen, E.; Boschker, H.T.S. Species-Specific Effects of Two Bioturbating Polychaetes on Sediment Chemoautotrophic Bacteria. *Mar. Ecol. Prog. Ser.* **2016**, *549*, 55–68. [CrossRef]
67. Andersen, T.J.; Jensen, K.T.; Lund-Hansen, L.; Mouritsen, K.N.; Pejrup, M. Enhanced Erodibility of Fine-Grained Marine Sediments by Hydrobia Ulvae. *J. Sea Res.* **2002**, *48*, 51–58. [CrossRef]

MDPI AG
Grosspeteranlage 5
4052 Basel
Switzerland
Tel.: +41 61 683 77 34

*Biology* Editorial Office
E-mail: biology@mdpi.com
www.mdpi.com/journal/biology

www.ingramcontent.com/pod-product-compliance
Lightning Source LLC
Chambersburg PA
CBHW041416210926
51850CB00013B/3286
* 9 7 8 3 7 2 5 8 4 3 5 3 4 *